Integrative Mechanobiology

The first of its kind, this comprehensive resource integrates cellular mechanobiology with micro-nano techniques to provide unrivalled in-depth coverage of the field, including state-of-the-art methods, recent advances, and biological discoveries.

Structured in two parts, the first offers detailed analysis of innovative micro-nano techniques including FRET imaging, electron cryomicroscopy, micropost arrays, nanotopography devices, laser ablation, and computational image analysis. The second part of the book provides valuable insights into the most recent technological advances and discoveries in areas such as stem cell, heart, bone, brain, tumor, and fibroblast mechanobiology.

Written by a team of leading experts and well-recognized researchers, this is an essential resource for students and researchers in biomedical engineering.

Yu Sun is a Professor at the University of Toronto and is the Canada Research Chair in Micro and Nanoengineering Systems. He was inducted Fellow of ASME, IEEE, and CAE for his work on micro-nano devices and robotic systems. His awards include the 2010 IEEE Robotics and Automation Society Early Career Award and an NSERC E.W. R. Steacie Memorial Fellowship in 2013.

Deok-Ho Kim is a Professor in the Department of Bioengineering at the University of Washington. His awards include the Samsung Humantech Thesis Award (2009), the Harold M. Weintraub Award in Biological Sciences (2010), the American Heart Association National Scientist Development Award (2012), the KSEA Young Investigator Award (2013), and the BMES-CMBE Rising Star Award (2013).

Craig A. Simmons is a Professor at the University of Toronto. His research in cell mechanobiology, tissue engineering, and microtechnologies has been recognized with the Canada Research Chair in Mechanobiology, the McCharles Prize (2010), and the McLean Award (2012).

Integrative Mechanobiology

Micro- and Nano- Techniques in Cell Mechanobiology

EDITED BY

YU SUN

University of Toronto

DEOK-HO KIM

University of Washington

CRAIG A. SIMMONS

University of Toronto

CAMBRIDGE
UNIVERSITY PRESS

University Printing House, Cambridge CB2 8BS, United Kingdom

Cambridge University Press is part of the University of Cambridge.

It furthers the University's mission by disseminating knowledge in the pursuit of education, learning and research at the highest international levels of excellence.

www.cambridge.org
Information on this title: www.cambridge.org/9781107078390

First published 2015

Printed in the United Kingdom by TJ International Ltd. Padstow Cornwall

A catalogue record for this publication is available from the British Library

ISBN 978-1-107-07839-0 Hardback

Additional resources for this publication at www.cambridge.org/9781107078390

Contents

Contributors

Thomas H. Barker
Georgia Institute of Technology

Kyle G. Battiston
University of Toronto

Agamoni Bhattacharyya
University of California, San Diego

Jason A. Burdick
University of Pennsylvania

Dwight M. Chambers
Georgia Institute of Technology

Yu Suk Choi
University of Western Australia

Eddie Y. Chung
University of California, San Diego

Seok Chung
Korea University

Junsang Doh
Pohang University of Science and Technology

Mitsuhiro Ebara
National Institute for Materials Science (NIMS), Japan

Rodrigo Fernandez-Gonzalez
University of Toronto

Vincent F. Fiore
Georgia Institute of Technology

Elliot Fisher
University of Washington

Jianping Fu
University of Michigan

Murat Guvendiren
Rutgers University

Sewoon Han
University of California, Berkeley

Dorit Hanein
Sanford Burnham Medical Research Institute

Andrew W. Holle
Max Planck Institute for Intelligent Systems

Hyo Eun Jeong
Korea University

Jessie S. Jeon
Korea Advanced Institute of Science and Technology

Deok-Ho Kim
University of Washington

Hong-Nam Kim
Korea Institute of Science and Technology (KIST)

Hye Mi Kim
Pohang University of Science and Technology

Jeong Ah Kim
Korea Basic Science Institute

Hao-Chih Lee
Carnegie Mellon University

Andrea Leonard
University of Washington

Chwee Teck Lim
National University of Singapore

Francis Lin
University of Manitoba

Chao Liu
University of Toronto

Shaoying Lu
University of California, San Diego

Luke A. MacQueen
University of Toronto

Kevin Middleton
University of Toronto

Christopher Moraes
McGill University

Jerel Mueller
Virginia Tech

Yasaman Nematbakhsh
National University of Singapore

Qin Qin
University of California, San Diego

Marita L. Rodriguez
University of Washington

J. Paul Santerre
University of Toronto

Chang Ho Seo
University of Tokyo

Yue Shao
University of Michigan

Yoojin Shin
Korea University

Craig A. Simmons
University of Toronto

Nathan J. Sniadecki
University of Washington

Yu Sun
University of Toronto

William Tyler
Arizona State University

Koichiro Uto
University of Washington

Niels Volkmann
Sanford Burnham Medical Research Institute

Yingxiao Wang
University of California, San Diego

Shinuo Weng
University of Michigan

Jiandong Wu
University of Manitoba

Ge Yang
Carnegie Mellon University

Lidan You
University of Toronto

Jennifer L. Young
Max Planck Institute for Intelligent Systems

Teresa Zulueta-Coarasa
University of Toronto

Preface

Mechanical forces in cell microenvironments direct cellular and multicellular form and function. Cells sense the mechanical characteristics of their microenvironment and translate the mechanical cues to intracellular biochemical signals that regulate several cellular and molecular processes important in development, homeostasis, and disease. Much of our understanding of the molecular mechanisms underlying the ability of cells to sense and react to mechanical stimuli is largely based on traditional macroscale tissue culture assays. However, novel micro- and nanoscale techniques for investigating cellular mechanobiological processes in normal and pathophysiological contexts have been under intense development in recent years. These approaches are providing new insights into cell mechanotransduction and mechanobiological responses, leading to improved fundamental understanding of cell biology and new strategies for cell-based regenerative therapies. This book highlights many of those recent advances in integrative cellular mechanobiology, including new discoveries and micro- and nanoengineered technologies.

The two related themes in the book are "Micro-Nano Techniques in Cell Mechanobiology" and "Recent Progress in Cell Mechanobiology."

- Chapter 1 reviews micropatterning technologies for controlling input signals in cellular mechanosensing and FRET live-cell imaging for visualizing molecular networks.
- Chapter 2 discusses electron microscopy and three-dimensional single-particle analysis for discerning the molecular underpinnings of cellular processes.
- Chapter 3 describes stretchable micropost array cytometry for quantitative control and real-time measurement of both mechanical stimuli and cellular biomechanical responses with a high spatiotemporal subcellular resolution.
- Chapter 4 discusses micro-device arrays for applying dynamically controlled mechanical stimuli to cells cultured in 2-D and 3-D arrayed environments.
- Chapter 5 reviews micro-nano topographies mimicking native extracellular matrices for the regulation of cellular behaviors.
- Chapter 6 describes hydrolytically degradable hydrogels, protease-sensitive hydrogels with cell-controlled properties, and stimuli-responsive hydrogels with user-controlled properties.
- Chapter 7 reviews microengineered electrotaxis devices for directing cell migration.

- Chapter 8 reviews laser ablation in severing cellular structures and its application to the measurement of physical forces with minimal disruption of the cellular microenvironment.
- Chapter 9 discusses computational image analysis techniques that are used to analyze dynamic fluorescence microscopy images for cell mechanobiology applications.
- Chapter 10 highlights micro- and nanotools for investigating cell mechanics and mechanobiology with a focus on cancer cells.
- Chapter 11 reviews several types of stimuli-responsive polymeric materials that have been developed for directing cell fate.
- Chapter 12 discusses in vivo tissue level mechanical stimuli and in vitro stem cell mechanosensing.
- Chapter 13 discusses how mechanobiological stimulation aids in engineering vascular and valvular tissues, and the role for micro- and nanoscale technologies.
- Chapter 14 highlights micro- and nanoscaled in vitro testing platforms to better mimic the environment of bone cells and their applications to studying bone cell mechanobiology.
- Chapter 15 explains molecular mechanotransduction mechanisms involved in fibroblast mediation of wound healing and tissue fibrosis.
- Chapter 16 highlights recent studies using micropost array technologies to measure and manipulate the contractile response of stem cell derived cardiomyocytes.
- Chapter 17 reviews several micro-nano fabricated platforms applied to addressing T cell activation.
- Chapter 18 discusses microfluidics assays for investigating cellular morphogenesis in 3-D and tumor angiogenesis.
- Chapter 19 discusses recent progress and future directions regarding cellular mechanobiology as applied to neuronal function.

Many of the molecular mechanisms by which cells sense and respond to mechanobiological stimuli require further elucidation. It is likely that major advances to this end will come from powerful micro- and nanoengineered platforms and the creation of more physiologically relevant in vitro models of mechanotransduction. It is certain that we will witness more transformative techniques and intriguing new findings in cellular mechanobiology in the near future. We thank all of the chapter authors and reviewers, and hope this book contributes to the education of next-generation mechanobiologists and becomes a useful reference in the area of integrative cellular mechanobiology.

Yu Sun, Deok-Ho Kim, and Craig Simmons

Part I

Micro-nano techniques in cell mechanobiology

1 Nanotechnologies and FRET imaging in live cells

Eddie Y. Chung, Qin Qin, Agamoni Bhattacharyya, Shaoying Lu, and Yingxiao Wang

Live cells can sense the mechanical characteristics of the microenvironment and translate the mechanical cues to intracellular biochemical signals in physiology and disease. To investigate intracellular signaling transduction during mechanosensing, nanotechnologies, and FRET live-cell imaging technologies have been developed to visualize the output signals in real time, such as intracellular molecular activity. Meanwhile, micropatterned technologies have been applied to modulate the physical and mechanical environment surrounding the cell to fine-tune the input signals in cellular mechanosensing. These advanced technologies can join forces and shed new light into the molecular networks that control mechanotransduction in normal conditions and disease.

1.1 Introduction

Mechanical force plays crucial roles in regulating pathophysiological processes such as atherosclerosis and cancer metastasis (Makowski and Hotamisligil 2004; Bissell and Hines 2011). For example, shear stress without a clear direction can lead to endothelial cell (EC) dysfunction and atherogenesis (Bao, Lu, and Frangos 1999; Bao, Clark, and Frangos 2000; Gimbrone and Garcia-Cardena 2013). However, there is a lack of understanding on how the cells can perceive the spatiotemporal cues and transduce them into biochemical activities to regulate cellular functions. It has been hypothesized that the great spatiotemporal heterogeneity of force distribution under atheroprone-disturbed flows contributes to the pathophysiological modulation of EC responses in mechanotransduction (DePaola et al. 1992; Frangos, Huang, and Clark 1996; Davies 1997; Tardy et al. 1997; Nagel et al. 1999; Bao et al. 2000; Butler et al. 2000; Li et al. 2002). The hypotheses will need to be investigated through novel imaging approaches with high spatiotemporal resolution in single live cells, such as the fluorescence proteins, fluorescence resonance energy transfer (FRET) biosensors, and enabling nanotechnologies.

Fluorescence proteins have been widely applied in single live cell imaging, a practice that has revolutionized the research field of cell biology, including mechanotransduction and mechanobiology. FRET-based biosensors have been engineered and applied to visualize molecular activities in single live cells (Miyawaki et al. 1997; Mochizuki et al. 2001; Ting et al. 2001; Zhang et al. 2001; Violin et al. 2003; Kunkel et al. 2005; Wang et al. 2005; Zhang et al. 2005; Pertz et al. 2006; Buranachai and Clegg 2008). The FRET biosensors contain a pair of donor and acceptor fluorescence proteins. The ratiometric measurement of the biosensor signals utilizes the donor-to-acceptor intensity ratio to represent the activity of target molecules. This measurement is self-normalizing and independent of the heterogeneous biosensor expression level among different cells (Liao et al. 2012). The genetically engineered FRET biosensors also allow subcellular localization to cytosol, plasma membrane, or organelles, which can provide measurement of local molecular activities with high spatial resolution. Therefore, FRET-based biosensors provide a versatile tool to monitor molecular signals in live cells in real time (Wang, Shyy, and Chien 2008; Song, Madahar, and Liao 2011).

Over the past 10 years, nanotechnologies have had a huge impact on scientific research. Novel nanomaterials, as well as nano- and microtechnologies and ensuing imaging methods, have enabled researchers to investigate molecular networks inside the cell with high functionality and high fidelity. Specifically, nanosized quantum dots (QDs) have been applied as FRET donors for QD FRET biosensors to monitor molecular activities in live cells. In this chapter, we will introduce the development of FRET biosensors based on fluorescence proteins and nanosized quantum dots, as well as their applications in live cell imaging under controlled microenvironment, for studies in mechanotransduction and cancer.

1.2 FRET biosensors

It has become well known that signaling molecules function in a nonlinear network with pathways that are largely dependent on subcellular localizations. Therefore, it is essential to develop and utilize imaging tools to monitor spatiotemporal activation of different signaling elements in this network. FRET biosensors provide ideal tools for this purpose. For example, the FRET-based kinase biosensors have wide applications in systematic understanding of the functional roles played by Src kinase and focal adhesion kinase (FAK) in mechanosensing and mechanobiology (Wang et al. 2005; Ouyang et al. 2008; Seong et al. 2011; Seong et al. 2013). In addition, a FRET reporter for a mechanosensitive canonical transient receptor potential channel 6 (TRPC6) has been developed to visualize real-time local calcium influx mediated at TRPC6 in different lipid regions (Lei et al. 2014). Two FRET biosensors with distinct fluorescence spectrums have also been developed to simultaneously monitor both the activities of Src kinase and MT1-MMP protease within the same live cells (Ouyang et al. 2010). All these arrays of FRET biosensors will provide a colorful landscape of functional roles played by different molecular activities in live cells during mechanotransduction.

1.3 Src FRET biosensors and mechanotransduction

The Src FRET biosensor was developed by Wang et al. (2005) to visualize Src kinase activity in live cells. The biosensor contains a cyan fluorescence protein (CFP), a SH2

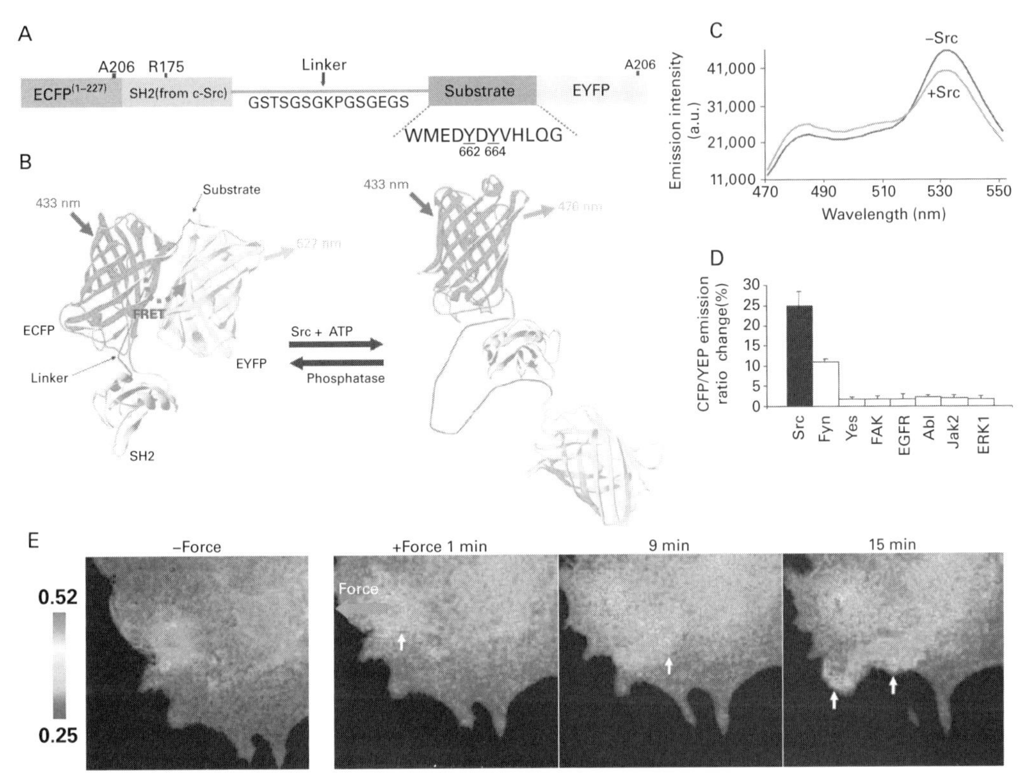

Figure 1.1 **The schematics and characterizations of the Src reporter, and observed wave-propagation in a live cell.**
(A) The Src reporter is composed of CFP, the SH2 domain, a flexible linker, the Src substrate peptide, and YFP. (B) The drawing illustrates the FRET effect of the Src reporter upon the actions of Src kinase or phosphatase. (C) Emission spectra of the Src reporter before (black) and after (red) phosphorylation by Src. (D) In vitro emission ratio changes (mean + s.d.) of the Src reporter in response to Src and other kinases. (E) Directional wave propagation (white arrow) of Src activation was observed after mechanical stimulation by bead traction (red arrow).
This research was originally published in the journal *Nature*. Wang et al. "Visualizing the mechanical activation of Src." *Nature* 2005, 434(7036): 1040–1045.

domain, a flexible linker, a Src substrate, and a yellow fluorescence protein (YFP, Fig. 1.1A, B). After the Src reporter is phosphorylated by Src kinase at the substrate tyrosine sites, it shows higher emissions of CFP and lower emissions of FRET (emission of YFP upon CFP excitation), which can be represented by a loss of FRET and an increase of CFP/FRET ratio caused by Src activation (Fig. 1.1C). The increase of CFP/FRET ratio was fully activated by c-Src, while only weakly activated by c-Fyn, and not activated by c-Yes or other kinases in mammalian cells (Fig. 1.1D). These results prove that the Src reporter is specifically activated by Src kinase in mammalian cells.

Fibronectin-coated beads can bind to integrins and thus couple with cytoskeleton (Miyamoto et al. 1995). These functionalized beads were then applied to human umbilical vein endothelial cells (HUVECs) and resulted in a local FRET response of the Src reporter around the beads (Arias-Salgado et al. 2003). However, if the beads were coated with noninteractive polylysine, then there was no significant FRET response. This result infers that specific integrin-cytoskeleton coupling is necessary for mechanotransduction. The authors then targeted the monomeric Src reporter to the plasma membrane (Thomas and Brugge 1997; Zacharias et al. 2002; Arias-Salgado et al. 2003). The FRET response of this membrane-targeted reporter was reversed and prevented by PP1, a general inhibitor of Src family kinase, which proves that the Src reporter has specificity toward Src kinase activity.

Upon applying mechanical force by laser tweezers on fibronectin-coated beads adhering to the cell surface, a robust FRET-signal change was observed, which indicated a rapid distal Src activation and a slower directional wave propagation of Src activation in live cells (Fig. 1.1E). The Src activation serves as a mechanism for cells to adapt to the new mechanical environment. Therefore, monitoring dynamic signal transduction processes by FRET biosensors enables the visualization and quantification of mechanotransduction process in live cells. This result marked the first time when the mechanotransduction process was visualized in live cells.

1.4 Focal adhesion kinase (FAK) FRET biosensors

Focal adhesions are molecular complexes at the contact sites at the interface between the cell and its extracellular matrix (ECM). Mature focal adhesions contain transmembrane receptor integrins, associated intracellular signaling proteins, and adapter proteins (Luo, Carman, and Springer 2007). Signaling proteins at focal adhesions include Src and FAK tyrosine kinases and receptor-like tyrosine phosphatase α (RPTP-α). In newly assembled focal adhesions, integrin ligates with ECM proteins, which subsequently recruit many intracellular structural and signaling proteins to the focal adhesion sites.

Since integrin lacks enzymatic activity, the signaling proteins in focal adhesions are crucial to transfer extracellular mechanical information toward the inside of the cells (Mitra, Hanson, and Schlaepfer 2005; Mitra and Schlaepfer 2006). For example, Src and FAK coordinate and regulate downstream signals in focal adhesions (Seong 2011; Seong et al. 2013). With a FAK FRET biosensor designed with a similar strategy as that of the Src biosensor, Seong et al. (2011) reported that growth-factor-induced FAK activation is mediated and maintained by Src activity, while FAK activation on cell adhesion is in fact independent of Src activation (Seong et al. 2011). FAK also interacts with different integrin receptors and extracellular matrix proteins to sense the mechanoenvironment (Fig. 1.2) (Seong et al. 2013). In this study, Seong et al. (2013) found that matrix protein fibronectin (FN)-mediated FAK activation is dependent on mechanical tension, while FAK can be sufficiently activated on type I collagen independent of tension. Therefore, different ECM proteins can differentially transmit or shield mechanical forces from the environment to the functional molecules in the cell and regulate cellular functions (Fig. 1.2) (Seong et al. 2013). In this case, FRET biosensors provided a precise picture of in situ molecular signals at real time in single

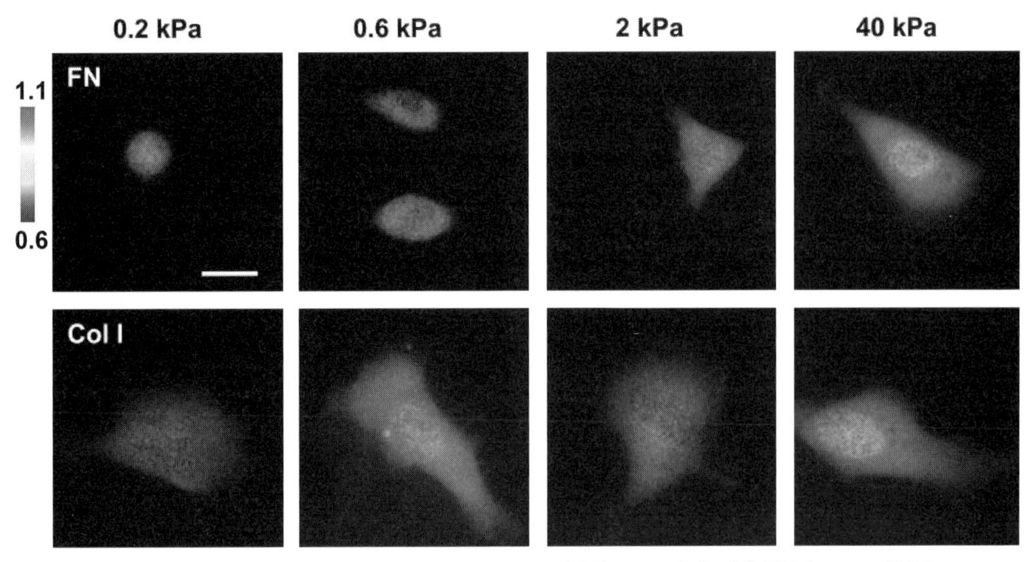

Figure 1.2 **FAK activation is dependent on substrate rigidity coupled with FN, but not Col I.** The representative ECFP/FRET ratio images of FAK biosensor in HT1080 cells cultured on the FN- (Top Panels) or Col I-coated (Bottom Panels) PA gels with different stiffness, as indicated. The color bar on the left shows the ECFP/YPet ratio values, with cold and hot colors representing low and high ratio values, as indicated.

This research was originally published in the journal *PNAS*. Seong et al. "Distinct biophysical mechanisms of FAK mechanoactivation by different extracellular matrix proteins." *PNAS* 2013, 110(48):19372–19377.

cells, which is especially important for the investigation of mechanosensing and mechanotransduction during cell-ECM interaction at focal adhesions (Bershadsky, Balaban, and Geiger 2003; Geiger, Spatz, and Bershadsky 2009).

1.5 Micropatterns, FRET, and fluorescence proteins (FPs)

To investigate how cells perceive and interpret the microenvironment to coordinate intracellular molecular signals and ultimately physiological functions, many technologies have been developed to manipulate the microenvironment around the cells and to study the effect of mechanical cues on cell behaviors. Among these, Kim et al. (2009) has combined micropatterning technologies with genetically encoded FRET biosensors to study spatiotemporal RhoA and Src activities in different microenvironments. A micropattern via micromolding in capillaries has been applied on glass coverslips on which cells attach (Fig. 1.3). After applying epidermal growth factor (EGF) on cells expressing Src or RhoA FRET biosensors, it has been shown that the microenvironment may regulate the spatiotemporal signaling network of RhoA and Src in live cells (Kim et al. 2009). Briefly, EGF induced a decrease of RhoA and an increase of Src activities with biphasic time courses in cells seeded on micropatterned glass, while the induced decrease of RhoA and increase of Src activities are relatively monophasic in

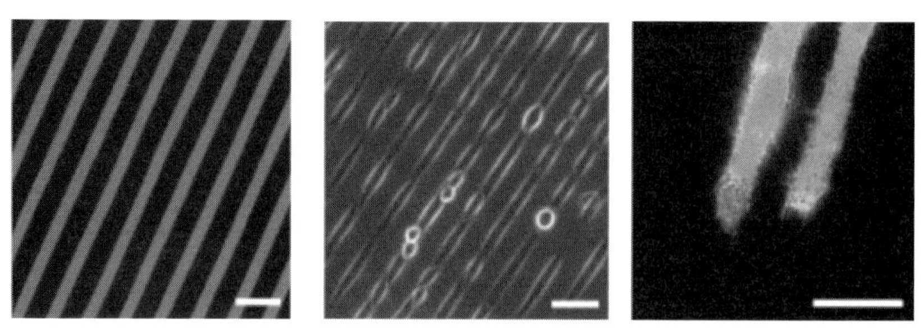

Figure 1.3 **Micropatterning on the silanized glass.**
Left: Fluorescence image of a pattern of NHS-rhodamine-conjugated Fn on silanized glass; Middle: Phase-contrast image (40x) of transfected HeLa cells cultured on the patterned surface; Right: Emission ratio images (100x) of the RhoA biosensor in transfected HeLa cells cultured on the patterned surface. Scale bars: 40 μm.
This research was originally published in the journal *Small*. Kim et al. "Visualizing the effect of microenvironment on the spatiotemporal RhoA and Src activities in live cells by FRET." *Small* 2009, 5(12): 1453–1459.

cells on glass without pattern. These results indicate that the cells may be capable of probing and adapting to the mechanical environment in response to stimuli. Further experiments show that the inhibition of Src activity abolishes this biphasic RhoA response toward EGF in cells on patterns. These results suggest that the observed micropattern effect on the biphasic RhoA activation in the cells is mediated by Src (Kim et al. 2009).

The spatial distribution of molecular signals within cells is crucial for cellular functions. Polarized molecular activities can guide the cell toward persistent and directional migration. Micropatterned strips of cell-adhesive extracellular matrix protein fibronectin separated by the nonadhesive copolymer, pluronic 127, have been developed to study the polarized molecular activities in live cells (Ouyang et al. 2008; Lu et al. 2011). In this system, cells are constrained to the fibronectin strips, which are about 10 μm in width. When two cells form a junction, they are stably polarized with one end connecting to a neighboring cell (the junction end) and the other end free of cell-cell contact (the free end) (Lu et al. 2011; Ouyang et al. 2013). Therefore, this system provides an ideal microenvironment in which to study polarized molecular activities and interactions in live cells.

Integrating the polarized micropatterns with the newly discovered high-efficiency FRET pair of enhanced CFP (ECFP) and YPet, the authors were able to visualize highly polarized Rac activity concentrating at the free end, which was activated by the growth factor stimulation and mediated by the Src-family kinase expression. In contrast to the polarized Rac activation, a global activation of Src was detected in these cells, which was modulated by the well-balanced endogenous Rac activity. These results demonstrate that the activations of Rac and Src have different spatial patterns, which mutually regulate each other (Ouyang et al. 2008).

Furthermore, the one-dimensional feature of the striped micropattern provides an ideal system for quantitative analysis of the observed fluorescence signals. For this purpose, an image analysis software package, Fluocell, was developed to quantify and analyze the dynamic regulation of signaling molecules in cells on pattern (Lu et al. 2011; Ouyang et al. 2013). In these studies, cells on micropattern were transfected with PH-Akt-GFP (the PH domain of Akt was fused to a green fluorescence protein, GFP), or a PAK-PBD-YFP (the PAK domain of PBD fused to a YFP), to monitor the intracellular PI3 K and Rac activities, respectively. Briefly, the cells with a junction end and a free end were stimulated with the growth factor and observed with polarized distribution of PH-Akt-GFP and PAK-PBD-YFP. The recorded fluorescence intensity images were aligned along the stripe direction, mapped to 1D, quantified, and normalized, so that different cells have a normalized uniform length and total fluorescence intensity at the longitudinal direction. This analytical approach allows the statistical analysis of image data from multiple cells, as well as the quantitative comparison of the sequential activation of PI3 K and Rac in the cells on pattern. These results revealed different polarization patterns of PI3 K and Rac1 activities induced by growth factor, suggesting that the initiation of edge extension occurred before PI3 K activation, which led to a stable extension of the free end followed by Rac activation (Lu et al. 2011). Further substantial experiments and analysis showed that PI3 K and Rac polarization depend specifically on the N-cadherin-p120-catenin complex at the cell-cell junction, whereas myosin II light chain and actin filament polarization depend on the junctional N-cadherin-β-catenin complex (Ouyang et al. 2013). Therefore, the integration of micropattern, fluorescence live cell imaging, and computational analysis provides a powerful tool to quantitatively investigate the spatiotemporal regulation of molecular signals in live cells under different microenvironment conditions.

1.6 Quantum dots as FRET donors

FRET technology has allowed single-cell imaging with high spatiotemporal resolution for applications in biology and medicine. The brightness of current available FRET pairs has limited broad usage, however, and a relatively high concentration of biosensors is needed to obtain meaningful FRET signals. The relatively broad spectrums of contemporary dyes also limit their scientific applications as FRET donors and acceptors. Quantum dots are inorganic semiconductor nanocrystals that have sharp, size-tunable symmetric emission spectra. By changing the size of quantum dots, it is possible to obtain specific emission wavelengths (Fig. 1.4), which therefore can be combined with almost every dye to form a FRET pair (Alivisatos, Gu, and Larabell 2005; Michalet et al. 2005; Smith et al. 2008). Most importantly, quantum dots have an order of magnitude higher extinction coefficient and quantum yield than conventional dyes, which make it suitable for molecular imaging (Alivisatos et al. 2005; Resch-Genger et al. 2008). Quantum dots, with their special core-shell inorganic composition, are also resistant to photobleaching in comparison to organic dyes (Reiss, Bleuse, and Pron 2002; Sukhanova et al. 2004). With these attractive features, they have been widely used in live cells, and even for in vivo imaging (Akerman et al. 2002; Dubertret et al. 2002; Ballou et al. 2004).

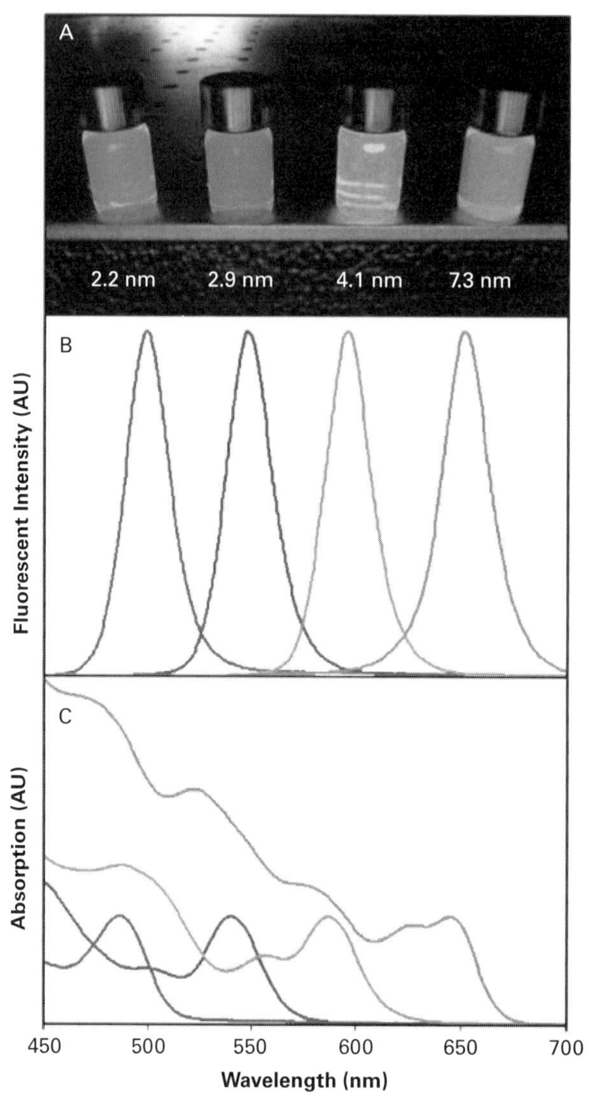

Figure 1.4 **Size-dependent optical properties of cadmium selenide QDs dispersed in chloroform, illustrating quantum confinement and size-tunable fluorescence emission.**
(A) Fluorescence image of four vials of monodisperse QDs with sizes ranging from 2.2 nm to 7.3 nm in diameter. This image was obtained with ultraviolet illumination. (B) Fluorescence spectra of the same four QD samples. Narrow emission bands (23–26 nm FWHM or full-width at half-maximum) indicate narrow particle size distributions. (C) Absorption spectra of the same four QD samples. Notice that the absorption spectra are very broad, allowing a broad wavelength range for excitation. Both the absorption and emission intensities are plotted in arbitrary units (AU).

Reprinted from Smith et al. "Bioconjugated quantum dots for in vivo molecular and cellular imaging." *Journal of Advanced Drug Delivery Reviews* 2008, 60: 1226–1240. Copyright (2008), with permission from Elsevier.

In 2003, Mauro's group had reported utilizing quantum dots as a FRET donor for the first time. By self-assembly of quantum dots with Escherichia coli maltose-binding protein (MBP), two quantum-dot-based FRET maltose sensors with different mechanisms have been proposed. In the first approach, when maltose was added to the sensor that has a fluorescence quencher at the saccharide-binding site, molecular quenchers were substituted and quantum-dot fluorescence increased in a systematic manner. In the second mechanism, the quantum dot–Cy3-Cy3.5 two-step FRET biosensor was presented. When maltose was detected, it substituted Cy3.5 and changed the sensor emission from Cy3.5 to Cy3 (Medintz et al. 2003).

In order to conjugate quantum dots to dye-labeled peptides, Medintz's group has demonstrated two linkage chemistries to form quantum-dot-based FRET pairs. The first one is by carbodiimide chemistry to form covalent bonds on terminal carboxyl groups of DHLA-PEG coated quantum dots to form a caspase 3 proteolytic biosensor. The other approach is by self-assembly between hexahistidine (His6)-appended peptides and ZnS shells of quantum dots to form tight noncovalent conjugation to function as a calcium FRET biosensor (Prasuhn et al. 2010).

1.7 Prospective

In summary, the integration of FRET biosensors, micropattern-controlled environments, and quantum dots can illuminate a bright future for mechanosensing and mechanobiology. Future development in this field may bring high-throughput library-screening approaches to the optimization of biosensors, sophisticatedly designed 3-D micro/nano-patterned environment, and better QDs for fluorescent imaging and FRET biosensors.

References

Akerman, M. E., W. C. Chan, P. Laakkonen, S. N. Bhatia, and E. Ruoslahti. (2002). "Nanocrystal targeting in vivo." *Proc Natl Acad Sci USA* **99**(20): 12617–12621.

Alivisatos, A. P., W. Gu, and C. Larabell. (2005). "Quantum dots as cellular probes." *Annu Rev Biomed Eng* **7**: 55–76.

Arias-Salgado, E. G., S. Lizano, S. Sarkar, J. S. Brugge, M. H. Ginsburg, and S. J. Shattil. (2003). "Src kinase activation by direct interaction with the integrin beta cytoplasmic domain." *Proc Natl Acad Sci USA* **100**(23): 13298–13302.

Ballou, B., B. C. Lagerholm, L. A. Ernst, M. P. Bruchez, and A. S. Waggoner. (2004). "Noninvasive imaging of quantum dots in mice." *Bioconjug Chem* **15**(1): 79–86.

Bao, X., C. B. Clark, and J. A. Frangos. (2000). "Temporal gradient in shear-induced signaling pathway: involvement of MAP kinase, c-fos, and connexin43." *Am J Physiol Heart Circ Physiol* **278**(5): H1598–1605.

Bao, X., C. Lu, and J. A. Frangos. (1999). "Temporal gradient in shear but not steady shear stress induces PDGF-A and MCP-1 expression in endothelial cells: role of NO, NF kappa B, and egr-1." *Arterioscler Thromb Vasc Biol* **19**(4): 996–1003.

Bershadsky, A. D., N. Q. Balaban, and B. Geiger. (2003). "Adhesion-dependent cell mechanosensitivity." *Annu Rev Cell Dev Biol* **19**: 677–695.

Bissell, M. J. and W. C. Hines. (2011). "Why don't we get more cancer? A proposed role of the microenvironment in restraining cancer progression." *Nat Med* **17**(3): 320–329.

Buranachai, C. and R. M. Clegg. (2008). "Fluorescence lifetime imaging in living cells." In *Methods in Molecular Biology Fluorescent Proteins: Methods and Applications*, edited by J. Rothnagel (Totowa, NJ: Humana Press).

Butler, P. J., S. Weinbaum, S. Chien, and D. E. Lemons. (2000). "Endothelium-dependent, shear-induced vasodilation is rate-sensitive." *Microcirculation* **7**(1): 53–65.

Davies, P. F. (1997). "Overview: temporal and spatial relationships in shear stress-mediated endothelial signalling." *J Vasc Res* **34**(3): 208–211.

DePaola, N., M. A. Gimbrone, Jr., P.F. Davies, and C. F. Dewey. (1992). "Vascular endothelium responds to fluid shear stress gradients." *Arterioscler Thromb* **12**(11): 1254–1257.

Dubertret, B., P. Skourides, D. J. Norris, V. Noireaux, A. H. Brivanlou, and A. Libchaber. (2002). "In vivo imaging of quantum dots encapsulated in phospholipid micelles." *Science* **298**(5599): 1759–1762.

Frangos, J. A., T. Y. Huang, and C. B. Clark. (1996). "Steady shear and step changes in shear stimulate endothelium via independent mechanisms–superposition of transient and sustained nitric oxide production." *Biochem Biophys Res Commun* **224**(3): 660–665.

Geiger, B., J. P. Spatz, and A. D. Bershadsky. (2009). "Environmental sensing through focal adhesions." *Nat Rev Mol Cell Biol* **10**(1): 21–33.

Gimbrone, M. A., Jr. and G. Garcia-Cardena. (2013). "Vascular endothelium, hemodynamics, and the pathobiology of atherosclerosis." *Cardiovasc Pathol* **22**(1): 9–15.

Kim, T. J., J. Xu, R. Dong, S. Lu, R. Nuzzo, and Y. Wang. (2009). "Visualizing the effect of microenvironment on the spatiotemporal RhoA and Src activities in living cells by FRET." *Small* **5**(12): 1453–1459.

Kunkel, M. T., Q. Ni, R. Y. Tsien, J. Zhang, and A. C. Newton. (2005). "Spatio-temporal dynamics of protein kinase B/Akt signaling revealed by a genetically encoded fluorescent reporter." *J Biol Chem* **280**(7): 5581–5587.

Lei, L., S. Lu, Y. Wang, T. Kim, D. Mehta, and Y. Wang. (2014). "The role of mechanical tension on lipid raft dependent PDGF-induced TRPC6 activation." *Biomaterials* **35**(9): 2868–2877.

Li, S., P. Butler, Y. Wang, Y. Hu, D. C. Han, S. Unami, J.-L. Guan, et al. (2002). "The role of the dynamics of focal adhesion kinase in the mechanotaxis of endothelial cells." *Proc Natl Acad Sci USA* **99**(6): 3546–3551.

Liao, X., S. Lu, Y. Zhuo, C. Winter, W. Xu, and Y. Wang. (2012). "Visualization of Src and FAK activity during the differentiation process from hMSCs to osteoblasts." *PLoS One* **7**(8): e42709.

Lu, S., T. J. Kim, C. E. Chen, M. Ouyang, J. Seong, X. Liao, and Y. Wang. (2011). "Computational analysis of the spatiotemporal coordination of polarized PI3 K and Rac1 activities in micro-patterned live cells." *PLoS One* **6**(6): e21293.

Luo, B. H., C. V. Carman, and T. A. Springer. (2007). "Structural basis of integrin regulation and signaling." *Annu Rev Immunol* **25**: 619–647.

Makowski, L. and G. S. Hotamisligil. (2004). "Fatty acid binding proteins–the evolutionary crossroads of inflammatory and metabolic responses." *J Nutr* **134**(9): 2464S–2468S.

Medintz, I. L., A. R. Clapp, H. Mattoussi, E. R. Goldman, B. Fisher, and J. M. Mauro. (2003). "Self-assembled nanoscale biosensors based on quantum dot FRET donors." *Nat Mater* **2**(9): 630–638.

Michalet, X., F. F. Pinaud, L. A. Bentolilia, J. M. Tsay, S. Doose, J. J. Li, et al. (2005). "Quantum dots for live cells, in vivo imaging, and diagnostics." *Science* **307**(5709): 538–544.

Mitra, S. K., D. A. Hanson, and D. D. Schlaepfer. (2005). "Focal adhesion kinase: in command and control of cell motility." *Nat Rev Mol Cell Biol* **6**(1): 56–68.

Mitra, S. K. and D. D. Schlaepfer. (2006). "Integrin-regulated FAK-Src signaling in normal and cancer cells." *Curr Opin Cell Biol* **18**(5): 516–523.

Miyamoto, S., H. Teramoto, O. A. Coso, J. S. Gutkind, S. K. Akiyama, and K. M. Yamada. (1995). "Integrin function: molecular hierarchies of cytoskeletal and signaling molecules." *J Cell Biol* **131**(3): 791–805.

Miyawaki, A., J. Llopis, R. Helm, J. M. McCaffery, J. A. Adams, M. Ikura, and R. Y. Tsien. (1997). "Fluorescent indicators for Ca2+ based on green fluorescent proteins and calmodulin." *Nature* **388**(6645): 882–887.

Mochizuki, N., S. Yamashita, K. Kurokawa, Y. Obha, T. Nagai, A. Miyawaki, and A. Matsuda. (2001). "Spatio-temporal images of growth-factor-induced activation of Ras and Rap1." *Nature* **411**(6841): 1065–1068.

Nagel, T., N. Resnick, C. F. Dewey, and M. A. Gimbrone. (1999). "Vascular endothelial cells respond to spatial gradients in fluid shear stress by enhanced activation of transcription factors." *Arterioscler Thromb Vasc Biol* **19**(8): 1825–1834.

Ouyang, M., H. Huang, N. C. Shaner, A. G. Remacle, S. A. Shiryaev, A. Y. Strongin, R. Y. Tsien, et al. (2010). "Simultaneous visualization of protumorigenic Src and MT1-MMP activities with fluorescence resonance energy transfer." *Cancer Res* **70**(6): 2204–2212.

Ouyang, M., S. Lu, T. Kim, C. E. Chen, J. Seong, D. E. Leckband, F. Wang, et al. (2013). "N-cadherin regulates spatially polarized signals through distinct p120ctn and beta-catenin-dependent signalling pathways." *Nat Commun* **4**: 1589.

Ouyang, M., J. Sun, S. Chien, and Y. Wang. (2008). "Determination of hierarchical relationship of Src and Rac at subcellular locations with FRET biosensors." *Proc Natl Acad Sci USA* **105**(38): 14353–14358.

Pertz, O., L. Hodgson, R. L. Klemke, and K. M. Hahn. (2006). "Spatiotemporal dynamics of RhoA activity in migrating cells." *Nature* **440**(7087): 1069–1072.

Prasuhn, D. E., A. Feltz, J. Blanco-Canosa, K. Susumo, M. H. Stewart, B. C. Mei, A. V. Yakoviev, et al. (2010). "Quantum dot peptide biosensors for monitoring caspase 3 proteolysis and calcium ions." *ACS Nano* **4**(9): 5487–5497.

Reiss, P., J. Bleuse, and A. Pron. (2002). "Highly Luminescent CdSe/ZnSe Core/Shell Nanocrystals of Low Size Dispersion." *Nano Lett* **2**(7): 781–784.

Resch-Genger, U., M. Grabolle, S. Cavaliere-Jaricot, R. Nitschke, and T. Nunn. (2008). "Quantum dots versus organic dyes as fluorescent labels." *Nat Methods* **5**(9): 763–775.

Seong, J., M. Ouyang, T. Kim, J. Sun, P.-C. Wen, et al. (2011). "Detection of focal adhesion kinase activation at membrane microdomains by fluorescence resonance energy transfer." *Nat Commun* **2**: 406.

Seong, J., A. Tajik, J. Sun, J.-L. Guan, M. J. Humphries, S. E. Craig, A. Shekaran, et al. (2013). "Distinct biophysical mechanisms of focal adhesion kinase mechanoactivation by different extracellular matrix proteins." *Proc Natl Acad Sci USA* **110**(48): 19372–19377.

Smith, A. M., H. Duan, A. M. Mohs, and S. Nie. (2008). "Bioconjugated quantum dots for in vivo molecular and cellular imaging." *Adv Drug Deliv Rev* **60**(11): 1226–1240.

Song, Y., V. Madahar, and J. Liao. (2011). "Development of FRET assay into quantitative and high-throughput screening technology platforms for protein-protein interactions." *Ann Biomed Eng* **39**(4): 1224–1234.

Sukhanova, A., J. Devy, L. Venteo, H. Kaplan, M. Artemyev, V. Oleinikov, D. Klinov, et al. (2004). "Biocompatible fluorescent nanocrystals for immunolabeling of membrane proteins and cells." *Anal Biochem* **324**(1): 60–67.

Tardy, Y., N. Resnick, M. A. Gimbone, and C. F. Dewey. (1997). "Shear stress gradients remodel endothelial monolayers in vitro via a cell proliferation-migration-loss cycle." *Arterioscler Thromb Vasc Biol* **17**(11): 3102–3106.

Thomas, S. M. and J. S. Brugge. (1997). "Cellular functions regulated by Src family kinases." *Annu Rev Cell Dev Biol* **13**: 513–609.

Ting, A. Y., K. H. Kain, R. L. Klemke, and R. Y. Tsien. (2001). "Genetically encoded fluorescent reporters of protein tyrosine kinase activities in living cells." *Proc Natl Acad Sci USA* **98**(26): 15003–15008.

Violin, J. D., J. Zhang, R. Y. Tsien, and A. C. Newton. (2003). "A genetically encoded fluorescent reporter reveals oscillatory phosphorylation by protein kinase C." *J Cell Biol* **161**(5): 899–909.

Wang, Y., E. L. Botvinick, Y. Zhao, M. W. Berns, S. Usami, R. Y. Tsien, and S. Chien. (2005). "Visualizing the mechanical activation of Src." *Nature* **434**(7036): 1040–1045.

Wang, Y., J. Y. Shyy, and S. Chien. (2008). "Fluorescence proteins, live-cell imaging, and mechanobiology: seeing is believing." *Annu Rev Biomed Eng* **10**: 1–38.

Zacharias, D. A., J. D. Violin, A. C. Newton, and R. Y. Tsien. (2002). "Partitioning of lipid-modified monomeric GFPs into membrane microdomains of live cells." *Science* **296**(5569): 913–916.

Zhang, J., C. J. Hupfeld, S. S. Taylor, J. M. Olefsky, and R. Y. Tsien. (2005). "Insulin disrupts beta-adrenergic signalling to protein kinase A in adipocytes." *Nature* **437**(7058): 569–573.

Zhang, J., Y. Ma, S. S. Taylor, and R. Y. Tsien. (2001). "Genetically encoded reporters of protein kinase A activity reveal impact of substrate tethering." *Proc Natl Acad Sci USA* **98**(26): 14997–15002.

2 Electron microscopy and three-dimensional single-particle analysis as tools for understanding the structural basis of mechanobiology

Niels Volkmann and Dorit Hanein

2.1 Introduction

Mechanobiology focuses on how physical forces and changes in cell mechanics contribute to physiology, development, and disease. Major challenges in the field are the understanding of mechanotransduction – the mechanism by which cells sense and respond to mechanical signals – and how cells regulate and initiate changes in mechanical properties. Ultimately, both phenomena depend on molecular interactions between macromolecules. One of the key components in processes relevant to mechanobiology is the actin cytoskeleton, which is involved in cell motility, cell-substrate interactions, transport processes, cytokinesis, and the establishment and maintenance of cell morphology. This multiplicity of functions is the consequence of the diversity of filamentous actin (F-actin) structures mediated by different actin-binding proteins (ABPs), each with unique and functionally significant mechanochemical properties (Blanchoin et al. 2014; Le Clainche and Carlier 2008).

For example, actin bundles and actin networks support and stabilize cellular protrusions, invaginations, or domains of the plasma membrane and thus have profound effects on cellular shape, division, adhesion, motility, and signaling. Also, during migration cells respond to chemical and mechanical cues by adapting their shape, dynamics, and adhesion to the extracellular matrix (ECM). In this process, matrix adhesions play a critical role (Gardel et al. 2010; Parsons et al. 2010). Matrix adhesions are multicomponent complexes linking the actin cytoskeleton to integrin receptors. Changes in the mechanical environment (force) induce changes in these mechanosenstive assemblies, which in turn alter the signals they generate. These alterations in signaling regulate migration as well as proliferation, gene expression, and cell survival. The signals play a central role in homeostatic processes like inflammation and in pathological processes such as cancer and thrombosis. A small number of structural proteins are thought to link the ECM to the actin cytoskeleton via integrin receptors, and a much larger number of proteins are involved in adhesion-based signaling processes. The detailed structure of this mechanosensitive connection is under intense study but the complex, multicomponent, and dynamic nature of matrix adhesions severely limits the use of direct atomic-resolution structural approaches such as X-ray crystallography and nuclear magnetic resonance (NMR). It is clear that ABPs act as mechanosensitive switches to control the mechanical coupling between the matrix adhesions and the actin cytoskeleton, and they control the assembly of actin bundles and actin networks. However, the detailed molecular mechanisms for many of the key components have remained elusive.

Major contributions to unraveling some of the underlying mechanistic underpinnings have been provided by electron microscopy in combination with three-dimensional (3-D) single-particle analysis. Instead of providing a comprehensive description of all molecular components involved in mechanobiology that have been investigated by using these techniques, this article focuses on the contribution to the understanding of a few key components in mechanotransduction, and in the determination of mechanical properties of cells, to demonstrate the utility of electron microscopy and single-particle analysis.

2.2 Electron microscopy

Due to dramatic improvements in experimental methods, computational techniques, and equipment (Kühlbrandt 2014; Frank 2013), electron microscopy has matured into a powerful and diverse collection of methods that allow the visualization of an extraordinary range of biological assemblies at resolutions ranging from atomic (0.3 nm) to molecular (about 2–3 nm). Many of the restrictions of X-ray crystallography or NMR spectroscopy do not apply to electron microscopy. Crystalline order is helpful but not necessary, there is no upper size limit for the structures studied, the quantities of sample needed are relatively small, and cryo-methods enable the observation of molecules in their native aqueous environment (Dubochet et al. 1988). The general underlying principle of the technology is the collection of projection images of the structure under scrutiny. For successful 3-D reconstruction, a good coverage of all possible orientations is necessary. If the particles are randomly oriented or obey helical symmetry, no active tilting of the sample is required to achieve this. Owing to the low signal-to-noise ratio, several thousands of projection images are collected per sample. Image processing of these projections – primarily involving iterative orientation determination, alignment, and averaging – is then used to reassemble the underlying three-dimensional structure (Frank 2006b; Volkmann and Hanein 2009; Penczek 2010; Scheres 2012). The exact steps of image analysis and image acquisition vary according to the symmetry and nature of the specimen. The following three reconstruction strategies were particularly useful in the current context.

2.2.1 Single particle reconstruction

In principle, virtually any isolated single macromolecular assembly is eligible for reconstruction by electron microscopy and single-particle approaches. Briefly, these methods take advantage of the fact that molecular assemblies often exist as many copies in the specimen, visible as isolated particles, distinguished only by their orientations. A sufficiently large number of particle images will thus cover the complete angular range of possible orientations. Even a small number of electron microscope images (micrographs), each of a different field of view, often contain enough particles to reconstruct the assembly in three dimensions. The averaging of many copies of the structure reduces the noise and carries potential for reaching high resolution, sometimes up to near atomic (Bai et al. 2013), even for particles without symmetry. Computationally, the most challenging task is to determine and refine the particle orientations. Orientation analysis is much more straightforward for biochemically homogeneous samples. In order to generate a high-resolution reconstruction, an initial starting model at moderate resolution (~3 nm) is acquired as a first step. Methods for doing that include the "random conical tilt" method (Radermacher 1988), "angular reconstitution" (van Heel 1987), the use of models of related structures, or electron tomography. The second stage involves cyclic model-based refinement. At each stage, more accurate values for the orientations for each particle are obtained by matching it against projections of the current model. Then, a refined reconstruction is calculated and the procedure is iterated exhaustively until convergence. Imaging contrast limitations affecting the fidelity of particle alignment lead to a lower size limit for the application of the technology – currently at about 150–250 kDa. For more detailed information on single-particle reconstruction see the textbook by Frank (Frank 2006b). The attainable resolution for a particular project depends on several factors including the homogeneity of the sample, the number of particles in the data set, the accuracy of the orientation parameters obtained, and the quality of the projection images. Recent advances in computational classification make it now possible to efficiently separate subpopulations of heterogeneous samples *in silico* (Scheres 2012; Spahn and Penczek 2009; Xu et al. 2011).

2.2.2 Processing of helical assemblies

Many biological assemblies important in the context of mechanobiology occur naturally in helical form, in particular cytoskeleton filaments. These filamentous structures are especially attractive targets for electron microscopy as they are not usually amenable to crystallization due to their natural tendency to polymerize into filaments. Filamentous structures were traditionally targeted by specifically developed helical reconstruction approaches that make explicit use of the helical symmetry in Fourier space (DeRosier and Moore 1970). Diffraction from a helix occurs on a set of layer lines related to the pitch repeat. Possible complications that arise during this type of helical analysis include partial decoration of the helix under study, bending of the helix in the plane of the image and perpendicular to the plane, or imperfect helical symmetry. To overcome these shortcomings, hybrid techniques that are primarily based on addition of aspects from single-particle analysis to the reconstruction process (Egelman 2007; Egelman 2000; Holmes et al. 2003) have gained popularity in recent years. Helical structures that were previously inaccessible to image analysis for these reasons can now be processed (Egelman 2007). Briefly, the helical projection images are subdivided into short overlapping segments that are then treated as single-particle images and processed accordingly. This strategy makes the whole gambit of single-particle processing methodology, most notably methods for separating heterogeneous classes from each other, available to processing of helices. Helical symmetry is then determined and applied using these single-particle reconstructions, significantly enhancing the fidelity of the technique.

2.2.3 Electron tomography and sub-tomogram averaging

Electron tomography is the most general method for obtaining three-dimensional information by electron microscopy (Frank 2006a; Lucic et al. 2005). In this technique, a series of images is taken of a single specimen as the specimen is tilted over a wide range of angles. The main advantage of this approach is the ability to generate 3-D structures of individual assemblies without the need for averaging, opening the way to identify unique conformations and single assemblies. On the other hand, electron tomography data acquisition schemes limit the attainable resolution. Today, with the use of computer-controlled microscopes and the availability of direct electron detection cameras, it has become possible to image large-scale structures at a resolution of better than 4 nm. While the main area of application for electron tomography are large, multicomponent objects, structures of purified single macromolecules can also be determined. In this case, the three-dimensional, noisy tomograms of the single macromolecular assemblies are aligned and averaged in three dimensions. In order to generate an averaged, high-quality three-dimensional structure, only a few hundred particles are required using tomographic techniques (Koster et al. 1997), an advantage of using electron tomography instead of single-particle analysis that requires several thousand particles for structure determination. The resolution currently achievable by this tomographic single-particle approach can exceed 1 nm for purified molecules (Bartesaghi et al. 2012; Schur et al. 2013) and can approach 2 nm for particles extracted from more complex environments.

2.3 Mechanical properties of cells

To a significant degree, the mechanical properties of cells are determined by the properties of the higher-order F-actin assemblies in the cell. The properties of these assemblies, in turn, are primarily determined by ABPs that arrange actin filaments into structures such as tight, stiff bundles or into more elastic actin filament networks. Two prominent examples – the Arp2/3 complex and fimbrin – have been studied with great success by electron microscopy and various single-particle analysis approaches.

2.3.1 Arp2/3 complex

The Arp2/3 complex is a key component of many of the cells' actin filament networks by virtue of its ability to initiate actin filament branches (daughter filaments) at an angle on the sides of preexisting "mother" filaments (Pollard 2007). Conserved among eukaryotes, Arp2/3 complex consists of two actin-related proteins, Arp2 and Arp3, and five additional subunits named ARPC1-5. In the presence of adenosine triphosphate (ATP) and nucleation promoting factors (NPFs) the complex binds to the side of a mother filament and initiates a daughter filament, which grows at its free barbed end. Although actin filaments are flexible, branch junctions appear to be very rigid, thus significantly impacting the mechanical properties of the underlying networks (Blanchoin et al. 2014). The initial structural models of Arp2 and Arp3 were based on their sequence homology with actin and led to the hypothesis that the Arp2/3 complex nucleates actin filaments through an interaction between the barbed end surface of Arp2 and Arp3 and the pointed ends of actin monomers (Kelleher et al. 1995). The structure of inactive Arp2/3 complex has been well established by X-ray crystallography (Robinson et al. 2001; Nolen et al. 2004; Nolen and Pollard 2007; Nolen et al. 2009; Luan and Nolen 2013) but shows Arp2 and Arp3 in an arrangement incompatible with actin filament growth. Crystallization attempts of activated Arp2/3 complex until now were unsuccessful, and the only structure of Arp2/3 complex with bound NPFs shows essentially the same conformation as the inactive form of the complex, exposing a secondary, low-affinity NPF binding site (Ti et al. 2011).

Electron microscopy in conjunction with various single-particle analysis approaches provided essential contributions to further our understanding of Arp2/3-mediated actin branch formation, in particular that of the primary NPF-bound and the active conformations of Arp2/3 complex. Single-particle reconstructions of Arp2/3 complex and projection averages of branch junctions (Volkmann et al. 2001) showed clearly that a significant conformational change takes place during branch formation. Comparison of single-particle reconstructions of wild-type Arp2/3 complex and mutants with modified activation behavior showed clear hints for conformational changes in the complex before binding to the mother filament occurs (Martin et al. 2005), most likely triggered by binding of NPFs to the complex. A major breakthrough was the reconstruction of the entire Arp2/3 branch junction at 2.6 nm resolution (Fig. 2.1A, B), obtained by electron tomography and sub-tomogram averaging (Rouiller et al. 2008). This reconstruction enabled the construction of an atomic model of the entire branch junction including the Arp2/3 complex, twelve mother filament protomers, and four daughter filament protomers. Through the use of statistical fitting approaches (Volkmann and Hanein 1999; Volkmann 2009), conformational changes in the complex itself as well as in the mother filament could be deduced and quantified (Rouiller et al. 2008). All subunits of the Arp2/3 complex are involved in interactions with the mother filament, which is significantly altered in two of the participating protomers. Arp2 and Arp3 adopt a filament-like conformation presenting the barbed ends of the Arps as a template for filament growth to the daughter filament. The atomic model was later used as a basis for molecular dynamics

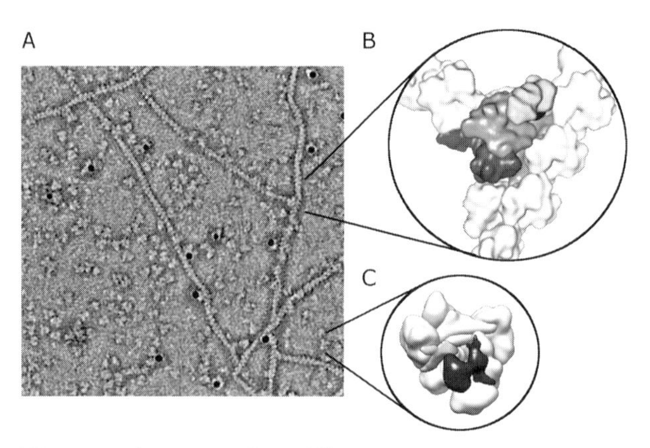

Figure 2.1 **Electron microscopy of Arp2/3 complex.**
(A) A field of few of actin filaments with branch junctions mediated by Arp2/3 complex with unbound Arp2/3 complexes in the background.
(B) Low-resolution representation of the model of the Arp2/3-mediated branch junction. Actin protomers are shown in solid white representation, Arp2/3 complex subunits are shown in different shades of gray.
(C) Representation of the model of Arp2/3 complex (white) with bound NPF (dark gray).

simulations of the entire branch junction, generating further insights about the dynamic behavior of the branch junction (Pfaendtner et al. 2012). Single-particle reconstructions of Arp2/3 complex with several bound NPFs (Fig. 2.1C) shed further light on the activation process (Xu et al. 2011). Interestingly, the primary binding site of NPFs overlaps with the mother-filament binding site, revealing an additional, unexpected level of regulation, consistent with a multiple-switch mechanism necessary to finely tune the kinetics of branch formation in vivo (Higgs and Pollard 1999; Marchand et al. 2001). Sub-tomogram averages of Arp2/3 complexes bound to the pointed ends of actin filaments reveal that a conformation similar to that in the branch junction, providing evidence for the existence of two distinct activation pathways for Arp2/3 complex, one in the context of branch formation, one in the context of pointed-end capping, with essentially the same conformational end point (Volkmann et al. 2014).

2.3.2 Fimbrin

Fimbrin is a member of the calponin homology (CH) domain superfamily of actin crosslinking proteins, which also includes alpha-actinin, beta-spectrin, and filamin. The superfamily is characterized by two tandemly arranged 125 residue CH domains. The majority of these actin crosslinking proteins form noncovalent homodimers, resulting in bivalent molecules supporting the crosslinking of actin filaments. The specific features of the dimerization domains control the separation and relative orientation of the actin binding domains (ABDs) and function as critical determinants of the mechanical properties of the resulting actin networks (Matsudaira 1991). In addition to crosslinking actin filaments, CH-domain-containing proteins also bind intermediate filaments and some proteins involved in signal transduction, linking them to the actin cytoskeleton. Fimbrin, also known as plastin, is unique among the CH-domain cross-linking proteins because it contains two tandem repeats of the ABD within a single polypeptide chain. Sequence homology and analysis of biochemical properties show that fimbrin is highly conserved from yeast to humans, indicating its important role in cellular function. Owing to the close proximity of its ABDs, fimbrin induces the formation of tightly bundled actin filament assemblies. Fimbrin is involved in the formation of actin bundles in intestinal microvilli, the hair cell stereocilia of the inner ear (Tilney et al. 1983; Tilney et al. 1995) and in the formation of fibroblast filopodia (Chafel et al. 1995). The overall domain organization of the molecule has been defined by X-ray crystallography (Klein et al. 2004), revealing a substantial amount of plasticity between the two ABDs.

Actin filaments or F-actin assemblies cannot be crystallized because the helical symmetry of the filament is incompatible with any possible crystal symmetry. Thus, structural studies of interfaces of ABDs – including fimbrins – with F-actin are not accessible by X-ray crystallography. For single-particle approaches and electron microscopy, on the other hand, helical symmetry is not only a factor that does not hamper the reconstruction process, it is in fact helpful by providing all views necessary for successful 3-D reconstruction (DeRosier and Moore 1970) and by providing symmetry constraints for averaging (Egelman 2007). Full advantage of these facts was taken to derive reconstructions of F-actin with bound fimbrin ABDs (Hanein et al. 1998; Hanein et al. 1997; Galkin et al. 2008), enabling the fitting of atomic models and analysis of the interfaces (Fig. 2.2A, B). Interestingly, the two ABDs, which were analyzed as separate domains bound to actin, employ significantly different binding interfaces on actin. Both domains induce significant changes in F-actin when binding. Comparison with the crystal structures of the fimbrin core (Klein et al. 2004) indicates that conformational changes within the core need to occur to allow formation of a crosslink. Electron microscopy was used to investigate the crosslinking rules of two-dimensional arrays formed by fimbrin (Fig. 2.2C), which suggest a cooperative coupling between actin polymerization, fimbrin binding, and crossbridge formation (Volkmann et al. 2001).

2.4 Mechanotransduction

Forces are transmitted from the ECM at matrix adhesion through integrins and adaptor proteins to the actin cytoskeleton. Two of the major components in this connection – vinculin and talin – have been extensively studied by electron microscopy and single-particle analysis approaches to provide essential information on the molecular mechanisms underlying matrix adhesion formation.

2.4.1 Vinculin

Vinculin is a conserved component and an essential regulator of cell morphology and migration (Zamir et al. 2000), which is central to brain and heart development

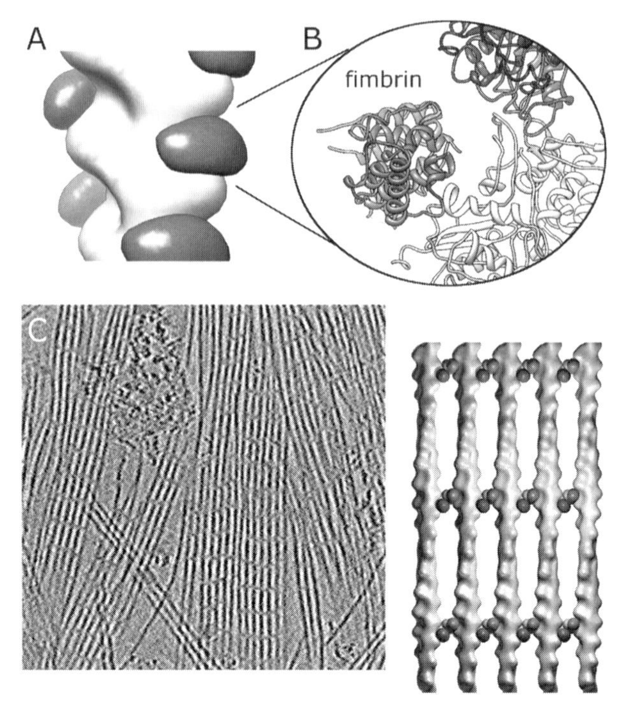

Figure 2.2 **Electron microscopy of fimbrin assemblies.**
(A) Reconstruction of F-actin (white) with bound fimbrin ABD1 (gray).
(B) Atomic model of the interface between fimbrin ABD1 (gray) and F-actin. Fimbrin ABD1 interacts with two adjacent actin protomers (white and dark gray) in the filament.
(C) Slice through a tomogram of crosslinked actin filaments. The actin filaments are visible as long vertical lines, the crosslinks appear as horizontal striations. A low-resolution representation of the derived model for actin filaments (light gray) crosslinked by fimbrin (dark gray) is shown on the right.

(Xu et al. 1998), and preservation of normal muscle function (Barstead and Waterston 1991). Hearts of vinculin-null mice are predisposed to stress-induced cardiomyopathy and mutations and deletion of metavinculin, a vinculin splice variant, were correlated with idiopathic dilated cardiomyopathy. There is ample evidence that vinculin functions in transducing force across cell membranes (Grashoff et al. 2010), in regulating matrix adhesion turnover (Saunders et al. 2006), in adhesion dynamics at the leading edge of migrating cells (Thievessen et al. 2013), and in controlling cell survival (Subauste et al. 2004). Vinculin contains five helical bundle domains (Bakolitsa et al. 2004), four of which making up the globular head domain (Vh), which is attached to a C-terminal helical bundle (Bakolitsa et al. 1999), the vinculin tail (Vt), by a flexible linker (Winkler and Taylor 1996). It is believed that in the cytosol the molecule adopts the compact autoinhibited form observed in the crystal structure (Bakolitsa et al. 2004), in which Vh and Vt form intermolecular interactions. When these interactions are relieved, the binding sites for several molecules – including talin – are exposed (Jockusch and Rudiger 1996; Zamir et al. 2000; DeMali et al. 2002). Vinculin activation is mediated by multiple mechanisms, including ligand binding to Vh and Vt, mechanical force, and phosphorylation (Peng et al. 2011). Binding to talin recruits vinculin to matrix adhesions (Burridge and Mangeat 1984). Vinculin's actin-binding site resides in Vt and is deactivated in the autoinhibited form of the full-length molecule (Johnson and Craig 1995). Full-length vinculin was shown to promote actin bundling (Jockusch and Isenberg 1981) and the existence of a cryptic dimerization site in Vt, which is activated upon binding to F-actin, was demonstrated (Johnson and Craig 2000). Förster (or fluorescence) resonance energy transfer (FRET) experiments were used to demonstrate conformational changes in full-length vinculin upon actin binding (Chen et al. 2005) and established the importance of this actin-binding conformation of vinculin for regional cellular dynamics in the context of living cells.

Similar to fimbrin, the interface structure of Vt and F-actin is not accessible by X-ray crystallography owing to the helical symmetry constraints of F-actin. Electron microscopy and single-particle approaches that take advantage of the helical symmetry were used to reconstruct F-actin with bound Vt (Janssen et al. 2006). Statistics-based fitting allowed characterization of the interface between F-actin and Vt (Fig. 2.3A) and revealed that the F-actin interface of Vt is partially blocked by Vh in the crystal structure of the inactive form, explaining why full-length vinculin does not bind strongly to F-actin and showing how head-tail dissociation relieves this steric hindrance. Vinculin contacts two adjacent F-actin protomers, providing an explanation why vinculin does

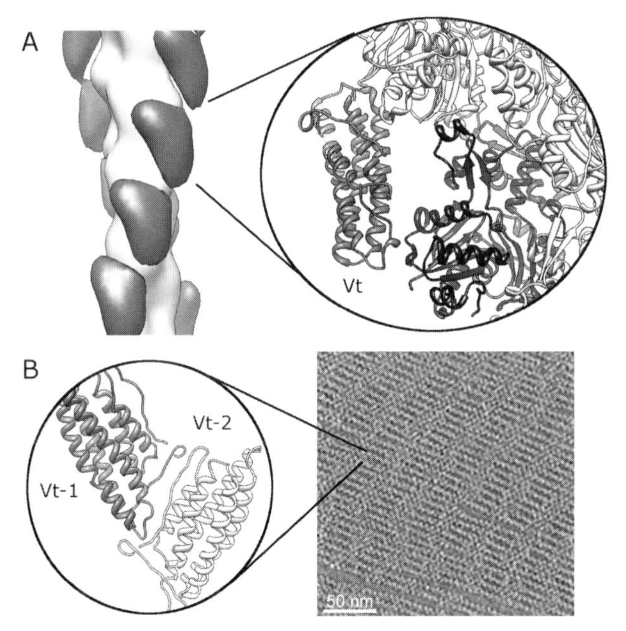

Figure 2.3 **Electron microscopy of vinculin assemblies.**
(A) Reconstruction of F-actin (white) with bound vinculin tail (gray). The right-hand side shows the atomic model of the vinculin tail (gray) bound to two adjacent actin protomers (white and dark gray) in the filament.
(B) A slice through a tomographic reconstruction of actin filaments crosslinked by vinculin tail. The left shows the atomic model of the vinculin tail dimer (monomers Vt-1 and Vt-2) derived form the data.

not bind to actin monomers. Electron tomography data from actin-filament arrays crosslinked by Vt (Janssen et al. 2006) allowed providing the molecular details of actin bundling through fitting of adjacent actin filaments with bound Vt to elucidate dimer formation (Fig. 2.3B). The analysis indicated that actin binding triggers a local conformational change in Vt that exposes cryptic dimerization surfaces in Vt, potentiating the formation of crosslinks when two Vt monomers on apposing filaments are presented in the correct position and orientation. Reconstructions of the tail domain of the muscle-specific splice variant metavinculin (MVt) bound to F-actin were also obtained by electron microscopy and single-particle analysis of helical segments (Janssen et al. 2012). The reconstruction showed that MVt binds to F-actin in the same manner as Vt and that the unique 68-residue insert that distinguishes MVt from Vt is positioned to block the dimerization site providing an explanation why MVt does not bundle actin filaments. Interestingly, MVt induces severing of actin filaments instead, most likely through the involvement of a gelsolin-like motif within the insert (Janssen et al. 2012). This ability of MVt may be essential for modulating compliance of vinculin-induced actin bundles when exposed to rapidly increasing external forces.

2.4.2 Talin

Talin is one of the most prominent cytosolic proteins that have been shown to interact with vinculin and is a key player in coupling the integrin family of cell adhesion molecules to the actin cytoskeleton at matrix adhesions (Critchley and Gingras 2008; Critchley 2009). There are two talin isoforms, talin-1 and talin-2, with 74% sequence identity (Debrand et al. 2009). Disruption of the talin-1 gene compromises cell spreading and matrix adhesion assembly in embryonic stem cells and talin-1-null fibroblasts derived from these cells cannot support the initial weak 2 pN linkage between fibronectin/integrin complexes and actomyosin (Jiang et al. 2003) or support the assembly of matrix-adhesion structures in response to external force (Giannone et al. 2003). Talin is comprised of an N-terminal head (residues 1–400) linked to a large flexible rod (residues 482 to the C-terminus) by a flexible linker. The talin head domain contains the primary binding site for the beta-integrin cyto-domains (Anthis et al. 2009; Wegener et al. 2007). The crystal structure of the talin-1 head shows that it adopts a rather extended structure (Elliott et al. 2010). The talin rod contains sixty-two amphipathic helices that are organized into thirteen successive helix bundles (Goult et al. 2013) and harbors two actin-binding sites (Hemmings et al. 1996) and numerous binding sites for vinculin (Gingras et al. 2005).

The C-terminal helix of the talin rod contains a dimerization domain, which is necessary to form the biologically relevant talin homodimer. The best-characterized actin-binding site is located in the helix bundle next to the dimerization domain and requires the formation of a homodimer for actin binding. Electron microscopy and single-particle analysis in conjunction with Small Angle X-ray Scattering (SAXS) and X-ray crystallography (Gingras et al. 2007) were used to derive a model of the dimeric construct of the terminal helices and the preceding helix bundles bound to F-actin (Fig. 2.4). The analysis revealed an asymmetric pincer-like arrangement of the

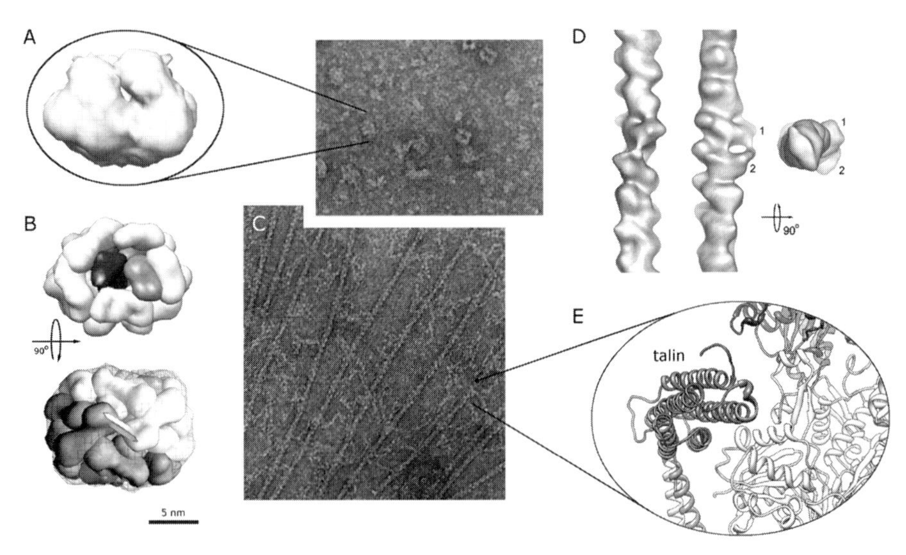

Figure 2.4 **Electron microscopy of talin assemblies.**
(A) Reconstruction of the inactive full-length talin-1 homodimer.
(B) Model of the domain arrangement in the talin-1 dimer. The two heads are shown in different shades of gray. The rod of one monomer is shown in white, the rod of the second in light gray. The dimerization helices are symbolized as arrows.
(C) Field of actin filaments with bound talin rods.
(D) Actin filament with single bound C-terminal talin ABD dimer (monomers are indicated by numbering).
(E) Atomic model of one talin ABD (gray) bound to two adjacent actin protomers (white and dark gray) in the filament.

two helix bundles on the filament with the dimerization domain holding the arrangement in place. A second binding site of talin, which resides in the talin head, was characterized by electron microscopy and helix-based single-particle analysis (Lee et al. 2004), allowing the identification of the F-actin site involved in the interaction. Owing to its high degree of flexibility, the structure of the full-length talin molecule has so far not been accessible to X-ray crystallography. Electron microscopy and single-particle approaches were used to derive a three-dimensional reconstruction of the inactive talin-1 homodimer (Goult et al. 2013), revealing a compact globular shape (Fig. 2.4A). Because of the relatively low resolution of the reconstruction (2 nm), fitting of the individual thirteen helical rod bundles and the talin head into the reconstruction is highly ambiguous. However, by including inter-bundle V-angle constraints derived from SAXS and NMR analyses of overlapping talin-1 polypeptides (Goult et al. 2013), a unique fit can be deduced (Fig. 2.4B). This fit allowed characterization of the general arrangement of the talin-1 domains within the inactive dimer. The two heads are arranged side-by-side with the two rods wrapping around this central feature, burying the actin binding sites and presumably a large fraction of the vinculin binding sites. Upon activation and opening of dimer, the actin binding sites and many of the vinculin binding sites become exposed, allowing the assembly of a talin-based linkage between integrins and the actin cytoskeleton (Fig. 2.5).

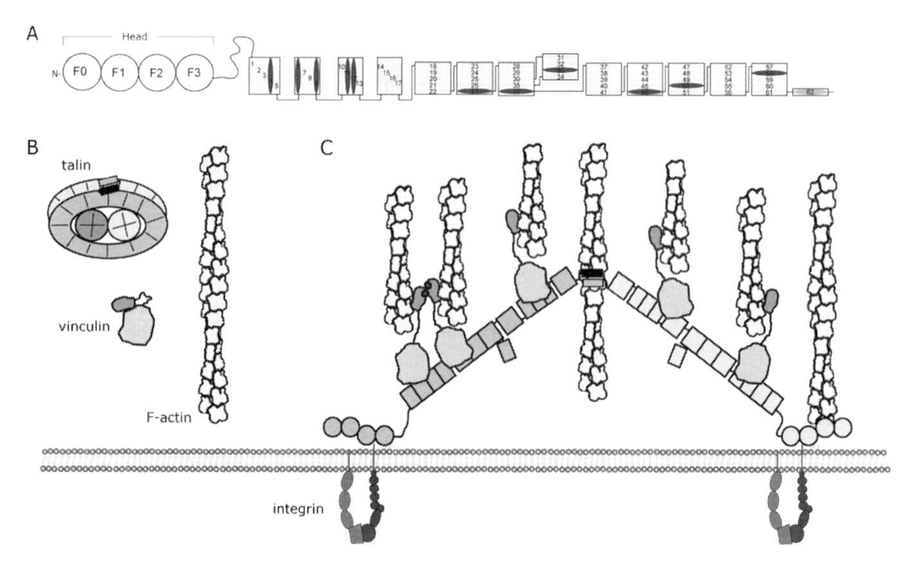

Figure 2.5 **Schematic drawing of the integrin-actin linkage.**
(A) Domain organization of talin. The four domains in the talin head are denoted F0-F3. Helix bundles in the talin rod are shown as squares, helices are shown in light gray with consecutive numbers. Vinculin binding sites are indicated in dark gray.
(B) An actin filament and the inactive forms of talin and vinculin.
(C) After activation, talin opens up and can bind directly to integrin and actin. Vinculin can bind to talin and actin filaments and can form crosslinks between actin filaments.

2.5 Outlook

Electron microscopy and single-particle analysis approaches have shown great utility as tools for understanding the molecular underpinnings of cellular processes relevant to mechanobiology. With the current surge in equipment and software advances (Kühlbrandt 2014; Frank 2013), it can be expected that these techniques will continue to provide indispensable structural understanding to the field. Furthermore, the ideal positioning of these approaches for combining information from atomic-resolution techniques through in-situ electron tomography (Volkmann 2012) with dynamic information from light microscopy (Hanein and Volkmann 2011), will likely ensure a key role in the future of mechanobiology research, not only for the understanding of the underlying biological processes but also for the understanding of molecular-level consequences of manipulations introduced into the cellular environment by complementary mechanobiological approaches.

Acknowledgments

This work was supported by National Institutes of Health NIH Program Project Grants P01-GM098412 and P01-GM066311 to DH and NV.

References

Anthis, N. J., K. L. Wegener, F. Ye, C. Kim, B. T. Goult, E. D. Lowe, I. Vakonakis, et al. (2009). "The structure of an integrin/talin complex reveals the basis of inside-out signal transduction." *EMBO J* **28**: 3623–3632.

Bai, X. C., I. S. Fernandez, G. McMullan, and S. H. Scheres. (2013). "Ribosome structures to near-atomic resolution from thirty thousand cryo-EM particles." *elife* **2**: e00461.

Bakolitsa, C., D. M. Cohen, L. A. Bankston, A. A. Bobkov, G. W. Cadwell, L. Jennings, D. R. Critchley, et al. (2004). "Structural basis for vinculin activation at sites of cell adhesion." *Nature* **430**: 583–586.

Bakolitsa, C., J. M. de Pereda, C. R. Bagshaw, D. R. Critchley, and R. C. Liddington. (1999). "Crystal structure of the vinculin tail suggests a pathway for activation." *Cell* **99**: 603–613.

Barstead, R. J., and R. H. Waterston. (1991). "Vinculin is essential for muscle function in the nematode." *J Cell Biol* **114**: 715–724.

Bartesaghi, A., F. Lecumberry, G. Sapiro, and S. Subramaniam. (2012). "Protein secondary structure determination by constrained single-particle cryo-electron tomography." *Structure* **20**: 2003–2013.

Blanchoin, L., R. Boujemaa-Paterski, C. Sykes, and J. Plastino. (2014). "Actin dynamics, architecture, and mechanics in cell motility." *Physiol Rev* **94**: 235–263.

Burridge, K., and P. Mangeat. (1984). "An interaction between vinculin and talin." *Nature* **308**: 744–746.

Chafel, M. M., W. Shen, and P. Matsudaira. (1995). "Sequential expression and differential localization of I-, L-, and T-fimbrin during differentiation of the mouse intestine and yolk sac." *Dev Dyn* **203**: 141–151.

Chen, H., D. M. Cohen, D. M. Choudhury, N. Kioka, and S. W. Craig. (2005). "Spatial distribution and functional significance of activated vinculin in living cells." *J Cell Biol* **169**: 459–470.

Critchley, D. R. (2009). "Biochemical and structural properties of the integrin-associated cytoskeletal protein talin." *Annu Rev Biophys* **38**: 235–254.

Critchley, D. R., and A. R. Gingras. (2008). "Talin at a glance." *J Cell Sci* **121**: 1345–1347.

Debrand, E., Y. El Jai, L. Spence, N. Bate, U. Praekelt, C. A. Pritchard, S. J. Monkley, and D. R. Critchley. (2009). "Talin 2 is a large and complex gene encoding multiple transcripts and protein isoforms." *FEBS J* **276**: 1610–1628.

DeMali, K. A., C. A. Barlow, and K. Burridge. (2002). "Recruitment of the Arp2/3 complex to vinculin: coupling membrane protrusion to matrix adhesion." *J Cell Biol* **159**: 881–891.

DeRosier, D. J., and P. B. Moore. (1970). "Reconstruction of three-dimensional images from electron micrographs of structures with helical symmetry." *J Mol Biol* **52**: 355–369.

Dubochet, J., M. Adrian, J.-J. Chang, J.-C. Homo, J. Lepault, A. W. McDowall, and P. Schultz. (1988). "Cryo-electron microscopy of vitrified specimens." *Q Rev Biophys* **21**: 129–228.

Egelman, E. H. (2000). "A robust algorithm for the reconstruction of helical filaments using single-particle methods." *Ultramicroscopy* **85**: 225–34.

Egelman, E. H. (2007). "Single-particle reconstruction from EM images of helical filaments." *Curr Opin Struct Biol* **17**: 556–561.

Elliott, P. R., B. T. Goult, P. M. Kopp, N. Bate, J. G. Grossmann, G. C. Roberts, D. R. Critchley, et al. (2010). "The structure of the talin head reveals a novel extended conformation of the FERM domain." *Structure* **18**: 1289–1299.

Frank, J. (2006a). *Electron tomography: methods for three-dimensional visualization of structures in the cell*. New York: Springer.

Frank, J. (2006b). *Three-dimensional electron microscopy of macromolecular assemblies: visualization of biological molecules in their native state*. Oxford: Oxford University Press.

Frank, J. (2013). "Story in a sample – the potential (and limitations) of cryo-electron microscopy applied to molecular machines." *Biopolymers* **99**: 832–836.

Galkin, V. E., A. Orlova, O. Cherepanova, M. C. Lebart, and E. H. Egelman. (2008). "High-resolution cryo-EM structure of the F-actin-fimbrin/plastin ABD2 complex." *Proc Natl Acad Sci USA* **105**: 1494–1498.

Gardel, M. L., I. C. Schneider, Y. Aratyn-Schaus, and C. M. Waterman. (2010). "Mechanical integration of actin and adhesion dynamics in cell migration." *Annu Rev Cell Dev Biol* **26**: 315–333.

Giannone, G., G. Jiang, D. H. Sutton, D. R. Critchley, and M. P. Sheetz. (2003). "Talin1 is critical for force-dependent reinforcement of initial integrin-cytoskeleton bonds but not tyrosine kinase activation." *J Cell Biol* **163**: 409–419.

Gingras, A. R., N. Bate, B. T. Goult, L. Hazelwood, I. Canestrelli, J. G. Grossmann, H. Liu, et al. (2007). "The structure of the C-terminal actin-binding domain of talin." *EMBO J* **27**: 458–469.

Gingras, A. R., W. H. Ziegler, R. Frank, I. L. Barsukov, G. C. Roberts, D. R. Critchley, and J. Emsley. (2005). "Mapping and consensus sequence identification for multiple vinculin binding sites within the talin rod." *J Biol Chem* **280**: 37217–37224.

Goult, B. T., X. P. Xu, A. R. Gingras, M. Swift, B. Patel, N. Bate, P. M. Kopp, et al. (2013). "Structural studies on full-length talin1 reveal a compact auto-inhibited dimer: Implications for talin activation." *J Struc Biol* **184**: 21–32.

Grashoff, C., B. D. Hoffman, M. D. Brenner, R. Zhou, M. Parsons, M. T. Yang, M. A. McLean, et al. (2010). "Measuring mechanical tension across vinculin reveals regulation of focal adhesion dynamics." *Nature* **466**: 263–266.

Hanein, D., P. Matsudaira, and D. J. DeRosier. (1997). "Evidence for a conformational change in actin induced by fimbrin (N375) binding." *J Cell Biol* **139**: 387–396.

Hanein, D., and N. Volkmann. (2011). "Correlative light-electron microscopy." *Adv Protein Chem Struct Biol* **82**: 91–99.

Hanein, D., N. Volkmann, S. Goldsmith, A. M. Michon, W. Lehman, R. Craig, D. DeRosier, et al. (1998). "An atomic model of fimbrin binding to F-actin and its implications for filament crosslinking and regulation." *Nat Struct Biol* **5**: 787–792.

Hemmings, L., D. J. Rees, V. Ohanian, S. J. Bolton, A. P. Gilmore, B. Patel, H. Priddle, et al. (1996). "Talin contains three actin-binding sites each of which is adjacent to a vinculin-binding site." *J Cell Sci* **109**: 2715–2726.

Higgs, H. N., and T. D. Pollard. (1999). "Regulation of actin polymerization by Arp2/3 complex and WASp/Scar proteins." *J Biol Chem* **274**: 32531–3254.

Holmes, K. C., I. Angert, F. J. Kull, W. Jahn, and R. R. Schroder. (2003). "Electron cryo-microscopy shows how strong binding of myosin to actin releases nucleotide." *Nature* **425**: 423–427.

Janssen, M. E., E. Kim, H. Liu, L. M. Fujimoto, A. Bobkov, N. Volkmann, and D. Hanein. (2006). "Three-dimensional structure of vinculin bound to actin filaments." *Mol Cell* **21**: 271–281.

Janssen, M. E., H. Liu, N. Volkmann, and D. Hanein. (2012). "The C-terminal tail domain of metavinculin, vinculin's splice variant, severs actin filaments." *J Cell Biol* **197**: 585–593.

Jiang, G., G. Giannone, D. R. Critchley, E. Fukumoto, and M. P. Sheetz. (2003). "Two-piconewton slip bond between fibronectin and the cytoskeleton depends on talin." *Nature* **424**: 334–337.

Jockusch, B. M., and G. Isenberg. (1981). "Interaction of alpha-actinin and vinculin with actin: opposite effects on filament network formation." *Proc Natl Acad Sci USA* **78**: 3005–3009.

Jockusch, B. M., and M. Rudiger. (1996). "Crosstalk between cell adhesion molecules: vinculin as a paradigm for regulation by conformation." *Trends Cell Biol* **6**: 311–315.

Johnson, R. P., and S. W. Craig. (1995). "F-actin binding site masked by the intramolecular association of vinculin head and tail domains." *Nature* **373**: 261–264.

Johnson, R. P., and S. W. Craig. (2000). "Actin activates a cryptic dimerization potential of the vinculin tail domain." *J Biol Chem* **275**: 95–105.

Kelleher, J. F., S. J. Atkinson, and T. D. Pollard. (1995). "Sequences, structural models, and cellular localization of the actin– related proteins Arp2 and Arp3 from Acanthamoeba." *J Cell Biol* **131**: 385–397.

Klein, M. G., W. Shi, U. Ramagopal, Y. Tseng, D. Wirtz, D. R. Kovar, C. J. Staiger, et al. (2004). "Structure of the actin crosslinking core of fimbrin." *Structure* **12**: 999–1013.

Koster, A. J., R. Grimm, D. Typke, R. Hegerl, A. Stoschek, J. Walz, and W. Baumeister. (1997). "Perspectives of molecular and cellular electron tomography." *J Struct Biol* **120**: 276–308.

Kühlbrandt, W. (2014). "Cryo-EM enters a new era." *elife* **3**: e03678.

Le Clainche, C., and M. F. Carlier. (2008). "Regulation of actin assembly associated with protrusion and adhesion in cell migration." *Physiol Rev* **88**: 489–513.

Lee, H.-S., R. M. Bellin, D. L. Walker, B. Patel, P. Powers, H. Liu, B. Garcia-Alvarez, et al. (2004). "Characterization of an actin-binding site within the talin FERM domain." *J Mol Biol* **343:** 771–784.

Luan, Q., and B. J. Nolen. (2013). "Structural basis for regulation of Arp2/3 complex by GMF." *Nat Struct Mol Biol* **20**: 1062–1068.

Lucic, V., T. Yang, G. Schweikert, F. Forster, and W. Baumeister. (2005). "Morphological characterization of molecular complexes present in the synaptic cleft." *Structure* **13**: 423–434.

Marchand, J. B., D. A. Kaiser, T. D. Pollard, and H. N. Higgs. (2001). "Interaction of WASP/Scar proteins with actin and vertebrate Arp2/3 complex." *Nat Cell Biol* **3**: 76–82.

Martin, A. C., X. P. Xu, I. Rouiller, M. Kaksonen, Y. Sun, L. Belmont, N. Volkmann, et al. (2005). "Effects of Arp2 and Arp3 nucleotide-binding pocket mutations on Arp2/3 complex function." *J Cell Biol* **168**: 315–328.

Matsudaira, P. (1991). "Modular organization of actin crosslinking proteins." *Trends Biochem Sci* **16**: 87–92.

Nolen, B. J., and T. D. Pollard. (2007). "Insights into the influence of nucleotides on actin family proteins from seven structures of Arp2/3 complex." *Mol Cell* **26**: 449–457.

Nolen, B. J., R. S. Littlefield, and T. D. Pollard. (2004). "Crystal structures of actin-related protein 2/3 complex with bound ATP or ADP." *Proc Natl Acad Sci USA* **101**: 15627–15632.

Nolen, B. J., N. Tomasevic, A. Russell, D. W. Pierce, Z. Jia, C. D. McCormick, J. Hartman, et al. (2009). "Characterization of two classes of small molecule inhibitors of Arp2/3 complex." *Nature* **460**: 1031–1034.

Parsons, J. T., A. R. Horwitz, and M. A. Schwartz. (2010). "Cell adhesion: integrating cytoskeletal dynamics and cellular tension." *Nat Rev Mol Cell Biol* **11**: 633–643.

Penczek, P. A. (2010). "Image restoration in cryo-electron microscopy." *Methods Enzymol* **482**: 35–72.

Peng, X., E. S. Nelson, J. L. Maiers, and K. A. Demali. (2011). "New insights into vinculin function and regulation." *Int Rev Cell Mol Biol* **287**: 191–231.

Pfaendtner, J., N. Volkmann, D. Hanein, P. Dalhaimer, T. D. Pollard, and G. A. Voth. (2012). "Key structural features of the actin filament Arp2/3 complex branch junction revealed by molecular simulation." *J Mol Biol* **416**: 148–161.

Pollard, T. D. (2007). "Regulation of actin filament assembly by Arp2/3 complex and formins." *Annu Rev Biophys Biomol Struct* **36**: 451–477.

Radermacher, M. (1988). "Three-dimensional reconstruction of single particles from random and nonrandom tilt series." *J Electron Microsc Tech* **9**: 359–94.

Robinson, R. C., K. Turbedsky, D. A. Kaiser, J. B. Marchand, H. N. Higgs, S. Choe, and T. D. Pollard. (2001). "Crystal structure of Arp2/3 complex." *Science* **294**: 1679–184.

Rouiller, I., X. P. Xu, K. J. Amann, C. Egile, S. Nickell, D. Nicastro, R. Li, et al. (2008). "The structural basis of actin filament branching by Arp2/3 complex." *J Cell Biol* **180**: 887–895.

Saunders, R. M., M. R. Holt, L. Jennings, D. H. Sutton, I. L. Barsukov, A. Bobkov, R. C. Liddington, E. A. Adamson, G. A. Dunn, and D. R. Critchley. (2006). "Role of vinculin in regulating focal adhesion turnover." *Eur J Cell Biol* **85**: 487–500.

Scheres, S. H. (2012). "A Bayesian view on cryo-EM structure determination." *J Mol Biol* **415:** 406–418.

Schur, F. K., W. J. Hagen, A. de Marco, and J. A. Briggs. (2013). "Determination of protein structure at 8.5 Å resolution using cryo-electron tomography and sub-tomogram averaging." *J Struc Biol* **184**: 394–400.

Spahn, C. M., and P. A. Penczek. (2009). "Exploring conformational modes of macromolecular assemblies by multiparticle cryo-EM." *Curr Opin Struct Biol* **19**: 623–631.

Subauste, M. C., O. Pertz, E. D. Adamson, C. E. Turner, S. Junger, and K. M. Hahn. (2004). "Vinculin modulation of paxillin-FAK interactions regulates ERK to control survival and motility." *J Cell Biol* **165**: 371–381.

Thievessen, I., P. M. Thompson, S. Berlemont, K. M. Plevock, S. V. Plotnikov, A. Zemljic-Harpf, et al. (2013). "Vinculin-actin interaction couples actin retrograde flow to focal adhesions, but is dispensable for focal adhesion growth." *J Cell Biol* **202**: 163–177.

Ti, S. C., C. T. Jurgenson, B. J. Nolen, and T. D. Pollard. (2011). "Structural and biochemical characterization of two binding sites for nucleation-promoting factor WASp-VCA on Arp2/3 complex." *Proc Natl Acad Sci USA* **108**: E463–E471.

Tilney, L. G., E. H. Egelman, D. J. DeRosier, and J. C. Saunder. (1983). "Actin filaments, stereocilia, and hair cells of the bird cochlea. II. Packing of actin filaments in the stereocilia and in the cuticular plate and what happens to the organization when the stereocilia are bent." *J Cell Biol* **96**: 822–834.

Tilney, L. G., M. S. Tilney, and G. M. Guild. (1995). "F actin bundles in Drosophila bristles. I. Two filament cross-links are involved in bundling." *J Cell Biol* **130**: 629–638.

van Heel, M. (1987). "Angular reconstitution: a posteriori assignment of projection directions for 3D reconstruction." *Ultramicroscopy* **21**: 111–124.

Volkmann, N. (2009). "Confidence intervals for fitting of atomic models into low-resolution densities." *Acta Crystallogr D Biol Crystallogr* **65**: 679–689.

Volkmann, N. (2012). "Putting structure into context: fitting of atomic models into electron microscopic and electron tomographic reconstructions." *Curr Opin Cell Biol* **24**: 141–147.

Volkmann, N., and D. Hanein. (1999). "Quantitative fitting of atomic models into observed densities derived by electron microscopy." *J Struc Biol* **125**: 176–184.

Volkmann, N., and D. Hanein. (2009). "Electron microscopy in the context of systems biology." *In Structural Bioinformatics*, J. Gu and P. E. Bourne, eds. New York: Wiley-Blackwell, 143–170.

Volkmann, N., C. Page, R. Li, and D. Hanein. (2014). "Three-dimensional reconstructions of actin filaments capped by Arp2/3 complex." *Eur J Cell Biol* **93**: 179–183.

Volkmann, N., K. J. Amann, S. Stoilova-McPhie, C. Egile, D. C. Winter, L. Hazelwood, J. E. Heuser, et al. (2001). "Structure of Arp2/3 complex in its activated state and in actin filament branch junctions." *Science* **293**: 2456–2459.

Wegener, K. L., A. W. Partridge, J. Han, A. R. Pickford, R. C. Liddington, M. H. Ginsberg, and I.D. Campbell. (2007). "Structural basis of integrin activation by talin." *Cell* **128**: 171–182.

Winkler, H., and K. A. Taylor. (1996). "Three-dimensional distortion correction applied to tomographic reconstructions of sectioned crystals." *Ultramicroscopy* **63**: 125–132.

Xu, W., H. Baribault, and E. D. Adamson. (1998). "Vinculin knockout results in heart and brain defects during embryonic development." *Development* **125**: 327–337.

Xu, X. P., I. Rouiller, B. D. Slaughter, C. Egile, E. Kim, J. R. Unruh, X. Fan, et al. (2011). "Three-dimensional reconstructions of Arp2/3 complex with bound nucleation promoting factors." *EMBO J* **31**: 236–247.

Zamir, E., M. Katz, Y. Posen, N. Erez, K. M. Yamada, B. Z. Katz, S. Lin, et al. (2000). "Dynamics and segregation of cell-matrix adhesions in cultured fibroblasts." *Nat Cell Biol* **2**: 191–196.

3 Stretchable micropost array cytometry

A powerful tool for cell mechanics and mechanobiology research

Yue Shao, Shinuo Weng, and Jianping Fu

It has become increasingly appreciated that living mammalian cells are not just complex biochemical reactors but also sophisticated biomechanical systems that can adapt their mechanical properties to various signals and perturbations from the extracellular space, and integrate with intracellular signaling events through a process called mechanotransduction, to regulate cell behaviors. To gain fundamental insights into such biomechanical nature of mammalian cells, many biomechanical tools have been developed with unprecedented spatiotemporal resolutions covering both molecular and cellular length scales. In this chapter, we describe a recently developed biomechanical tool, termed "stretchable micropost array cytometry" (SMAC), which is capable of quantitative control and real-time measurements of both mechanical stimuli and cellular biomechanical responses with a high spatiotemporal subcellular resolution. We further discuss implementations of the SMAC for characterizing cell cytoskeletal contractile force, cell stiffness, and cell adhesion signaling and dynamics at both whole-cell and subcellular scales in real time. We conclude with remarks regarding future improvements and applications of the SMAC for cell mechanics and mechanobiology studies.

3.1 Introduction

Conventionally, cells constituting a living organism have been admired as complex biochemical systems governing what we call life via a myriad of biochemical reactions, driving cellular processes such as metabolism, gene expression, and protein synthesis. However, strong evidence has emerged in recent years suggesting cells are not only biochemical reactors but also sophisticated biomechanical systems exhibiting correlated physical properties that are adaptive to different signals and perturbations in the extracellular space and play an essential and integral role in many physiological and pathological processes (Bao and Suresh 2003; Engler et al. 2006; Fu et al. 2010; Harris et al. 1980; McBeath et al. 2004; Wang et al. 1993; Wozniak and Chen 2009).

As a most prominent example, mammalian adherent cells exert a mechanical contractile force through the actin cytoskeleton (CSK) against the extracellular matrix (ECM) and adjacent cells (Harris et al. 1980; Liu et al. 2010; Maruthamuthu et al. 2011). Spatiotemporally regulated CSK contractile force is critical for developmental processes such as neural tube closure (Nishimura et al. 2012), epithelial morphogenesis (Zhang et al. 2011), mesenchymal condensation (Mammoto et al. 2011), etc. Abnormal and dysregulated CSK contractile force is a hallmark of many human diseases, including but not limited to cardiovascular diseases (Hahn and Schwartz 2009), tumor stromal remodeling (stiffening) (Butcher et al. 2009; Levental et al. 2009; Wirtz et al. 2011), and loss of hemostasis (Ono et al. 2008). In addition, CSK contractile force has been leveraged in tissue engineering and regenerative medicine for engineering stem cell fate in vitro, owing to its role in mechanotransduction to regulate relevant intracellular signaling pathways (Fu et al. 2010; Sun et al. 2014). Correlated with CSK contractile force, other physical properties of mammalian cells, such as cell shape (Kilian et al. 2010), size (Khetan et al. 2013; McBeath et al. 2004), and mechanical stiffness (Suresh 2007; Suresh et al. 2005), have also been documented as important cell phenotypes characterizing multiple developmental events such as epithelial morphogenesis (Aragona et al. 2013) for example, and cell fate patterning (Ruiz and Chen 2008), and human disease progression such as cancer and malaria.

Although cell mechanical properties are emergent phenotypes resulting from intracellular molecular events, they can in turn regulate a myriad of molecular events inside cells to influence intracellular molecular machineries and signaling to dictate cell behaviors, fates, and functions (Chen et al. 2004; Chen 2008; DuFort et al. 2011; Wang et al. 2009). Altogether, characterizing, manipulating, and fundamental understanding of cell mechanics can significantly improve our knowledge of cell biology and provide potential phenotypical markers for disease profiling and cell engineering.

Recent development of micro- and nanotechnology has provided scientists and engineers a broad range of tools for cell mechanics and mechanotransduction studies. These tools can serve as miniaturized mechanical actuators to perturb cells or microscale force sensors monitoring dynamic subcellular CSK contractile force (Kim et al. 2009; Shao and Fu 2014). To join in and contribute toward this effort, our research group has recently developed a versatile technique, termed "stretchable micropost array cytometry" (SMAC), which can for the first time allow for quantitative control and real-time measurements of both mechanical stimuli and cellular biomechanical responses with a high spatiotemporal subcellular resolution. Specifically, the SMAC integrates sensitive, high-resolution elastomeric micropost force sensors with a global cell-stretching mechanism using a stretchable elastomeric membrane. Thus, using a computer-controlled vacuum, SMAC can be activated to apply stretching forces to adherent cells attached onto the micropost force sensors. In this chapter, we describe the development of SMAC and then highlight examples demonstrating its implementation for quantifications of cellular CSK contractile force, cell stiffness, and cell adhesion signaling and dynamics (Lam et al. 2012; Mann et al. 2012; Shao et al. 2014). We conclude with remarks regarding future improvements and applications of SMAC for cell mechanics and mechanobiology studies.

3.2 Fabrication and functionalization of SMAC

The SMAC system is essentially a direct improvement and functional extension of the elastomeric micropost array device (mPAD), which was originally developed as a force

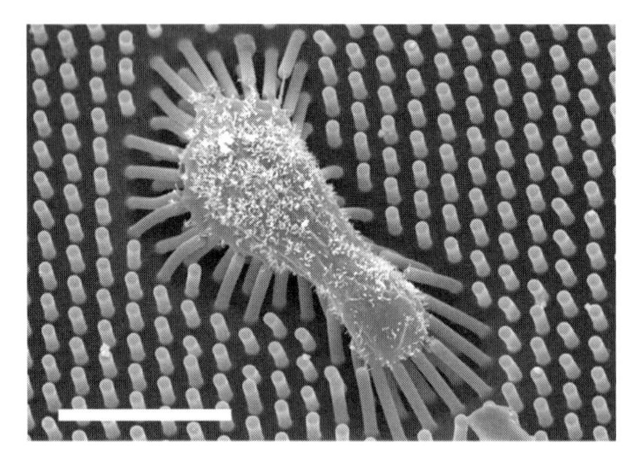

Figure 3.1 **A scanning electron microscopy (SEM) image of a cell on the PDMS micropost array.** The micropost bent under the tension applied by the cell. The deflection of the micropost tips was used to calculate the magnitude of the cell tension. Scale bar: 20 μm.

Reprinted from the journal *Integrative Biology* 2014, 6: 300–311, Shao et al. "Global architecture of the F-actin cytoskeleton regulates cell shape-dependent endothelial mechanotransduction." Reproduced with permission from the Royal Society of Chemistry. Copyright 2014.

sensor with a high spatiotemporal subcellular resolution to report CSK contractile forces using regularly arranged, vertical microscale cantilevers made out of polydimethylsiloxane (PDMS) (Fig. 3.1) (du Roure et al. 2005; Fu et al. 2010; Shao et al. 2014; Tan et al. 2003). Real-time imaging of deflections of PDMS micropost tips allows direct quantification of subcellular CSK contractile forces applied on each micropost using the mPAD. To extend the mPAD technique and make it suitable for quantitative control and real-time measurement of both mechanical stimuli and cellular biomechanical responses, we developed a protocol to fabricate and functionalize the PDMS micropost array onto a thin stretchable PDMS film (Figs. 3.2 and 3.3), which could then be integrated with a cell-stretching device to apply basal mechanical stretches to cells cultured on the micropost array (Fig. 3.4). Below we discuss the detailed three-step experimental procedure for SMAC.

(1) Fabrication of the PDMS micropost array on a thin PDMS film. Fig. 3.2 illustrates the procedure for generating the stretchable PDMS micropost array. Briefly, PDMS prepolymer (usually with the PDMS base to curing agent ratio of 10:1) is first spun coated onto a clean coverslip pretreated with (tridecafluoro-1,1,2,2,-tetrahydrooctyl)-1-trichlorosilane before thermally cured. Silanization renders the coverslip superhydrophobic, facilitating subsequent peeling of the PDMS film from the coverslip. In parallel, a negative PDMS template containing an array of negative features (holes) is generated by casting PDMS prepolymer over the Si micropost master fabricated by photolithography and deep reactive ion etching (DRIE). The negative PDMS template is oxidized with oxygen plasma and then silanized before being applied to fabricate a positive PDMS micropost array on the PDMS thin film covering the coverslip. After curing, the coverslip and the PDMS thin film holding the positive PDMS micropost array are altogether gently peeled off the negative template. Collapse of PDMS microposts induced during the peeling process can be rescued

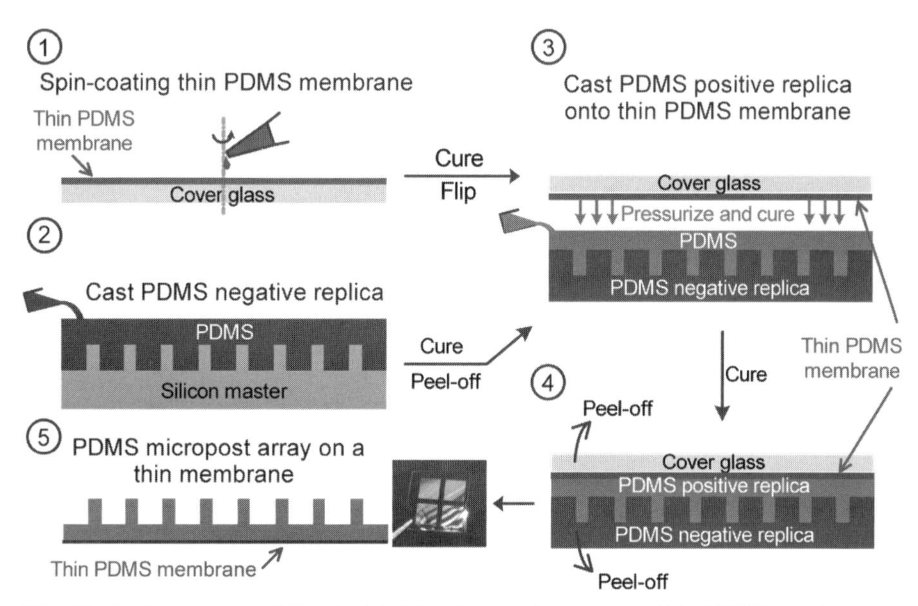

Figure 3.2 **The fabrication process of the stretchable micropost array on a thin PDMS membrane.** Reprinted from the journal *Integrative Biology* 2014, 6: 300–311, Shao et al. "Global architecture of the F-actin cytoskeleton regulates cell shape-dependent endothelial mechanotransduction." Reproduced with permission from the Royal Society of Chemistry. Copyright 2014.

by using CO_2 critical point drying. Finally, the stretchable PDMS thin film containing the PDMS micropost array can be peeled off the coverslip and handled as a flexible, free standing membrane (Fig. 3.3).

(2) Functionalization of the PDMS micropost array. Functionalization of the PDMS microposts is achieved using microcontact printing, which is performed while the stretchable micropost array membrane is still held on the coverslip. As shown in Fig. 3.3, a PDMS stamp (either flat, featureless, or with patterned surface structures) is first inked with a protein solution. Various soluble ECM proteins such as fibronectin (FN) and collagen (Col) are compatible with microcontact printing. Due to hydrophobic interaction, hydrophobic segments of solute ECM proteins are repelled from the watery environment, resulting in the ECM proteins to adsorb onto the PDMS stamp surface. The PDMS stamp is then rinsed and blown dry. Finally, the stretchable micropost array membrane is activated by UV ozone treatments, before being brought into a conformal contact with the PDMS stamp to transfer ECM proteins to the micropost tops.

(3) Integration of the PDMS micropost array membrane onto a stretchable PDMS membrane. Plasma-assisted PDMS-PDMS bonding is utilized for bonding the PDMS micropost array membrane onto a stretchable PDMS membrane. It should be noted that since any exposure to oxygen plasma is destructive to ECM proteins, during the plasma treatment the PDMS micropost array membrane is flipped over with the PDMS microposts brought into a conformal contact with a freshly prepared PDMS block. The backside of the PDMS micropost array membrane and the top

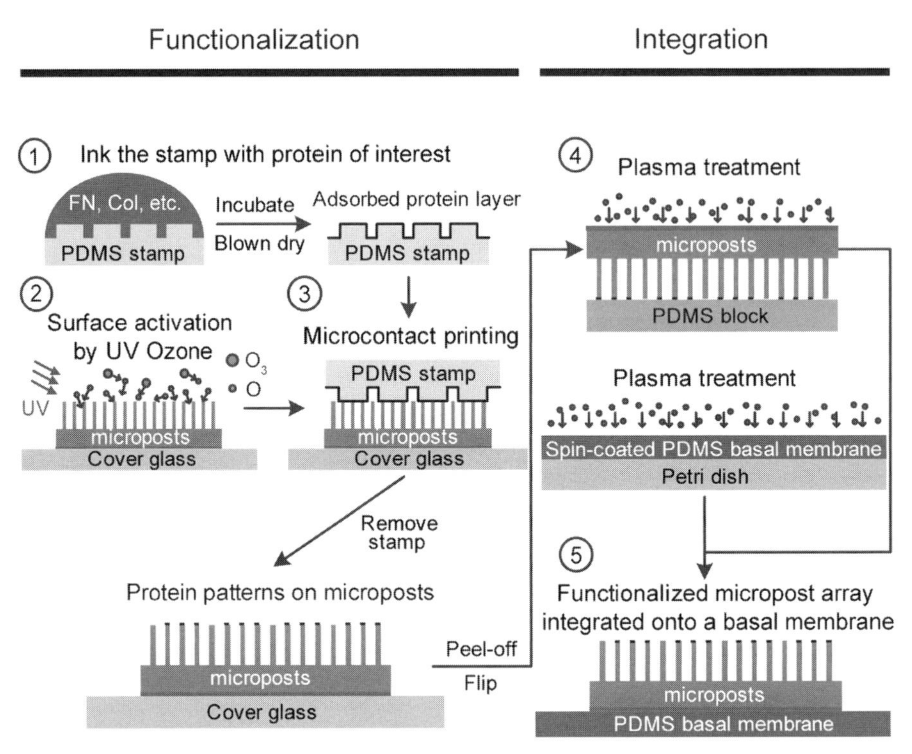

Figure 3.3 **Step-by-step schematics of the functionalization and integration of the stretchable micropost array.**

Reprinted from the journal *Integrative Biology* 2014, 6: 300–311, Shao et al. "Global architecture of the F-actin cytoskeleton regulates cell shape-dependent endothelial mechanotransduction." Reproduced with permission from the Royal Society of Chemistry. Copyright 2014.

surface of a PDMS basal membrane are both treated briefly with oxygen plasma before they are bonded permanently.

To visualize the PDMS microposts under fluorescence microscopy, the PDMS microposts are fluorescently labeled with a Δ9-DiI solution. Passivation of shafts and the bottom surface of the PDMS micropost array is achieved by incubating the array with a Pluronic F-127 solution, rendering these areas resistive to protein adsorption.

The SMAC has been implemented with a custom-designed cell-stretching device (CSD) to generate controlled mechanical stretches to the PDMS micropost array membrane using a computer-controlled vacuum (Fig. 3.4A). The CSD contains a circular viewing aperture surrounded by an annular vacuum chamber. The CSD is activated for cell stretch by applying vacuum to draw the periphery of the PDMS base membrane into the vacuum chamber, causing the central area of the PDMS micropost array membrane holding the PDMS micropost array to stretch equibiaxially. Importantly, by using simulation-assisted design and rationally dividing the annular vacuum chamber into multiple compartments, the CSD can generate different patterns of stretch on the PDMS micropost array membrane. For example, we can modify the internal topology of the CSD annular vacuum chamber by placing two identical

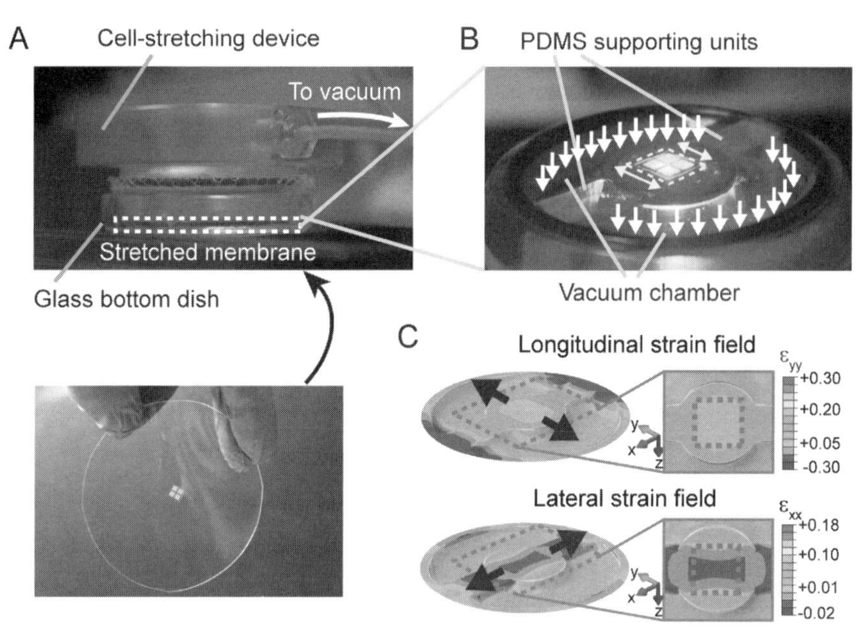

Figure 3.4 Implementation of SMAC by integrating the stretchable micropost array with a customized cell-stretching device.
(A) Experimental setup. The stretchable micropost array attached to a PDMS basal membrane was mounted onto a vacuum-driven cell-stretching device. The device was inverted and placed onto a glass bottom dish for high-resolution live-cell imaging using an inverted microscope. (B) Customization of the device for uniaxial cell-stretching. Two identical PDMS supporting units were inserted symmetrically in the annular vacuum chamber and divided the chamber into two compartments. A bipolar suction (single-headed arrows) generated by the vacuum created a uniaxial stretching field (double-headed arrows) in the central region of the PDMS basal membrane, to which the stretchable micropost array was attached. (C) Finite element simulations verified that a uniform longitudinal strain field and a much smaller transverse strain field was generated using the design in (B).

Reprinted from the journal *Integrative Biology* 2014, 6: 300–311, Shao et al. "Global architecture of the F-actin cytoskeleton regulates cell shape-dependent endothelial mechanotransduction." Reproduced with permission from the Royal Society of Chemistry. Copyright 2014.

PDMS supporting units at two opposite locations in the chamber to divide the chamber into two exact vacuum compartments. Such simple modification of the CSD setup allows convenient generations of uniaxial stretches on the PDMS micropost array membrane (Figs. 3.4B and C).

3.3 Applications of SMAC for studying contractile responses of vascular cells to external stretch

As the first important application of the SMAC, we have studied stretch-mediated contractile responses of vascular cells – including both vascular smooth muscle cells (VSMCs) and endothelial cells (ECs) – which is critically relevant to vascular

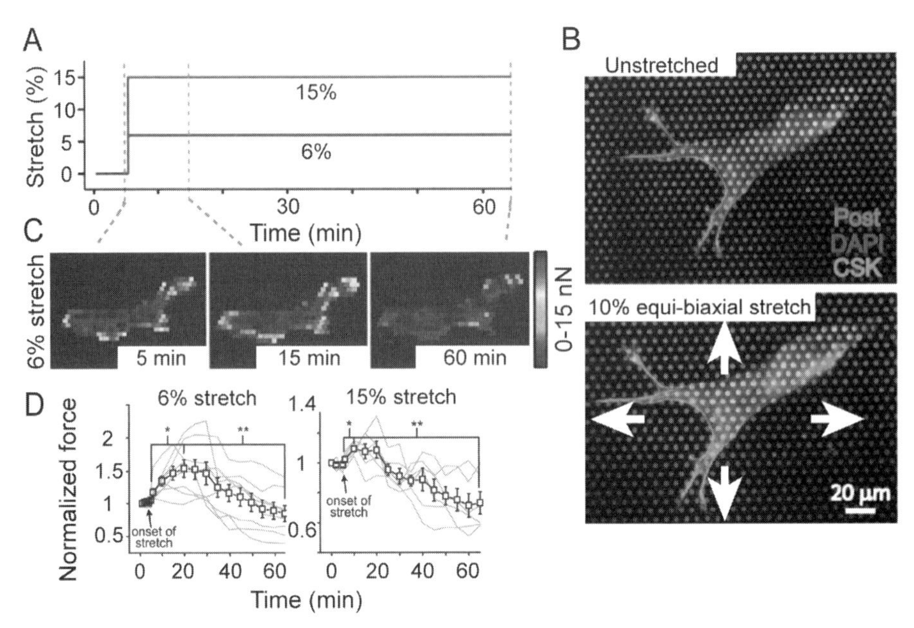

Figure 3.5 **Time-dependent biphasic response of VSMC contractility to a step-function equibiaxial stretch.**
(A) The stretch signal applied in experiments. (B) Immunofluorescence images of single VSMCs before and after the equibiaxial stretch. (C) Subcellular mapping of CSK contractile force and its time-dependent response under 6% stretch. (D) Normalized whole-cell contractility and its time-dependent trajectory in response to 6% and 15% stretch, respectively.

Reprinted from the journal *Lab Chip* 2012, 12: 731–740, Mann et al. "A silicone-based stretchable micropost array membrane for monitoring live-cell subcellular cytoskeletal response." Reproduced with permission from the Royal Society of Chemistry. Copyright 2012.

physiology and pathology. Specifically, we have examined live-cell subcellular dynamic responses of CSK contractile force in VSMCs to a sustained static equibiaxial cell stretch (Mann et al. 2012). By applying a step-function equibiaxial stretch to single VSMCs cultured on the stretchable micropost array membrane (Fig. 3.5A, B), we have observed that VSMCs regulate their CSK contractility in a biphasic manner: they first acutely enhance their contraction to resist rapid cell deformation ("stiffening") before they allow slow adaptive inelastic CSK reorganization to release their contractility ("softening") (Fig. 3.5C, D). Importantly, the contractile response across entire single VSMCs is mainly contributed from cell adhesion sites on cell peripheral regions (Fig. 3.5C). The biphasic contractile response of VSMCs to a sustained static equibiaxial cell stretch is independent of the magnitude of the stretch (Fig. 3.5D). However, stretch magnitude can affect the extent of both the stiffening and softening phases (Fig. 3.5D). Under a medium stretch (6%), the total CSK contractile force of VSMCs peaks at 20 minutes after the onset of stretch before gradually decreasing to a level comparable to the initial CSK contractile force – a phenomenon called tensional homeostasis (Chien 2007) – whereas overstretched (15%) VSMCs demonstrate shorter and less significant stiffening before decreasing to a contractile level much less than the initial CSK contractile state. Such observation suggests that VSMCs are sensitive to the magnitude of mechanical stretches and as such may respond differently to physiological and pathological stretch conditions generated by healthy and abnormal blood pressures, respectively.

Recently, we have applied the SMAC to study mechanotransduction of ECs (Fig. 3.6) (Shao et al. 2014). In particular, although cell shape of ECs has long been considered as a result from external mechanical forces such as laminar shear flow and basal lateral stretch, there has been little experimental data directly answering whether and how cell shape can in turn regulate EC mechanotransduction. To address this, we have utilized SMAC to exert static uniaxial cell-stretching forces to single ECs (Fig. 3.4B) and quantified the resulting instantaneous CSK contractile response of ECs (Fig. 3.6A). Interestingly, we have observed a consistent increase of CSK contractility in ECs with a low aspect ratio cell shape, for example circular and square-shaped cells (Fig. 3.6B–D). Such tensional reinforcement is independent of cell stretch direction (Fig. 3.6C, D). However, for ECs with a high aspect ratio cell shape, as with rectangular shapes, longitudinal stretches can still result in increases in CSK contractility in ECs, whereas a transverse stretch is unable to induce a significant CSK contractile response (Fig. 3.6E, F). Thus, our data suggests that cell shape alone can serve as a regulatory factor in EC mechanotransduction via CSK contractility (Fig. 3.6G). Importantly, a spatiotemporally coordinated, uniform subcellular response of CSK contractility has been observed in ECs with low aspect ratio cell shapes (Fig. 3.6H), suggesting that there may be a global force-sensing and responding mechanism underlying EC mechanotransduction. Incorporating the F-actin CSK into our study, we have identified that a globally interconnected actin network is responsible for the spatiotemporal coordination of subcellular CSK contractile forces in ECs. Furthermore, through a combination of experiments and theoretical modeling, we have identified the cell shape-dependence of F-actin architecture responsible for the differential sensitivity of EC mechanotransduction under a basal directional stretch. Our study using the SMAC directly demonstrates the biomechanical process underlying EC mechanotransduction and implies the role of cell shape and the F-actin CSK in controlling EC properties through CSK contractility and CSK contractility-mediated cell signaling events.

3.4 Application of SMAC for mapping subcellular cell stiffness

Cell stiffness is another important cell mechanical property correlated with CSK contractility and dynamics and involved in mechanotransduction (Chiang et al. 2013; Chowdhury et al. 2010; Tee et al. 2011). In addition, cell stiffness is correlated with progression of diseases such as cancer and malaria (Suresh 2007; Suresh et al. 2005). However, it is still unclear how cell stiffness changes in different subcellular regions when subjected to external mechanical perturbations, which is important for understanding spatial variation and coordination in mechanoresponsiveness at the subcellular level. To address this knowledge gap, we have recently applied the SMAC in conjunction with a finite element method to achieve whole-cell cell stiffness measurements with a subcellular spatial resolution (Lam et al. 2012).

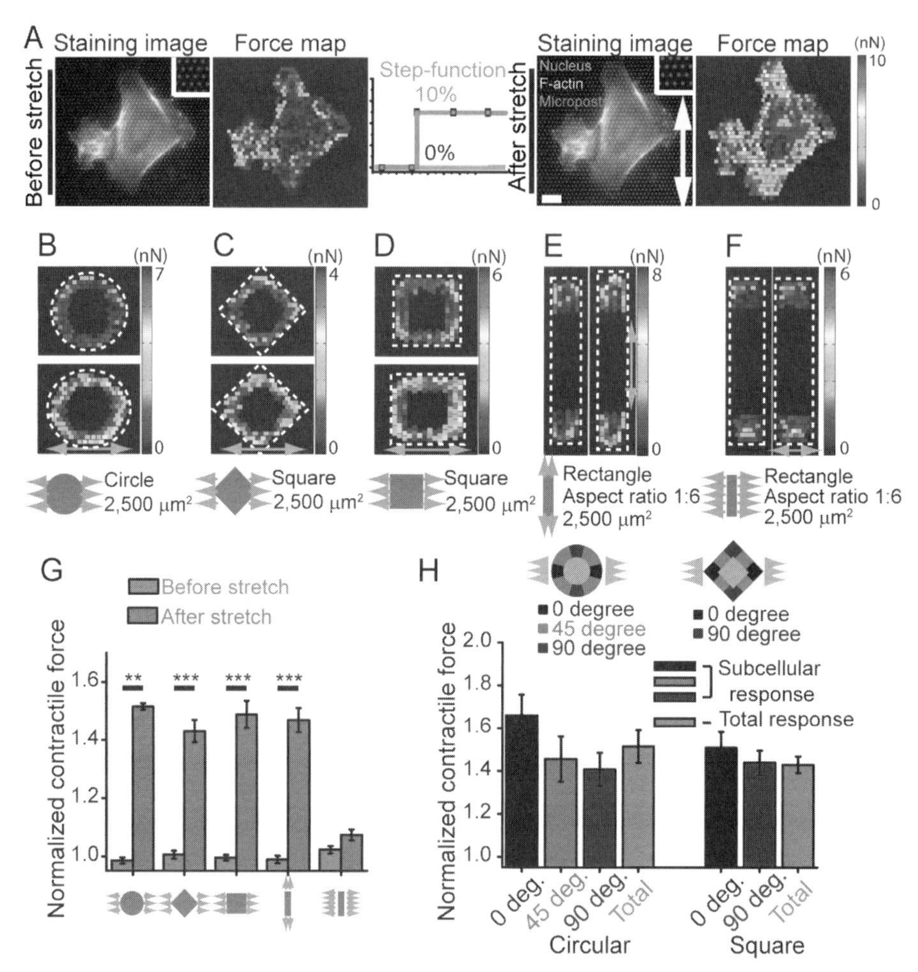

Figure 3.6 **Cell shape directly regulates EC mechanotransduction via cell tension.**
(A) Immunofluorescence images and subcellular CSK forces maps of ECs before (left panel) and after (right panel) a uniaxial step-function stretch (middle panel) was applied. Direction of the stretch was along the double-headed arrow. (B–F) Subcellular CSK force maps of cells of different shapes before and after the stretch, respectively. The stretched cell and the direction of the stretch were identified with a double-headed arrow. (G) Statistical analysis of the cell shape-dependent responses of total CSK contractility under uniaxial stretch. (H) Statistical analysis of the subcellular CSK contractile force in response to the uniaxial stretch, demonstrated spatially coordinated responses throughout the cell of low aspect-ratio.

Reprinted from the journal *Integrative Biology* 2014, 6: 300–311, Shao et al. "Global architecture of the F-actin cytoskeleton regulates cell shape-dependent endothelial mechanotransduction." Reproduced with permission from the Royal Society of Chemistry. Copyright 2014.

The finite element method leverages the intrinsic discrete nature of the PDMS micropost array that is arranged in a regular hexagonal formation. Thus, individual cells adherent on the PDMS micropost array are automatically discretized into subcellular triangular grids formed by the PDMS microposts underlying the cells (Fig. 3.7A). Using the SMAC, the initial force and displacement vectors, $\boldsymbol{f}_{\text{ini}}$ and $\boldsymbol{u}_{\text{ini}}$, on each node of the grid throughout the

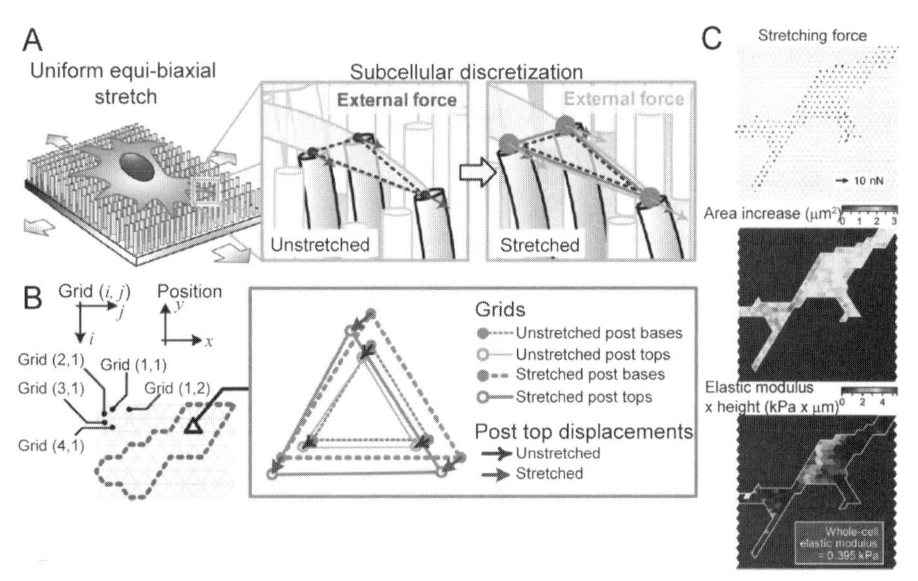

Figure 3.7 **A SMAC-based finite element method for mapping subcellular mechanical stiffness in live single cells.**
(A) Schematic of the external mechanical stretch, micropost array-based subcellular discretization, and the change of nodal force and grid area in response to the stretch. (B) Terminology used in the SMAC-based finite element method. (C) Results showing the change of nodal forces (top) and grid area (middle), and the calculated subcellular stiffness (bottom) in a single live VSMC.

Reprinted from the journal *Integrative Biology* 2012, 4: 1289–1298, Lam et al. "Live-cell subcellular measurement of cell stiffness using a microengineered stretchable micropost array membrane." Reproduced with permission from the Royal Society of Chemistry. Copyright 2012.

entire cell area, can be acquired. Immediately after the onset of an equibiaxial stretch applied to the cell, a new set of vectors, $\boldsymbol{f}_{\mathrm{str}}$ and $\boldsymbol{u}_{\mathrm{str}}$, can be obtained. Thus, by subtracting the initial force and displacement vectors $\boldsymbol{f}_{\mathrm{ini}}$ and $\boldsymbol{u}_{\mathrm{ini}}$ from $\boldsymbol{f}_{\mathrm{str}}$ and $\boldsymbol{u}_{\mathrm{str}}$, respectively, we can obtain changes of CSK contractile force and post displacement, $\boldsymbol{f} = \boldsymbol{f}_{\mathrm{str}} - \boldsymbol{f}_{\mathrm{ini}}$ and $\boldsymbol{u} = \boldsymbol{u}_{\mathrm{str}} - \boldsymbol{u}_{\mathrm{ini}}$, on each node, which can then be used for calculation of mechanical stiffness of the small triangular grid of subcellular area using the finite element format of linear elasticity (Fig. 3.7A). Specifically, the mathematical finite element format for a grid containing node 1, 2, and 3, with combined force vector $\boldsymbol{f} = [f_{1x}, f_{1y}, f_{2x}, f_{2y}, f_{3x}, f_{3y}]$ and displacement vector $\boldsymbol{u} = [u_{1x}, u_{1y}, u_{2x}, u_{2y}, u_{3x}, u_{3y}]$, reads (Fig. 3.7B):

$$\begin{bmatrix} f_{1x} \\ f_{1y} \\ f_{2x} \\ f_{2y} \\ f_{3x} \\ f_{3y} \end{bmatrix} = \frac{AE_h}{1-v^2} B^T \begin{bmatrix} 1 & v & 0 \\ v & 1 & 0 \\ 0 & 0 & (1-v)/2 \end{bmatrix} B \begin{bmatrix} u_{1x} \\ u_{1y} \\ u_{2x} \\ u_{2y} \\ u_{3x} \\ u_{3y} \end{bmatrix} \tag{3.1}$$

where

$$B = \frac{1}{2A}\begin{bmatrix} y_2 - y_3 & 0 & y_3 - y_1 & 0 & y_1 - y_2 & 0 \\ 0 & x_3 - x_2 & 0 & x_1 - x_3 & 0 & x_2 - x_1 \\ x_3 - x_2 & y_2 - y_3 & x_1 - x_3 & y_3 - y_1 & x_2 - x_1 & y_1 - y_2 \end{bmatrix} \quad (3.2)$$

and

$$A = \frac{1}{2}\begin{bmatrix} 1 & 1 & 1 \\ x_1 & x_2 & x_3 \\ y_1 & y_2 & y_3 \end{bmatrix} \quad (3.3)$$

E_h stands for the product of the grid's Young's modulus and its height, and v is the Poisson's ratio of the grid. We can combine the relations described by Eq. (3.1) for all grid elements within the cell area into a global matrix form as

$$F = S(E_h, v)U \quad (3.4)$$

where the global force matrix $\boldsymbol{F}$ and the global displacement matrix $\boldsymbol{U}$ are acquired through SMAC, and the matrices of $\boldsymbol{E}_h$ and $\mathbf{v}$ are the unknown parameters we will solve numerically. It should be noted that as $\boldsymbol{E}_h$ and $\mathbf{v}$ are defined on each grid, the total number of unknowns is 2 m, where m is the number of grids. Yet, the total number of equations in Eq. (3.4) is twice the number of nodes, i.e., 2 n. As m usually is greater than n, it needs to add additional constraints to make Eq. (3.4) numerically solvable. Hereby, we assume that the grids at $i = 2\,n, j$ and $i = 2\,n - 1, j$ share the same cell stiffness and Poisson's ratio, where i and j denote the position of the grid as illustrated in Fig. 3.7B. Now the stiffness and Poisson's ratio matrices $\boldsymbol{E}_h$ and $\mathbf{v}$ become numerically solvable.

Using the SMAC in conjunction with the finite element method, we have for the first time successfully mapped the subcellular heterogeneous distribution of cell stiffness throughout a live single cell. For example, Fig. 3.7C shows maps of contractile force change and cell area change after cell stretch and the corresponding subcellular cell stiffness distribution. It is noticeable that the perinuclear area has a great E_h. In addition, we have observed a positive correlation between whole-cell cell stiffness and CSK contractility, confirming that these two cell mechanical properties are coupled traits during mechanosensing.

3.5 Application of SMAC to study FA dynamics and morphogenesis

To study the intricate connection between CSK contractility and subcellular force-dependent molecular events (Hoffman et al. 2011; Moore et al. 2010; Oakes and Gardel 2014), we have recently combined the SMAC with live-cell imaging using adherent mechanosensitive cells expressing fluorescently labeled focal adhesion (FA) proteins to study force-mediated FA dynamics. As shown in Fig. 3.8A, fluorescent images recorded in real time for both Δ9-DiI–labeled PDMS microposts and YFP-paxillin containing FAs have provided information needed for extracting subcellular distributions

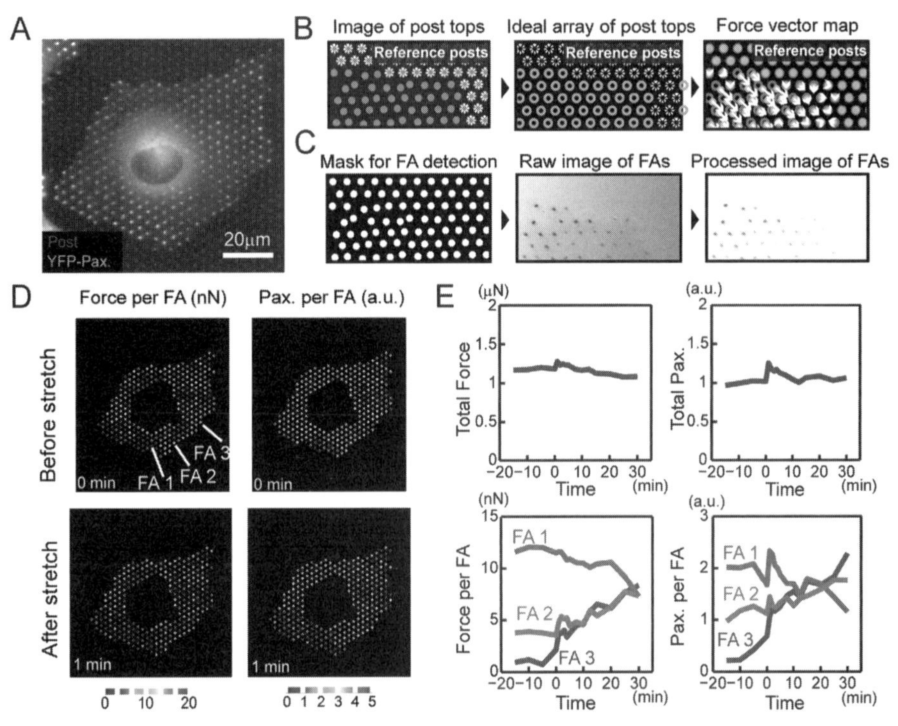

Figure 3.8 **A multiplex SMAC for simultaneously studying biomechanical and biochemical dynamics at both cellular and subcellular scales.**
(A) A fluorescent live-cell image visualizing both cell tension (by fluorescently labeled micropost tops) and individual FAs (by YFP-paxillin). (B) The image analysis method to calculate the FA-mediated CSK contractile force (yellow arrows) by registering the deflected micropost tops with undeformed micropost bases. The marked reference posts, which located outside the YFP-expressing cell, were also considered undeformed and irrelevant, and were excluded from the calculation. (C) The image analysis method to map and quantify individual FA intensity throughout a single cell, by using the micropost tops as a mask to detect individual FAs and then using rolling-ball algorithm to remove the uneven background. (D) Subcellular mapping of CSK contractile force and FA intensity before and after a step-function equibiaxial stretch was applied to the cell. Three single FAs were selected and the time-lapse date was provided for each, as shown in E. (E) Time-lapse tracking of the total CSK force and FA intensity (top panel) as well as single FA force and intensity (bottom panel) in a single cell, demonstrating that the single-cell biomechanical and biochemical homeostasis was driven by distinct, compensatory subcellular dynamics.

and dynamics of CSK contractile force and FA morphogenesis and dynamics. Specifically, subcellular CSK contractile force was calculated using micropost deflection measured by registering the image of micropost tops against that of micropost bases (Fig. 3.8B). For analysis of FA morphogenesis, the image of micropost tops was used as an in situ mask for detecting individual FAs localized on individual microposts. Raw images of FAs were further processed to subtract the uneven background using a rolling-ball algorithm, yielding a clean image resolving the intensity of each single FA (Fig. 3.8C).

Combining the SMAC with live-cell imaging, we have successfully demonstrated simultaneous mapping and tracking of subcellular CSK contractile force and FA dynamics in response to equibiaxial stretches (Fig. 3.8D, E). Interestingly, we have observed biphasic dynamics featuring single-cell homeostasis on both CSK contractility and FA morphogenesis, suggesting robust biomechanical and biochemical stability under mechanical perturbations (Fig. 3.8E). However, we have also observed a highly heterogeneous, yet compensatory dynamics among FAs (Fig. 3.8E). Specifically, while FAs sustaining high initial CSK contractile force switched to a relaxing phase after an initial reinforcement, those bearing low initial CSK contractile force exhibited a continuous reinforcement of both CSK contractile force and FA formation. Such differential subcellular responses can compensate each other and eventually converge into a steady state, comparable to the homeostatic state at the whole cell level.

3.6 Concluding remarks

As a unique feature of the SMAC, it can apply global, uniform mechanical perturbations to live single cells, while acquiring cellular responses with a subcellular resolution in real time. By comparing whole-cell responses to subcellular activities, it may help elucidate how single-cell-level behaviors emerge from complex yet organized subcellular dynamics. Leveraging this unique capability of the SMAC, we have successfully characterized CSK contractile force, cell stiffness, and FA dynamics at both whole-cell and subcellular scales in real time.

Although the applications of SMAC demonstrated here have been carried out using a custom-designed cell-stretching device for live single cells, SMAC is versatile to be integrated with many other cell mechanics tools for investigating different cell mechanics and mechanobiology questions in various contexts. The SMAC technique can also be implemented with other types of tissue cultures to study mechanobiology beyond the cellular scale. For example, an epithelial monolayer model can be applied with the SMAC to study the mechanotransduction in collective cell migration, and a VSMC/EC bilayer tissue model can be used to study mechanobiology of vascular smooth muscle/endothelium interactions. In addition, it will also be exciting to combine the SMAC with advanced live-cell imaging such as total internal reflection fluorescence (TIRF) and fluorescence resonance energy transfer (FRET) microscopy and dynamic molecular manipulation tools such as optogenetics (Fenno et al. 2011), to examine dynamic signaling events at the molecular level involved in mechanotransduction.

Acknowledgments

We acknowledge financial support for our research from the National Science Foundation (CMMI 1129611 and CBET 1149401), the National Institutes of Health (1R21HL114011), and the American Heart Association (12SDG12180025). We also thank Jennifer M. Mann and Raymond H. W. Lam for developing the SMAC system and its cell mechanics and mechanobiology applications.

References

Aragona, M., et al. (2013). "A mechanical checkpoint controls multicellular growth through yap/taz regulation by actin-processing factors." *Cell* **154**: 1047–1059.

Bao, G. and S. Suresh. (2003). "Cell and molecular mechanics of biological materials." *Nature Mater* **2**: 715–725.

Butcher, D., et al. (2009). "A tense situation: Forcing tumour progression." *Nat Rev Cancer* **9**: 108–122.

Chen, C. S. (2008). "Mechanotransduction – a field pulling together?" *J Cell Sc* **121**: 3285–3292.

Chen, C. S., et al. (2004). "Mechanotransduction at cell-matrix and cell-cell contacts." *Annu Rev Biomed Eng* **6**: 275–302.

Chiang, M., et al. (2013). "Relationships among cell morphology, intrinsic cell stiffness and cell-substrate interactions." *Biomaterials* **34**: 9754–9762.

Chien, S. (2007). "Mechanotransduction and endothelial cell homeostasis: the wisdom of the cell."*Am J Physiol Heart Circ Physiol* **292**:H1209–H1224.

Chowdhury, F., et al. (2010). "Material properties of the cell dictate stress-induced spreading and differentiation in embryonic stem cells." *Nature Mater* **9**: 82–88.

du Roure, O., et al. (2005). "Force mapping in epithelial cell migration." *Proc Natl Acad Sci USA* **102**: 2390–2395.

DuFort, C. C., et al. (2011). "Balancing forces: architectural control of mechanotransduction." *Nat RevMol Cell Biol* **12**: 308–319.

Engler, A., et al. (2006). "Matrix elasticity directs stem cell lineage specification." *Cell* **126**: 677–689.

Fenno, L., et al. (2011). "The development and application of optogenetics." *Annu Rev Neurosci* **34**: 389–412.

Fu, J., et al. (2010). "Mechanical regulation of cell function with geometrically modulated elastomeric substrates." *Nat Methods* **7**: 733–736.

Hahn, C. and M. Schwartz. (2009). "Mechanotransduction in vascular physiology and atherogenesis." *Nat Rev Mol Cell Biol* **10**: 53–62.

Harris, A. K., et al. (1980). "Silicone-rubber substrata – new wrinkle in the study of cell locomotion." *Science* **208**: 177–179.

Hoffman, B., et al. (2011). "Dynamic molecular processes mediate cellular mechanotransduction." *Nature* **475**: 316–323.

Khetan, S., et al. (2013). "Degradation-mediated cellular traction directs stem cell fate in covalently crosslinked three-dimensional hydrogels." *Nature Mater* **12**: 458–465.

Kilian, K., et al. (2010). "Geometric cues for directing the differentiation of mesenchymal stem cells." *Proc Natl Acad Sci USA* **107**: 4872–4877.

Kim, D.-H., et al. (2009). "Microengineered platforms for cell mechanobiology." *Annu Rev Biomed Eng* **11**: 203–233.

Lam, R., et al. (2012). "Live-cell subcellular measurement of cell stiffness using a microengineered stretchable micropost array membrane." *Integr Biol* **4**: 1289–1298.

Levental, K., et al. (2009). "Matrix crosslinking forces tumor progression by enhancing integrin signaling." *Cell* **139**: 891–906.

Liu, Z., et al. (2010). "Mechanical tugging force regulates the size of cell-cell junctions." *Proc Natl Acad Sci USA* **107**: 9944–9949.

Mammoto, T., et al. (2011). "Mechanochemical control of mesenchymal condensation and embryonic tooth organ formation." *Dev Cell* **21**: 758–769.

Mann, J., et al. (2012). "A silicone-based stretchable micropost array membrane for monitoring live-cell subcellular cytoskeletal response." *Lab Chip* **12**: 731–740.

Maruthamuthu, V., et al. (2011). "Cell-ECM traction force modulates endogenous tension at cell-cell contacts." *Proc Natl Acad Sci USA* **108**: 4708–4713.

McBeath, R., et al. (2004). "Cell shape, cytoskeletal tension, and RhoA regulate stem cell lineage commitment." *Dev Cell* **6**: 483–495.

Moore, S., et al. (2010). "Stretchy proteins on stretchy substrates: The important elements of integrin-mediated rigidity sensing." *Dev Cell* **19**: 194–206.

Nishimura, T., et al. (2012). "Planar cell polarity links axes of spatial dynamics in neural-tube closure." *Cell* **149**: 1084–1097.

Oakes, P. and M. Gardel. (2014). "Stressing the limits of focal adhesion mechanosensitivity." *Curr Opin Cell Biol* **30C**: 68–73.

Ono, A., et al. (2008). "Identification of a fibrin-independent platelet contractile mechanism regulating primary hemostasis and thrombus growth." *Blood* **112**: 90–99.

Ruiz, S. and C. Chen. (2008). "Emergence of patterned stem cell differentiation within multicellular structures." *Stem Cells* **26**: 2921–2927.

Shao, Y. and J. P. Fu. (2014). "Integrated micro/nanoengineered functional biomaterials for cell mechanics and mechanobiology: a materials perspective." *Adv Mater* **26**, 1494–1533.

Shao, Y., et al. (2014). "Global architecture of the F-actin cytoskeleton regulates cell shape-dependent endothelial mechanotransduction." *Integr Biol* **6**: 300–311.

Sun, Y. B., et al. (2014). "Hippo/YAP-mediated rigidity-dependent motor neuron differentiation of human pluripotent stem cells." *Nature Mater* **13**: 599–604.

Suresh, S. (2007). "Biomechanics and biophysics of cancer cells." *Acta Biomater* **3**: 413–438.

Suresh, S., et al. (2005). "Connections between single-cell biomechanics and human disease states: gastrointestinal cancer and malaria." *Acta Biomater* **1**: 15–30.

Tan, J., et al. (2003). "Cells lying on a bed of microneedles: an approach to isolate mechanical force." *Proc Natl Acad Sci USA* **100**: 1484–1489.

Tee, S.-Y., et al. (2011). "Cell shape and substrate rigidity both regulate cell stiffness." *Biophys J* **100**: L25–L27.

Wang, N., et al. (1993). "Mechanotransduction across the cell-surface and through the cytoskeleton." *Science* **260**: 1124–1127.

Wang, N., et al. (2009). "Mechanotransduction at a distance: mechanically coupling the extracellular matrix with the nucleus." *Nat Rev Mol Cell Biol* **10**: 75–82.

Wirtz, D., et al. (2011). "The physics of cancer: the role of physical interactions and mechanical forces in metastasis." *Nat Rev Cancer* **11**: 512–522.

Wozniak, M. A. and C. S. Chen. (2009). "Mechanotransduction in development: a growing role for contractility." *Nat Rev Mol Cell Biol* **10**: 34–43.

Zhang, H. M., et al. (2011). "A tension-induced mechanotransduction pathway promotes epithelial morphogenesis." *Nature* **471**: 99–103.

4 Microscale generation of dynamic forces in cell culture systems

Christopher Moraes, Luke A. MacQueen, Yu Sun, and Craig A. Simmons

Dynamic mechanical forces play a critical role in modulating cellular function, and inclusion of these extracellular stimuli in culture systems may improve the relevance and utility of biological results. In this work, we discuss recent advances made by our research group in applying dynamically controlled mechanical stimuli to cells cultured in 2-D and 3-D arrayed environments. Advantages in throughput and precision arising from microengineering such systems are demonstrated, with illustrative examples of potential biological applications. Engineering challenges associated with building these culture systems are explored, and the design and fabrication strategies that we have developed are discussed. Finally, the ability to incorporate additional sensing technologies into these dynamic screening platforms is explored.

4.1 Introduction

Mechanical forces play a critical role in regulating cellular function. As early as 1892, Julius Wolff described bone remodeling, in which bones change shape, density, and stiffness when mechanical loading conditions are altered (Wolff 1986). For this reason, astronauts experience severe bone loss after extended periods in microgravity (Holick 2000). Similarly, muscles develop as a result of exercise, and skin forms hard calluses under repeated wear. Only recently however, have these organ-level phenomena been traced to cellular function, and findings that cells are exquisitely sensitive to mechanical forces have prompted great interest in how mechanical forces broadly regulate tissue homeostasis, disease progression, and cell fate and function. Understanding fundamental mechanisms of how cells interact with the mechanical microenvironment could eventually result in predictive control of cell fate and function, which will enable rational design strategies for tissue engineering and identify avenues for cell-based regenerative therapies (Moraes et al. 2011).

Currently, typical strategies to manipulate cell function for these purposes include presenting selected microenvironmental cues based on prior knowledge and informed reasoning. However, this approach has only met with limited success, and significant advances in predictive control will likely arise from an improved understanding of the integrated response of cells to multiple regulatory stimuli in the microenvironment (Zandstra 2004). As such, an emerging trend in regenerative medicine strategies is high-throughput screening using cell-based arrays (Simon and Lin-Gibson 2011). By systematically manipulating biomaterials (Anderson et al. 2005), matrix proteins

(Flaim et al. 2005), cell sources and soluble factors (Gómez-Sjöberg et al. 2007) in an array-based format, researchers are able to determine how cells respond to various stimuli in their environment. Such assays have been developed with the hopes of recreating organ-level functionality by simulating realistic cell culture environments (Moraes et al. 2012), and thereby accelerating drug discovery in the pharmaceutical industry. However, this approach as well has only met with limited success. One factor that could account for the low success rates of these approaches is the inability of current state-of-the-art high-throughput screening platforms to incorporate physiologically relevant mechanical forces in the screening parameters. Dynamic mechanical forces exist in vivo and are known to play a pivotal role in driving cell fate and function (Moraes et al. 2011). In the case of load-bearing cells, mechanical stimuli can supplant other factors (McBeath et al. 2004). Furthermore, mechanical stimuli can alter cellular response to other stimuli (MacKenna et al. 1998; Kong et al. 2005), and are hence critical features of the microenvironment. The inability of high-throughput screening systems to capture mechanical features of the in vivo microenvironment may be the critical limiting factor in the discovery of positive "hits" that are then translatable to therapeutic applications. Thus, there is a need to develop high-throughput screening technologies that incorporate dynamic mechanical stimulation to expedite scientific discovery and innovation.

Commercial systems currently available for mechanical stimulation of cells tend to be relatively large in size due to limitations in manufacturing technologies, and are expensive, cumbersome, require large quantities of expensive reagents, and are low throughput. These factors have limited the widespread use of mechanically dynamic bioreactors in screening applications. Despite the demonstrated importance of mechanical forces in regulating and modulating cell fate and function, it is only recently that systems capable of screening for mechanical stimulation in cell-based assays have been developed. This deficiency has precluded systematic screening for multiple *combinations* of mechanical, chemical, and matrix parameters.

The past decade has seen the rapid emergence of microfabrication technologies for biological and biomedical applications. Microfabricated systems are designed and executed with precise control over feature sizes with micron-level resolution. Microsystems provide a number of advantages over conventional technologies including the ability to exploit favorable scaling laws at smaller length scales; increases in portability and throughput via system miniaturization; and the potential to integrate multiple components on a single chip. This "lab-on-a-chip" approach is quite promising for biomedical applications, and we have already seen the development and deployment of microfabricated point-of-care diagnostic tools and miniaturized high-throughput bioreactor systems (Figallo et al. 2007). Recent progress in such fabrication technologies could alleviate some of the manufacturing concerns in developing array-based platforms to precisely apply mechanical forces to cells in culture. In addition, the substantial decrease in equipment footprint can dramatically increase screening throughput, thereby offering the possibility of systematically screening multiple combinations of parameters in the cellular microenvironment. *Combinatorially* probing cellular response to various types of stimuli using conventional techniques generates an unfeasibly large number of experimental conditions, and would require alternative approaches. Microfabricated systems also require greatly reduced quantities of reagents, can speed up experimental times and efficiency, can improve precision, and because they are batch fabricated could eventually be inexpensively manufactured to produce disposable testing platforms. Hence, microfabrication technologies may have the ability to accelerate scientific discovery and innovation, as compared to more conventional macroscale techniques.

In this chapter, we review engineered microdevices developed in our lab that have been designed to apply dynamic mechanical forces to cell/biomaterial systems. We discuss design strategies that have been particularly useful, and direct specific attention to microsystems that enable studies that would not be possible with conventional techniques. In particular, we limit discussion to devices that offer precise control of applied mechanical forces in the cellular microenvironment, while simultaneously enhancing experimental throughput, as these core criteria are essential for developing next-generation increased-throughput screening platforms.

4.2 Simple 2-D stretching mechanisms

Cell culture on flat, two-dimensional static substrates has remained the most commonly used cell culture system in biological laboratories. As such, the ability to incorporate dynamic mechanical forces into two-dimensional environments is particularly important in order to compare biological results with the greatest variety of existing culture systems. Applying dynamic mechanical forces to cells cultured on flat surfaces is most simply realized by culturing cells on a flexible substrate, and deforming the substrate using candidate actuation mechanisms. In this section we briefly review potential actuation mechanisms for such a purpose, highlight simple systems utilized to apply mechanical loads to cells cultured on substrates, and discuss some of the challenges associated with these approaches.

4.2.1 Selection of actuation mechanisms

While commercial cell-stretching systems utilize primarily pneumatic actuation mechanisms, engineering cell-stretching systems at the microscale opens new possibilities for alternative actuation mechanisms. For example, electrostatic forces may be used to drive small deformations on the microscale, due to favorable scaling conditions associated with microengineering. Though impractically large voltages are necessary to electrostatically drive motion in large devices, the relative increase of interacting surface areas available by engineering miniaturized comb structures enables deformations of tens of microns with kilovolts of applied potential. Designing such systems to stretch cells serves the dual purpose of both actuating the deformation mechanism, as well as sensing the resulting displacements using the same comb structures connected in a capacitive sensing configuration. While this approach has been utilized to apply mechanical forces to cells by a number of research groups (Mukundan et al. 2012;

Scuor et al. 2006), severe limitations exist in utilizing this method for most cell culture applications. First, the incompatible needs for conductive liquid media and high voltages force additional design constraints on such devices (Mukundan et al. 2012). Second, driving large voltage loads requires specialized equipment and techniques not readily available in most wet-labs. Third, displacements achieved with this approach make this technique suitable to stretch single cells, and require a large device footprint to generate enough force to stimulate a cell. Given the broad heterogeneity in biological systems, sequentially probing single cells is unlikely to adequately capture the ensemble average response of a cell population. Though useful for characterizing single-cell mechanics, this approach has yet to be successfully utilized to probe more complex biochemical cellular responses to mechanical stimulation.

Other actuation mechanisms such as magnetically driven deformation have been used to apply forces directly to cells via magnetic twisting cytometry (Sniadecki 2010), but poor scaling factors prevent the use of this approach in driving deformation of cell-supporting surfaces at the microscale. In contrast, pneumatic- or hydrostatic-driven architectures are independent of scale when the pressure is supplied off-chip, and can provide the highest forces and power densities possible at the microscale (Volder and Reynaerts 2010). Applied pressures can be controlled within 1 Pa by simply controlling the height of a reservoir in a hydrostatically driven system. Maximum pressures are limited only by the size of available pumps, and mechanisms to dynamically control pressure generation and application are readily available. Furthermore, easily fabricated microfluidic systems allow the uniform distribution of pneumatic pressures to multiple locations on a chip. Hence, the simplicity and versatility of pneumatically driven systems are particularly attractive for high-throughput cell stretching applications.

4.2.2 Pneumatic "bulge" mechanisms

Simple deformation of inflatable diaphragms provides the simplest method to translate applied pressure differentials to deform a substrate, and have recently been used by several research groups to apply forces to cultured cells. In this approach, cells cultured on thin elastomeric diaphragms are stretched as the inflatable diaphragms expand (Fig. 4.1A). Tan et al. (2008) first demonstrated this approach in 2008, in which free-standing polydimethylsiloxane (PDMS) films containing topographical alignment cues were inflated, forcing cells to align along the topography and respond to the applied stimulation (Tan et al. 2008). A similar system was more recently used to measure calcium signaling in response to applied dynamic stretch (Kim and Jeong 2012). This approach can be used with both air pressure and injection of liquid using a programmable syringe pump, as demonstrated by Huang et al. (2013) who used a hydraulic actuation scheme to investigate interactions between lung cancer and myofibroblast cells in a mechanically dynamic environment. Furthermore, by manipulating the size of the inflatable pressure cavities, a single pressure source may be used to generate a variety of strains produced across the chip (Shimizu et al. 2011). This is particularly important, as the primary failure mode of most microfluidic devices is in

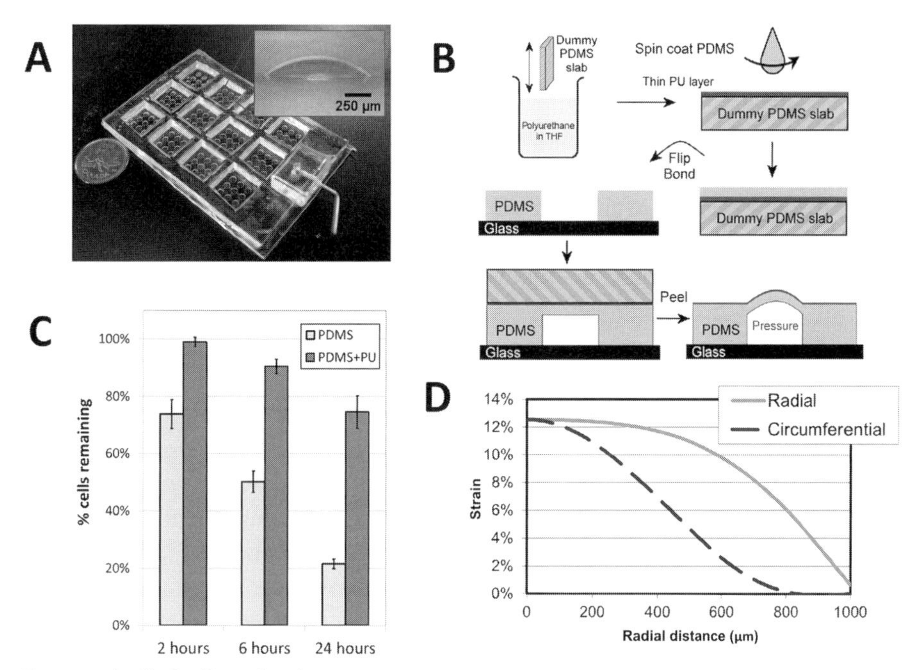

Figure 4.1 **Pneumatic "bulge" mechanisms.**
(A) High-throughput platform to apply dynamic mechanical stimulation to cells cultured on 2-D surfaces. Pneumatic pressure inflates a silicone diaphragm, stretching cells cultured on the surface. (B) Fabrication process to incorporate polyurethane films into a cell stretching platform. (C) Use of composite polyurethane/PDMS films significantly improves cell adhesion over conventional PDMS substrates. (D) Analytical model of radial and circumferential strains across the surface of an inflating diaphragm. Strains are typically not uniform across the device surface.
(A) and (B) reprinted from the journal *Integrative Biology* 2013, 5: 673–680; reproduced with permission from Moraes et al. "Microdevice array-based identification of distinct mechanobiological response profiles in layer-specific valve interstitial cells." Data for (C) and (D) previously reported in Moraes et al. (2013).

their world-to-chip interconnects, and reducing the number of these connections greatly improves the robustness of the device in operation. Manipulation of cavity geometries may also be used to alter the strain fields generated on such a device (Zhou and Niklason 2012). While circular diaphragms can be used to create axisymmetric biaxial strain fields, alternative geometries may be also used to adjust strains along each of the principal strain axes.

4.2.3 Design of peripheral equipment

While many configurations of peripheral equipment are possible, we briefly describe the setup used in recent work from our lab, in which we utilize a pneumatic actuation program to inflate microfabricated diaphragms and screen for combinatorial mechanobiological effects on disease progression in an aortic valve fibrosis (Moraes et al. 2013).

Briefly, an array of bulging diaphragms was microfabricated in soft polymeric materials, and cyclically inflated using an external pump (Fig. 4.1A). In order to control the applied external pressure, a pulse-width modulation voltage controller was designed to vary the root-mean-square voltage applied to a 5 V eccentric diaphragm micropump (SP 500 EC-LC, Schwarzer Precision, Germany). Pressure output was measured and calibrated using a simple differential digital output pressure sensor (allsensors.com). The pressure source was then connected to the chip via a solenoid valve and manifold (Pneumadyne, Plymouth, MN). The valve is designed to either connect the device to the applied pressure source, or to vent the device to atmospheric pressure, and is controlled by application of a 5 V power source. To generate cyclic pressure waveforms, a simple timing circuit was setup using a standard IC 555 chip to set the period and duty cycle of the desired square-pressure waveform. Inexpensive timer kits to generate these waveforms are readily available commercially. To simulate the heartbeat in cardiovascular microenvironments, we selected a 50% duty cycle applied at 1 Hz and maintained this stimulation profile for 24 hours. While this setup is the simplest we are aware of, more complex peripheral equipment may be set up in LabView or with microcontrollers to dynamically vary duty cycles, pressure levels, and frequency over the course of an experiment.

4.2.4 Advantages at the microscale

In this application, the use of microscale stretching systems affords two critical advantages over conventional techniques. First, reducing the lateral size of the diaphragm reduces the vertical displacement necessary to generate strains comparable to conventional systems. This minimized perturbation reduces differential hydrostatic normal forces that can occur between stretched and relaxed conditions in macroscale systems (Brown 2000), decoupling these two confounding factors in cell culture. Second, as primary cells often need to be seeded at a minimum density to ensure healthy attachment, spreading, and proliferation, the miniaturized system size reduces the number of primary cells required for a study. As cells typically display combinatorial responses to applied mechanobiological cues, reducing system size enables probing multiple combinations of culture conditions, which can be particularly challenging when primary cells are scarce or challenging to obtain.

Cells in the heart valve leaflet exist in a mechanically complex microenvironment, which can play an important role in driving disease progression (Chen and Simmons 2011; Yip and Simmons 2011). In order to identify the effects of these mechanical cues on disease progression, primary cells were isolated from either side of the aortic valve leaflet, and cultured under dynamic mechanical stimulation conditions for 24 hours, and then assayed for incorporation of α-smooth muscle actin (SMA) into stress fibers in the cytoskeleton. This marker is the defining indicator of differentiation toward the myofibroblast phenotype (Tomasek et al. 2002; Hinz 2010), a highly contractile cell type associated with valvular fibrosis. Obtaining large numbers of primary cells is challenging, as prolonged expansion culture on tissue culture plastic itself prompts differentiation toward the myofibroblast phenotype. Hence, given the limited availability and sensitivity of these cells, high-throughput microfabricated methods are required to combinatorially probe this biological system. Using the described platform, we identified distinctive mechanobiological response profiles of each cell type to combinations of mechanical, biochemical, and matrix protein cues. As these cell types were previously considered to be from the same population, these quantitative screens have relevance in understanding disease progression in the valve, and in developing tissue-engineered replacements. Given the limited availability of these cells, these findings would not be possible to observe using conventional macroscale equipment.

4.2.5 Maintaining cell adhesion

Polydimethylsiloxane (PDMS) is the material most commonly used in building prototype cell culture microdevices, for a variety of reasons. It is elastomeric, transparent, gas permeable, chemically stable, nontoxic, can be cured at low temperatures, and retains micro- and nano-patterned features necessary for device operation (Duffy et al. 1998). Though some issues with culturing cells on PDMS have been identified (Regehr et al. 2009) and microengineers are gradually exploring more conventional materials such as polystyrene (Berthier et al. 2012), the options for mechanically dynamic culture platforms are limited. Material toughness, elasticity and compliance become important factors, in addition to more broadly applicable design criteria such as optical transparency, biocompatibility, and amenability to processing; and PDMS fills these design requirements extremely well.

However, proteins are typically adsorbed onto the PDMS surface by noncovalent, nonspecific binding, patterns of which have been shown to degrade after just four days in static culture (Moraes, Kagoma, et al. 2009). It is hence reasonable to assume that under mechanical loading, matrix proteins that interface between the cell culture substrate and the cell will detach, resulting in significant loss of cellular adhesion from the surface, a finding demonstrated in a simple cell-stretching system (Wipff et al. 2009).

To address this critical issue, several groups have developed techniques to covalently bind extracellular matrix proteins to PDMS surfaces (Wipff et al. 2009; Bellin et al. 2009), using glutaraldehyde-based crosslinking chemistry. As chemical modification may influence protein activity, these treatments may have subtle, undefined effects on cell culture. To circumvent this issue and provide clinically relevant substrates for these mechanically dynamic in vitro culture models, we developed a technique to integrate polyurethane (PU) culture substrates onto PDMS surfaces (Moraes, Kagoma, et al. 2009), and then used this technique to develop composite material diaphragms for mechanically dynamic cell culture (Moraes et al. 2013). Briefly, dummy PDMS slabs were dip-coated in polyurethane (Tecoflex SG-80A; Lubrizol Corporation; Wickliffe, OH) and allowed to dry vertically. Drying the polyurethane in an enclosed environment with minimal air flows resulted in an optically transparent, smooth film less than one micron in thickness. A thin film of PDMS was then spin-coated and partially cured onto this film. Curing the PDMS on the polyurethane film formed a tighter bond than between precured PDMS and polyurethane, enabling the transfer of the composite culture film to the pneumatic device (Fig. 4.1B) (Moraes, Kagoma, et al. 2009). In this way, we integrate polyurethane, a well-studied and clinically relevant biomaterial, into the PDMS fabrication process. Although polyurethane is a low-toughness material and is prone to fatigue and mechanical creep, designing these composite material diaphragms allows us to bolster polyurethane's poor mechanical properties with the robustness of the underlying PDMS layer while maintaining the desirable biochemical properties of polyurethane.

Simple adhesion assays of primary cells on mechanically dynamic culture films indicate that polyurethane coatings significantly improve cell adhesion to the underlying layer, even over relatively long-term stimuli protocols (Fig. 4.1C). These values may be further improved using existing methods to modify Tecoflex polyurethane to achieve several-fold increases in cell adhesion (Dennes and Schwartz 2008).

4.2.6 Analysis of diaphragm stress

The "bulging" systems described here do not produce a uniform strain field, but generate strains that vary across the surface. While finite element methods are typically used to analyze surface strains, we present an analytical approach to approximate radial and circumferential strains across an inflating circular diaphragm by measuring vertical deflection of the diaphragm (Moraes et al. 2013). Based on work by Suhir et al. (Suhir 1991) this model provides a method to rapidly identify and obtain design parameters based on quick characterization experiments. The following equations relate peak radial and circumferential stresses and strains in an axisymmetric model:

$$\varepsilon_r(r) = \frac{1}{E}(\sigma_r - v\sigma_\theta) \tag{4.1}$$

$$\varepsilon_\theta(r) = \frac{1}{E}(\sigma_\theta - v\sigma_r) \tag{4.2}$$

where ε_r and ε_θ are the radial and circumferential strains respectively, and are functions of the radial distance r. E and v are the Young's modulus and Poisson's ratio of the diaphragm material. As most culture systems are designed in PDMS, the diaphragms may be considered incompressible ($v = 0.5$).

The radial and circumferential stresses can be determined using the following equations:

$$\sigma_r = -\frac{6D}{h^2}\left(\frac{d^2w}{dr^2} + \frac{v}{r}\frac{dw}{dr}\right) + \frac{1}{r}\frac{d\phi}{dr} \tag{4.3}$$

$$\sigma_\theta = -\frac{6D}{h^2}\left(\frac{1}{r}\frac{dw}{dr} + v\frac{d^2w}{dr^2}\right) + \frac{d^2\phi}{dr^2} \tag{4.4}$$

where D is the flexural rigidity of the circular film and is defined by

$$D = \frac{Eh^3}{12(1 - v^2)} \tag{4.5}$$

The functions $w(r)$ and $\Phi(r)$ are defined as

$$w(r) = f\left[1 - \left(\frac{r}{a}\right)^2\right]^2 \tag{4.6}$$

$$\phi(r) = \frac{f^2 E}{12}\left[\left(\frac{5 - 3v}{1 - v}\right)\left(\frac{r}{a}\right)^2 - \frac{1}{4}\left(\frac{r}{a}\right)^4 + \frac{1}{9}\left(\frac{r}{a}\right)^6 - \frac{1}{48}\left(\frac{r}{a}\right)^8\right] \tag{4.7}$$

where f is the measured deflection of the diaphragm, a is the diaphragm radius, and h is the thickness of the diaphragm. For the composite material diaphragms developed in our lab (Moraes et al. 2013), a "rule of mixtures" approach was used to estimate the homogenized E based on a ratio between the thicknesses of the constituent materials and their individual moduli.

Experimental verification of vertical displacement in conjunction with this model demonstrates that for a 2 mm diameter diaphragm having a thickness of 45 μm and under an applied pressure of 11 kPa, the radial and circumferential strains deviate significantly across the membrane surface (Fig. 4.1D). This presents a significant concern in culturing cells on the surface of such membranes: cells in different regions of the device experience distinct mechanical strain conditions. This is an inherent issue with inflatable diaphragm systems. To compensate for the strain heterogeneity, only these cells adhered within the central, relatively uniform region of the device could be analyzed and included in reported datasets. However, these uniformly stimulated cell populations may still interact via soluble signals with mechanically static populations, which may potentially confound biological results. More importantly, as cells are known to be extremely sensitive to even small differences in strain magnitude (Moraes, Chen, et al. 2010), this approach may not provide sufficient strain uniformity for some applications. Hence, for some investigations, alternative approaches will likely be necessary to improve the strain uniformity of microscale cell stretching systems.

4.3 Advanced stretching configurations

Uniformity of strains can be achieved by including deformation mechanisms that limit substrate stretch to in-plane deformations. One possible method is through the use of in-plane stretching mechanisms, such as those demonstrated by Huh et al. (2010), in which a thin PDMS film is stretched by two symmetrically deforming walls. Alternatively, a design approach proposed by the Flexcell Corporation has been to include a loading post across which a flexible substrate can be stretched (Fig. 4.2A). This approach, demonstrated by us (Moraes, Chen, et al. 2010) and by others (Simmons et al. 2011) has been utilized to generate precisely controlled deformation over multiple experimental units on a microfabricated array. We have successfully used this technology to create arrays of miniaturized stretching units

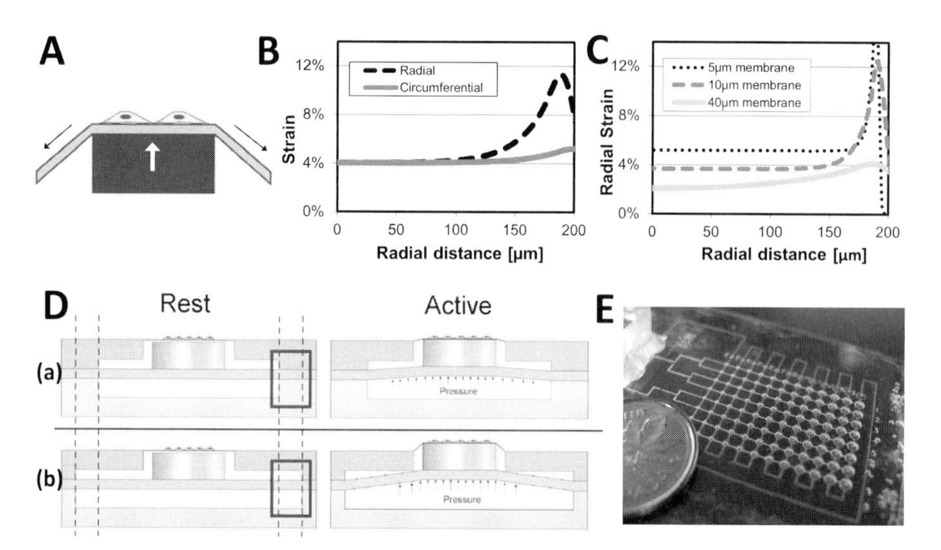

Figure 4.2 **Advanced stretching configurations.**

(A) Stretching a culture film over a loading post creates in-plane deformations that generate relatively uniform equibiaxial strains for a circular loading post. (B) Finite element simulations demonstrate regions of uniform equibiaxial radial and circumferential strains in the central region of the loading post. (C) Strain uniformity is strongly influenced by thickness of the cell culture diaphragm in a microfabricated format. Thicker diaphragms create bending stresses that reduce strain uniformity across the loading post. (D) Microfabricated device schematic cross-section, in which geometry is used to modulate surface strains for a single pressure source. The red frame highlights differences between geometries across the array: increased size of the cavity beneath the loading post results in increased vertical deflections and therefore, increased surface strains. (E) Microfabricated system in which equibiaxial strains are modulated from 2 to 15% across the device array.

Reprinted from the journal *Lab on a Chip* 2010, 10: 227, Moraes, Chen et al. "Microfabricated arrays for high-throughput screening of cellular response to cyclic substrate deformation." Reproduced with permission from Moraes, Chen, et al. (2010).

that simultaneously generate uniform equibiaxial strains ranging in magnitude from 2% to 15% across the array. Using this system, 108 dynamic stretching experiments can be performed simultaneously on a 6.5 cm^2 chip. This high-throughput system was then used to identify a novel time- and strain-magnitude dependent activation of the Wnt signaling pathway in primary valve interstitial cells (Moraes, Chen, et al. 2010). However, as discussed in the following sections, designing these uniform strain generating systems poses significant challenges, particularly when engineering these systems in microfabricated, array-based formats.

4.3.1 Incorporating posts to improve strain uniformity

Out-of-plane deflection of a thin film by raising a circular loading post produces an area of uniform equibiaxial strain in the central region of the post. While analytical solutions for strains in the central region of the loading post are possible (Brown 2000), capturing the nonuniform stresses away from the central region may be challenging, and hence finite element models present the simplest method to analyze strain profiles across the device surface (Fig. 4.2B) (Moraes, Chen, et al. 2010). Finite element simulations indicate that strain profiles along the radial and circumferential directions are equal and uniform for a significant percentage of the area overlying the loading post. The size of this equibiaxially uniform region is dependent on the thickness of the cell culture diaphragm: bending stresses in thicker cell culture films lead to greater stress nonuniformities at the edges of the loading post (Fig. 4.2C). Hence, minimizing thickness of the cell-culture diaphragm is a critical parameter in maintaining uniformity of strain in a scaled down system.

Modifying the post shape may be used to control the relative magnitudes of strains at different regions of the stretching unit to generate anisotropic strain profiles (Gopalan et al. 2003), but this has yet to be demonstrated using microfabricated strain arrays. The geometry of the system may also be used to modulate the strain magnitude of equibiaxial strains for a fixed applied pressure load by adjusting the ratio of culture film distention to culture film area. While others have achieved this by changing the diameter of the loading post, such that larger loading posts decrease the strains generated (Simmons et al. 2011), this approach creates culture surfaces of different diameters across a screening array. As cell function is modulated by the number of cells within a colony, this introduces a confounding factor that may influence cell function independent of applied strain magnitude. In contrast, we chose an alternate method to make the available culture surface constant across the array, in which the loading post is actively raised to distend the cell culture diaphragm. Pneumatic pressure applied beneath the loading post raises the post, and the geometry of the cavity underlying the loading post modulates the displacement of the post for a constant applied pressure (Fig. 4.2D, E) (Moraes, Chen, et al. 2010). In this way, our design minimizes confounding parameters that may influence biological results.

4.3.2 Challenges in fabrication

Two core challenges arise in fabricating the devices proposed in Fig. 4.2D: (1) alignment of multiple micropatterned layers, and (2) addressing shrinkage-induced alignment registration errors between the layers.

Multilayer soft lithography (MSL) is a technique in which multiple layers of PDMS are aligned, stacked, and bonded to create complex three-dimensional device architectures (Unger et al. 2000). To reliably align the layers, a simple custom-built alignment mechanism was designed to maintain a small distance between the two samples, while allowing freedom of movement in the x, y, and theta directions (Fig. 4.3A) (Moraes, Sun, et al. 2010). Briefly, a vacuum chuck was designed to hold a 3" × 2" glass slide, and attached to the arm of a micromanipulator (Siskiyou; Mission Viejo, CA). This system holds a layer of micropatterned PDMS over another micropatterned PDMS piece mounted on a rotary stage (Newmark; Grants Pass, OR). Alignment is observed through a zoom scope (Navitar, 12x zoom system; Rochester, NY), and the micromanipulator is adjusted until the features line up. The two substrates are then brought into contact and

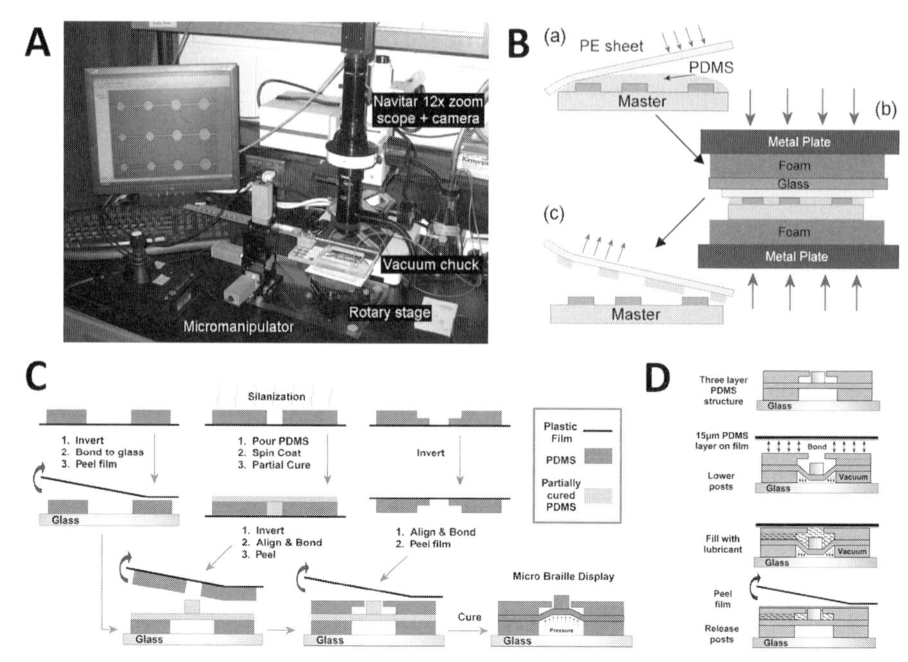

Figure 4.3 **Customized fabrication techniques.**
(A) Custom-designed alignment system to build multilayer structures. A micromanipulator is designed to move a PDMS structure over a second PDMS layer mounted on a rotary stage. (B) Sandwich mold fabrication is used to eliminate shrinkage of the PDMS film by curing PDMS between an inkjet transparency and a mold under compression. (C) Fabrication process to create micro braille displays in which a loading post can be vertically actuated. (D) Active fabrication methodology to add a cell culture diaphragm and lubricate the cavity between the loading post and diaphragm.
(A)–(C) reprinted from the *Journal of Micromechanics and Microengineering*, 2009, 19: 065015, Moraes, Sun, et al. "Solving the shrinkage-induced PDMS alignment registration issue in multilayer soft lithography." Reproduced with permission from Moraes, Sun, et al. (2009).
(D) Reprinted from the journal *Lab on a Chip* 2010, 10: 227, Moraes, Chen et al. "Microfabricated arrays for high-throughput screening of cellular response to cyclic substrate deformation." Reproduced with permission from Moraes, Chen, et al. 2010.

bonded via oxygen plasma-induced bonding. Once bonded, the vacuum seal on the upper substrate is broken, releasing the stack of bonded PDMS layers.

However, PDMS has a tendency to shrink during curing. The amount of shrinkage in each layer is very sensitive to a number of parameters, including cure time and temperature, layer thickness and PDMS component ratios (Lee and Lee 2008). While varied shrinkage across multiple layers may not make a significant difference in fabricating a single 3-D microstructure, even a very small difference in shrinkage ratios will affect the ability to align an array of microstructures over a relatively large area. Since it is particularly challenging to perfect fabrication process parameters in a general multipurpose lab environment, simply calibrating and rescaling the device designs to accommodate shrinkage ratios can be challenging to implement. Furthermore, it will not be possible to obtain a perfectly accurate shrinkage ratio, thereby limiting the size of the cell-stretching array that can be produced, unless the tolerance for registration between aligned layers is quite large.

In order to address this issue, we utilized sandwich mold fabrication, a technique by which thin films of PDMS are fabricated by squeezing the uncured polymer between a micropatterned mold and a flexible plastic transparent sheet (Fig. 4.3B) (Jo et al. 2000). Once cured, the PDMS micropatterned sheet and attached transparency is then gently peeled from the mold, placed into the alignment system, and brought into contact with the underlying layer. The mechanical stability of the attached transparency prevents the PDMS sheet from undergoing shrinkage, and hence the spot-to-spot distances between structures on the array remain constant (Moraes, Sun, et al. 2009). Once the PDMS layer is firmly bonded to the lower layer in the stack, the transparent film is gently peeled away. In this way, multiple layers can be fabricated independent of shrinkage considerations, and the fabrication process outlined in Fig. 4.3C can be used to construct arrays of distensible loading post structures (Moraes, Sun, et al. 2009).

4.3.3 Challenges in operation

The reduction in component sizes of this system increases the problem of stiction between the loading post and the deforming culture diaphragm. "Stiction" or "static friction," refers to the problem frequently encountered in MEMS, in which the increased surface area of the miniaturized system relative to the volume of the components increases surface-scaling adhesive forces between components. Small parts tend to stick together, and detachment requires substantial applied forces. This problem is rarely encountered in PDMS devices, as design for most applications can include reasonably thick PDMS diaphragms that provide sufficient spring-generated forces to separate components temporarily attached together by stiction. However, as previously discussed, reducing the thickness of the cell culture film is critical in providing reasonably large areas of uniform equibiaxial stimulation. Hence, spring constants for these thin films (<10 μm) are by necessity reduced, and stiction becomes a significant issue between the interacting loading post and culture film surfaces.

To reduce the friction between the two surfaces, a lubricant was designed to ensure device operation. Standard oil-based lubricants were unsuitable because the hydrophobic molecules migrate through PDMS (Regehr et al. 2009), where they can cause swelling and were in some cases toxic to cultured cells. The lubricant needs to maintain effectiveness on PDMS surfaces over time. This is particularly challenging as the chemical surface moieties of PDMS changes over time, as the plasma-oxidized, hydrophilic PDMS layer that is created on the surface to bond layers together is replaced by hydrophobic migratory chains from within the PDMS bulk (Duffy et al. 1998). Glycerol is biocompatible, does not swell PDMS (Lee et al. 2003), and has been shown to have a low coefficient of friction between hydrophilic oxidized PDMS surfaces (Bongaerts et al. 2007). In order to maintain the hydrophilic character of the PDMS surfaces, deionized water was added to the lubricant, as oxidized surfaces have been shown to remain stable for up to two weeks when in contact with water (Anderson et al. 2000).

The ratio between glycerol and water was arbitrarily selected based on ease of filling the microfluidic delivery channels, as experimental characterization indicated no significant differences between 30% and 90% solutions of glycerol in water.

In order to prevent the culture film from sticking to the loading post during fabrication, an "actuated fabrication methodology" was required to ensure that the system was well lubricated during the fabrication process (Fig. 4.3D). During this process, negative pressure is applied beneath the loading post, increasing the distance between the loading post and the cell culture film. The culture film is then plasma-oxidized and bonded to the surface, and the channel filled with lubricant. Once the culture film is securely bonded, the loading post is released, and device fabrication can continue. Cells plated on the culture film surface can then be dynamically stimulated using a similar peripheral equipment setup as described earlier.

4.4 Extending to three-dimensional systems

Three-dimensional (3-D) cell culture has emerged as an important aspect of recreating the cellular environment and thereby providing realistic cues to cultured cells. Known to modulate cellular response to chemical factors in the microenvironment (Hwang et al. 2006; Weaver et al. 1997; Smalley et al. 2006), and to alter cellular mechanisms of probing environmental mechanics (Huebsch et al. 2010), 3-D culture systems are expected to be the next generation of high-throughput screening systems. However, these culture systems present unique technical challenges and opportunities. While applying mechanical forces to 3-D culture is performed regularly using conventional equipment and techniques, substantial work remains to be done in miniaturizing such platforms to increase screening throughput. While the simplest modes of mechanical stimulation in three-dimensional structures are tension and compression, few examples exist in which dynamic mechanical tension is applied to tissue constructs. Recent work in our lab has focused on dynamic compression of cell-laden hydrogels, and in this section, we discuss the technical requirements and challenges associated with developing microscale compression culture systems.

4.4.1 Potential compression modes

Compression of macroscale three-dimensional biomaterials cannot be modeled using simple solid mechanical approaches. As most biomaterials are hydrogels, dynamic compressive stimuli causes a coupled fluid-solid interaction, in which fluid pressure initially spikes and is gradually normalized as water is expelled from the deforming hydrogel, and mechanical deformation gradually increases to equilibrium values. This fluid-driven time dependency complicates mechanical analysis of deforming hydrogel systems. However, in addition to scaling up the throughput of experimental platforms, microfabricated 3-D compression systems may also significantly decrease the mechanical equilibration kinetics of the system. Because the miniaturized hydrogels have a significantly greater surface area to volume ratio, fluid can be expelled rapidly, reducing the fluid pressure wave time constant within the system. Though this assumption needs to be rigorously proven through simulations that consider specific material parameters of the system being studied, we assume that simple solid mechanical deformation analysis is sufficient to characterize deformation of microfabricated hydrogels.

Compression of 3-D materials typically follows two schemes: confined compression and unconfined compression. Under confined compression, the hydrogel is encased in a porous material to allow fluid flow, and a close-fitting piston is used to apply compressive stresses to the material. This results in uniform stresses generated throughout the material, but due to precise machining requirements, can be quite challenging to miniaturize. Under unconfined compression, a free hydrogel is compressed between porous plates, which can be significantly easier to implement in a miniaturized format, but results in varied strain profiles through the hydrogel. In recent work, we present a third alternative to these compression schemes, in which hydrogels are "semiconfined" (Moraes et al. 2011), This approach balances ease of fabrication with generation of uniform strains in the central region of the hydrogel.

4.4.2 Building increased throughput compression systems

In recent work, we have developed microfabricated screening arrays for hydrogel deformation based on unconfined (Moraes, Wang, et al. 2010) and semiconfined (Moraes et al. 2011) compression (Fig. 4.4A, B). To achieve unconfined compression,

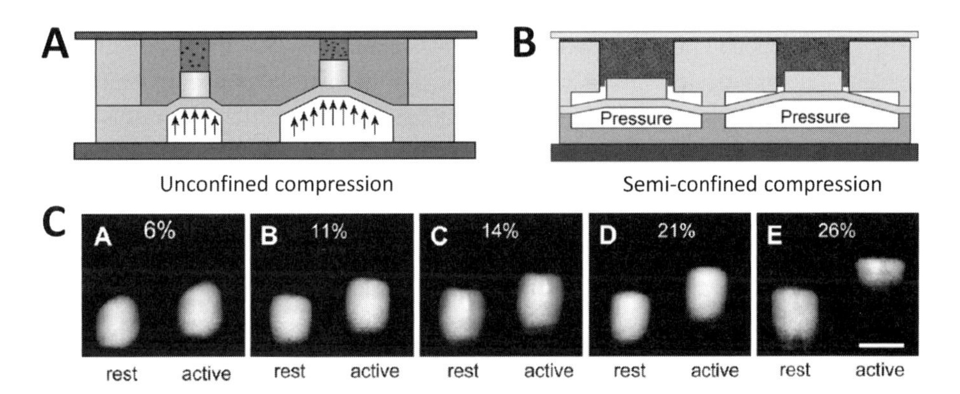

Figure 4.4 **Microfabricated screening arrays for hydrogel deformation.**
Schematic diagrams for microdevice arrays for (A) unconfined, and (B) semi-confined compression schemes.

(A) and (B) reprinted from the journal *Biomaterials* 2010, 31(8): 577–584, Moraes, Wang, et al. "A microfabricated platform for high-throughput unconfined compression of micropatterned biomaterial arrays." Reproduced with permission from Moraes, Wang, et al. (2010) and C. Moraes et al. (2011).

Cell-laden hydrogels are polymerized in situ and compressed by the application of pressure beneath the loading posts. (C) Confocal images of cells at rest and under a range of applied deformations. Cells exhibit a highly non-linear deformation profile when compared to deformation of the surrounding matrix.

(C) Reproduced with permission from Moraes, Wang, et al. (2010).

photopolymerizable poly ethylene glycol (PEG) hydrogels were micropatterned directly over the vertically actuated loading posts used in the previous studies. Once the loading post is actuated, the cell-laden hydrogel is compressed against a fixed glass surface. Using this approach, we produced a simple 5 × 5 array of compressive units. Using similar geometries as previously discussed, global compressive strains were varied from 6% to 26% across the array, using a single pressure source. Tuning this range of strains is easily accomplished by changing the height of the polymerization chamber: increasing the height reduces the global strain, while making the polymerization chamber thinner can be used to increase the applied strains.

This approach is conceptually simple, but a few technical issues need to be addressed. First, this system requires photopolymerization of aligned biomaterial constructs. Hence, a careful alignment procedure needs to be designed to place a photomask directly over the loading posts prior to polymerization. Alignment was achieved using the custom alignment system described in Section 4.4.1. Second, the use of a photopolymerizable biomaterial places inherent limitations on the system. UV-driven, free-radical photopolymerization chemistry can often be toxic to cells, as is the case in our previous work (Moraes, Wang, et al. 2010). In order to ensure cell viability, the photopolymerization system requires careful optimization of polymerization times and intensities. Third, free-radical polymerization systems are often quenched by the presence of oxygen. PDMS is an oxygen-rich material, and photopolymerization within this environment often requires long exposure times (and consequent increased cell death) to compensate for oxygen quenching. In order to address this issue, we coated a thin layer of Parylene-C over the PDMS posts. Parylene-C acts as a barrier to oxygen contained within the PDMS, thereby improving polymerization times. Furthermore, a conformal layer of Parylene can be deposited by vapor phase processing, making it particularly attractive for the complex geometries needed in this particular microtechnology. Finally, requiring photopolymerizable materials places significant restrictions on the variety of materials that can be incorporated into this device.

Semiconfined compression presents an attractive alternative to these challenges. Briefly, a polymerizable biomaterial is poured over the device and allowed to form a hydrogel. Surface tension between the hydrogel and trapped air within the device prevents the gel from forming in the crevices beside the loading posts. This approach provides significant advantages over the unconfined compression system. First, as previously mentioned, strain uniformities are improved within the central region of the hydrogel. Second, once the PDMS microdevice is fabricated, no further alignment steps are necessary, making the device usable by a broad spectrum of end users. Third, the system will work with any polymerizable hydrogel system, including natural extracellular matrix components such as collagen and matrigel. Since scaffold material plays a critical role in regulating cell function, the versatility of this approach greatly improves the potential utility of this screening platform. Finally, time-consuming optimization protocols for cell viability are not as stringent, as the gelation chemistries that can be employed typically produce significantly reduced cytotoxicities.

4.4.3 Transmission of mechanical stresses

Characterization of matrix stiffness is critically important in designing microfabricated compression systems. In order to ensure that the biomaterial does undergo compression, the PDMS loading post array must be designed to deliver the appropriate displacements with sufficient force. In the case of stiffer biomaterials, this may mean using stiffer PDMS compositions or designing thicker support layers for the vertically actuated loading post. An increase in applied pressure will then be necessary to achieve sufficient displacement.

However, a more subtle effect can occur in the transmission of mechanical deformations to cells embedded within the matrix. Cells embedded within the photopolymerized PEG hydrogels under unconfined compression displayed an unusual nonlinear response to applied mechanical deformation (Fig. 4.4C). Confocal images of cells embedded within the matrix reveals that cells do not deform under applied loads up to 21% applied strain. At strains greater than 21%, cells undergo a dramatic deformation. This phenomenon arises from nonlinear transfer of loads within the hydrogel matrix. If the matrix is softer than the embedded cells, deformation of the matrix shields the cells from applied loads. However, at high compressive strains, the matrix pores collapse, increasing the stiffness of the matrix relative to the cell. Once this occurs, strains are more effectively transferred to the embedded cells. This nonlinear transfer of loads can significantly influence cellular response to applied deformation: if the biomaterial shields the cell from applied forces, would the cell experience a distinct mechanical environment? While subtle forces may be sufficient to trigger cellular responses, increased throughput screening in mechanically well-defined systems is necessary to identify if cells respond to subtle applied forces as well as applied displacements.

4.5 Integrating advanced technologies into dynamic mechanical stimulation platforms

Incorporating additional technologies into the microfabricated strain generation platforms discussed here can significantly improve the utility and capabilities of these devices. As the devices are largely fabricated in PDMS, they are compatible with the wide variety of technologies being developed by the microfabrication and biomedical communities. These technologies can broadly be classified as sensory or stimulatory technologies. Sensory technologies involve developing integrated systems to provide more information about the system under study, while stimulatory technologies are designed to incorporate additional controlled mechanical and chemical factors into the screening platforms.

While a variety of technologies exist for these purposes, current work in our lab has focused on incorporating strain sensors directly into a 3-D compression system, in order to measure stiffness of a tissue construct that has been conditioned with mechanical compressive stimulation (MacQueen et al. 2012). To demonstrate this technology, strain sensors are integrated into the deforming PDMS layer, which compresses soft PDMS

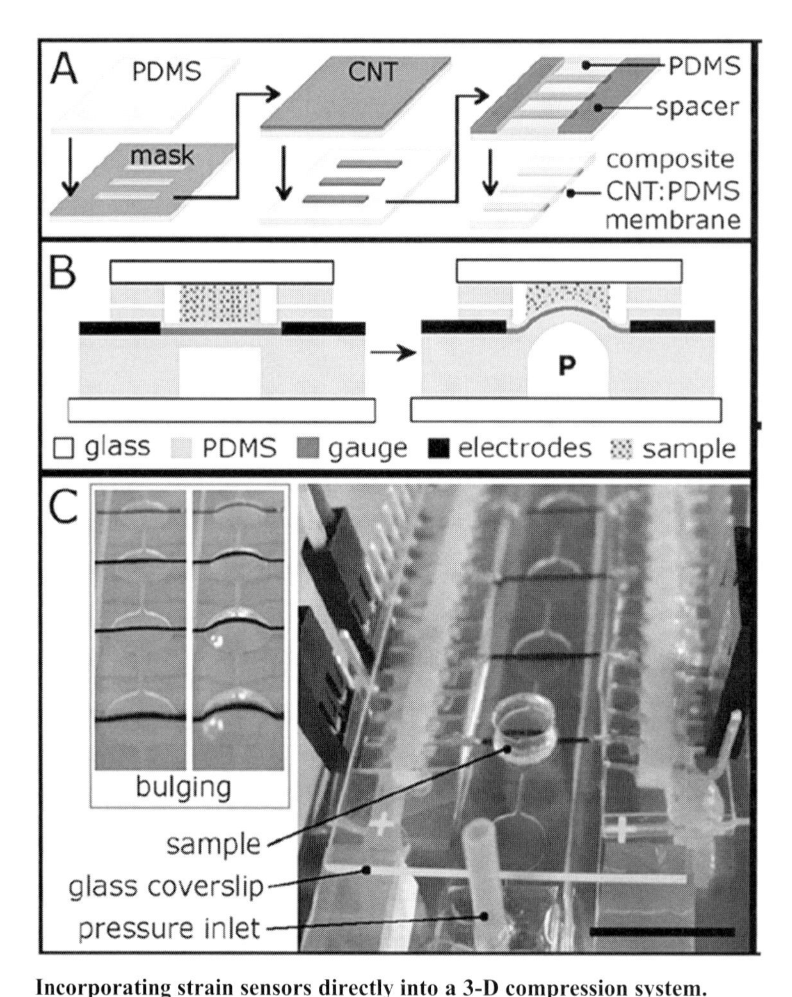

Figure 4.5 **Incorporating strain sensors directly into a 3-D compression system.**
(A) Schematic diagram of the fabrication process involved in integrating carbon nanotube / PDMS composite sensors into a compression system. (B) Hydrogel samples undergoing compression in the microfabricated system. (C) Actuated devices compress biomaterial constructs.

Reprinted from the journal *Lab on a Chip* 2102, 12(20): 4178–4184, MacQueen et al. "Miniaturized platform with on-chip strain sensors for compression testing of arrayed materials." Reproduced with permission from MacQueen et al. (2012).

constructs (Fig. 4.5). To fabricate the strain sensors, multi-wall carbon nanotubes (CNTs) are mixed in with PDMS prior to curing, and screen-printed in strips on the PDMS membrane. The composite CNT-PDMS material alters resistance under applied strains, and a simple ammeter is sufficient to measure these changes for the range of strain values generated in the device. Once cured the CNT-PDMS is encapsulated in a second layer of PDMS to create a strain sensor embedded within the deforming membrane. Resistance to membrane deflection by biomaterials of different stiffness can hence be measured in real-time to monitor evolving construct stiffness in mechanically dynamic environments.

4.6 Conclusions

The ability to incorporate dynamic mechanical cues into both two-dimensional and three-dimensional culture environments may significantly improve the ability of high-throughput screening tools to accurately recreate realistic cell culture environments. The use of microfabrication technologies enables end users to study biological systems in a high-throughput manner that may lead to biological results and insights that would be challenging, if not impossible to obtain with conventional approaches. Microengineering also significantly improves the precision with which mechanical stimuli can be applied, resulting in cell culture platforms with minimized confounding factors. Furthermore, the ability to incorporate additional novel sensing and stimulation technologies will further improve the utility of such systems. Although such technologies have yet to achieve mainstream usage, these strategies represent an important step forward in designing mechanically and physiologically realistic cell culture screening platforms.

References

Anderson, D. G. et al. (2005). "Biomaterial microarrays: rapid, microscale screening of polymer–cell interaction." *Biomaterials* **26**(23): 4892–4897.
Anderson, J. R. et al. (2000). "Fabrication of topologically complex three-dimensional microfluidic systems in PDMS by rapid prototyping." *Analytical Chemistry* **72**(14): 3158–3164.
Bellin, R. M. et al. (2009). "Defining the role of syndecan-4 in mechanotransduction using surface-modification approaches." *Proceedings of the National Academy of Sciences* **106**(52): 22102–22107.
Berthier, E., E. W. K. Young, and D. Beebe. (2012). "Engineers are from PDMS-land, Biologists are from Polystyrenia." *Lab on a Chip* **12**(7): 1224–1237.
Bongaerts, J. H. H., K. Fourtouni, and J. R. Stokes. (2007). "Soft-tribology: lubrication in a compliant PDMS–PDMS contact." *Tribology International* **40**(10–12): 1531–1542.
Brown, T. D. (2000). "Techniques for mechanical stimulation of cells in vitro: a review." *J Biomech* **33**(1): 3–14.
Chen, J. H. and C. A. Simmons. (2011). "Cell-matrix interactions in the pathobiology of calcific aortic valve disease: critical roles for matricellular, matricrine, and matrix mechanics cues." *Circ Res* **108**(12): 1510–1524.
Dennes, T. J. and J. Schwartz. (2008). "Controlling cell adhesion on polyurethanes." *Soft Matter* **4**(1): 86–89.
Duffy, D. C. et al. (1998). "Rapid prototyping of microfluidic systems in poly(dimethylsiloxane)." *Analytical Chemistry* **70**(23): 4974–4984.

Figallo, E. et al. (2007). "Micro-bioreactor array for controlling cellular microenvironments." *Lab on a Chip* **7**(6): 710–719.

Flaim, C. J., S. Chien, and S. N. Bhatia. (2005). "An extracellular matrix microarray for probing cellular differentiation." *Nature Methods* **2**(2): 119–125.

Gómez-Sjöberg, R. et al. (2007). "Versatile, fully automated, microfluidic cell culture system." *Analytical Chemistry* **79**(22): 8557–8563.

Gopalan, S. M. et al. (2003). "Anisotropic stretch-induced hypertrophy in neonatal ventricular myocytes micropatterned on deformable elastomers." *Biotechnology and Bioengineering* **81**(5): 578–587.

Hinz, B. (2010). "The myofibroblast: paradigm for a mechanically active cell." *Journal of Biomechanics* **43**(1): 146–55.

Holick, M. F. (2000). "Microgravity-induced bone loss—will it limit human space exploration?" *The Lancet* **355**(9215): 1569–1570.

Huang, J.-W. et al. (2013). "Interaction between lung cancer cell and myofibroblast influenced by cyclic tensile strain." *Lab on a Chip* **13**(6): 1114–1120.

Huebsch, N. et al. (2010). "Harnessing traction-mediated manipulation of the cell/matrix interface to control stem-cell fate." *Nature Materials* **9**: 518–526.

Huh, D. et al. (2010). "Reconstituting organ-level lung functions on a chip." *Science* **328**(5986): 1662–1668.

Hwang, N. S. et al. (2006). "Effects of three-dimensional culture and growth factors on the chondrogenic differentiation of murine embryonic stem cells." *Stem Cells* **24**: 284–291.

Jo, B. H. et al. (2000). "Three-dimensional micro-channel fabrication in polydimethylsiloxane (PDMS) elastomer." *Journal of Microelectromechanical Systems* **9**(1): 76–81.

Kim, T. K. and O. C. Jeong. (2012). "Fabrication of a pneumatically-driven tensile stimulator." *Microelectronic Engineering* **98**: 715–719.

Kong, H. J. et al. (2005). "Non-viral gene delivery regulated by stiffness of cell adhesion substrates." *Nature Materials* **4**(6): 460–464.

Lee, J. N., Park, C. and G. M. Whitesides. (2003). "Solvent compatibility of poly(dimethylsiloxane)-based microfluidic devices." *Analytical Chemistry* **75**(23): 6544–6554.

Lee, S. W. and S. S. Lee. (2008). "Shrinkage ratio of PDMS and its alignment method for the wafer level process." *Microsystem Technologies* **14**(2): 205–208.

MacKenna, D. A. et al. (1998). "Extracellular signal-regulated kinase and c-Jun NH2-terminal kinase activation by mechanical stretch is integrin-dependent and matrix-specific in rat cardiac fibroblasts." *Journal of Clinical Investigation* **101**(2): 301–310.

MacQueen, L. et al. (2012). "Miniaturized platform with on-chip strain sensors for compression testing of arrayed materials." *Lab on a Chip* **12**(20): 4178–4184.

McBeath, R. et al. (2004). "Cell shape, cytoskeletal tension, and RhoA regulate stem cell lineage commitment." *Developmental Cell* **6**(4): 483–495.

Moraes, C. et al. (2011). "Semi-confined compression of microfabricated polymerized biomaterial constructs." *Journal of Micromechanics and Microengineering* **21**, 054014.

Moraes, C. et al. (2012). "Organs-on-a-Chip: A Focus on Compartmentalized Microdevices." *Annals of Biomedical Engineering* **40**(6): 1211–1227.

Moraes, C. et al. (2013). "Microdevice array-based identification of distinct mechanobiological response profiles in layer-specific valve interstitial cells." *Integrative Biology* **5**, 673–680.
Moraes, C., J.-H. Chen, et al. (2010). "Microfabricated arrays for high-throughput screening of cellular response to cyclic substrate deformation." *Lab on a Chip* **10**, 227.
Moraes, C., Y. K. Kagoma, et al. (2009). "Integrating polyurethane culture substrates into poly(dimethylsiloxane) microdevices." *Biomaterials* **30**(28): 5241–5250.
Moraes, C., G. H. Wang, et al. (2010). A microfabricated platform for high-throughput unconfined compression of micropatterned biomaterial arrays. *Biomaterials* **31**(3): 577–584.
Moraes, C., Y. Sun, and C. A. Simmons (2009). "Solving the shrinkage-induced PDMS alignment registration issue in multilayer soft lithography." *Journal of Micromechanics and Microengineering* **19**, 065015.
Moraes, C., Y. Sun, and C. A. Simmons. (2010). "Microfabricated platforms for mechanically dynamic cell culture. journal of visualized experiments." *J Vis Exp* **46**: e2224.
Moraes, C., Y. Sun, and C. A. Simmons. (2011). "(Micro)managing the mechanical microenvironment." *Integrative Biology* **3**: 959–971.
Mukundan, V., W. J. Nelson, and B. L. Pruitt. (2012). "Microactuator device for integrated measurement of epithelium mechanics." *Biomedical Microdevices* **15**(1): 117–123.
Regehr, K. J. et al. (2009). "Biological implications of polydimethylsiloxane-based microfluidic cell culture." *Lab on a Chip* **9**: 2132.
Scuor, N. et al. (2006). "Design of a novel MEMS platform for the biaxial stimulation of living cells." *Biomedical Microdevices* **8**(3): 239–246.
Shimizu, K. et al. (2011). "Development of a biochip with serially connected pneumatic balloons for cell-stretching culture." *Sensors and Actuators B: Chemical* **156**(1): 486–493.
Simmons, C. S. et al. (2011). "Integrated strain array for cellular mechanobiology studies." *J Micromech Microeng* **21**(5): 54016–54025.
Simon, C. G. and S. Lin-Gibson. (2011). "Combinatorial and high-throughput screening of biomaterials." *Advanced Materials* **23**(3): 369–387.
Smalley, K. S. M., M. Lioni, and M. Herlyn. (2006). "Life isn't flat: taking cancer biology to the next dimension." *In Vitro Cellular & Developmental Biology – Animal* **42**: 242.
Sniadecki, N. J. (2010). "Minireview: a tiny touch: activation of cell signaling pathways with magnetic nanoparticles." *Endocrinology* **151**(2): 451–457.
Suhir, E., 1991. *Structural analysis in microelectronic and fiber-optic systems*. New York: Van Nostrand Reinhold.
Tan, W. et al. (2008). "Development and evaluation of microdevices for studying anisotropic biaxial cyclic stretch on cells." *Biomed Microdevices* **10**(6): 869–882.
Tomasek, J. J. et al. (2002). "Myofibroblasts and mechano-regulation of connective tissue remodelling." *Nat Rev Mol Cell Biol* **3**(5): 349–363.
Unger, M. A. et al. (2000). "Monolithic microfabricated valves and pumps by multilayer soft lithography." *Science* **288**(5463): 113–116.

Volder, M. D. and D. Reynaerts. (2010). "Pneumatic and hydraulic microactuators: a review." *Journal of Micromechanics and Microengineering* **20**(4): 043001.
Weaver, V. M. et al. (1997). "Reversion of the malignant phenotype of human breast cells in three-dimensional culture and in vivo by integrin blocking antibodies." *The Journal of Cell Biology* **137**(1): 231–245.
Wipff, P.-J. et al. (2009). "The covalent attachment of adhesion molecules to silicone membranes for cell stretching applications." *Biomaterials* **30**(9): 1781–1789.
Wolff, J. (1986). *Law of Bone Remodelling*. New York: Springer.
Yip, C. Y. Y. and C. A. Simmons. (2011). "The aortic valve microenvironment and its role in calcific aortic valve disease." *Cardiovascular Pathology* **20**(3): 177–182.
Zandstra, P. W. (2004). "The opportunity of stem cell bioengineering." *Biotechnology and Bioengineering* **88**(3): 263–263.
Zhou, J. and L. E. Niklason. (2012). "Microfluidic artificial 'vessels' for dynamic mechanical stimulation of mesenchymal stem cells." *Integrative Biology* **4**(12): 1487–1497.

5 Multiscale topographical approaches for cell mechanobiology studies

Koichiro Uto, Elliot Fisher, Hong-Nam Kim, Chang Ho Seo, and Deok-Ho Kim

Human tissues are sophisticated ensembles of various cell types embedded in the complex but defined structures of the extracellular matrix (ECM). ECM is configured in a hierarchical structure from nano- to microscale, with many biological molecules forming large scale configurations and textures with feature sizes up to macroscopic scale (several hundred microns). The physicochemical, biological and mechanostructural properties of native ECM play a critical role in constructing a microenvironment for cells and tissues. In conjunction with the rapid evolution of material science and its fabrication techniques, studies of the topography and elasticity of ECM and other materials have allowed advanced interrogation of cellular mechanotransduction and cellular responses to mechanostructural cues. By learning from and mimicking the highly organized ECM structures found in vivo, topography-guided approaches to regulate cell function and fate have been widely investigated in the last several decades. Here, we review recent efforts in mimicking the micro- and nanotopography of the native ECM in vitro for the regulation of cellular behaviors. We also discuss how these biomimetic topographical surfaces have been applied to fundamental cell mechanobiology studies into cell adhesions, migrations, and differentiation as well as toward efforts in tissue engineering.

5.1 Introduction

The microenvironment surrounding cells regulates various cellular responses, including adhesion, growth, migration, and differentiation. In addition to the biochemical properties of the extracellular matrix (ECM), mechanostructural cues are of great importance because cells reside permanently on their microenvironment, which has tissue-relevant viscoelasticity and topography. Evidence has consistently indicated that topography can induce decision of cell fate, suggesting that the interaction between adhered cells and their extracellular environment in the form of topography, which is often hierarchically ordered from the nanoscale to the macroscale, can directly modulate cell function. For example, the heart is a complex organ system composed of a highly diverse arrangement of muscle (i.e., cardiomyocyte) and nonmuscle cells. These cells are assembled into the specific three-dimensional structures of the mature heart. The structural organization of the heart tissue is evident hierarchically at multiple length scales, ranging from the nanometer scale of actin/myosin motors, to the micrometer scale of sarcomere assembly, up to the three-dimensional structure of heart organ (Fig. 5.1) (Chien et al. 2008). This assembly clearly indicates the importance of scaling, a design problem that spans several

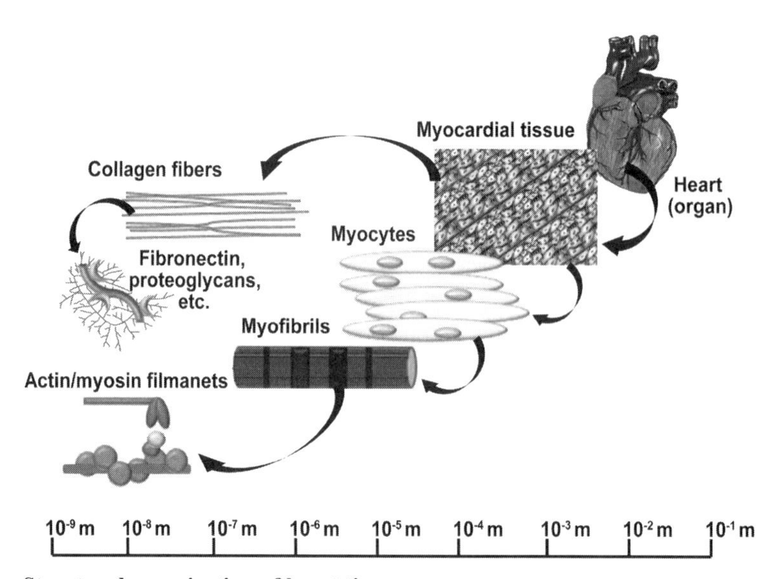

Figure 5.1 **Structural organization of heart tissue.**
The multiscale hierarchical organization of the heart is evident across several orders of spatial magnitude, on the molecular, subcellular, cellular, tissue, and organ levels.

orders of spatial magnitude, from the alignment of actin/myosin complexes within a sarcomere, to their alignment in myofibrils, to the organization of myofibrils in a myocyte, and further to the coupling between myocytes in anisotropic, laminar muscle.

Research has indicated that both tissue-specific function and in vivo regeneration can be driven by nanotopographical cues. This is because cells in their natural environment interact with ECM components such as collagen fibers on a nanometer scale, allowing them to arrange nanoscale structural units in the required configurations and textures for the generation of new functional tissues or organs. This occurs via the sophisticated spatial and mechanical harmonization of cells and ECM components. For example, mechanostructural cues imparted by the native ECM of the myocardium are known to influence the structural and functional properties of engineered cardiac tissue (Laflamme and Murry 2011; Lim et al. 2010). Many other cell types exhibit similar multiscale dependence on mechanostructural cues from their surrounding environment. The structure of the natural ECM in various tissues including not only heart, but also bone, nerve, and skin usually reveals highly oriented grooved fibrous nanoscale structures with various length scales (Fig. 5.2) (Kim et al. 2013). Concentric nanoscale-thick cylinders enhance mechanical properties of cortical bone, and the aligned collagen matrix in the dermis of skin layer leads to anisotropic bulk mechanical properties. Inspired by the ultrastructural character of natural ECM, the use of nanoengineering technology to develop a sophisticated matrix mimic in vivo has been attractive to engineers and biologists in the fields of classical cell biology, tissue engineering, and regenerative medicine.

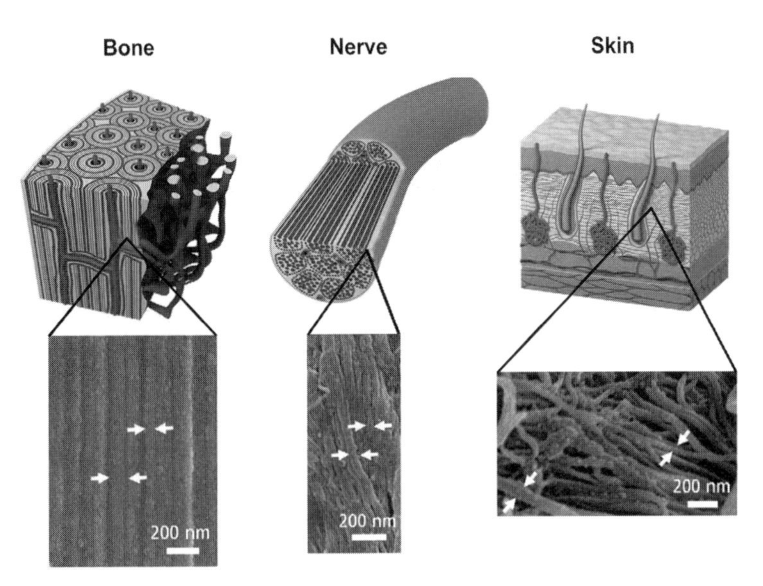

Figure 5.2 **Graphical illustrations and SEM images of ex vivo human bone, nerve and skin, showing well aligned nanostructures with various fibrous features.**
Cortical bone shows aligned brick and mortar structure. Nerves show structural anisotropy that aids signal conduction. Skin has organized ECM matrix fibers in the dermis, which aids in wound healing and confers bulk mechanical properties. White arrows indicate single nanoscale fibrous features. Reproduced with permission from Kim et al. 2013.
Reprinted with permission from Elsevier.

In this chapter, we focus on recent efforts to construct microenvironments using nanoengineering techniques and an understanding of the dependence of cell-specific function on topographical features. Specifically, we address the issues of how substrate topography influences cell adhesion (Section 5.2), migration (Section 5.3) and differentiation (Section 5.4). In addition, we discuss how the use of topographical substrates has been applied to tissue engineering (Section 5.5). This chapter reviews studies examining the relationship between cells and artificial mechanostructural environments with micro- and nanoscale features, and presents an in-depth understanding of current integrative mechanobiology.

5.2 Topography-guided cell adhesions

Application of lithographic, etching, and patterning techniques developed by the silicon microelectronics industry to soft matters, including biomolecules and artificial polymers, has facilitated investigations into the intricate role of micro- and nanoscale topography in scaffold materials on all aspects of cellular behavior and cell fate.

In vitro, characteristic structural functions of complex engineered tissues are heavily influenced by how well cell shape and polarity resemble those of cells in vivo (Chen et al. 1997; McBeath et al. 2004). Substrate topography has been shown to strongly influence the shape and polarity of many different cell types through a process known as contact guidance (Dunn and Ebendal 1978; Clark et al. 1991). Specifically, Teixeira et al. (2003 and 2006) reported that quantitative analysis of corneal epithelial cells showed that the extent of cellular alignment varied with the specific dimensions of substrate topographies, such as depth and pitch of surface grooves, as well as with changes to culture media, suggesting a synergistic interaction between topographic features (mechanostructural factors) and soluble biochemical factors. Although these data strongly suggested that surface structural features such as surface topography regulate the cell adhesion process, nanotopographic features found in vivo are still poorly understood.

In cardiac cell biology and tissue engineering it has been well known that the nanoscale arrangement of topographical and molecular patterns generated by collagen fibers play an important role in cardiac function. Substrates with micrometer feature size in both two-dimensional (2-D) and three-dimensional (3-D) culture systems have been employed to direct cardiomyocytes into anisotropic arrangements for electrophysiological and mechanical characterization (Fast et al. 1996; Bursac et al. 2002; Gopalan et al. 2003). To investigate whether finer, nanoscale, control over the cell–material interface facilitates the creation of truly biomimetic cardiac tissue constructs that recapitulate the structural and functional aspects of in vivo ventricular myocardial phenotype, Kim et al. (2010) has used UV-assisted capillary force lithography (CFL) to fabricate well-defined nanogrooved polyethylene glycol (PEG) substrates, which mimic ECM fiber dimensions. To account for the scale of the ECM cues and variability in the diameter of collagen fibrils, they designed an anisotropically nanofabricated substrate (ANFS) with various sizes, ranging from a ridge width of 150 nm, a spacing of 50 nm, and a height of 200 nm, to a ridge width of 800 nm, a spacing of 800 nm, and a depth of 500 nm. Neonatal rat ventricular myocytes (NRVMs) cultured on these surfaces showed elongation and alignment in the direction of the nanoscale cues and formed a dense monolayer. Transmission electron microscopy (TEM) also revealed alignment of myofilaments (Mf) in the direction of ANFS, with actin fiber bundles closely following individual grooves (Fig. 5.3A, white arrows) and focal adhesions (FA) often forming at the sides of the ridges and grooves (Fig. 5.3A, white arrow heads), suggesting that anisotropic cell-artificial matrix interaction induced extensive topographical alteration in FA assembly and cytoskeletal alignment. TEM analysis also revealed that on the 400 nm (1:1 ridge and groove) substrate, cells appeared to extend downward toward the groove floor, but this "sagging" was incomplete (Fig. 5.3B). By contrast, the sagging of the cells into the grooves of the 800 nm (1:1 ridge and groove) pattern appeared to fill the grooves more completely, with FAs present throughout the cell–groove interface (Fig. 5.3C), suggesting a more extensive cell–ANFS adhesion surface. Thus, a significant difference between small and large groove structure – even on the same nanoscale – was observed. Furthermore, the nanoscale width of grooves can regulate the level of cell penetration into the groove, thus modulating the strength of cell-substrate adhesion. Taken together, these findings suggest that nanotopographic control by the ANFS mediates the structural orientation of FA

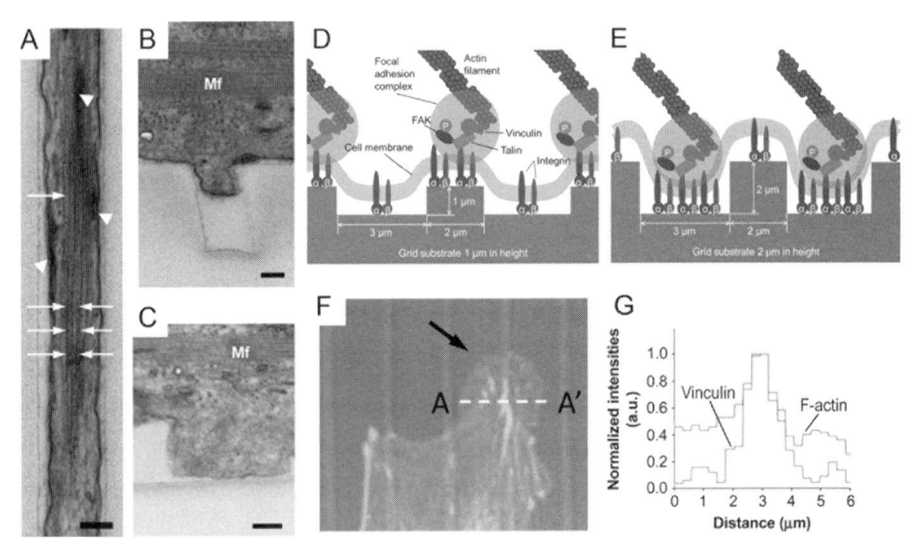

Figure 5.3 **The influence of anisotropically nanofabricated substrates on cell mechanobiology is mediated by pattern dimensions.**

(A) An enlarged view of actin bundles (white arrows) and focal adhesions (dark and thick lines indicated by white arrowheads) preferentially formed in parallel to the individual ridges and grooves of the ANFS. (B and C) Representative cross-sectional view of the PEG sidewalls showing the lower extent of cell protrusion into (B) a 400-nm-wide groove than of that into (C) an 800-nm-wide groove. [Scale bar: 200 nm] Reproduced with permission from Kim et al. 2010. (D and E) A schematic describing the different sites of focal adhesion induced by different heights of the grid micropatterns. (D) FAs on the grid substrate 1 μm high developed at the cell periphery on the top surface of the grid. However, (E) FAs on the grid micropatterned substrate of 2 μm or 4 μm high formed at the bottom square pits. FAK, focal adhesion kinase; α, integrin alpha subunit; β, integrin beta subunit; P, phosphorylation of FAK at tyrosine 397. Reproduced with permission from Seo et al. 2013. (F and G) Spatial co-localization of the actin cytoskeleton and FAs in the protruding lamelipodium aligned proximal to the individual ridge. Reproduced with permission from Kim 2009.

(D), (E), (F), and (G) reprinted with permission from Elsevier.

assembly, which further recruits cytoskeletal molecules into aligned FA sites. As an interesting extension, Kim et al. later employed thermoresponsive nanopatterned substrata (TNFS) to demonstrate that murine myoblast (C2C12) cell monolayers maintain anisotropic alignment engendered by substratum topography even after detachment in response to temperature change (Jiao et al. 2014).

Recent reports have revealed that the cell senses a variety of mechanostructural cues from the microenvironments through FA, which is a robust molecular complex that consists of proteins such as vinculin, talin, paxillin, and focal adhesion kinase (Riviline et al. 2001; Hoffman 2011). Although the complex molecules and underlying mechanism on the FAs have been intensively investigated for many years (Calderwood and Ginsburg 2003; Izaguirre et al. 2001; DeMali et al. 2001), the role of mechanostructural cues including topography and stiffness of substrates on the regulation of adhesion dynamics is not fully understood.

In this regard, Seo and Ushida et al. examined the spatial properties of FAs by changing the height of micropatterns in microgrid and post structures as well as by modulating the stiffness of the substrates (Seo et al. 2013). They designed two kinds of PDMS substrates with grid (2 μm wide with 3 μm interval) and post (3 μm wide and spaced 2 μm apart) micropatterns of different heights (1, 2, and 4 μm), and cultured mouse mesenchymal stem cells (C3H10T1/2 cells) on the fibronectin-coated PDMS micropatterned substrates. These experiments revealed that the localization of FA is highly regulated by topographical variation: that is, by the height of grid micropatterns, but not by the stiffness of substrates or the function of actin cytoskeleton, although the latter strongly influenced the FA size or area. The focal adhesion sites switching (FASS), which is the point that FA sites switched from top grid area to the bottom square area by a mere increase in the height of the micropattern, was observed by the increase of height (Fig. 5.3D and E). FAs on the 1-μm high substrate were formed on the top areas (Fig. 5.3D); whereas they formed on the bottom areas on the 2-μm high substrate (Fig. 5.3E). In other words, this demonstrated that FA localization on substrates with well-defined structure is critically determined by the topographical variation in substrate surface.

Kim, Han et al. (2009) has also reported the use of UV-assisted CFL to create a model substrate of anisotropic micro- and nanotopographic pattern arrays with variable local density for the analysis of cell-substrate interactions. They have reported the fabrication of a single cell adhesion substrate with constant ridge width (1 μm), and depth (400 nm), and variable groove widths (1–9.1 μm), which allowed them to characterize the dependence of NIH3T3 fibroblast responses to the anisotropy and local density of the variable surface pattern. As discussed above, many previous studies reported the importance of substrate topography on the orientation of the cytoskeleton and cells themselves through the organization and regulation of FAs (Teixeira et al. 2003; Kim, Seo et al. 2009). Kim et al. (2009) newly addressed the issue of how the topographical pattern density of substrates influences the formation of FAs and cytoskeleton remodeling. They clearly demonstrated that individual NIH3T3 fibroblasts aligned their actin cytoskeleton along the direction of the ridges, established focal contacts proximal to the ridges, and extended lamellipodia preferentially along more densely spaced ridges (the range of groove widths: 1–3.8 μm) (Fig. 5.3F and G). Cells on sparsely spaced one-dimensional (1-D) ridge arrays (the range of groove widths: 8.1–9.1 μm) also aligned their stress fibers, but their focal contacts were more randomly distributed, similar to those on the flat continuous surfaces. These results surely indicate that the topographical pattern density plays a powerful role in defining cell shape and orientation through the adhesion process.

5.3 Topography-guided cell migration

Cell migration plays an important role in numerous physiological and pathological processes. These include morphogenesis, wound healing, and tumor metastasis

(Lo et al. 2000). To achieve appropriate physiological outcomes, cell movement must maintain a defined direction and adequate speed in response to environment cues (Petrie et al. 2009). In terms of cell migration behavior, cells in various tissues respond differently according to diverse factors, such as gradient of electric fields (electrotaxis) (Erickson and Nuccitelli 1984), light intensity (phototaxis) (Saranak and Foster 1997), gravitational potential (geotaxis) (Lowe 1997), and dissolved (chemotaxis) or surface-immobilized chemicals (haptotaxis) (Carter 1965; Carter 1967; Aznavoorian et al. 1900). In addition to these chemical or biochemical cues, physical topography or rigidity of substrates also influences cell migration (Lo et al. 2000). In contrast to those cultured on flat substrates, on which the trajectory of migration is essentially random, corneal epithelial cells migrate parallel to the direction of nanoridges when they are present (Diehl 2005). This cellular behavior on artificial nanoengineered substrates is consisted with in vivo findings with carcinoma cell migration along aligned ECM (Provenzano et al. 2006; Provenzano et al. 2008). In addition, neutrophils migrate more rapidly on grooved surfaces than on flat surfaces (Tan and Saltzman 2002). The significant findings presented in these studies collectively indicate how nanotopography acts as a regulator of directional migration.

Kim et al. demonstrated that cell migration is also sensitive to variation in the density of nanotopographical cues. NIH3T3 fibroblasts appeared to be very sensitive not only to the local density of the topographic features but also to their gradients (Fig. 5.4A and B)

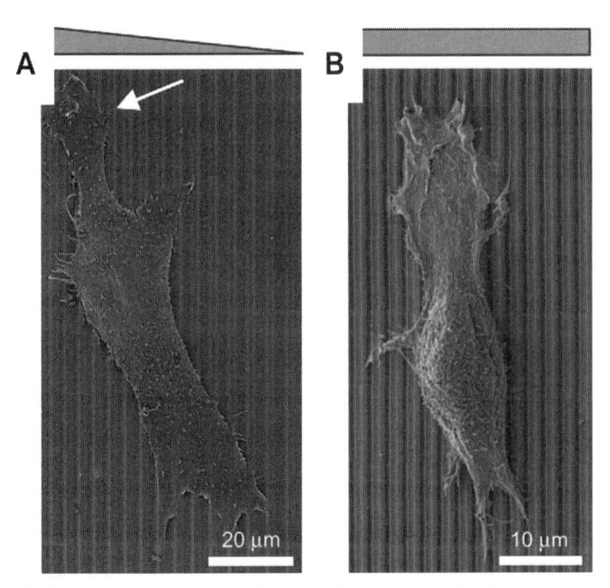

Figure 5.4 **Cell migrations are sensitive to the density of substrate nanopatterns.**
(A and B) SEM images of critical point NIH3T3 fibroblasts plated for 14 h (A) on variable ridge pattern arrays with graded spacing and (B) on regularly spaced topographic pattern arrays with 1 µm wide ridges and 1 µm wide grooves. The white arrow indicates membrane protrusion extending toward the more densely spaced ridges. Reproduced with permission from Kim, Han et al. (2009).

Reprinted with permission from Elsevier.

(Kim, Han et al. 2009). The migration speed of fibroblasts was found to be dependent on the size of the interval between nanopatterned ridges, that is, on the groove width. Interestingly, they found that cell migration speed depended biphasically on the topographic density, with the faster migration (~40 μm/h) occurring at an intermediated ridge pattern density with the range of groove width of 5.6–6.9 μm, suggesting a complex interplay between cell sensing of microscale and nanoscale features. As is indicated by the trajectories of migrating fibroblasts toward regions of denser patterns (Fig. 5.4A), cell migration can be guided by gradients in nanoengineered substrates. This behavior has been termed "topotaxis," which can be used to describe both individual and collective cell migration behavior (Kim, Provenzano et al. 2012).

By contrast, it is well known that fibroblast can naturally migrate to areas of less-organized matrix as ECM is repaired and the density of organized matrix is increased, thus ensuring self-organized repair propagation, as seen in the wound healing process. From this perspective, Kim, Seo et al. (2009) has also designed topographically tailored substrates of variable local density. Inspired by in vivo ECM architectures, they created substrates with 600×600 μm^2 patterns and a lattice network of perpendicular ridges of different local densities. They observed that cells on the topographically patterned areas of graded lattice arrays indeed migrated preferentially toward and accumulated at zones of optimal topographic density. These textured substrates with variable local density and anisotropy serve as strong structural and mechanical mimics of in vivo ECM that can be used to guide cell migration. Using this platform, it is conceivable that systematic variation of the local density of a 2-D patterned substrate could give rise to a planar assembly of cells in a specified location.

While these studies investigated the effect of topographic cues in a way that relied on the assumption that extracellular tissue structures encountered by migrating cells are either 2-D or 3-D tissue networks, most cells experience 3-D interactions that alter cell migration behavior, most of which are not recapitulated in 2-D cell culture (Even-Ram and Yamada 2005). In fact, Cukierman et al. (2001) demonstrated a major discrepancy between 2-D and 3-D culture modalities in that fibroblast readily migrate along ligand-dense ECM fibers in 3-D at rates ~1.5 times faster than in 2-D systems. Similarly, Doyle et al. (2009) demonstrated that a single, nearly 1-D micropatterned line can mimic many aspects of the phenotype induced by cell-derived oriented 3-D matrices. They prepared multiple patterns to test the roles of different ECM topographies on fibroblast migration by using microphotopatterning of hydrophilic polyvinyl alcohol (PVA). When adhered to single 1.5 μm wide fibrillar lines, fibroblasts demonstrated rapid spreading, polarization, motility, and uniaxial phenotype. This result is similar to fibroblast behavior observed in cell-derived 3-D matrices. In striking contrast, fibroblasts attaching to dot arrays (~5 μm spacing; 3.4 μm^2 area per ECM island) showed multiple points of protrusive activity with slow migration. Doyle et al. also addressed the effect of ligand density on cell migration rate by coating 1-D lines and 2-D surfaces with fibronectin of different concentrations. They observed a biphasic effects on migration rate in 2-D surfaces, whereas 1-D migration was observed to be significantly faster than 2-D velocities at all concentrations, suggesting that fundamental aspects of 3-D cell migration in vivo may be successfully studied in more reductionist 1-D and 2-D cell culture systems so long as they provide relevant architecture.

5.4 Topography-guided stem cell differentiation

The term "stem cell" refers to the cell types that have multi- or pluripotent ability, and can be differentiated into specific cell lineages in response to external stimuli. Importantly, stem cells in the tissue niche also respond to surrounding microenvironment as has been discussed above, usually as indicated by differential proliferation, migration, and ultimately differentiation (Kim et al. 2013). Among the numerous cues affecting differentiation (soluble factors, ECM composition, etc.), the micro- and nanotopography of substrates plays an important role in deciding stem cell function and fate (Kshitiz et al. 2011; Kshitiz et al. 2012). Mechanical topography is believed to control large cell populations in a minimally invasive manner. Furthermore, micro- and nanotopographies can be used not only for the differentiation, but also for the maintenance of stemness via the suitable design of geometry (McMurray et al. 2011; Chen et al. 2012). Therefore, the manipulation of topographical cues has strong potential as a strategy for stem cell engineering and regenerative medicine. Although there are a few exceptions depending on the materials and surface chemistries, many studies reported that stem cell fate induced by topographical cues showed strong correlation with cell adhesion morphologies including spreading and elongation. For example, in the case of mesenchymal stem cells, less-spread, rounded, and less-adhered cells show strong correlation with adipogenic lineage (differentiation into fat cells), while characteristics such as being highly-spread, flat, and strongly adhering displayed strong correlation with osteogenic lineage (differentiation into bone cells) (Dalby et al. 2014). Furthermore, several reports indicated that soft microenvironments induce adipogenic or neurogenic lineages, while rigid microenvironment directs osteogenic commitment (Engler et al. 2006; Fu et al. 2010). Intracellular tension has also been observed to be important in the determination of stem cell fate (Guilak et al. 2009).

One important factor in micro- and nanotopography is feature size, which is the most often-controlled factor in manipulating stem cell differentiation. Oh et al. (2009) emulated the effect of surface roughness on stem cell function at the surface of orthopedic implant by engineering the diameter of TiO_2 nanotubes from 30 to 100 nm by anodizing titanium substrates. When human mesenchymal stem cells (hMSCs) were cultured in the absence of osteogenic induction media, the differentiation into osteoblast-like cells, genes responsible for osteogenesis, and cytoskeletal tension were increased as the diameter of TiO_2 nanotube increased, a trend that was correlated with cell elongation (~10-fold increase). The size of topography can also be used in the preservation of stemness. For this purpose, Chen et al. prepared glass substrates having nanoscale roughness (1, 70, and 150 nm) by reactive ion etching technique (Chen et al. 2012). When human embryonic stem cells (hESCs) were cultured on these substrates, hESCs on the smooth surface (roughness of 1 nm) maintained stemness in long-term culture of seven days, while those on rough surfaces spontaneously differentiated. Furthermore, stemness was decreased as the nanoroughness increases from 1 to 150 nm (93.6%, 41%, and 36.6%, respectively). These studies each indicate that the differentiation of stem cells increases as the feature scale increases on nanoscale topographies (<200 nm).

Another important factor in micro- and nanotopography is topographical density, which can be varied and optimized depending on the pattern design, the type of surface patterns, such as pillar and groove, the stiffness of the material, and any surface coatings that may be applied. Studies controlling topographical spacing have been widely conducted using CFL due to its high reliability and reproducibility (Suh et al. 2009; Ahn et al. 2014). Ahn et al. (2014) demonstrated that in the case of nanopost patterns, when the spacing of nanopost was controlled as 1.2, 2.4, 3.6, and 5.6 μm, adipogenesis showed a biphasic trend, with maximum at 2.4 μm, but osteogenesis showed a monotonic increment as the spacing increased (Fig. 5.5A). According to quantitative reverse transcription-polymerase chain reaction (qRT-PCR) analysis, osteogenesis was strongly correlated with genes related to cytoskeletal changes such as actin, RhoA and adhesion proteins such as integrin, and focal adhesion kinase (FAK). Additionally, hMSC elastic modulus was found to vary monotonically with substratum post density with maximum at 5.6 μm and minimum at 2.4 μm (Fig. 5.5B). These results suggest that substrate topography can regulate cytoskeleton mechanical stiffness, an important differentiation characteristic. Kim et al. demonstrated that in the case of grooves structures, hMSCs cultured on the 550-nm-wide nanogrooves with spacing of 550, 1650, and 2750 nm showed maximum osteo- and neurogenesis in the intermediate spacing of 1650 nm rather than dense (550 nm) and sparse (2750 nm) cases (Kim ... Park et al. 2013). The biphasic trend of

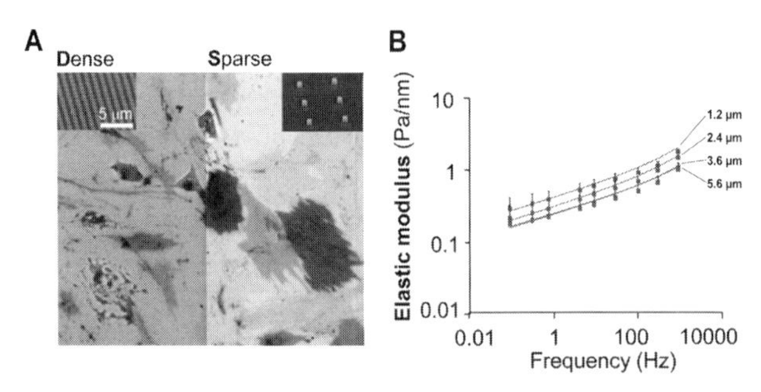

Figure 5.5 **The density of substrate nanoscale topographic features regulates cytoskeleton mechanical stiffness, an important differentiation characteristic.**
(A) Stained cells cultured on dense (1.2 μm, left) and sparse (5.6 μm, right) micropost arrays showing substrate dependent morphologies. Scanning electron microscopy (SEM) images of nanopatterned arrays with varying post densities. All the tested patterns, with a flat control, were fabricated on a single glass coverslip to reduce experimental variation. (B) Reprinted with permission from Elsevier. Representative elastic moduli of hMSC cultured on micropost substata with several post-to-post distances; hMSC cultured on substrates with 5.6 μm post-to-post distances were the stiffest, and those on 2.4 μm substrates were the least stiff.

Reproduced with permission from Ahn et al. (2014).

differentiation was strongly correlated with cell-shape index, cell-substrate interaction (integrin), and cell-cell interaction (N-cadherin). Furthermore, such a biphasic trend of differentiation was maintained similarly in the presence and absence of endothelial cells (Kim ... Gar et al., 2013). These results collectively indicate that there might be an optimum spacing in the designing of topographical density for maximized differentiation.

The importance of ordered-ness, which means the homogeneity (or heterogeneity), of surface features can be seen in the disordered hexagonal arrangement of type X collagen observed in bone tissue (Kwan et al. 1991; Stephens et al. 1992). Inspired by such disordered morphologies, Dalby et al. (2007) has fabricated hexagonal and square arrays of nanopits with ordered and disordered arrangements. Interestingly, hMSCs demonstrated higher differentiation into a mesodermal lineage on the slightly disordered square array, which is consistent with the disordered arrangement observed in vivo. Inversely, the precisely arranged square nanopit array could be utilized in the maintenance of stemness of MSCs (McMurray et al. 2011). Such ordered-ness of nanotopography seems more important when the feature size is around 100 nm or less. For example, in the case of nanoscale features with size of ~70 nm, the arrangement of topography strongly affects cell adhesion and integrin clustering. The MSCs on RGD patterns show reduced adhesion and clustering in the precise arrangement case, presumably due to the effective clustering distance required Huang et al. 2009; Malmström et al. 2011).

5.5 Topography-induced mechanobiology in tissue engineering

The creation of tissues in vitro that closely mimic their native analogues requires faithful biomimicry of the complex structures of the native extracellular environment found in vivo. Thus, the advancement of the understanding and implementation of micro- and nanotopographical platforms to influence cell mechanobiology is a salient frontier in the field of tissue engineering. Ultimately, anisotropically nanopatterned substrates can be integrated into a wide variety of tissue engineering platforms and devices for cell culture in two, quasi-three, and three dimensions. In the following broad overview, efforts in developing and implementing such devices and scaffolds are articulated with respect to several tissue types, beginning with pertinent efforts in cardiac tissue engineering.

The focused application of nanotopographical methodologies in cardiac tissue engineering has shown great and mounting promise in recent research efforts. The natively anisotropic alignment and elongation of cardiomyocytes is a critical feature of heart tissue that is known to contribute substantially to contractile and electro-active functionalities. By manipulating characteristics of nanotopographical substrates, various measurable aspects of tissue function can be influenced. In the engineering of cardiac tissue in vitro, the degree of physiological resemblance is characterized by observing structural and anisotropic properties. Such properties include the rate of action potential propagation, anisotropy and frequency of synchronous contractions, sarcomere striation, actin cytoskeleton alignment, connexin43 expression at cell-cell contacts, and the formation of monolayers and 3-D constructs of high cell density (Shimizu et al. 2002; Sekine et al. 2013; Sakaguchi et al. 2013).

The use of transplanted nanopatterned cell patches has been demonstrated by Kim et al. (2012) to promote myocardial regeneration in vivo in the rat model. When cardiosphere-derived cells (CDCs) were plated on both unpatterned and anisotropically nanofabricated substrates (ANFS) made from PEG, it was found that confluent cell monolayers formed fully in just two days on the patterned substrate, and exhibited a high degree of anisotropic alignment with the nanopattern ridges, whereas CDCs plated on unpatterned substrates showed random orientation and no mutual alignment. Cell adhesion, as quantified by substrate attachment retention times under perfusion flow conditions of trypsin-EDTA solution (0.05%) in the direction of the ANFS, was found to be improved by a factor of two for cells plated on patterned substrates relative to those plated on unpatterned substrates. Motility and proliferation were similarly quantified and found to be improved for CDCs on ANFS. Grafting of monolayered ANFS-based CDC patches onto the infarcted hearts of adult Wistar-Kyoto rats showed improved tissue repair as characterized by thicker infarct wall formation in rats that received ANFS-based CDC patches (Fig. 5.6A).

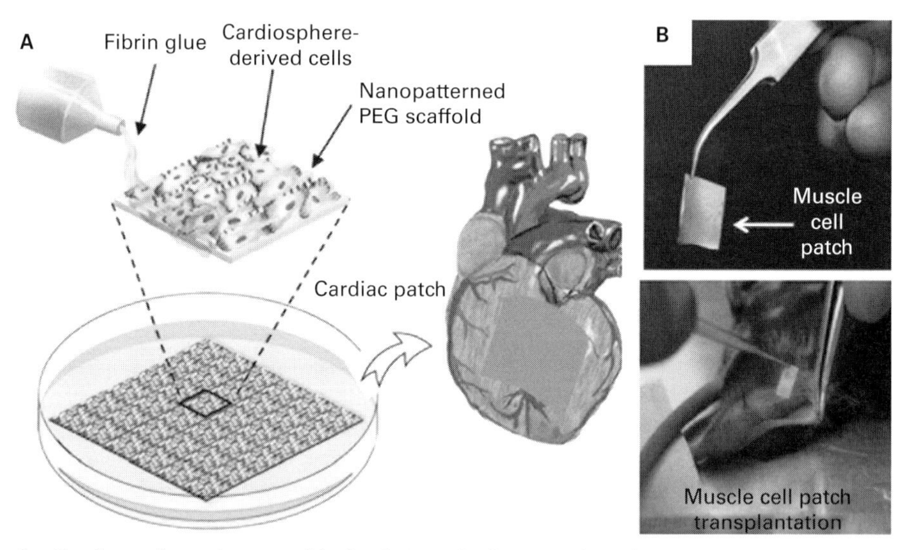

Figure 5.6 **Applications of nanotopographical substrate in tissue engineering.**
(A) Engraftment of a nanopatterned cardiac cell patch into the infarcted rat heart in vivo. Schematic outlining proposed strategy of integration of autologous stem cells with nanotopographical cues in development of tissue engineered cardiac patches. Reproduced with permission from Kim et al. (2012). (B) Transplantation of nanopatterned muscle cell patch into the quadriceps muscles of muscular dystrophy (*mdx*) mice. Reproduced with permission from Yang et al. (2014).

Reprinted with permission from Elsevier.

The imposition of structure on injectable cardiac therapies has been demonstrated to be beneficial in promoting neovascularization following myocardial infarction (MI). Lin et al. conducted a study in which they permanently ligated the left anterior descending (LAD) coronary artery to create MI in Sprague-Dawley (SD) rats, which they immediately followed with intramyocardial injection of treatments from six directions at the peri-infarct area. They reported that the injection of self-assembling peptide nanofibers in conjunction with vascular endothelial growth factor (NF/VEGF) led to a sustained release of up to fourteen days after delivery and improved cardiac performance twenty-eight days after MI (Lin et al. 2012). This has been attributed to recruitment of endogenous myofibroblasts for repair by the structured microenvironment imposed by injected NFs.

As a final representative example of the fruitful application of nanotopographically oriented tissue engineering methods to cardiac tissue engineering, Kim et al. has demonstrated that NRVMs cultured on ANFS made from PEG demonstrated monolayer formation, elongation, structural and synchronous contractile anisotropy, and upregulation of connexin43 as compared to those cultured on unpatterned substrates. This improvement was found to be dependent on nanotopographical dimensions, with increased performance as characterized above on ANFS with widths of 800 nm and spacing of 800 nm (Kim et al. 2010).

As with other tissue types, the organization of skeletal muscle tissue developed in vitro relies on the mimicry of this highly ordered, multiscale structure in the native tissue. The parallel array of striated multinucleate myotubes found spanning the length of skeletal muscles allows for the axial propagation of force throughout the tissue (Kim, Jiao et al. 2013). The maturation of myotubes into myofibers leads to the organization of myofiber bundles and the formation of the perimysium ECM layer, which contains both longitudinal and transverse collagen fiber structures (Pope et al. 2008). Yang et al. (2014) demonstrated that the use of nanotopographical substrates produced by CFL allows for the differentiation of primary mononucleated cells into mature muscle patches, as characterized by myotube formation, and increased expression of several myogenic regulatory factors. Transplantation of such engineered muscle patches in *mdx* mice models for Duchenne muscular dystrophy led to high dystrophin expression in new muscle fibers, signaling the success of this approach to dystrophin replacement and myogenesis in vivo (Fig. 5.6B).

Tissue engineering of other tissues has been also studied and introduced based on the surface topographical mechanobiology. The engineering of bone tissue is typically approached through the isolation of patient MSCs or osteoblasts and subsequent culture on a variety of 3-D bone substitute material scaffolds with the eventual goal of transplantation. The application of micro- and nanoscale topographies to scaffold design has shown promise in promoting organized osteogenesis (Dalby et al. 2006). The role of mechanobiological factors (mechanical stretching as detected by mechanosensing osteoblasts and osteocytes, cell shape) in bone growth and repair have been shown to be critically important to bone growth and organization (Tanaka et al. 2005), and have been shown to outweigh the role of chemical factors such as dexamethasone (Jagodzinski et al. 2004).

The physiological epidermis tissue layer is made up of epithelial cells, stratified on the basal lamina, which is made up of an ordered layer of ECM fibers, including type IV collagen, laminin, heparin sulfate proteoglycan, and fibronectin. The use of micro- and nanofabrication techniques in the development of skin tissue cultures shows promise in promoting tissue organization, function, and emulation of native tissues. Dermal fibroblasts have been demonstrated to attach to electrospun polymer NFs precoated with ECM molecules such as collagen (Venugopal et al. 2006; Venugopal and Ramakrishna 2005). Other natural biopolymers such as silk and chitin have also been employed as nanostructured scaffolds, and both have exhibited good keratinocyte and fibroblast attachment (Min et al. 2004; Noh et al. 2006).

The native environment of neurons and supporting glial cells is also known to support cellular structure and function in a number of ways, including the provision of soluble chemical and bounded chemical cues, and micro- and nanotopographical cues imparted by the ECM or cell membrane structures, which have been shown to promote neuronal growth even in the absence of chemical factors (Moore et al. 2011; Yurchenko and Wadsworth 2004; Tsiper and Yurchenko 2002). Mimicry of the topographical features of native brain tissue using ANFS has shown boons in neuronal differentiation, neurite growth, cell alignment, and cell migration, which may be useful for regeneration of neural tissue (Yim et al. 2007; Lee et al. 2010). These effects have been primarily observed with nanoscale architectures, with no significant differences observed between microscale grooved substrates and unpatterned substrates (Recknor et al. 2006).

As with many other cell types, tissue-engineering efforts using both endothelial cells (ECs) and smooth muscle cells (SMCs) have benefited from the application of nanotopographical substrate design. Nanotopography is thought to influence ECs through integrin-mediated attachment of cells to their substrate through FAs. Several studies have shown that ECs grow and express integrin proteins and FAs anisotropically in parallel with the nanopatterned substrates on which they are plated, even without surface ECM coatings (Hwang et al. 2010; Liliensiek et al. 2010). Similarly, physiological SMCs are embedded in helically organized collagen and elastin lamellae, and have been found to align and elongate anisotropically in the pattern direction when plated on micro- and nanotopographical cell culture substrates (Hu at al. 2005; Sarkar et al. 2005). Nanotopography has been integrated with studies of cell detachment and vascular grafting, where it has given rise to improvements in the mechanical properties of cultured tissues as compared to those of tissues cultured on flat substrates (Zorlutuna et al. 2009; Isenberg et al. 2012). In an attempt to mimic the 3-D structures of physiological vascular tissue, layer-by-layer stacking of nanotopographically cultured cell sheets has been performed by several researchers. Observation of SMC differentiation and SMC contractile markers after transplantation of such multilayer scaffolds has indicated that such methods may hold therapeutic promise with continued development (Sarkar et al. 2008; Tan and Desai 2005; Feng et al. 2007).

5.6 Conclusion

As described above, micro- and nanotopography influence cell functions including not only adhesion and migration but also stem cell lineage commitment. Furthermore, as an opposing aspect of the same effect, engineered topographies can be used to maintain stemness while letting stem cells proliferate. Both of these aspects have important implications for the advancement of tissue engineering and regenerative medicine, and demonstrate the utility of surface topography as a promising engineering tool. In spite of the advances reported here, much work remains in the study and expanded application of micro- and nanoscale topographies. For example, the effect of topography could be further enhanced through the use of surface treatments or coatings, which are known to influence signaling pathways depending on chemical moieties or binding receptors (Seong et al. 2013). In addition, the biochemical signals that mediate mechanotransduction need to be further investigated because signaling molecules and proteins have the capacity to enhance topographical cues synergistically. A recent study found that YAP (Yes-associated protein) and TAZ (transcriptional coactivator with PDZ-binding motif) signaling plays a role in the transportation of mechanical signals to the nucleus as a biochemical transporter, a finding that has potential applications in stem cell engineering (Dupont et al. 2011). In looking to the future of stem cell engineering, it is essential to consider the incorporation of materials engineering. For example, recent efforts toward the development of dynamically tunable polymers and hydrogels may allow researchers to unravel many of the biological mechanisms of stem cell differentiation in situ (Burdick and Murphy 2012).

From the numerous investigations described herein and elsewhere, it is clear that the nanotopography of the ECM plays a critical role in regulating cell behavior. These sophisticated hierarchical interactions, which occur at an extremely small-length scale, have often been overlooked when experiments are conducted in the context of larger-length scales. However, specific tools to study the influence of intricate nanoscale features of the ECM are now available, and scientists should now be able to take full advantage of such tools to gain additional insight into fundamental mechanisms driving cell behavior.

Although no in vitro system can perfectly recapitulate the characteristics of the in vivo environment, many in vitro systems hold advantages in their reduced, but relevant, complexity. The nanotopographically defined cell culture models described in this chapter provide a unique middle ground, maintaining the simplicity of traditional in vitro systems while mimicking small-scale 3-D features of the ECM down to the molecular level. However, many challenges still remain with respect to materials. The systems developed to date generally allow only for static control, which is far from the dynamic environment found in vivo. Current in vitro tissue engineering technology is yet far from achieving complete physiological mimicry of all the key properties that would allow for simultaneous, independent, and modular dynamic control of different cellular characteristics (Kim and Hayward 2012). Ultimately, it is evident that our understanding of underlying micro- and nanotopographically induced mechanobiology will not only help to answer basic biological questions about how cells sense and respond to various cues in their surrounding microenvironment, but also will blaze new trails in the advancement of functional tissue engineering and regenerative medicine.

References

Ahn, E. H., Y. Kim, Kshitiz, S. S. An, J. Afzal, S. Lee, et al. (2014). "Spatial control of adult stem cell fate using nanotopographic cues." *Biomaterials* **35**(8): 2401–2410.

Aznavoorian S, M. L. Stracke, H. Krutzsch, E. Schiffmann, and L. A. Liotta. (1990). "Signal transduction for chemotaxis and haptotaxis by matrix molecules in tumor cells." *J Cell Biol* **110**(4): 1427–1438.

Burdick, J. A., and W. L. Murphy. (2012). "Moving from static to dynamic complexity in hydrogel design." *Nat Commun* **3**: 1269.

Bursac, N., K. K. Parker, S. Iravanian, and L. Tung. (2002). "Cardiomyocyte cultures with controlled macroscopic anisotropy: a model for functional electrophysiological studies of cardiac muscle." *Circ Res* **91**(12): e45–54.

Calderwood, D. A., and M. H. Ginsberg. (2003). "Talin forges the links between integrins and actin." *Nat Cell Biol* **5**(8): 694–697.

Carter, S. B. (1965). "Principles of cell motility: the direction of cell movement and cancer invasion." *Nature* **208**(5016): 1183–1187.

Carter, S. B. (1967). "Haptotaxis and the mechanism of cell motility." *Nature* **213**(5073): 256–260.

Chen, C. S., M. Mrksich, S. Huang, G. M. Whitesides, and D. E. Ingber. (1997). "Geometric control of cell life and death." *Science* **276**(5317): 1425–1428.

Chen, W., L. G. Villa-Diaz, Y. Sun, S. Weng, J. K. Kim, R. H. W. Lam, et al. (2012). "Nanotopography influences adhesion, spreading, and self-renewal of human embryonic stem cells." *ACS Nano* **6**(5): 4094–4103.

Chien, K. R., I. J. Domian, and K. K. Parker. (2008). "Cardiogenesis and the complex biology of regenerative cardiovascular medicine." *Science* **322**(5907): 1494–1497.

Clark, P., P. Connolly, A. S. Curtis, J. A. Dow, and C. D. Wilkinson. (1991). "Cell guidance by ultrafine topography in vitro." *J Cell Sci* **9**(Pt 1): 73–77.

Cukierman, E., R. Pankov, D. R. Stevens, and K. M. Yamada (2001). "Taking cell-matrix adhesions to the third dimension." *Science* **294**(5547): 1708–1712.

Dalby, M. J., N. Gadegaard, and G. Curtis. (2007). "Nanotopographical control of human osteoprogenitor differentiation." *Curr Stem Cell Res Ther* **2**(2): 129–138.

Dalby, M. J., N. Gadegaard, and R. O. C. Oreffo. (2014). "Harnessing nanotopography and integrin-matrix interactions to influence stem cell fate." *Nat Mater* **13**(6): 558–569.

Dalby, M. J., D. McCloy, M. Robertson, C. D. W. Wilkinson, and R. O. C. Oreffo. (2006). "Osteoprogenitor response to defined topographies with nanoscale depths." *Biomaterials* **27**(8): 1306–1315.

DeMali, K. A., C. A. Barlow, and K. Burridge. (2002). "Recruitment of the Arp2/3 complex to vinculin: coupling membrane protrusion to matrix adhesion." *J Cell Biol* **159**(5): 881–891.

Diehl, K. A., J. D. Foley, P. F. Nealey, and C. J. Murphy. (2005). "Nanoscale topography modulates corneal epithelial cell migration." *J Biomed Mater* **75**(3): 603–611.

Doyle, A. D., F. W. Wang, K. Matsumoto, and K. M. Yamada. (2009). "One-dimensional topography underlies three-dimensional fibrillar cell migration." *J Cell Biol* **184**(4): 481–90.

Dunn, G. A., and T. Ebendal. (1978). "Contact guidance on oriented collagen gels." *Exp Cell Res* **111**(2): 475–479.

Dupont, S., L. Morsut, M. Aragona, E. Enzo, S. Giulitti, M. Cordenonsi, et al. (2011). "Role of YAP/TAZ in mechanotransduction." *Nature* **474**(7350): 179–183.

Engler, A. J., S. Sen, H. L. Sweeney, and D. E. Discher. (2006). "Matrix elasticity directs stem cell lineage specification." *Cell* **126**(4): 677–689.

Erickson, C. A., and R. Nuccitelli. (1984). "Embryonic fibroblast motility and orientation can be influenced by physiological electric fields." *J Cell Biol* **98**(1): 296–307.

Even-Ram, S., and K. M. Yamada. (2005). "Cell migration in 3D matrix." *Curr Opin Cell Biol* **17**(5): 524–532.

Fast, V. G., B. J. Darrow, J. E. Saffitz, and A. G. Kléber. (1996). "Anisotropic activation spread in heart cell monolayers assessed by high-resolution optical mapping. Role of tissue discontinuities." *Circ Res* **79**(1): 115–127.

Feng, J., M. B. Chan-Park, J. Shen, and V. Chan. (2007). "Quick layer-by-layer assembly of aligned multilayers of vascular smooth muscle cells in deep microchannels." *Tissue Eng* **13**(5): 1003–1012.

Fu, J., Y.-K. Wang, M. T. Yang, R. A. Desai, X. Yu, Z. Liu, et al. (2010). "Mechanical regulation of cell function with geometrically modulated elastomeric substrates." *Nat Methods* **7**(9): 733–736.

Gopalan S. M., C. Flaim, S. N. Bhatia, M. Hoshijima, R. Knoell, K. R. Chien, et al. (2003). "Anisotropic stretch-induced hypertrophy in neonatal ventricular myocytes micropatterned on deformable elastomers." *Biotechnol Bioeng* **81**(5): 578–587.

Guilak, F., D. M. Cohen, B. T. Estes, J. M. Gimble, W. Liedtke, and C. S. Chen. (2009). "Control of stem cell fate by physical interactions with the extracellular matrix." *Cell Stem Cell* **5**(1): 17–26.

Hoffman, B. D., C. Grashoff, and M. A. Schwartz. (2011). "Dynamic molecular processes mediate cellular mechanotransduction." *Nature* **475**(7356): 316–323.

Hu, W., E. K. F. Yim, R. M. Reano, K. W. Leong, and S. W. Pang. (2005). "Effects of nanoimprinted patterns in tissue-culture polystyrene on cell behavior." *J Vac Sci Technol* **23**(6): 2984–2989.

Huang, J., S. V. Grater, F. Corbellini, S. Rinck, E. Bock, R. Kemkemer, et al. (2009). "Impact of order and disorder in RGD nanopatterns on cell adhesion." *Nano Lett* **9**(3): 1111–1116.

Hwang, S. Y., K. W. Kwon, K.-J. Jang, M. C. Park, J. S. Lee, and K. Y. Suh. (2010). "Adhesion assays of endothelial cells on nanopatterned surfaces within a microfluidic channel." *Anal Chem* **82**(7): 3016–3022.

Isenberg, B. C., D. E. Backman, M. E. Kinahan, R. Jesudason, B. Suki, P. J. Stone, et al. (2012). "Micropatterned cell sheets with defined cell and extracellular matrix orientation exhibit anisotropic mechanical properties." *J Biomech* **45**(5): 756–61.

Izaguirre, G., L. Aguirre, Y. P. Hu, H. Y. Lee, D. D. Schlaepfer, B. J. Aneskievich, et al. (2001). "The cytoskeletal/non-muscle isoform of alpha-actinin is phosphorylated on its actin-binding domain by the focal adhesion kinase." *J Biol Chem* **276**(31): 28676–28685.

Jagodzinski, M., M. Drescher, J. Zeichen, S. Hankemeier, C. Krettek, U. Bosch, et al. (2004). "Effects of cyclic longitudinal mechanical strain and dexamethasone on osteogenic differentiation of human bone marrow stromal cells." *Eur Cell Mater* **7**: 35–41, 41.

Jiao, A., N. E. Trosper, H. S. Yang, J. Kim, J.H. Tsui, S. D. Frankel, et al. (2014). "Thermoresponsive nanofabricated substratum for the engineering of three-dimensional tissues with layer-by-layer architectural control." *ACS Nano* **8**(5): 4430–4439.

Kim, D.-H., K. Han, K. Gupta, K. W. Kwon, K.-Y. Suh, and A. Levchenko. (2009). "Mechanosensitivity of fibroblast cell shape and movement to anisotropic substratum topography gradients." *Biomaterials* **30**(29): 5433–5444.

Kim, D.-H., Kshitiz, R. R. Smith, P. Kim, E. H. Ahn, H. N. Kim, et al. (2012). "Nanopatterned cardiac cell patches promote stem cell niche formation and myocardial regeneration." *Integr Biol* **4**(9): 1019–1033.

Kim, D.-H., E. A. Lipke, P. Kim, R. Cheong, S. Thompson, M. Delannoy, et al. (2010). "Nanoscale cues regulate the structure and function of macroscopic cardiac tissue constructs." *Proc Natl Acad Sci USA* **107**(2): 565–570.

Kim, D.-H., C. H. Seo, K. Han, K. W. Kwon, A. Levchenko, and K.-Y. Suh. (2009). "Guided cell migration on microtextured substrates with variable local density and anisotropy." *Adv Funct Mater* **19**(10): 1579–1586.

Kim, H. N., A. Jiao, N. S. Hwang, M. S. Kim, D. H. Kang, D.-H. Kim, et al. (2013). "Nanotopography-guided tissue engineering and regenerative medicine." *Adv Drug Deliv Rev* **65**(4): 536–558.

Kim, J., and R. C. Hayward. (2012). "Mimicking dynamic in vivo environments with stimuli-responsive materials for cell culture." *Trends Biotechnol* **30**(8): 426–439.

Kim, J., H. N. Kim, K.-T. Lim, Y. Kim, S. Pandey, P. Garg, et al. (2013). "Synergistic effects of nanotopography and co-culture with endothelial cells on osteogenesis of mesenchymal stem cells." *Biomaterials* **34**(30): 7257–7268.

Kim, J., H. N. Kim, K.-T. Lim, Y. Kim, H. Seonwoo, S. H. Park, et al. (2013). "Designing nanotopographical density of extracellular matrix for controlled morphology and function of human mesenchymal stem cells." *Sci Rep* **3**: 3552.

Kim, D.-H., P. P. Provenzano, C. L. Smith, and A. Levchenko. (2012). "Matrix nanotopography as a regulator of cell function." *J Cell Biol* **197**: 351–360.

Kshitiz, D.-H. Kim, D. J. Beebe, and A. Levchenko. (2011). "Micro- and nanoengineering for stem cell biology: the promise with a caution." *Trends Biotechnol* **29**(8): 399–408.

Kshitiz, J. Park, P. Kim, W. Helen, A. J. Engler, A. Levchenko, et al. (2012). "Control of stem cell fate and function by engineering physical microenvironments." *Integr Biol* **4**(9): 1008–1018.

Kwan, A. P., C. E. Cummings, J. A. Chapman, and M. E. Grant. (1991). "Macromolecular organization of chicken type X collagen in vitro." *J Cell Biol* **114**(3): 597–604.

Laflamme, M. A., and C. E. Murry. (2011). "Heart regeneration." *Nature* **473**(7347): 326–335.

Lee, M. R., K. W. Kwon, H. Jung, H. N. Kim, K. Y. Suh, K. Kim, et al. (2010). "Direct differentiation of human embryonic stem cells into selective neurons on nanoscale ridge/groove pattern arrays." *Biomaterials* **31**(15): 4360–4366.

Liliensiek, S. J., J. A. Wood, J. Yong, R. Auerbach P. F. Nealey, and C. J. Murphy. (2010). "Modulation of human vascular endothelial cell behaviors by nanotopographic cues." *Biomaterials* **31**(20): 5418–5426.
Lin, Y.-D., C.-Y. Luo, Y.-N. Hu, M.-L. Yeh, Y.-C. Hsueh, M.-Y. Chang, et al. (2012). "Instructive nanofiber scaffolds with VEGF create a microenvironment for arteriogenesis and cardiac repair." *Sci Transl Med* **4**(146): 146ra109.
Lo, C. M., H. B. Wang, M. Dembo, and Y. L. Wang. (2000). "Cell movement is guided by the rigidity of the substrate." *Biophys J* **79**(1): 144–152.
Lowe, B. (1997). "The role of Ca2+ in deflection-induced excitation of motile, mechanoresponsive balancer cilia in the ctenophore statocyst." *J Exp Biol* **200**(Pt 11): 1593–1606.
Malmström, J., J. Lovmand, S. Kristensen, M. Sundh, M. Duch, and D. S. Sutherland. (2011). "Focal complex maturation and bridging on 200 nm vitronectin but not fibronectin patches reveal different mechanisms of focal adhesion formation." *Nano Lett* **11**(6): 2264–2271.
McBeath, R., D. M. Pirone, C. M. Nelson, and K. Bhadriraju. (2004). "Cell shape, cytoskeletal tension, and RhoA regulate stem cell lineage commitment." *Dev Cell* **6**(4): 483–495.
McMurray, R. J., N. Gadegaard, P. M. Tsimbouri, K. V. Burgess L. E. McNamara, R. Tare, et al. (2011). "Nanoscale surfaces for the long-term maintenance of mesenchymal stem cell phenotype and multipotency." *Nat Mater* **10**(8): 637–644.
Min, B.-M., G. Lee, S. H. Kim, Y. S. Nam, T. S. Lee, and W. H. Park. (2004). "Electrospinning of silk fibroin nanofibers and its effect on the adhesion and spreading of normal human keratinocytes and fibroblasts in vitro." *Biomaterials* **25**(7–8): 1289–1297.
Moore, S. W., and M.P. Sheetz. (2011). "Biophysics of substrate interaction: influence on neural motility, differentiation, and repair." *Dev Neurobiol* **7**(11): 1090–1101
Noh, H. K., S. W. Lee, J.-M. Kim, J.-E. Oh, K.-H. Kim, C.-P. Chung, et al. (2006). "Electrospinning of chitin nanofibers: degradation behavior and cellular response to normal human keratinocytes and fibroblasts." *Biomaterials* **27**(21): 3934–3944.
Oh, S., K. S. Brammer, Y. S. J. Li, D. Teng, A. J. Engler, S. Chien, et al. (2009). "Stem cell fate dictated solely by altered nanotube dimension." *Proc Natl Acad Sci USA* **106**(7): 2130–2135.
Petrie, R. J., A. D. Doyle, and K. M. Yamada. (2009). "Random versus directionally persistent cell migration." *Nat Rev Mol Cell Biol* **10**(8): 538–549.
Pope, A. J., G. B. Sands, B. H. Smaill, and I. J. LeGrice. (2008). "Three-dimensional transmural organization of perimysial collagen in the heart." *Am J Physiol Heart Circ Physiol* **295**(3): H1243–H1252.
Provenzano, P. P., K. W. Eliceiri, J. M. Campbell, D. R. Inman, J. G. White, and P. J. Keely. (2006). "Collagen reorganization at the tumor-stromal interface facilitates local invasion." *BMC Med* **4**(1): 38.
Provenzano, P. P., D. R Inman, K. W. Eliceiri, S. M. Trier, P. J. Keely. (2008). "Contact guidance mediated three-dimensional cell migration is regulated by Rho/ROCK-dependent matrix reorganization." *Biophys J* **95**(11): 5374–5384.
Recknor, J. B., D. S. Sakaguchi, S. K. Mallapragada. (2006). "Directed growth and selective differentiation of neural progenitor cells on micropatterned polymer substrates." *Biomaterials* **27**(22): 4098–4108.

Riveline, D., E. Zamir, N. Q. Balaban, U. S. Schwarz, T. Ishizaki, S. Narumiya, et al. (2001). "Focal contacts as mechanosensors: externally applied local mechanical force induces growth of focal contacts by an mDia1-dependent and ROCK-independent mechanism." *J Cell Biol* **153**(6): 1175–1186.

Sakaguchi, K., T. Shimizu, S. Horaguchi, H. Sekine, M. Yamato, M. Umezu, et al. (2013). "In vitro engineering of vascularized tissue surrogates." *Sci Rep* **3**: 1316.

Saranak, J., and K. W. Foster (1997). "Rhodopsin guides fungal phototaxis." *Nature* **387**(6628): 465–466.

Sarkar, S., M. Dadhania, P. Rourke, T. A. Desai, and J. Y. Wong. (2005). "Vascular tissue engineering: microtextured scaffold templates to control organization of vascular smooth muscle cells and extracellular matrix." *Acta Biomater* **1**(1): 93–100.

Sarkar, S., B. C. Isenberg, E. Hodis, J. B. Leach T. A. Desai, and J. Y. Wong. (2008). "Fabrication of a layered microstructured polycaprolactone construct for 3-D tissue engineering." *J Biomater Sci Polym Ed* **19**(10): 1347–1362.

Sekine, H., T. Shimizu, K. Sakaguchi, I. Dobashi, M. Wada, M. Yamato, et al. (2013). "In vitro fabrication of functional three-dimensional tissues with perfusable blood vessels." *Nat Commun* **4**: 1399.

Seo, C. H., H. Jeong, K. S. Furukawa, Y. Suzuki, and T. Ushida 2013. "The switching of focal adhesion maturation sites and actin filament activation for MSCs by topography of well-defined micropatterned surfaces." *Biomaterials* **34**(7): 1764–1771.

Seong, J., A. Tajik, J. Sun J.-L. Guan, M. J. Humphries, S. E. Craig, et al. (2013). "Distinct biophysical mechanisms of focal adhesion kinase mechanoactivation by different extracellular matrix proteins." *Proc Natl Acad Sci USA* **110**(48): 19372–19377.

Shimizu, T., M. Yamato, and T. Akutsu. (2002). "Electrically communicating three-dimensional cardiac tissue mimic fabricated by layered cultured cardiomyocyte sheets." *J Biomed Mater Res* **60**(10): 110–117.

Stephens, M., A. P. Kwan, M. T. Bayliss, and C. W. Archer. (1992). "Human articular surface chondrocytes initiate alkaline phosphatase and type X collagen synthesis in suspension culture." *J Cell Sci* **103**(Pt 4): 1111–1116.

Suh, K.-Y., M. C. Park, and P. Kim. (2009). "Capillary force lithography: a versatile tool for structured biomaterials interface towards cell and tissue engineering." *Adv Funct Mater* **19**(17): 2699–712.

Tan, J., and W. M. Saltzman. (2002). "Topographical control of human neutrophil motility on micropatterned materials with various surface chemistry." *Biomaterials* **23**(15): 3215–3225.

Tan, W., and T. A. Desai. (2005). "Microscale multilayer cocultures for biomimetic blood vessels." *J Biomed Mater Res* **72**(2): 146–160.

Tanaka, S. M., H. B. Sun, R. K. Roeder, D. B. Burr, C. H. Turner, and H. Yokota. (2005). "Osteoblast responses one hour after load-induced fluid flow in a three-dimensional porous matrix." *Calcif Tissue Int* **76**(4): 261–271.

Teixeira, A. I., G. A. Abrams, P. J. Bertics, C. J. Murphy, and P. F. Nealey. (2003). "Epithelial contact guidance on well-defined micro- and nanostructured substrates." *J Cell Sci* **116**(Pt 10): 1881–1892.

Teixeira, A. I., G. A. McKie, J. D. Foley, P. J. Bertics, P. F. Nealey, and C. J. Murphy. (2006). "The effect of environmental factors on the response of human corneal epithelial cells to nanoscale substrate topography." *Biomaterials* **27**(21): 3945–3954.

Tsiper, M. V., and P. D. Yurchenco. (2002). "Laminin assembles into separate basement membrane and fibrillar matrices in Schwann cells." *J Cell Sci* **115**(5): 1005–1015.

Venugopal, J., and S. Ramakrishna "Biocompatible nanofiber matrices for the engineering of a dermal substitute for skin regeneration." *Tissue Eng* **11**(5–6): 847–54.

Venugopal, J. R., Y. Zhang, and S. Ramakrishna. (2006). "In vitro culture of human dermal fibroblasts on electrospun polycaprolactone collagen nanofibrous membrane." *Artif Organs* **30**(6): 440–446.

Yang, H. S., N. Ieronimakis, J. H. Tsui, H. N. Kim, K.-Y. Suh, M. Reyes, et al. (2014). "Nanopatterned muscle cell patches for enhanced myogenesis and dystrophin expression in a mouse model of muscular dystrophy." *Biomaterials* **35**(5): 1478–86.

Yim, E. K. F., S. W. Pang, and K. W. Leong. (2007). "Synthetic nanostructures inducing differentiation of human mesenchymal stem cells into neuronal lineage." *Exp Cell Res* **313**(9): 1820–1829.

Yurchenco, P. D., and W. G. Wadsworth. (2004) "Assembly and tissue functions of early embryonic laminins and netrins." *Curr Opin Cell Biol* **16**(5): 572–579.

Zorlutuna, P., A. Elsheikh, V. Hasirci. (2009). "Nanopatterning of collagen scaffolds improve the mechanical properties of tissue engineered vascular grafts." *Biomacromolecules* **10**(4): 814–821.

6 Hydrogels with dynamically tunable properties

Murat Guvendiren and Jason A. Burdick

6.1 Introduction

Cells reside in a highly dynamic extracellular matrix (ECM) microenvironment, where mechanical, biochemical, and topographical cues are displayed in a spatiotemporal fashion (Discher et al. 2009). Cells sense their surrounding microenvironment by simply pulling and pushing it, and in response generate biochemical activity – including changes in their morphology, migration, proliferation, and differentiation – through a process known as mechanotransduction (Wang et al. 1993). Likewise, the ECM is responsive to cells, as cell secreted enzymes and growth factors can remodel and alter the surrounding ECM properties. Importantly, many cellular processes, including during development (Daley et al. 2008) and with diseases such as fibrosis and cancer (Georges et al. 2007; Huang and Ingber 2005) are dynamic in nature and involve spatiotemporal changes in matrix elasticity, organization, and biochemical activity. Thus, there is a growing interest in the development of material systems (e.g., hydrogels) with dynamic properties, where changes in material properties can be altered either by the user or the cell itself.

Hydrogels are three-dimensional (3-D) networks formed from water-soluble polymers with high water contents. Hydrogels mimic many biochemical and biophysical properties of soft tissues due to their flexibility in design, making them attractive for tissue engineering applications as carriers for cells or for the delivery of therapeutics (Grieshaber et al. 2011). For instance, a variety of natural (such as gelatin and collagen) and synthetic (such as polyacrylamide and polyethylene glycol) polymers have been used to mimic the elasticity of native tissues including brain (0.2–1 kPa), muscle (10–15 kPa), and precalcified bone (30–45 kPa) (Engler et al. 2006). Synthetic hydrogels are often functionalized (either chemically or by physical absorption) with a wide range of cell-adhesive moieties and/or growth factors to control cell attachment and function. However, most of these hydrogel systems possess static properties that do not capture the dynamic nature of native ECM. In this chapter, we focus on hydrogels with dynamic properties where features such as material degradation, mechanics, and display of biochemical cues are altered in time. Specifically, this review will focus on classes of materials including: hydrolytically degradable hydrogels, protease sensitive hydrogels with cell-controlled properties, and stimuli-responsive hydrogels with user-controlled properties via triggers such as light and temperature.

6.2 Hydrolytically degradable hydrogels

Hydrogels formed from biodegradable polymers are early examples of dynamic systems, where network mechanics gradually decrease with time by degradation, mainly via hydrolysis. Although hydrolysis was originally implemented to eliminate implantable materials with degradation, the dynamic mechanics observed with these systems may be useful to interrogate changing cellular environments. Nonetheless, these property changes must be considered in the context of cellular environments. These hydrolytically degradable hydrogels are also used to deliver chemical cues, such as growth factors, for a range of biomedical applications (Anseth et al. 2002). Examples of these hydrogel systems include hydrophilic polymers linked to poly(α-hydroxy esters) (Metters et al. 2000), polyvinyl alcohol (Martens et al. 2002), and polypropylene fumarates (Jo et al. 2001), each with tunable hydrolysis based on material design. Hydrolytic degradation can also be introduced into hydrogels from synthetically modified biopolymers by suitable chemistry and can be tuned by controlling the degradable link such as in the case of hyaluronic acid and dextran (de Groot et al. 2001; Sahoo et al. 2008).

While hydrolytically degradable hydrogels show some promise with their degree of tunability, they possess some drawbacks, such as that the timescale for degradation may not always coincide with cellular function, and degradation occurs globally instead of locally (not cell specific). Below, alternative approaches to induce spatiotemporal control over hydrogel degradation are discussed.

6.3 Protease sensitive hydrogels

Enyzmatically degradable materials are quickly evolving to exploit the activity of the cell to alter material properties. Many biopolymers, such as collagen and fibrin, are enzymatically degradable, and hydrogels formed using these biopolymers as building blocks usually exhibit this property (unless modified during processing). For instance, collagen hydrogels have inherently dynamic properties, and can be degraded and remodeled in response to cellular activity (Hadjipanayi et al. 2009). In some cases, synthetic modification of biopolymers may deteriorate their bioactivity (e.g., degradation properties), depending on the nature and extent of the modification. For instance, when biopolymers, including hyaluronic acid, dextran, and gelatin, were modified with methacrylates to form chemically crosslinked hydrogels, depending on the degree of methacrylation, they were observed to be nondegradable (Khetan et al. 2013; Nichol et al. 2010). For some biopolymers, synthetic modifications are necessary to form a hydrogel (Burdick and Prestwich 2011).

There is continued effort to introduce aspects of natural matrices (e.g., cell-triggered degradation) into synthetic biomaterials to form more biomimetic materials that can assist the tissue regeneration process. Enzyme-responsive molecules (peptides or proteins) are particularly attractive to induce cell-mediated dynamic properties in

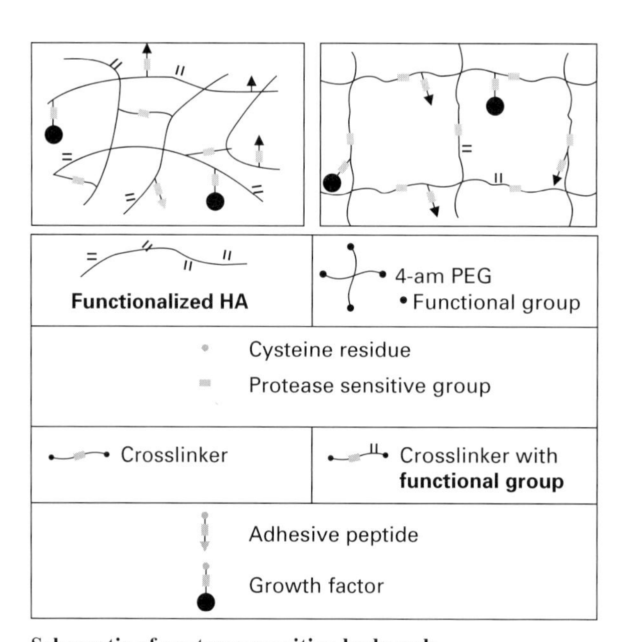

Figure 6.1 **Schematic of protease-sensitive hydrogels.**
Protease-sensitive hydrogels are sensitive to enzymatic cleavage of crosslinks and/or biochemical cues (e.g., adhesive peptides, growth factors). Two hypothetical cases include functionalized hyaluronic acid (HA) (top left) and end functionalized 4-arm PEG (top right) with functional groups (e.g., acrylates, maleimides) designed to react with cysteine residues.

hydrogels (Lutolf et al. 2003a), where the molecule is cleaved in the presence of cell-secreted enzymes, such as matrix-metalloproteases (MMP) or other enzymes like elastases and plasmin (Lutolf et al. 2003b; Patterson and Hubbell 2010; Gobin and West 2002; Kim and Healy 2003). There are multiple ways to incorporate these peptides into the polymer network, such as in the form of a pendant chain and/or a crosslinker (Fig. 6.1). The former is generally used to induce dynamic presentation of biochemical cues, such as cell-adhesive moieties, or growth factors (Salinas and Anseth 2008), whereas the latter creates dynamic crosslinks that can be used to control hydrogel degradation (which correlates with mechanics) and delivery of trapped molecules within the hydrogel (Zisch et al. 2003; Seliktar et al. 2004). In many cases, it has been observed that cell-mediated remodeling of hydrogels requires the presence of both cell-adhesive cues and degradation and the kinetics of cleavage are determined by crosslink density and peptide specificity (Patterson and Hubbell 2010). Cysteine is the most commonly used amino acid to tether peptides and proteins to synthetic polymers, as thiols can facilitate Michael-type addition reaction with various reactive groups such as maleimides, vinyl sulfones, and (meth)acrylates (Phelps et al. 2012; Metters and Hubbell 2005). Michael-type reactions are advantageous as compared to free radical polymerization (such as in the case of UV-induced polymerization) as they do not require free-radicals. However, reactions are improved with a nucleophilic buffering agent such as triethanolamine (TEA) (Pratt et al. 2004; Mather et al. 2006) that might also have cytotoxic effects on some cell types (Phelps et al. 2012; Shikanov et al. 2011).

In the former approach (cleavable tethers), the main idea is to use an enzyme-responsive molecule as a linker between the polymer backbone and the biochemical cue, enabling release of the biochemical cue with the upregulation of certain cell-secreted enzymes. The enzyme-responsive group is generally incorporated into the polymer prior to crosslinking, which occurs independently. As just one example, Salinas and Anseth employed this approach to study the role of cell-mediated temporal regulation of integrin binding peptides in hMSC (human mesenchymal stem cell) chondrogenesis within PEG hydrogels (Salinas and Anseth 2008). Their motivation was to introduce the adhesive cue at early times to promote hMSC survival and then remove it to eliminate issues in chondrogenesis due to adhesion persistence. Since the activity of MMP-13 in chondrogenic media conditions was found to increase significantly from culture day 9 to 14, the hydrogel was designed with an MMP-13 cleavable RGD peptide to enhance hMSC adhesion in the early stages of differentiation and then allow it to be cleaved at later times. This approach resulted in much higher glycosaminoglycan production and collagen type II deposition when compared to hydrogels with persistent RGD presentation. This study clearly shows the importance of temporal control of biochemical cues in stem cell differentiation.

The latter approach (cleavable crosslinks) requires a bifunctional peptide or a molecule with an enzyme-responsive group. Many protease degradable crosslinker peptides with cysteine functionality on both ends were developed to fabricate enzymatically degradable hydrogels (Raeber et al. 2005). Patterson and Hubbell conducted a detailed experimental study in vitro using a collection of protease degradable peptides both in soluble form and in the form of crosslinkers within PEG based hydrogels (Patterson and Hubbell 2010). Hydrogels were formed via Michael-type addition reactions between bifunctional thiol-containing peptides and 4-arm PEG-vinyl sulfone. Hydrogels degraded faster when the appropriate enzymes were matched to enzyme susceptibility of the crosslinker peptide, showing the specificity of degradation. This property can be used to specifically match hydrogel design to the wound environment and enzymatic activity of a specific cell type. For instance, cell-demanded release of vascular endothelial growth factor (VEGF) for vascularized tissue growth was achieved by conjugating RGD-containing peptides (G*C*GYG*RGDSP*G) to PEG macromers covalently conjugated with VEGF, and cross-linking the hydrogel with MMP-sensitive peptides (G*C*RDG*PQGIWGQ*DR*C*G). (Zisch et al. 2003). VEGF-RGD activated endothelial cells showed a significant increase in latent MMP-2 zymogen production and when transforming growth factor beta-1 (TGF-β1) was encapsulated into these hydrogels, release of the TGF-β1 with MMP-induced degradation converted the latent MMP zymogen into its active form, further enhancing hydrogel degradation (Seliktar et al. 2004). This approach is also powerful for the on-demand delivery of therapeutic agents (Kim and Yoo 2010; Tauro and Gemeinhart 2005), or even molecules to tune protease activity (Purcell et al. 2014) in diseases where protease activity is altered.

In these hydrogels the reaction kinetics and coupling efficiency of thiols differ significantly depending on the reactive group (e.g., maleimides, vinyl sulfones, or acrylates). For instance, Garcia et al. investigated the reaction kinetics of a thiol containing cell-adhesive peptide (GRGDSP*C*) and enzymatically degradable crosslinker peptides (G*C*RD*VPM*SMRGGDR*C*G) with 4-arm PEG macromers end functionalized with maleimides (PEG-4MAL), vinyl sulfones (PEG-4VS), or acrylates (PEG-4A) (Phelps et al. 2012). For the same reaction conditions (400 mM TEA buffer), MAL showed nearly 100% RGD incorporation for 1:1 molar ratio of MAL:peptide within 10 min, whereas a 4:1 molar ratio of VS:peptide was required for 60 min incubation for vinyl sulfone and 8:1 molar ratio of A:peptide was required for complete incorporation of peptide to acrylates at 60 min. Similarly, the time of gelation was significantly shorter for PEG-4MAL (1–5 min) compared to PEG-4VS (30 to 60 min) and PEG-4A (60 min). Gelation time is particularly important for homogenous cell distribution when the cells are encapsulated within the hydrogel.

6.4 Light responsive hydrogels

As an alternative to the above systems, light can be used to alter mechanical and biochemical properties of hydrogels. Importantly, this can occur in the presence of cells and at a user-defined time and space in a highly controlled manner. Precise control of light responsiveness on the scale of a single cell is also possible with the implementation of multi-photon microscopy (Kloxin et al. 2010b). Many biological processes are dynamic in nature, and light responsive hydrogels provide highly tunable 3-D cell-culture platforms to study cell response to dynamic changes in their microenvironments. With this regard, light responsive hydrogels are categorized into two main groups including light-initiated addition (photopolymerizable hydrogels) or removal (photolabile hydrogels) of chemical moieties into the hydrogel (schematic in Fig. 6.2). The former is used to introduce crosslinks and/or biochemical cues, whereas the latter is used to cleave crosslinks (photodegradable hydrogels) or remove biochemical cues and/or caging compounds to block biochemical activity.

6.4.1 Photopolymerizable hydrogels

Photopolymerizable hydrogels have been used for many tissue-engineering applications as they provide the potential for in situ polymerization via light (UV, visible, infrared) exposure (Guvendiren et al. 2012), including in the presence of cells. It is also possible to use the photopolymerization reaction as a secondary mechanism to incorporate biochemical cues and additional crosslinks spatiotemporally into already formed hydrogels. In order to guide cell behavior, biochemical and biomechanical patterning of 3-D hydrogels has been achieved via photopolymerization within PEG hydrogels (Hahn et al. 2006). Specifically, PEG-diacrylate (PEGDA) hydrogels, formed by photopolymerization and not to complete conversion of all acrylate groups, were swollen with a precursor solution containing an appropriate photoinitiator and acryloyl-PEG-peptide or lower molecular weight PEGDA. Light exposure induced chemical tethering of peptide moieties into the hydrogel or altered hydrogel

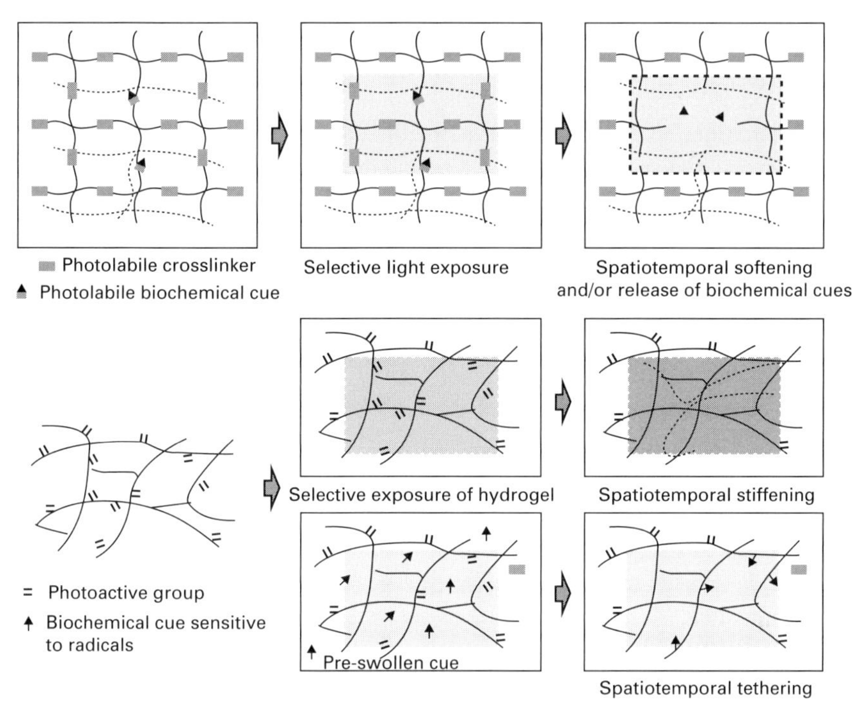

Figure 6.2 **Schematic of light-induced spatiotemporal changes within 3-D hydrogels, including light induced cleavage (top) and photopolymerization (bottom).**
Top: Hydrogel with photolabile crosslinks and/or biochemical cues becomes softer and/or releases biochemical cues when selected space within the hydrogel is exposed to light. Bottom: Hydrogel with photoactive groups can be stiffened or biochemical cues introduced spatially when the hydrogel exposed to light in the presence of a photoinitiator.

crosslinking density. The degree of spatial control was shown to increase significantly from conventional photomasks to single-photon laser to two-photon laser (Hahn et al. 2006). This technique was then used to create RGD patterns within proteolytically degradable hydrogels, where cells preferentially migrated and invade into only the RGD-containing regions in the hydrogel.

To spatially manipulate cell-mediated degradation, a similar approach was also reported using sequential click reactions, one for hydrogel formation and another for biochemical patterning within the hydrogel (DeForest et al. 2009). The hydrogel was formed via a step-growth polymerization reaction under physiological conditions between tetra-azide (-N_3) end groups in 4-arm PEG macromers and alkynes (-C≡C-) from a bis(DIFO3) di-functionalized crosslinker peptide. The peptide crosslinker also contained enzymatically degradable groups and a photoreactive group for chemical patterning. Cells remained rounded when encapsulated within hydrogels without RGD, whereas cells degraded the matrix and spread within RGD-coupled hydrogels. Using the same logic, cell-induced degradation, and hence, spreading, was spatially and temporally controlled by UV-induced coupling of RGD within the hydrogels (DeForest et al. 2009). Acrylated- and methacrylated-HA macromers were also designed that could undergo multiple modes of crosslinking including Michael-type addition and photopolymerization (Khetan et al. 2009; Khetan and Burdick 2010). Protease degradable hydrogels containing RGD peptides were formed under conditions to only consume a fraction of the acrylates in the HA macromers. A photoinitiator was then introduced and the hydrogel was exposed to UV light through a photomask, further crosslinking the network by consuming remaining acrylates and introducing multiple modes of degradation with spatial control. When cultured in a mixed adipogenic/osteogenic induction medium, hMSCs encapsulated within the UV-exposed regions remained rounded and underwent adipogenesis, whereas hMSCs degraded the matrix, spread, and underwent osteogenesis in the unexposed regions (Khetan and Burdick 2010).

A similar approach was used to investigate the underlying mechanism of cell-matrix adhesive interactions for cells encapsulated in hydrogels (Khetan et al. 2013). For this purpose, HA macromers were functionalized with maleimides for hydrogel formation via addition polymerization with protease degradable peptides and for incorporation of adhesive moieties, and methacrylates were included for light-induced crosslinking. This technique was used to fabricate 3-D hydrogels of equivalent elastic moduli that either permitted or restricted cell-mediated degradation. When cell-mediated degradation was permitted, cells exhibited high degrees of spreading and high tractions, favoring an osteogenic phenotype, whereas cells exhibited low degrees of spreading and low tractions, favoring an adipogenic phenotype, when cell-mediated degradation was restricted (Fig. 6.3). This study demonstrated that for covalently crosslinked hydrogels, hMSC differentiation is independent of cell morphology or matrix mechanics but directed by degradation-mediated cellular tractions. Furthermore, delayed secondary crosslinking was employed to induce a temporal switch of the degradable hydrogel to a restrictive state, which in turn resulted in a switch from greater osteogenesis to more adipogenesis.

Besides patterning biochemical cues or restricting degradation, photopolymerization can also be used as a secondary crosslinking mechanism to induce stiffening. In this regard, a step-wise approach was developed to crosslink hydrogels via an addition reaction (first) and then light-mediated crosslinking (second), to fabricate stiffening hydrogels (from ~3 kPa to 30 kPa) in the presence of cells (Guvendiren and Burdick 2012). Stiffening hydrogels were used to investigate short- and long-term hMSC responses to temporal stiffening. Cells were found to increase their area from 500 to 3000 μm^2, and exhibit greater traction from ~1 kPa to 10 kPa within 4 hours after stiffening. For long-term culture studies (14 days) in mixed adipogenic/osteogenic differentiation media, cells selectively differentiated based on the period of culture before and after stiffening, such that adipogenic differentiation was favored for later stiffening, and osteogenic differentiation was favored for earlier stiffening (see Fig. 6.3). This approach was also used for spatial manipulation of hydrogel mechanics to spatially control cellular behavior, including differentiation, from a single pool of cells (Marklein and Burdick 2010; Marklein et al. 2012; Guvendiren et al. 2014).

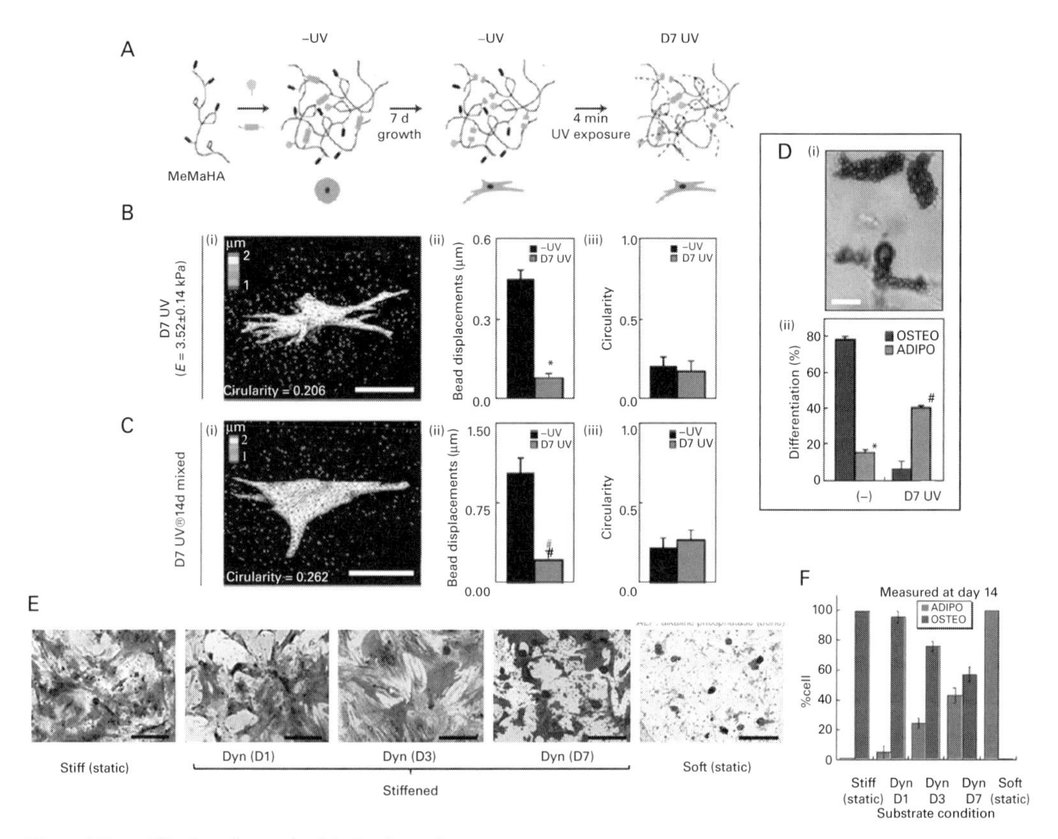

Figure 6.3 **Photopolymerizable hydrogels.**
(A) Hydrogels were formed from hyaluronic acid modified with both maleimide and methacrylate groups (MeMaHA) to entrap hMSCs to permit spreading (–UV) or restrict material degradation (UV) at a user-defined time. Turning off degradation (D7 UV compared to –UV) altered the generation of traction forces in spread cells initially (B) and after culture for another 14 days in differentiation media (C), and led to differences in cell differentiation (D), favoring adipogenesis in mixed adipogenic/osteogenic induction media when the degradation is restricted. (E, F) On 2-D stiffening culture systems, hMSC fate is controlled by the timing of stiffening, with longer periods on a soft substrate leading to more adipogenesis and longer times on a stiffer substrate leading to more osteogenesis.
(A)–(D), Khetan et al. (2013); (E)–(F), Guvendiren and Burdick (2012).

6.4.2 Photolabile hydrogels

Light has also been used to cleave crosslinks and chemical moieties within hydrogels to manipulate mechanical and biochemical properties of hydrogels. Nitrobenzyl ether-derived macromers are the most commonly used photocleavable crosslinkers to dynamically manipulate crosslinking density, and have been widely used in PEG hydrogels (Ifkovits and Burdick 2007). For example, photodegradable hydrogels were designed to tune mechanical properties of hydrogels (softening) in the presence of cells (Kloxin et al. 2010a) This hydrogel system was used to investigate cardiac valve fibrosis, indicated by tissue stiffening due to the persistent presence of myofibroblasts, differentiated from fibroblasts residing in valvular tissue (Wang et al. 2012; Kloxin et al. 2010a). However, rather than stiffening, the study focused on softening of the substrate from 32 kPa (mimicking diseased cardiac valve) to 7 kPa (mimicking the healthy tissue). The study showed that softening of the substrates led to de-activation of valvular myofibroblasts, which differentiated into dormant fibroblasts within 6 hours.

Alternatively, this approach was applied to fabricate dynamic hydrogels with light induced cleavage of biochemical cues. For instance, 3-D hydrogels were formed from polymers tethered with biomolecules through *o*-nitrobenzyl–based linkers, which enable cleavage of these biomolecules in response to UV light (Kloxin et al. 2009). These dynamic hydrogels were used to investigate the effect of temporal changes in cell adhesion molecules on hMSC chondrogenesis. In contrast to cell-mediated cleavage of RGD (as described above), a light responsive group enabled the user-defined display of RGD for temporal control over adhesion and this dynamic presentation altered chondrogenesis within the hydrogels.

Photodegradable hydrogels were also used to release encapsulated cells from hydrogels. To control the degradation rate, a wide range of ortho-nitrobenzyl group (light sensitive) containing macromers were synthesized, and used as crosslinkers to form hydrogels (Griffin and Kasko 2012). Degradation rate for the hydrogels increased with either decreasing number of aryl ethers or changing the functionality from primary to secondary at the benzylic site. To exploit the differences in degradation kinetics of crosslinkers, two different crosslinkers were used to form a bilayer hydrogel, such that the first gel was formed using one crosslinker and the second gel was formed in direct contact with the first one. To monitor cell release from the first and second layer, hMSCs expressing red or green fluorescent proteins were encapsulated within the hydrogels, respectively. When the bilayer hydrogel was exposed to UV light, cells were released faster from hydrogels crosslinked with faster degrading crosslinker as compared to slower degrading counterpart. These results show that it is possible to obtain user-defined sequential delivery of multiple cell populations from a single hydrogel composed of multiple compartments.

Alternatively, a photolabile hydrogel with photo-conjugation capability was designed that enabled chemical and mechanical (via degradation) patterning (DeForest and Anseth 2011). For this purpose a peptide crosslinker was designed to induce gelation at physiological conditions via strain-promoted azide-alkyne cycloaddition reaction. The peptide also contained a photodegradable link and free alkene. Mechanical patterning was achieved by UV induced photodegradation. Chemical patterning was obtained by using visible light induced radically mediated thiol-ene reaction between thiol-containing (such as cysteine) biochemical groups and alkenes (-C=C).

6.4.2.1 Caged compounds to form photolabile hydrogels

Caging compounds are chemical groups or protecting groups that are designed for photo-release applications (Ellis-Davies 2007). Photo-release technology enables light-controlled spatiotemporal activity or display of bioactive molecules. In the former case, a caging group is used to hinder bioactivity of a molecule of interest (caged molecule). In the latter approach, the cage is used to block the functional group, thus blocking tethering of biochemical cues to it. Caging groups can then be removed at user defined times with exposure to light, turning on the bioactivity of the molecule or reactivity of the chemical group. Uncaging has many advantages in regard to release time, location, and timing. These advantages make caging compounds particularly interesting for spatiotemporal presentation of bioactive molecules on or within hydrogels. For instance, photolabile hydrogels were formed from S-2-nitrobenzyl-cysteine (S-NBC) functionalized agarose (Luo and Shoichet 2004). S-NBC was released upon exposure to ultraviolet (UV) light to leave free sulphydryl groups, which were used to tether maleimide functionalized RGD (GRGDS). When primary rat dorsal ganglia cells were cultured on the surface of the agarose hydrogels patterned with RGD in 3-D, cells on the hydrogel surface invaded within the RGD patterned regions. Similarly, localizing the multiphoton beam in 3-D hydrogels from coumarin-modified thiolated agarose induced chemically differentiated patterns for immobilization of bioactive cues (Wosnick and Shoichet 2008). The photorelease of the cage in 3-D space created patterned volumes of free thiol groups, which were then used to selectively immobilize thiol-reactive biomolecules into the patterned volumes.

6.4.2.2 Photocaging of specific affinity interactions

There is a recent effort to spatially control bioactivity of hydrogels by localizing proteins and peptides via specific high-affinity binding interactions, such as streptavidin-biotin and barnase-barstar. High affinity binding interactions are suitable to functionalize surfaces, polymers, and hydrogels, as they form strong and specific binding. However, these reactions are spontaneous and generally very fast, which limits the ability of a user to control them spatially. Shoichet and colleagues applied photorelease technology using caging molecules to overcome this problem, and showed simultaneous immobilization of amino-terminal sonic hedgehog (SHH) and ciliary neurotrophic factor (CNTF) into 3-D agarose hydrogels (Wylie et al. 2011). In this technique, agarose hydrogels were functionalized with coumarin-caged thiols, which were uncaged locally by two-photon irradiation. These reactive thiols were then used to sequentially incorporate maleimide functionalized barnase and streptavidin (Fig. 6.4). Note that maleimides are reactive toward thiols via Michael-type addition reaction. The binding partner barstar (for barnase) and biotin (for streptavidin), functionalized with SHH and CNTF respectively, were introduced into the hydrogel, and simultaneously immobilized by barnase and biotin (Wylie et al. 2011). This approach requires special synthesis capabilities, long reaction times, and easy access to a multiphoton confocal microscope.

Lutolf and colleagues reported a similar approach using enzymatic crosslinking reactions for spatiotemporal control of biochemical cues (Mosiewicz et al. 2013). In this approach, PEG hydrogels were functionalized with caged (inactive) enzymatic

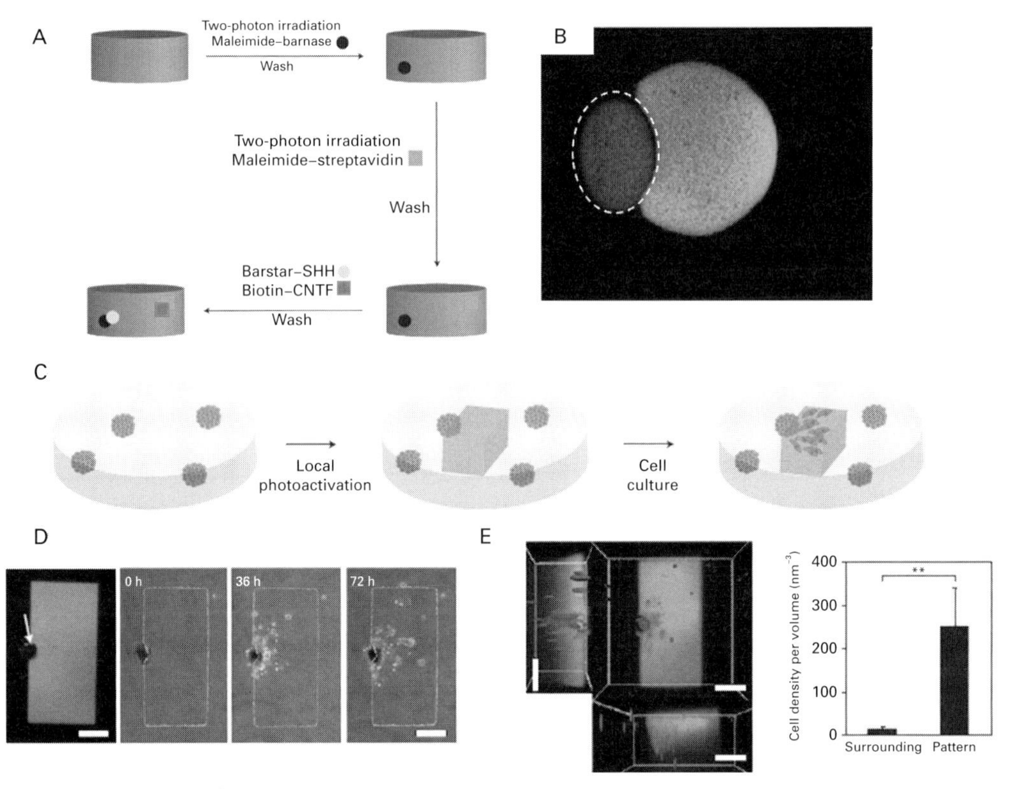

Figure 6.4 **Photolabile hydrogels.**
(A) Sequential tethering of orthogonal binding pairs (barnase-barstar, then streptavidin-biotin) via two-photon uncaging of reactive groups to incorporate one pair for each (barnase and streptavidin), allowing high-affinity binding of complementary pair (barstar and biotin) and spatial patterning of biochemical signals. (B) Demonstration of the approach with fluorescent molecules

Reprinted by permission from Macmillan Publishers Ltd: *Nature Materials.* Wylie et al. (2011), copyright (2011).

(C) Spatial presentation of RGD molecules using photoactivation allowed cells to invade an MMP sensitive hydrogel locally, with time lapse images of cellular invasion (D), and DAPI-stained cells with quantification of cell density (E). (Scale bars are 200 microns.)

Reprinted by permission from Macmillan Publishers Ltd: *Nature Materials*. Lutolf and Hubbell (2005), copyright (2013).

peptides, which were locally activated by exposure to light from a confocal laser. Highly localized tethering of biomolecules into the hydrogel was achieved using counter-reactive enzymes modified with a biomolecule of interest. They were able to spatially pattern vascular endothelial growth factor 121 ($VEGF_{121}$), fibronectin derived adhesion peptide arginine-glycine-aspartic acid (RGD), the recombinant fibronectin fragment $FN_{9\text{-}10}$, and platelet-derived growth factor B (PDGF-BB) using this approach. With RGD, patterning led to spatially controlled cellular behavior (Fig. 6.4).

6.5 Hydrogels from DNA-based reversible crosslinks

DNA is a biopolymer composed of adenine (A), guanine (G), cytosine (C), and thymine (T) nucleotides. DNA maintains a double-stranded helical structure due to hydrogen bonding between base pairs (Watson-Crick base pairs), such as A-T, and G-C. While the double helix structure makes DNA a suitable carrier for genetic information, the base pairing allows recognition of specific genes by DNA-binding proteins. In the early twenty-first century, DNA became an important building block for stimuli responsive materials and devices in nanotechnology and biotechnology, due to highly specific DNA interactions. (Yurke et al. 2000; Liu et al. 2011; Bath and Turberfield 2007; Dietz et al. 2009; Han et al. 2011; Hou and Jiang 2009; Krishnan and Simmel 2011; Liedl et al. 2007a; Liedl et al. 2007b; Liu and Balasubramanian 2003; Simmel and Dittmer 2005).

These favorable properties also fostered development of DNA-based hybrid hydrogels with tunable stimuli responsiveness (Liu et al. 2011; Xing et al. 2011; Nagahara and Matsuda 1996; Watson et al. 2001; Cheng et al. 2009; He et al. 2010; Kang et al. 2011; Murakami and Maeda 2005). Langrana et al. developed a technique to control substrate stiffness reversibly by using DNA crosslinked polyacrylamide (PAM) hydrogels (Lin et al. 2004). In this approach, single-stranded DNAs were immobilized in the PAM backbone to form acrylamide-modified DNA. For this purpose, Acrydite, a phosphoramidite, was conjugated to DNA via 5' end of oligonucleotide forming a methacrylated DNA. The authors copolymerized two sets of Acrydite-modified DNAs (designated by SA1 and SA2) with acrylamide. The DNA-modified acrylamides were then crosslinked by introducing a third DNA strand (referred as L2), which was designed to form base pairs with both SA1 and SA2. To enable reversible DNA crosslinking, the crosslinker DNA L2 was designed to have a "toehold" region to allow removal of the L2 when a removal strand was introduced. The removal strand (referred as CL2) with base sequence complementary to L2 displaced the L2 via branch migration. The mechanics of the DNA hydrogels were determined by the monomer concentration, and crosslinker-DNA length and concentration (Lin et al. 2005). Fibroblast growth (Jiang et al. 2010b) and neurite outgrowth (Jiang et al. 2010a) were investigated by using DNA crosslinked hydrogels with reversible stiffness between ~6 kPa to 23 kPa.

Some of the limitations to this system includes current limitation to 2-D cell substrates, inhomogeneity of the initially formed gel, and that an additional process involving melting followed by subsequent cooling is required to obtain uniform hydrogels (Lin et al. 2004; Lin et al. 2005). Also, delivery of exogenous DNA could cause potential problems in the presence of cells if the DNA sequence has similarity to the specific cell genome (Jiang et al. 2010a; Jiang et al. 2010b). Additionally, introducing free DNA strands could potentially lead to significant changes in the net charge of the hydrogel since free DNA is highly negatively charged (Lin et al. 2005; Gawel and Stokke 2011).

6.6 Two-component dynamic hydrogels

Hydrogels with reversible dynamic mechanics and degradation can be fabricated using a two-component polymer system, such that each component forms a distinct network without covalent bonding between them. These systems are referred to as interpenetrating polymer networks (IPNs). To achieve dynamic properties, one polymer component is designated as the structural component, whereas the other is the modular component. The structural component forms the hydrogel in the presence of cells and the modular component (not crosslinked) is then used to form another network via external stimuli – UV (Sun et al. 2012) or ion (Gillette et al. 2010) – increasing the hydrogel mechanics.

A two-component polymer system composed of type I collagen and alginate was used to induce reversible control of hydrogel mechanics (Gillette et al. 2010). In this case, collagen was used to form the structural component of the hydrogel whereas alginate was used as a modular component such that the crosslinking state of the alginate was controlled by adding or removing divalent cations, inducing stiffening or softening respectively. Alginate was crosslinked by pipetting the Ca^{2+} solution and uncrosslinked by adding citrate solution onto the membrane. Cellular spreading was controlled by switching the state of alginate crosslinking such that cell area increased significantly from 500 μm^2 to 2000 μm^2 by inducing alternating crosslinking and uncrosslinking conditions every 2 hours. Although a variety of cell types were encapsulated in Ca^{2+} crosslinked alginate hydrogels, and similar hydrogels were used for in vivo studies, it is possible for free Ca^{2+} ions to interfere with cellular signaling and influence cell phenotype (Barradas et al. 2012; Choi et al. 2013; Zhang et al. 2013b) and the long-term stability of these gels needs further investigation for use in cell culture.

In a recent study, the same two-component hydrogel system was used to switch human pluripotent stem cells (HPSCs) from self-renewal to differentiation by simply removing the modulatory component (alginate) from the hydrogel (Dixon et al. 2014). Here, the structural component, type I collagen, was identified as a differentiation-permissive microenvironment for the HPSCs, whereas alginate was identified as the self-renewal permissive microenvironment. The timing of this switch was also shown to regulate the lineage commitment to ectoderm (early switch) or mesoderm/endoderm (late switch). A similar approach was recently adapted to create a dynamic 3-D co-culture system composed of photocurable PEG hydrogels and Ca-alginate hydrogels (Sugiura et al. 2013). In this case, murine embryonic stem cells (mESCs) were encapsulated in micropatterned photocurable PEG hydrogels, which were then surrounded by Ca-alginate hydrogels encapsulated with human hepatocellular carcinoma cells (HepG2). After 4 days of co-culture, Ca-alginate gels were removed by sodium citrate solution. Authors reported that dynamic exposure of mESCs to HepG2 secreted paracrine factors induced higher expression of cardiac genes and proteins with increased beating after 16 days of culture. These types of approaches can be potentially useful to design dynamic 3-D tissue constructs to investigate particular cell-cell interactions in a spatially and temporally controlled manner.

6.7 Temperature and pH responsive hydrogels

Stimuli responsive hydrogels are a subgroup of hydrogels that can show substantial changes in material structure with alteration in some signal or stimulus. The most common stimuli to control hydrogel properties are temperature and pH, which could be used to control the presentation of biochemical cues on a substrate surface (Ebara et al. 2008; Hoffman and Stayton 2007; Hoffman et al. 1997). Temperature responsive polymers (most commonly poly(N-isopropyl acrylamide)) are generally used as a tissue culture platform for cell-sheet engineering (Klouda and Mikos 2008; Yamato and Okano 2004). In this approach, cells are cultured to a confluent cell-sheet at normal incubation temperatures (37°C), and spontaneous detachment of cell sheets are obtained when the temperature is reduced (~32°C), due to changes in the surface hydrophobicity (Yamato and Okano 2004). When compared to other techniques, such as use of proteolytic enzymes, this approach preserves the cell-cell interactions and deposited ECM within the sheet. However, the temperature and time of transition are determined by the polymer chemistry; for example, hydrogels based on 2-(diethylamino)ethyl acrylate require incubation of 30 min at 15°C (Zhang et al. 2013a).

ABA type triblock copolymer hydrogels were used to control substrate mechanics reversibly by changing the pH to evaluate the mechanosensing of mouse myoblast cells (C2C12) (Yoshikawa et al. 2011). A 40-fold increase in elastic modulus was obtained for thin films (3–4 μm) when the pH of the culture medium was switched from pH = 7 (E = 1.4 kPa) to pH = 8 (40 kPa). In response, the cell area increased from ~300 μm^2 to ~600 μm^2, which was reversible. Although this study shows the importance of dynamic changes in substrate mechanics in cellular spreading, significant changes in surface hydrophobicity and net charge limit applications of these approaches.

6.8 Summary

Dynamic hydrogels are emerging as important material platforms to capture the dynamic nature of native ECM toward either fundamental studies of cells in response to their microenvironment or for application in fields such as tissue engineering and regenerative medicine. These advances in our ability to control materials are further approaching the complexity that is observed in biological systems, such as during development or in disease. Spatiotemporal control over a wide range of biochemical and biophysical properties are now possible by engineering hydrogels that respond to cells or through user-defined signals, such as the introduction of light to either add or remove chemical moieties to the hydrogel. Ultimately, the complexity of such systems is increasing to open up previously unexplored research questions related to dynamic systems.

References

Anseth, K. S., Metters, A. T., Bryant, S. J., Martens, P. J., Elisseeff, J. H. and Bowman, C. N. (2002). "In situ forming degradable networks and their application in tissue engineering and drug delivery." *Journal of Controlled Release* **78**: 199–209.

Barradas, A. M. C., Fernandes, H. A. M., Groen, N., Chai, Y. C., Schrooten, J., van de Peppel, J., van Leeuwen, J., van Blitterswijk, C. A. and de Boer, J. (2012). "A calcium-induced signaling cascade leading to osteogenic differentiation of human bone marrow-derived mesenchymal stromal cells." *Biomaterials* **33**: 3205–3215.

Bath, J. and Turberfield, A. J. (2007). "DNA nanomachines." *Nature Nanotechnology* **2**: 275–284.

Burdick, J. A. and Prestwich, G. D. (2011). "Hyaluronic acid hydrogels for biomedical applications." *Advanced Materials* **23**: H41–H56.

Cheng, E. J., Xing, Y. Z., Chen, P., Yang, Y., Sun, Y. W., Zhou, D. J., Xu, L. J., Fan, Q. H. and Liu, D. S. (2009). "A pH-triggered, fast-responding DNA hydrogel." *Angewandte Chemie-International Edition* **48**: 7660–7663.

Choi, S., Yu, X. H., Jongpaiboonkit, L., Hollister, S. J. and Murphy, W. L. (2013). "Inorganic coatings for optimized non-viral transfection of stem cells." *Scientific Reports* **3**: 1587.

Daley, W. P., Peters, S. B. and Larsen, M. (2008). "Extracellular matrix dynamics in development and regenerative medicine." *Journal of Cell Science* **121**: 255–264.

de Groot, C. J., van Luyn, M. J. A., van Dijk-Wolthuis, W. N. E., Cadee, J. A., Plantinga, J. A., Den Otter, W. and Hennink, W. E. (2001). "In vitro biocompatibility of biodegradable dextran-based hydrogels tested with human fibroblasts." *Biomaterials* **22**: 1197–1203.

DeForest, C. A. and Anseth, K. S. (2011). "Cytocompatible click-based hydrogels with dynamically tunable properties through orthogonal photoconjugation and photocleavage reactions." *Nature Chemistry* **3**: 925–931.

DeForest, C. A., Polizzotti, B. D. and Anseth, K. S. (2009). "Sequential click reactions for synthesizing and patterning three-dimensional cell microenvironments." *Nature Materials* **8**: 659–664.

Dietz, H., Douglas, S. M. and Shih, W. M. (2009). "Folding DNA into twisted and curved nanoscale shapes." *Science* **325**: 725–730.

Discher, D. E., Mooney, D. J. and Zandstra, P. W. (2009). "Growth factors, matrices, and forces combine and control stem cells." *Science* **324**: 1673–1677.

Dixon, J. E., Shah, D. A., Rogers, C., Hall, S., Weston, N., Parmenter, C. D. J., McNally, D., et al. (2014). "Combined hydrogels that switch human pluripotent stem cells from self-renewal to differentiation." *Proceedings of the National Academy of Sciences USA* **111**: 5580–5585.

Ebara, M., Yamato, M., Aoyagi, T., Kikuchi, A., Sakai, K. and Okano, T. (2008). "The effect of extensible PEG tethers on shielding between grafted thermo-responsive polymer chains and integrin-RGD binding." *Biomaterials* **29**: 3650–3655.

Ellis-Davies, G. C. R. (2007). "Caged compounds: photorelease technology for control of cellular chemistry and physiology." *Nature Methods* **4**: 619–628.

Engler, A. J., Sen, S., Sweeney, H. L. and Discher, D. E. (2006). "Matrix elasticity directs stem cell lineage specification." *Cell* **126**: 677–689.

Gawel, K. and Stokke, B. T. (2011). "Logic swelling response of DNA-polymer hybrid hydrogel." *Soft Matter* **7**: 4615–4618.

Georges, P. C., Hui, J. J., Gombos, Z., Mccormick, M. E., Wang, A. Y., Uemura, M., Mick, R., et al. (2007). "Increased stiffness of the rat liver precedes matrix deposition: implications for fibrosis." *American Journal of Physiology-Gastrointestinal and Liver Physiology* **293**: G1147–G1154.

Gillette, B. M., Jensen, J. A., Wang, M. X., Tchao, J. and Sia, S. K. (2010). "Dynamic hydrogels: switching of 3D microenvironments using two-component naturally derived extracellular matrices." *Advanced Materials* **22**: 686–691.

Gobin, A. S. and West, J. L. (2002). "Cell migration through defined, synthetic extracellular matrix analogues." *Faseb Journal* **16**: 751–753.

Grieshaber, S. E., Jha, A. K., Farran, A. J. E. and Jia, X. Q. (2011). "Hydrogels in tissue engineering." In *Biomaterials for Tissue Engineering Applications: A Review of the Past and Future Trends*, J. A. Burdick and R. L. Mauck, eds. New York: Springer, 9–46.

Griffin, D. R. and Kasko, A. M. (2012). "Photodegradable macromers and hydrogels for live cell encapsulation and release." *Journal of the American Chemical Society* **134**: 13103–13107.

Guvendiren, M. and Burdick, J. A. (2012). "Stiffening hydrogels to probe short– and long-term cellular responses to dynamic mechanics." *Nature Communications* **3**: 792.

Guvendiren, M., Perepelyuk, M., Wells, R. G. and Burdick, J. A. (2014). "Hydrogels with differential and patterned mechanics to study stiffness-mediated myofibroblastic differentiation of hepatic stellate cells." *Journal of the Mechanical Behavior of Biomedical Materials* **38**: 198–208.

Guvendiren, M., Purcell, B. and Burdick, J. A. (2012)."Photopolymerizable systems." In *Polymer Science: A Comprehensive Reference*, vol. **9**, M. Krzysztof and M. Martin, eds. Amsterdam: Elsevier, 413–438.

Hadjipanayi, E., Mudera, V. and Brown, R. A. (2009). "Guiding cell migration in 3D: a collagen matrix with graded directional stiffness." *Cell Motility and the Cytoskeleton* **66**: 121–128.

Hahn, M. S., Miller, J. S. and West, J. L. (2006). "Three-dimensional biochemical and biomechanical patterning of hydrogels for guiding cell behavior." *Advanced Materials* **18**: 2679–2684.

Han, D. R., Pal, S., Nangreave, J., Deng, Z. T., Liu, Y. and Yan, H. (2011). "DNA origami with complex curvatures in three-dimensional space." *Science* **332**: 342–346.

He, X. J., Weiz, B. and Mi, Y. L. (2010). "Aptamer based reversible DNA induced hydrogel system for molecular recognition and separation." *Chemical Communications* **46**: 6308–6310.

Hoffman, A. S. and Stayton, P. S. (2007). "Conjugates of stimuli-responsive polymers and proteins." *Progress in Polymer Science* **32**: 922–932.

Hoffman, A. S., Stayton, P. S., Shimoboji, T., Chen, G. H., Ding, Z. L., Chilkoti, A., Long, C., et al. (1997). "Conjugates of stimuli-responsive polymers and biomolecules: Random and site-specific conjugates of temperature-sensitive polymers and proteins." *Macromolecular Symposia* **118**: 553–563.

Hou, X. and Jiang, L. (2009). "Learning from nature: building bio-inspired smart nanochannels." *Acs Nano* **3**: 3339–3342.

Huang, S. and Ingber, D. E. (2005). "Cell tension, matrix mechanics, and cancer development." *Cancer Cell* **8**: 175–176.

Ifkovits, J. L. and Burdick, J. A. (2007). "Review: photopolymerizable and degradable biomaterials for tissue engineering applications." *Tissue Engineering* **13**: 2369–2385.

Jiang, F. X., et al. (2010a). "Effect of dynamic stiffness of the substrates on neurite outgrowth by using a DNA-crosslinked hydrogel." *Tissue Engineering Part A* **16**: 1873–1889.

Jiang, F. X., et al. (2010b). "The relationship between fibroblast growth and the dynamic stiffnesses of a DNA crosslinked hydrogel." *Biomaterials* **31**: 1199–1212.

Jo, S., Shin, H. and Mikos, A. G. (2001). "Modification of oligo(poly(ethylene glycol) fumarate) macromer with a GRGD peptide for the preparation of functionalized polymer networks." *Biomacromolecules* **2**: 255–261.

Kang, H. Z., Liu, H. P., Zhang, X. L., Yan, J. L., Zhu, Z., Peng, L., Yang, H. H., et al. (2011). "Photoresponsive DNA-cross-linked hydrogels for controllable release and cancer therapy." *Langmuir* **27**: 399–408.

Khetan, S. and Burdick, J. A. (2010). "Patterning network structure to spatially control cellular remodeling and stem cell fate within 3-dimensional hydrogels." *Biomaterials* **31**: 8228–8234.

Khetan, S., Guvendiren, M., Legant, W. R., Cohen, D. M., Chen, C. S. and Burdick, J. A. (2013). "Degradation-mediated cellular traction directs stem cell fate in covalently crosslinked three-dimensional hydrogels." *Nature Materials* **12**: 458–465.

Khetan, S., Katz, J. S. and Burdick, J. A. (2009). "Sequential crosslinking to control cellular spreading in 3-dimensional hydrogels." *Soft Matter* **5**: 1601–1606.

Kim, H. S. and Yoo, H. S. (2010). "MMPs-responsive release of DNA from electrospun nanofibrous matrix for local gene therapy: in vitro and in vivo evaluation." *Journal of Controlled Release* **145**: 264–271.

Kim, S. and Healy, K. E. (2003). "Synthesis and characterization of injectable poly(N-isopropylacrylamide-co-acrylic acid) hydrogels with proteolytically degradable cross-links." *Biomacromolecules* **4**: 1214–1223.

Klouda, L. and Mikos, A. G. (2008). "Thermoresponsive hydrogels in biomedical applications." *European Journal of Pharmaceutics and Biopharmaceutics* **68**: 34–45.

Kloxin, A. M., Benton, J. A. and Anseth, K. S. (2010a). "In situ elasticity modulation with dynamic substrates to direct cell phenotype." *Biomaterials* **31**: 1–8.

Kloxin, A. M., Tibbett, M. W. and Anseth, K. S. (2010b). "Synthesis of photodegradable hydrogels as dynamically tunable cell culture platforms." *Nature Protocols* **5**: 1867–1887.

Kloxin, A. M., Kasko, A. M., Salinas, C. N. and Anseth, K. S. (2009). "Photodegradable hydrogels for dynamic tuning of physical and chemical properties." *Science* **324**: 59–63.

Krishnan, Y. and Simmel, F. C. (2011). "Nucleic acid based molecular devices." *Angewandte Chemie-International Edition* **50**: 3124–3156.

Liedl, T., et al. (2007a). "Controlled trapping and release of quantum dots in a DNA-switchable hydrogel." *Small* **3**: 1688–1693.

Liedl, T., Sobey, T. L. and Simmel, F. C. (2007b). "DNA-based nanodevices." *Nano Today* **2**: 36–41.

Lin, D. C., Yurke, B. and Langrana, N. A. (2004). "Mechanical properties of a reversible, DNA-crosslinked polyacrylamide hydrogel." *Journal of Biomechanical Engineering* **126**: 104–110.

Lin, D. C., Yurke, B. and Langrana, N. A. (2005). "Inducing reversible stiffness changes in DNA-crosslinked gels." *Journal of Materials Research* **20**: 1456–1464.

Liu, D. S. and Balasubramanian, S. (2003). "A proton-fuelled DNA nanomachine." *Angewandte Chemie-International Edition* **42**: 5734–5736.
Liu, D. S., Cheng, E. J. and Yang, Z. Q. (2011). "DNA-based switchable devices and materials." *NPG Asia Materials* **3**: 109–114.
Luo, Y. and Shoichet, M. S. (2004). "A photolabile hydrogel for guided three-dimensional cell growth and migration." *Nature Materials* **3**: 249–253.
Lutolf, M. P. and Hubbell, J. A. (2005). "Synthetic biomaterials as instructive extracellular microenvironments for morphogenesis in tissue engineering." *Nature Biotechnology* **23**: 47–55.
Lutolf, M. P., et al. (2003a). "Synthetic matrix metalloproteinase-sensitive hydrogels for the conduction of tissue regeneration: engineering cell-invasion characteristics." *Proceedings of the National Academy of Sciences USA* **100**: 5413–5418.
Lutolf, M. P., et al. (2003b). "Cell-responsive synthetic hydrogels." *Advanced Materials* **15**: 888–892.
Marklein, R. A. and Burdick, J. A. (2010). "Spatially controlled hydrogel mechanics to modulate stem cell interactions." *Soft Matter* **6**: 136–143.
Marklein, R. A., Soranno, D. E. and Burdick, J. A. (2012). "Magnitude and presentation of mechanical signals influence adult stem cell behavior in 3-dimensional macroporous hydrogels." *Soft Matter* **8**: 8113–8120.
Martens, P., Holland, T. and Anseth, K. S. (2002). "Synthesis and characterization of degradable hydrogels formed from acrylate modified poly(vinyl alcohol) macromers." *Polymer* **43**: 6093–6100.
Mather, B. D., Viswanathan, K., Miller, K. M. and Long, T. E. (2006). "Michael addition reactions in macromolecular design for emerging technologies." *Progress in Polymer Science* **31**: 487–531.
Metters, A. and Hubbell, J. (2005). "Network formation and degradation behavior of hydrogels formed by Michael-type addition reactions." *Biomacromolecules* **6**: 290–301.
Metters, A. T., Anseth, K. S. and Bowman, C. N. (2000). "Fundamental studies of a novel, biodegradable PEG-b-PLA hydrogel." *Polymer* **41**: 3993–4004.
Mosiewicz, K. A., Kolb, L., van der Vlies, A. J., Martino, M. M., Lienemann, P. S., Hubbell, J. A., Ehrbar, M., et al. (2013). "In situ cell manipulation through enzymatic hydrogel photopatterning." *Nature Materials* **12**: 1071–1077.
Murakami, Y. and Maeda, M. (2005). "DNA-responsive hydrogels that can shrink or swell." *Biomacromolecules* **6**: 2927–2929.
Nagahara, S. and Matsuda, T. (1996). "Hydrogel formation via hybridization of oligonucleotides derivatized in water-soluble vinyl polymers." *Polymer Gels and Networks* **4**: 111–127.
Nichol, J. W., Koshy, S. T., Bae, H., Hwang, C. M., Yamanlar, S. and Khademhosseini, A. (2010). "Cell-laden microengineered gelatin methacrylate hydrogels." *Biomaterials* **31**: 5536–5544.
Patterson, J. and Hubbell, J. A. (2010). "Enhanced proteolytic degradation of molecularly engineered PEG hydrogels in response to MMP-1 and MMP-2." *Biomaterials* **31**: 7836–7845.
Phelps, E. A., Enemchukwu, N. O., Fiore, V. F., Sy, J. C., Murthy, N., Sulchek, T. A., Barker, T. H., et al. (2012). "Maleimide cross-linked bioactive PEG hydrogel exhibits

improved reaction kinetics and cross-linking for cell encapsulation and in situ delivery." *Advanced Materials* **24**: 64–70.

Pratt, A. B., Weber, F. E., Schmoekel, H. G., Muller, R. and Hubbell, J. A. (2004). "Synthetic extracellular matrices for in situ tissue engineering." *Biotechnology and Bioengineering* **86**: 27–36.

Purcell, B. P., Lobb, D., Charati, M. B., Dorsey, S. M., Wade, R. J., Zellars, K. N., Doviak, H., et al. (2014). "Injectable and bioresponsive hydrogels for on-demand matrix metalloproteinase inhibition." *Nature Materials* **13**: 653–661.

Raeber, G. P., Lutolf, M. P. and Hubbell, J. A. (2005). "Molecularly engineered PEG hydrogels: a novel model system for proteolytically mediated cell migration." *Biophysical Journal* **89**: 1374–1388.

Sahoo, S., Chung, C., Khetan, S. and Burdick, J. A. (2008). "Hydrolytically degradable hyaluronic acid hydrogels with controlled temporal structures." *Biomacromolecules* **9**: 1088–1092.

Salinas, C. N. and Anseth, K. S. (2008). "The enhancement of chondrogenic differentiation of human mesenchymal stem cells by enzymatically regulated RGD functionalities." *Biomaterials* **29**: 2370–2377.

Seliktar, D., Zisch, A. H., Lutolf, M. P., Wrana, J. L. and Hubbell, J. A. (2008). "MMP-2 sensitive, VEGF-bearing bioactive hydrogels for promotion of vascular healing." *Journal of Biomedical Materials Research* **68A**: 704–716.

Shikanov, A., Smith, R. M., Xu, M., Woodruff, T. K. and Shea, L. D. (2011). "Hydrogel network design using multifunctional macromers to coordinate tissue maturation in ovarian follicle culture." *Biomaterials* **32**: 2524–2531.

Simmel, F. C. and Dittmer, W. U. (2005). "DNA nanodevices." *Small* **1**: 284–299.

Sugiura, S., Cha, J. M., Yanagawa, F., Zorlutuna, P., Bae, H. and Khademhosseini, A. (2013). "Dynamic three-dimensional micropatterned cell co-cultures within photocurable and chemically degradable hydrogels." *Journal of Tissue Engineering and Regenerative Medicine*: n/a-n/a.

Sun, J., Xiao, W. Q., Tang, Y. J., Li, K. F. and Fan, H. S. (2012). "Biomimetic interpenetrating polymer network hydrogels based on methacrylated alginate and collagen for 3D pre-osteoblast spreading and osteogenic differentiation." *Soft Matter* **8**: 2398–2404.

Tauro, J. R. and Gemeinhart, R. A. (2005). "Matrix metalloprotease triggered delivery of cancer chemotherapeutics from hydrogel matrixes." *Bioconjugate Chemistry* **16**: 1133–1139.

Wang, H., Haeger, S. M., Kloxin, A. M., Leinwand, L. A. and Anseth, K. S. (2012). "Redirecting valvular myofibroblasts into dormant fibroblasts through light-mediated reduction in substrate modulus." *PloS One* **7**(7): e39969.

Wang, N., Butler, J. P. and Ingber, D. E. (1993). "Mechanotrunsduction across the cell surface and through the cytoskeleton." *Science* **260**: 1124–1127.

Watson, K. J., Park, S. J., Im, J. H., Nguyen, S. T. and Mirkin, C. A. (2001). "DNA-block copolymer conjugates." *Journal of the American Chemical Society* **123**: 5592–5593.

Wosnick, J. H. and Shoichet, M. S. (2008). "Three-dimensional chemical patterning of transparent hydrogels." *Chemistry of Materials* **20**: 55–60.

Wylie, R. G., Ahsan, S., Aizawa, Y., Maxwell, K. L., Morshead, C. M. and Shoichet, M. S. (2011). "Spatially controlled simultaneous patterning of multiple growth factors in three-dimensional hydrogels." *Nature Materials* **10**: 799–806.

Xing, Y. Z., Cheng, E. J., Yang, Y., Chen, P., Zhang, T., Sun, Y. W., Yang, Z. Q., et al. (2011). "Self-assembled DNA hydrogels with designable thermal and enzymatic responsiveness." *Advanced Materials* **23**: 1117–1121.

Yamato, M. and Okano, T. (2004). "Cell sheet engineering." *Materials Today* **7**: 42–47.

Yoshikawa, H. Y., Rossetti, F. F., Kaufmann, S., Kaindl, T., Madsen, J., Engel, U., Lewis, A. L., et al. (2011). "Quantitative evaluation of mechanosensing of cells on dynamically tunable hydrogels." *Journal of the American Chemical Society* **133**: 1367–1374.

Yurke, B., Turberfield, A. J., Mills, A. P., Simmel, F. C. and Neumann, J. L. (2000). "A DNA-fuelled molecular machine made of DNA." *Nature* **406**: 605–608.

Zhang, R., et al. (2013a). "A thermoresponsive and chemically defined hydrogel for long-term culture of human embryonic stem cells." *Nature Communications* **4**.

Zhang, W. J., et al. (2013b). "The synergistic effect of hierarchical micro/nano-topography and bioactive ions for enhanced osseointegration." *Biomaterials* **34**: 3184–3195.

Zisch, A. H., Lutolf, M. P., Ehrbar, M., Raeber, G. P., Rizzi, S. C., Davies, N., Schmokel, H., et al. (2003). "Cell-demanded release of VEGF from synthetic, biointeractive cell-ingrowth matrices for vascularized tissue growth." *FASEB Journal* **17**: 2260–2262.

7 Microengineered tools for studying cell migration in electric fields

Jiandong Wu and Francis Lin

The migratory ability of various cell types contributes to cell functions, physiological processes, and disease pathologies. Among the diverse environmental guiding mechanisms for cell migration, the electric field is a long-known important guiding cue. The electric field–directed cell migration, termed "electrotaxis," can mediate processes that are important for human health such as wound healing, immune responses, and cancer metastasis. The growing interest in better understanding electrotaxis has motivated technological developments to enable more advanced electrotaxis studies. In particular, various microengineered devices have been developed and applied to studying electrotaxis over recent years. In general these new experimental tools can better control electric field application in cell migration experiments, whereas each developed tool offers its own features. Successful applications of the new devices have been demonstrated for studying electrotaxis of various cell types such as cancer cells, lymphocytes, animal models, and tissue cells related to wound healing, as well as for investigating electric field–mediated orientation responses in stem cells and yeast cells. In this chapter, we will provide the background information in directed cell migration, electrotaxis, and cell migration assays. We follow with a survey of fabrication and assembly methods of various microengineered electrotaxis devices and experimental setup and analysis methods, as well as their applications for cell studies. Finally, we conclude the chapter with our perspective on the issues challenging this research area and on the proposed directions for future development.

7.1 Introduction

7.1.1 Directed cell migration and electrotaxis

Directed cell migration critically contributes to physiological processes such as immune responses and tissue regeneration (Baggiolini 1998; Zhao et al. 2006). In addition, directed cell migration is a key player for mediating various diseases such as autoimmune diseases and cancers (Muller et al. 2001; Luster et al. 2005). Diverse chemical and physical guiding signals can direct the movement of various cell types. Chemical concentration gradient guided cell migration, termed "chemotaxis" or "haptotaxis" depending on the soluble or surface bound nature of the chemical gradient, enables many highly regulated physiological behaviors of motile cells in tissues including immune cell homing and cancer metastasis, to name a few (Campbell and Butcher 2000; Muller et al. 2001). The ability of immune cells to sense minute chemical concentration differences over their small body (typically ~10 μm) and drive persistent migration toward the chemical source is remarkable (Zigmond 1977). On the other hand, tumor cell demands higher chemical gradient strength, which is sometimes accompanied with co-stimulating factors, to provide a suitable physiological environment for promoting the cell's directional locomotion (Wang et al. 2004; Mosadegh et al. 2008). Beyond chemotaxis, various physical factors ranging from mechanical stress to temperature difference and to electromagnetic fields can also direct cell motion (Hsu et al. 2005; McCormick et al. 2011; DeLong et al. 1993). Among them, electric field (EF) guided cell migration, termed "electrotaxis" or "galvanotaxis," is a long-known and important mechanism (Zhao 2009; McCaig et al. 2005; Mycielska and Djamgoz 2004; Cortese et al. 2014).

Endogenous direct current electric fields (dcEF) are found in many physiological systems including developing embryo, healing wounds, and cancers (McCaig et al. 2005). One of the best known examples is the wound-generated electric field that attracts the surrounding tissue cells such as epithelial cells for wound reconstruction (McCaig et al. 2005). Electrotaxis of various wound-healing related tissue cell types have been well demonstrated (McCaig et al. 2005). Application of an external electric field to the wound can either accelerate or suppress the healing rate depending on the direction of the applied electric field relative to the field produced by the wound (Zhao et al. 2006). Our work showed that most lymphocytes and monocytes from peripheral blood as well as motile lymphocytes in skin tissues are also capable of migrating toward an applied electric field, suggesting the possible role of electrical guiding mechanisms for immune cell trafficking during the wound-healing process (Lin et al. 2008; Li et al. 2011). In cancerous environments, an electric field can be established between the tumor and the surrounding tissues, which favors the spreading of metastatic cancer cells to other secondary sites (Djamgoz et al. 2001; Pu et al. 2007). The electrotaxis response depends on the cancer cell types, and it often correlates with the cell's metastatic potential (Huang et al. 2009; Pu et al. 2007). In contrast to chemotaxis, specific cell surface sensors and downstream signaling pathways for detecting an electric field is not commonly identified in electrotaxis. On the other hand, several electrotactic signaling mechanisms have been proposed, such as electromigration of cell surface components and calcium signaling (Wu and Lin 2011; Zhao et al. 1999); and some downstream signaling pathways that are common to cell motility such as PI3 Kinase and PTEN signaling have been demonstrated to be important for electrotaxis as well (Zhao et al. 2006).

7.1.2 Conventional cell migration and electrotaxis assays

In vitro cell migration experimentation is an important research approach for characterizing directed migration of different cells under various defined stimulating conditions. Historically, the two main focuses of in vitro cell migration studies have been on high-throughput screening at the cell population level and detailed cell migration characterizations at the single cell level. Chemotaxis studies require gradient-generating assays. Boyden chamber or transwell assay is the most commonly used cell motility and chemotaxis assay in both basic science and applied biomedical research (Boyden 1962). It can produce relatively long-term but unstable chemical gradients across a membrane insert, which separates the upper and bottom compartments of the transwell, and allows measurements of the number of migrated cells to the attractant source in the bottom compartment at the end of the experiments by different cell-counting methods. The transwell assay can be performed in a multiwell plate enabling high-throughput parallel experiments. However, the readout of the transwell assay is based on large cell numbers at the end point. In a different direction, various single cell–based real-time visualization assays coupled with time-lapse optical microscopy have been employed largely for cell biology–driven studies. The Zigmond chamber (or its circular geometry variant, the Dunn chamber), which produces a linear chemical gradient over a bridge between the chemical source and sink wells made in glass (Zigmond 1977); the under agarose assay, which generates chemical gradients by diffusion of chemicals from the reagent reservoirs through agarose gel (Nelson et al. 1975); and micropipette-based assay, which releases the attractant molecules to cells seeded in a petri dish from a micropipette tip (Lohof et al. 1992), are some well-known examples of two-dimensional (2-D) real-time visualization assays. These assays offer the ability to characterize cell migration behaviors at the single cell level while suffering from reduced experimental throughput, and they require specialized live cell imaging facilities. In general, these conventional cell chemotaxis assays are limited for their ability to control and manipulate chemical gradients.

Different from chemotaxis assays, in vitro electrotaxis studies basically follow a gold standard dish-based assay or its variants (McCaig et al. 1994). The dish-based assay builds a glass chamber for cell seeding and electric field application in a petri dish. Electric field is applied using Ag/AgCl electrodes through salt bridges to the cell chamber. Then cell migration in response to the electric field is monitored by time-lapse microscopy and further analyzed by cell tracking. Electrotaxis studies using the dish-based assay have achieved great success in characterizing the robust dynamical cell migration responses to electric fields in many different cell systems (Sato et al. 2007; Djamgoz et al. 2001; Zhao et al. 2006; Song et al. 2007; Gamboa et al. 2010). We in a previous study modified the transwell assay to assess electrotaxis of different human peripheral blood leukocyte subsets to electric fields by inserting platinum electrodes to the upper and bottom well of the transwell (Lin et al. 2008). This assay inherits the high-throughput feature of the transwell assay for fast screening studies but suffers from nonuniform and less-controlled electric field applications. Similar to conventional chemotaxis assays, the relatively large size of the electrotaxis assays described above is the main limiting factor to control electric field application.

7.1.3 Microengineered devices for cell migration studies

Advanced cell migration investigation requires precise configurations of the cellular microenvironments and the ability to manipulate it when needed in in vitro cell migration systems. In the same time, experimental throughput, reagent consumption, and the ability for sophisticated cell migration analysis are other key demands for cell migration assays by the researchers. Microengineering technologies enable the development of new cell migration devices that help address the above-discussed issues (Whitesides 2006). Microfluidic devices found their playground in cell migration studies over the past fifteen years or so (Li and Lin 2011). Particularly, various gradient-generating microfluidic devices and systems have been developed for chemotaxis studies. In our most recent review, we identified over a hundred published chemotaxis studies using microfluidic devices in the past six years (Wu et al. 2013b). These studies demonstrated the unique strength of lab-on-chip analysis for a broad range of chemotaxis studies including high-throughput parallel experimentation, flexible gradient generation, and cell migration testing in 2-D and 3-D environments, cell migration studies in coexisting gradients, biomimicking of transmigration process using co-culture chips, effect of spatial confinement on cell migration and chemotaxis, and finally development of integrated microfluidic chemotaxis analysis systems more suitable for end research and clinical users.

Here we turn our attention to the microengineered electrotaxis assays. In recent years, researchers found the benefits of on-chip electrotaxis analysis. In general, the electrotaxis chips can produce well-defined electric fields and reduce the joule heating effect in the microchannels while allowing real-time cell tracing for quantitative cell migration analysis (Wu and Lin 2014). In the early development, the electrotaxis chips with a simple straight channel design were fabricated in different materials such as plastic and elastic polymer (Lin et al. 2008; Li et al. 2011). These chips were successfully used for studying electrotactic migration or orientation of different cell types and animal models such as T lymphocytes, yeast cells, and *Caenorhabditis elegans* (*C. elegans*) (Lin et al. 2008; Minc and Chang 2010; Rezai et al. 2010a). Recently, new devices with more advanced designs and new experimental strategies were developed to improve experimental throughput and controls. These new developments enabled a range of new electrotaxis studies and applications (Wu and Lin 2014). To name a few, the microfluidic chips allow electrotaxis experiments in more physiologically relevant 3-D environments (Sun et al. 2012b); multiple electric field strengths can be configured on a single chip for parallel comparison of cancer cell electrotaxis (Huang et al. 2009); we have developed a microfluidic device to study lymphocyte migration in competing chemical gradients and electric fields (Li et al. 2012a); polydimethylsiloxane (PDMS) pillar substrate has been integrated to the electrotaxis assay for cell traction measurement during collective epithelial cell electrotaxis (Li et al. 2012b); narrow microfluidic channels were used to study the confinement effect on fibroblast electrotaxis (Huang et al. 2013); finally, strategies have been developed for efficient on-chip sorting of *C. elegans* based on electrotaxis (Rezai et al. 2012; Manière et al. 2011; Han et al. 2012).

In the rest of this chapter, we will review the fabrication methods of different microengineered electrotaxis devices, experimental setup and analysis methods, as well as their applications for cell studies. The chapter concludes with our personal perspective on the issues of this research area and proposed directions for future development.

7.2 Microengineered devices for electrotaxis analysis

The conventional electrotaxis chamber is prepared in a petri dish (Zhao 2009). The chamber is fabricated by placing two stripes of glass in parallel on the bottom surface of a petri dish as the walls. Then another piece of cover slip is placed on the top of the walls to cover the chamber. Despite its wide use, the petri dish–based assays are limited in miniaturization and electric field control. Microfluidic devices have become popular tools for cell migration studies due to advantages in their small size, cellular microenvironmental control, and low reagent consumption (Li and Lin 2011). In recent years, microfluidic devices have been increasingly applied for studying cell migration in electric fields. In this section, we will briefly survey different fabrication methods of microfluidic electrotaxis devices, the general setup for electrotaxis experiment and data analysis methods.

7.2.1 Device fabrication and assembly

Microfluidic devices have been developed using different materials for electrotaxis studies including plastic, glass, and PDMS (Fig. 7.1A–C). The plastic device can be fabricated by cutting out a cell culture channel in a double-sided adhesive plastic film (Fig. 7.1A). Then the plastic film is sandwiched between the cell culture substrate and a top cover (Lin et al. 2008; Huang et al. 2009). This type of devices is suitable for mass fabrication at low cost. However, the adhesive can be toxic to cells and thus is less suitable for long-term experiments. In some studies, a polymethylmethacrylate (PMMA) assembly was used to connect the cell culture chamber and the reagent ports (Huang et al. 2009; Sun et al. 2012b). The glass device is fabricated by etching the microfluidic channel into a glass substrate, and the channel is sealed by bonding it to a glass cover (Li et al. 2011). The glass device has defined rigid channels with integrated on-chip electrodes for electric field application (Fig. 7.1B). The drawbacks include more complicated fabrication procedures, relatively high cost, and the use of corrosive chemicals. PDMS is widely used for microfluidic device fabrication due to its advantages in fast prototyping, channel dimension control, compatibility for long-term cell culture, and optical transparency (Fig. 7.1C). The PDMS devices are fabricated using the soft lithography technique (Li et al. 2011). It is basically done by PDMS molding to a photoresist master followed by bonding the PDMS replica to a substrate to seal the channel.

7.2.2 Electric field application

Introducing electric field into the cell chamber is critical for doing electrotaxis experimentation. This is typically done by making two medium reservoirs at the two ends of the cell channel for insertion of electrodes (e.g., platinum electrodes) (Lin et al. 2008). The electrodes are connected to an adjustable power supply. To avoid direct contact of electrodes to cell culture, agar salt bridges are used to connect the medium reservoirs with inserted Ag/AgCl electrodes and the medium wells in the device

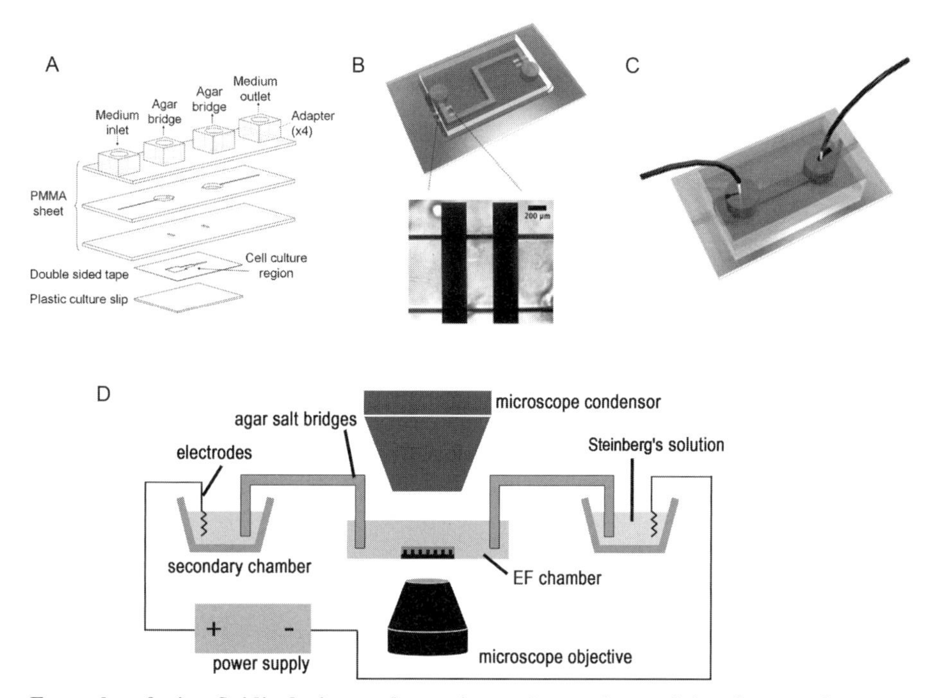

Figure 7.1 **Examples of microfluidic devices and experimental setup for studying electrotaxis.** (A) Plastic microfluidic device. (B) Glass microfluidic device. (C) PDMS microfluidic device. (D) Illustration of experimental setup.
(A) reproduced with permission from Elsevier (Huang et al. 2009); (B) and (C) reproduced with permission from The Royal Society of Chemistry (Li et al. 2011); (D) reproduced with permission from Springer (Li et al. 2012b).

(Fig. 7.1D) (Li et al. 2012b). In a recent study, a feedback loop with a computer-controlled voltage source is used to maintain the desired electrical voltage applied to the device (Song et al. 2013).

7.2.3 Electrotaxis experiment setup

To observe cell migration in electric field, the microfluidic device is placed under a microscope. A typical setup is shown in Fig. 7.1D. If required, a fresh medium can be continuously infused into the device. For long-term observation, a microscope incubation chamber is used to maintain the temperature and CO_2 injection for the electrotaxis device. Then cell migration in the device in response to the applied electric field is recorded by time-lapse microscopy.

7.2.4 Data analysis methods

The positions of cells from the time-lapse images or videos are tracked using specialized software such as ImageJ and MetaMorph, and the tracking data are used for quantitative

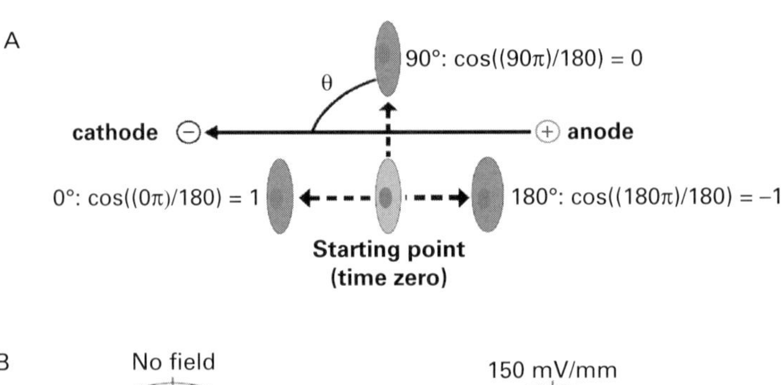

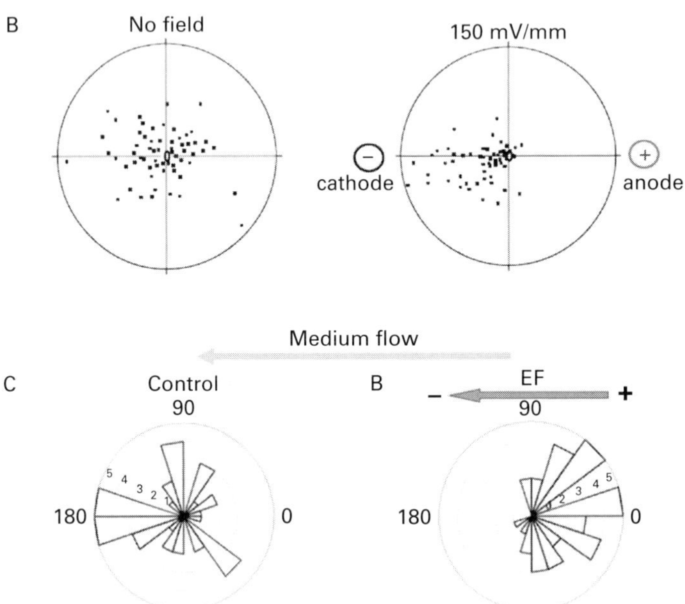

Figure 7.2 **Illustration of data analysis methods.**
(A) Definition of directedness; (B) scatter plot; (C) rose diagram.
(A) and (B) reproduced with permission from Springer (Tai et al. 2009); (C) reproduced with permission from Springer (Wu et al. 2013a).

electrotaxis analysis. The migration rate and directedness are two widely used parameters. The migration rate (or speed) is defined as the cell displacement (or distance) divided by the experiment time. The directedness is defined as the average cosine of the cell migration angle with respect to the direction of the applied electric field (Fig. 7.2A). The directedness is −1 for a cell moving toward the anode, and +1 for a cell moving toward the cathode. For a group of randomly migrating cells, the directedness is 0. The electrotactic index, which is the ratio of the displacement of the cells toward the electric field to the total migration distance, is also used to measure directional migration to electric field (Wu et al. 2013a; Li et al. 2011). The scatter graph that plots the distribution of the cells' final positions with respect to a common origin of their starting positions can be used to visualize the trend of cell migration in electric field (Fig. 7.2B) (Tai et al. 2009). Similarly, the rose diagram, which plots the distribution of cell migration angles, can be used to analyze directionality (Fig. 7.2C) (Wu et al. 2013a).

7.3 Applications of microengineered devices for electrotaxis studies

7.3.1 Cancer cell electrotaxis studies

Physiological electric fields can arise between tumors and the surrounding tissues, and these electric fields can potentially promote and guide metastatic cancer cell migration (Djamgoz et al. 2001). Previous studies have demonstrated electrotaxis of different types of cancer cells in vitro using conventional electrotaxis assays (Pu et al. 2007). Huang et al. (2009) developed a plastic microfluidic device, which can generate multiple electric fields, to study long-term electrotaxis of two different lung cancer cell lines (Fig. 7.3A) Wang et al. (2011) used a similar device and observed the biased growth of filopodia of lung cancer cells towards the cathode. This asymmetric filopodium growth was found to be related to the cathodal-biased distribution of epidermal growth factor receptors (EGFR). To better mimic the 3-D microenvironment in vivo, Sun et al. (2012b) incorporated 3-D scaffold in a microfluidic device and studied the electrotactic response of lung cancer cells in the 3-D scaffolds (Fig. 7.3B). The authors demonstrated that cell migration under EF is different in 2-D and 3-D environments, possibly due to different cell morphology and substrate stiffness. To investigate the mechanism underlying cancer cell electrotaxis, Tsai et al. (2013) developed a large-scale dcEF stimulation device, which can collect a large amount of cellular proteins upon electrical stimulation for protein-based analysis to evaluate electrotactic signaling for lung adenocarcinoma cells. We used a PDMS microfluidic device to study electrotaxis of MDA-MB-231 cells, a human metastatic breast cancer cell line (Fig. 7.3C) (Wu et al. 2013a). Our results showed the anode-directing migration of the MDA-MB-231 cells and that dcEF can induce EGFR polarization and increase intracellular calcium ions in these cells.

7.3.2 Immune cell electrotaxis studies

Previous studies using conventional electrotaxis assays have shown that the applied electric field within the physiological range can induce directional migration of immune cells such as mouse neutrophils (Zhao et al. 2006). Using a plastic microfluidic device, we showed that memory T cells from human peripheral blood migrate toward the cathode of an applied electric field (Lin et al. 2008). In a later study, we further developed two microfluidic devices, one in PDMS and the other in glass, to study electrotaxis of activated human blood T cells (Li et al. 2011). Our results showed that activated T cells migrate toward the cathode of applied DC electric fields (Fig. 7.4A). These findings suggest electrotaxis as a novel guiding mechanism

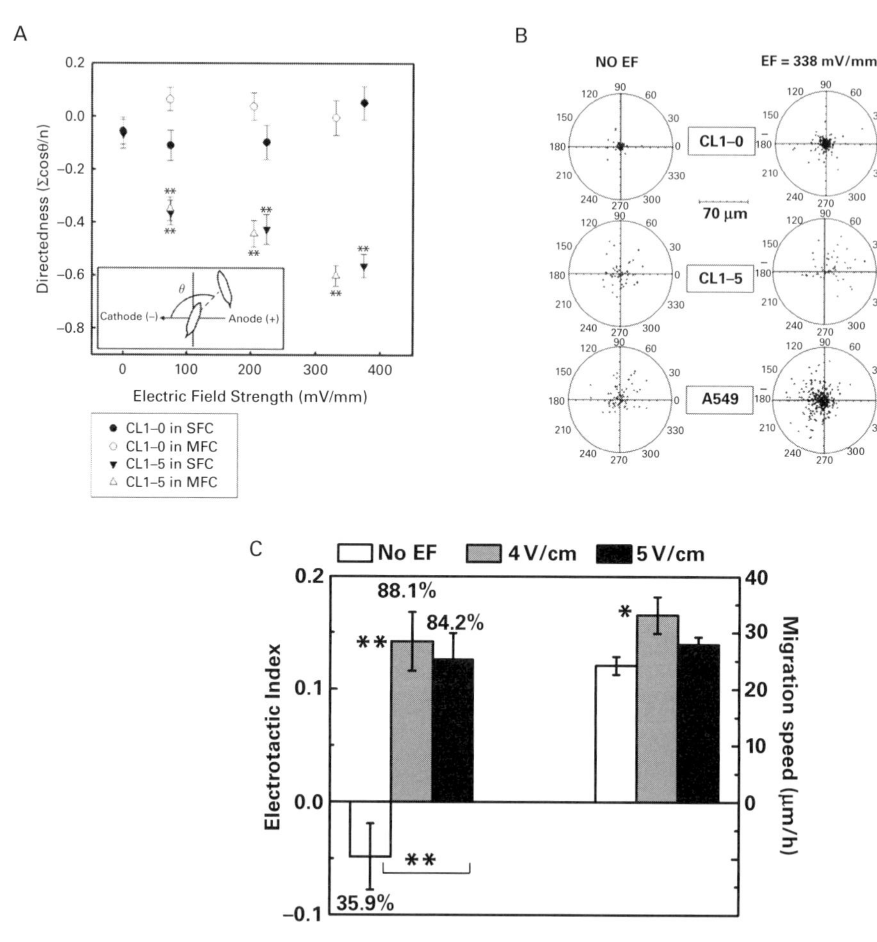

Figure 7.3 **Representative cancer cell electrotaxis studies.**
(A) Directedness of CL1–0 and CL1–5 cells after 2 hrs of EF stimulation in a single field chip (SFC) and a multifield chip (MFC). (B) Polar plots of cancer cells (CL1–0, CL1–5, and A549) migration after 2 hrs with and without the applied EF in a 3-D device. (C) Comparison of electrotactic index and speed of MDA-MB-231 cells in microfluidic devices. ** $p < 0.01$, and * $p < 0.05$.
(A) reproduced with permission from Elsevier (Huang et al. 2009); (B) reproduced with permission from American Institute of Physics (Sun et al. 2012b); (C) reproduced with permission from Springer (Wu et al. 2013a).

for immune cell migration. To study the competition of chemotaxis and electrotaxis in directing immune cell migration, we developed a PDMS microfluidic device that can generate well-controlled co-existing chemical gradients and dcEF (Li et al. 2012a). Our results showed a reduction of both chemotaxis and electrotaxis of activated human blood T cells in competing chemical and electric fields. Furthermore, more cells migrated toward the cathode of a dcEF in the presence of a competing chemokine gradient, supporting the potential of electric field overriding other guiding cues for immune cell migration (Fig. 7.4B).

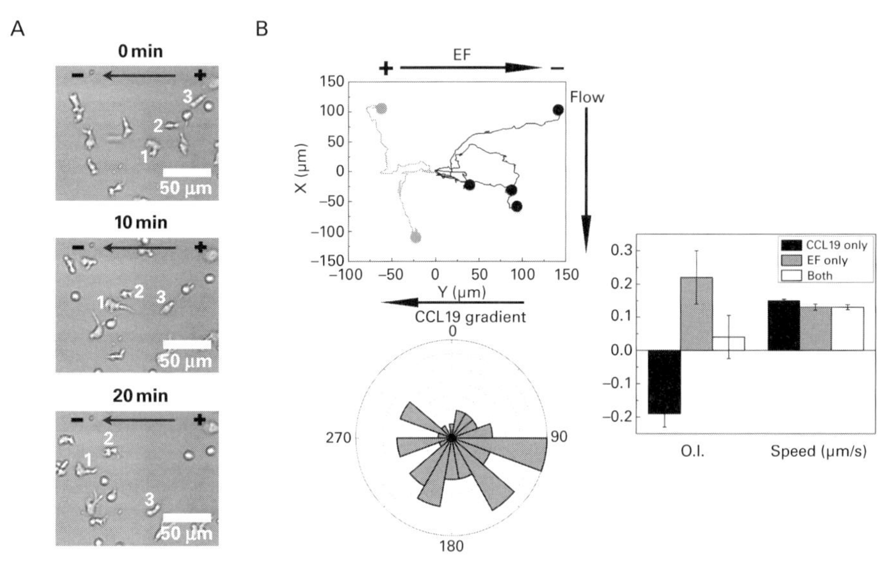

Figure 7.4 **Representative immune cell electrotaxis studies.**
(A) Electrotaxis of activated T cells from human blood using a PDMS microfluidic chip.
(B) Activated T cell migration in competing CCL19 gradients and dcEF in the microfluidic device.
(A) reproduced with permission from The Royal Society of Chemistry (Li et al. 2011);
(B) reproduced with permission from American Institute of Physics (Li et al. 2012a).

7.3.3 Tissue cell electrotaxis studies

Many tissue cells such as fibroblasts and epithelial cells contribute to the wound-healing process. Physiological electric fields mediate recruitment of these cells to the injured tissues for wound repair. In vitro studies have demonstrated that both dermal fibroblast and different epithelial cells can migrate toward an applied dcEF within the range of physiological strength (Guo et al. 2010; Zhao et al. 1996; Zhao et al. 2006). Beside the wound-produced dcEF, shear stress can be induced by interstitial fluid loss toward the wound. Song et al. (2013) developed a PDMS microfluidic device to study the collaborative effects of electric field and fluid shear stress on fibroblast migration in single, simultaneous, and sequential modes. Simultaneous dcEF and flow stimulations enhance directional cell migration. When dcEF and flow are applied sequentially, cell migration is affected by the current stimulation as well as the previous stimulating conditions. Huang et al. (2013) used a microfluidic chip to study the effect of spatial confinement on electrotaxis and motility of fibroblasts (Fig. 7.5A). The results show that confinement leads to increased cell velocity with or without an electric field. Beside single cell electrotaxis, collective migration of tissue cells in response to electrical stimuli is of growing interest to researchers over recent years. Li et al. (2012b) studied collective electrotaxis of large epithelial sheets. The study demonstrated that cells in monolayer migrated more efficiently and directionally toward the electric field than cells in isolation, and the collective migration is E-cadherin dependent. Furthermore, a PDMS pillar substrate was integrated into

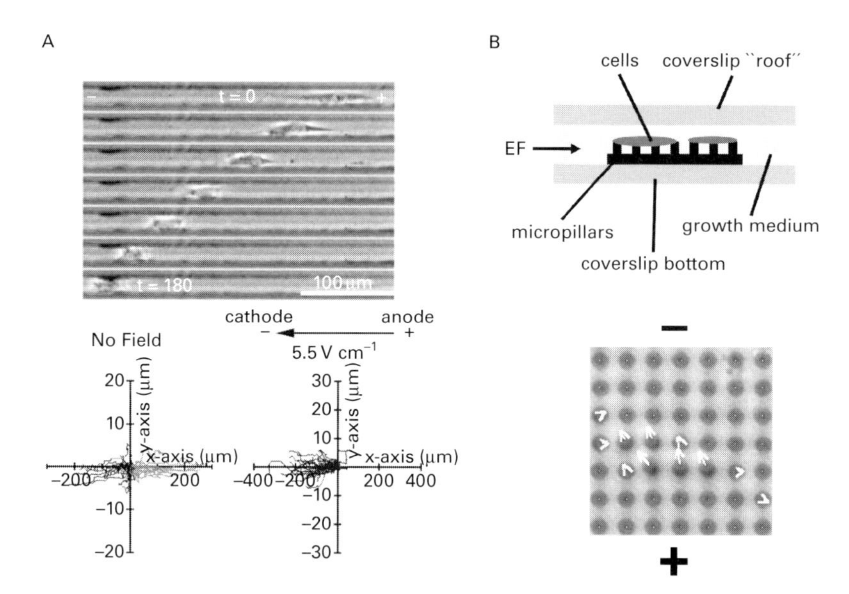

Figure 7.5 **Representative tissue cell electrotaxis studies.**
(A) Confined fibroblast electrotaxis in 20 μm channels. (B) Traction force measurement for collective electrotaxis of an epithelial cell monolayer sheet using an electrotaxis chamber integrated with a PDMS pillar substrate.
(A) reproduced with permission from PLOS (Huang et al. 2013); (B) reproduced with permission from Springer (Li et al. 2012b).

the electrotaxis assay for measuring coordinated traction forces generated by cells in collective electrotaxis (Fig. 7.5B). To better mimic the wound microenvironment, Sun et al. (2012a) integrated the barrier wound-healing assay with a microfluidic chip to study the migration of fibroblasts under different conditions such as serum, electric field, and wound-healing promoting drugs.

7.3.4 *Caenorhabditis elegans* electrotaxis studies

Caenorhabditis elegans is a widely utilized model organism in biology research owing to its useful features such as small size, transparency, short life cycle, and amenability for genetic manipulations (Li and Lin 2011). Electrotaxis of *C. elegans* has been widely studied using microfluidic devices. A PDMS microfluidic device was developed by Rezai et al. (2010a) to study the behavior of *C. elegans* in constant dcEF. The results demonstrated robust and sensitive movement of the worms toward the cathode without physiological and behavioral side effects. In a later study, a similar device was used to study the worms' movement in response to alternating electric fields (Fig. 7.6A) (Rezai et al. 2010b). The authors found that 1 Hz and higher frequency of the electric field can effectively localize worms in the channel. In

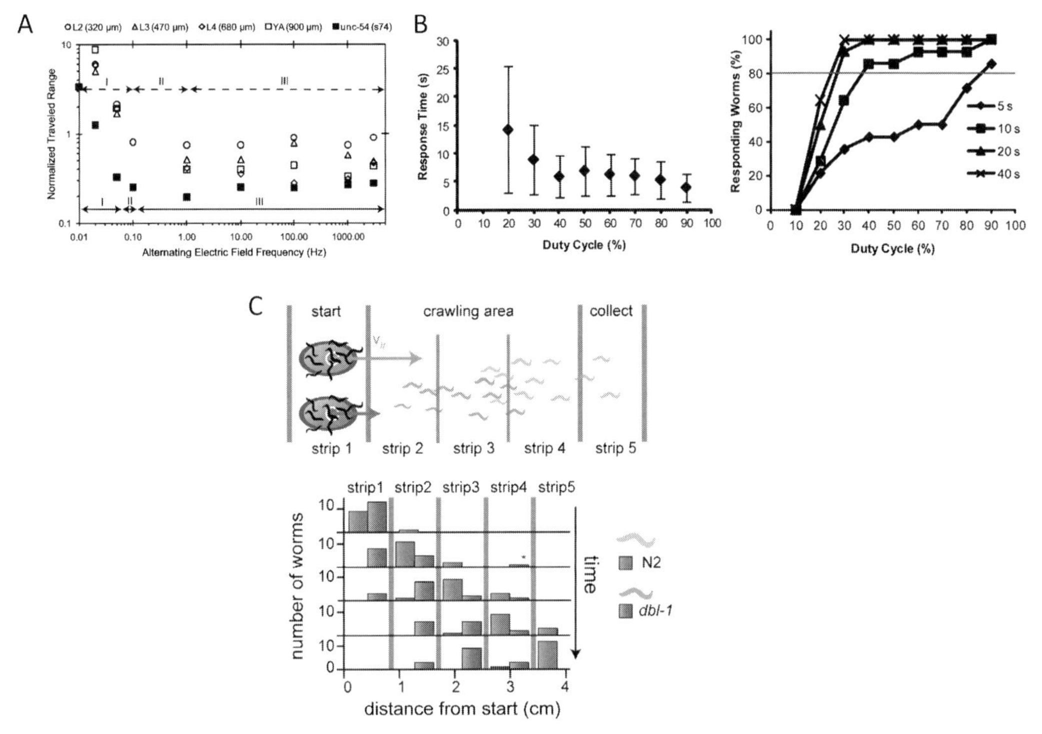

Figure 7.6 **Representative *C. elegans* electrotaxis studies.**
(A) Electrotaxis of *C. elegans* in different stages to alternating EF. (B) Effect of duty cycle on the turning response time and percentage of responding *C. elegans* to pulse dcEF. (C) Population sorting of a mixture of wild-type and mutant worms.
(A) reproduced with permission from American Institute of Physics (Rezai et al. 2010b); (B) reproduced with permission from American Institute of Physics (Rezai et al. 2011); (C) reproduced with permission from PLOS (Manière et al. 2011).

another study, the worms' migration in response to a pulse dcEF was demonstrated (Fig. 7.6B) (Rezai et al. 2011). Comparing to electrotaxis in stable dcEF, electrotaxis in the pulse dcEF has comparable speed, but is different in other parameters such as turning response time and the number of responding worms. Other studies used microfluidic devices to sort *C. elegans* of different ages, stages, or mutations based on their corresponding different electrotactic motility (Fig. 7.6C) (Rezai et al. 2012; Manière et al. 2011; Han et al. 2012).

7.3.5 Stem cell and yeast cell electrotactic orientation studies

Instead of electrotactic migration, some cell types showed orientation responses to an applied electric field. Microengineered devices have been applied to demonstrate the electrotactic orientation in different cells and to elucidate the mechanisms. We used a simple PDMS electrotaxis device to study the electrotactic orientation of rat adipose-derived stem cells (ASCs) in response to dcEF (Wadhawan et al. 2012). Over two hours

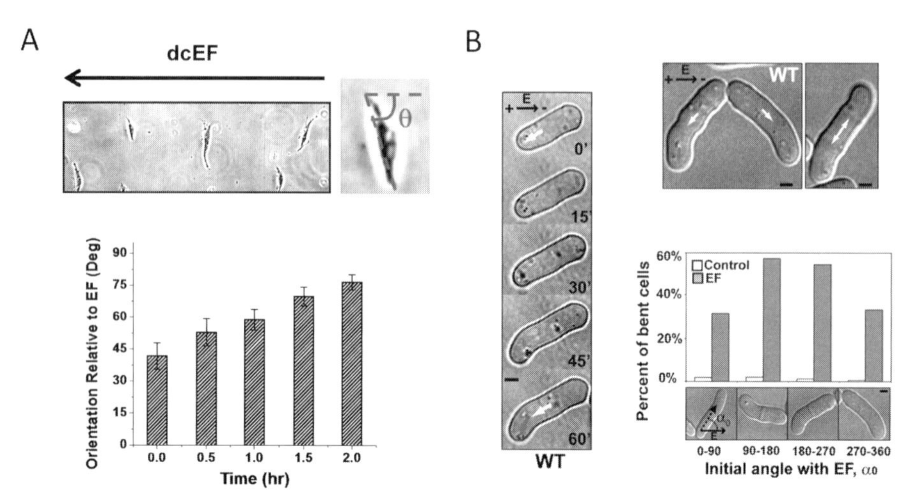

Figure 7.7 **Stem cell and yeast cell electrotactic orientation studies.** (A) Rat adipose-derived stem cell orientation in an applied dcEF in a PDMS microfluidic device. (B) Reorientation of fission yeast cell in a dcEF in a PDMS microfluidic device. (A) reproduced with permission from The Royal Society of Chemistry (Wadhawan et al. 2012); (B) reproduced with permission from Elsevier (Minc and Chang 2010).

of dcEF exposure, the majority of the ASCs oriented perpendicular to the direction of the dcEF (Fig. 7.7A). In an earlier study, Minc and Chang (2010) used a similar device to study dcEF-induced cell polarization in fission yeast *Schizosaccharomyces pombe.* They found the rod-shaped cells reorient in the orthogonal direction to the applied dcEF and produce a bent morphology (Fig. 7.7B).

7.4 Concluding remarks

In this chapter, we review the growing development of microengineered tools for studying cell migration in electric fields. We have no doubt that such development will continue to grow in the future, and that these new microdevices and systems will enable various advanced studies to answer interesting and important questions in electric field–directed cell migration.

With the rapid development of this research area, issues were identified and solved on an ongoing basis, while some long-standing issues remain to be addressed. Although multifield electrotaxis chips were developed (Huang et al. 2009), higher experimental throughput is required to meet the requirement of parallel electrotaxis testing under different conditions with repeats on a single chip. Such a high-throughput electrotaxis device is in principle doable but requires careful design to ensure well-controlled parallel electric field applications to multiple channels on the chip.

The three-dimensional extracellular matrix (ECM) has been incorporated to conventional electrotaxis assays to better mimic the physiological cell migration environments (Zhao et al. 2013). To date, many 3-D microfluidic devices have been developed for cell migration studies (Saadi et al. 2007; Abhyankar et al. 2008; Haessler et al. 2009; Sudo et al. 2009; Zervantonakis et al. 2010; Amadi et al. 2010). Therefore, it is anticipated that more advanced 3-D microfluidic electrotaxis chips will be developed in the near future. On the other hand, the complex composition of 3-D ECM will raise the difficulty for controlling electric field applications, which needs to be taken into consideration in designing the 3-D devices.

One of the long-standing issues in electrotaxis studies is that it is difficult to decouple the direct effect of electric fields in directing cell migration from other accompanying influencing factors due to bioelectrochemical changes of the cell environments. In this regard, careful considerations are needed to minimize the side effects of electric field applications such as by-products of electrodes and cells as well as pH changes in the medium for on-chip electrotaxis analysis. In addition, coexisting electric fields and other guiding cues are expected in tissue microenvironments, and understanding how cells respond to multiple coexisting guiding signals is fundamentally important for directed cell migration. Coexisting field microfluidic devices have been developed to study the competition of electrotaxis against chemotaxis and flowtaxis (Song et al. 2013; Li et al. 2012a). However, more accurate control of coexisting fields is needed to better examine the coordination by multiple guiding factors.

As we pointed out previously (Li and Lin 2011), many microfluidic electrotaxis devices are similar in their design and operating principle, involving a simple microfluidic channel wherein defined electric fields are applied through external or on-chip electrodes (Li et al. 2011; Minc and Chang 2010). Therefore, it is feasible and beneficial to standardize the format of microfluidic electrotaxis devices. On the other hand, advanced applications in electrotaxis studies will require more flexible chip designs to better control the electrical environments for cell migration and to configure complex cellular guiding environments.

We believe research enabled by the microengineered tools will lead to a better understanding of electric field–directed cell migration. The improved knowledge of electrotaxis will inspire new noninvasive medical interventions for treating physiological problems and diseases such as wound healing and cancers by electric field applications.

In conclusion, studying electric field–directed cell migration using microengineered tools is an emerging research area, which holds great promise to advance the science of directed cell migration and to facilitate development of new therapeutic applications.

Acknowledgments

The development of this manuscript is supported by a Discovery Grant from the Natural Sciences and Engineering Research Council of Canada (NSERC) to Francis Lin. Lin thanks the Winnipeg Rh Institute Foundation and the University of Manitoba for a Rh Award; Jiandong Wu thanks the Manitoba Health Research Council (MHRC) for a graduate fellowship.

References

Abhyankar, V. V., Toepke, M. W., Cortesio, C. L. Lokuta, M. A., Huttenlocher, A. and Beebe, D. J. (2008). "A platform for assessing chemotactic migration within a spatiotemporally defined 3D microenvironment." *Lab on a Chip* **8**: 1507–1515.

Amadi, O., Steinhauser, M., Nishi, Y., Chung, S., Kamm, R., Mcmahon, A. and Lee, R. (2010). "A low resistance microfluidic system for the creation of stable concentration gradients in a defined 3D microenvironment." *Biomedical Microdevices* **12**: 1027–1041.

Baggiolini, M. (1998). "Chemokines and leukocyte traffic." *Nature* **392**: 565–568.

Boyden, S. (2008). "The chemotactic effect of mixtures of antibody and antigen on polymorphonuclear leucocytes." *J Exp Med* **115**: 453–466.

Campbell, J. and Butcher, E. (2000). "Chemokines in tissue-specific and microenvironment-specific lymphocyte homing." *Curr Opin Immunol* **12**: 336–41.

Cortese, B., Palamà, I. E., D'amone, S. and Gigli, G. (2014). "Influence of electrotaxis on cell behaviour." *Integrative Biology* **6**: 817–830.

Delong, E. F., Frankel, R. B. and Bazylinski, D. A. (1993). "Multiple evolutionary origins of magnetotaxis in bacteria." *Science* **259**: 803.

Djamgoz Mba, Mycielska, M., Madeja, Z., Fraser, S. and Korohoda, W. (2001). "Directional movement of rat prostate cancer cells in direct-current electric field: involvement of voltagegated Na+ channel activity." *J Cell Sci* **114**: 2697–2705.

Gamboa, O. L. Pu, J., Townend, J., Forrester, J. V., Zhao, M., McCaig, C. and Lois, N. (2010). "Electrical estimulation of retinal pigment epithelial cells." *Exp Eye Res* **91**: 195–204.

Guo, A., Song, B., Reid, B., Gu, Y., Forrester, J. V., Jahoda, C. A. B. and Zhao, M. (2010). "Effects of physiological electric fields on migration of human dermal fibroblasts." *Journal of Investigative Dermatology* **130**: 2320–2327.

Haessler, U., Kalinin, Y., Swartz, M. A. and Wu, M. (2009). "An agarose-based microfluidic platform with a gradient buffer for 3D chemotaxis studies." *Biomed Microdevices* **11**: 827–835.

Han, B., Kim, D., Hyun Ko, U. and Shin, J. H. (2012). "A sorting strategy for C. elegans based on size-dependent motility and electrotaxis in a micro-structured channel." *Lab on a Chip* **12**: 4128–4134.

Hsu, S., Thakar, R., Liepmann, D. and Li, S. (2005). "Effects of shear stress on endothelial cell haptotaxis on micropatterned surfaces." *Biochemical and Biophysical Research Communications* **337**: 401–409.

Huang, C., Cheng, J., Yen, M. and Young, T. (2009). "Electrotaxis of lung cancer cells in a multiple-electric-field chip." *Biosens Bioelectron* **24**: 3510–3516.

Huang, Y.-J., Samorajski, J., Kreimer, R. and Searson, P. C. (2013). "The influence of electric field and confinement on cell motility." *PLoS One* **8**: e59447.

Li, J., et al. (2012a). "Microfluidic device for studying cell migration in single or co-existing chemical gradients and electric fields." *Biomicrofluidics* **6**: 024121.

Li, L., et al. (2012b). "E-cadherin plays an essential role in collective directional migration of large epithelial sheets." *Cellular and Molecular Life Sciences* **69**: 2779–2789.

Li, J., and Lin, F. (2011). "Microfluidic devices for studying chemotaxis and electrotaxis." *Trends in Cell Biology* **21**: 489–497.

Li, J., Nandagopal, S., Wu, D., Romanuik, S. F., Paul, K., Thomson, D. J. and Lin, F. (2011). "Activated T lymphocytes migrate toward the cathode of DC electric fields in microfluidic devices." *Lab on a Chip* **11**: 1298–1304.

Lin, F., Baldessari, F., Gyenge, C. C., Sato, T., Chambers, R. D., Santiago, J. G. and Butcher, E. C. (2008). "Lymphocyte electrotaxis in vitro and in vivo." *The Journal of Immunology* **181**: 2465–2471.

Lohof, A., Quillan, M., Dan, Y. and Poo, M. (1992). "Asymmetric modulation of cytosolic cAMP activity induces growth cone turning." *J. Neurosci* **12**: 1253–1261.

Luster, A., Alon, R. and von Andrian, U. (2005). "Immune cell migration in inflammation: present and future therapeutic targets." *Nat Immunol* **6**: 1182–90.

Manière, X., Lebois, F., Matic, I., Ladoux, B., Di Meglio, J.-M. and Hersen, P. (2011). "Running worms: C. elegans self-sorting by electrotaxis." *PLoS One* **6**: e16637.

McCaig, C. D., Allan, D. W., Erskine, L. Rajnicek, A. M. and Stewart, R. (1994). "Growing nerves in an electric field." *Neuroprotocols* **4**: 134–141.

McCaig, C. D., Rajnicek, A., Song, B. and Zhao, M. (2005). "Controlling cell behavior electrically: current views and future potential." *Physiol Rev* **85**: 943–978.

Mccormick, K. E., Gaertner, B. E., Sottile, M., Phillips, P. C. and Lockery, S. R. (2011). "Microfluidic devices for analysis of spatial orientation behaviors in semi-restrained Caenorhabditis elegans." *PLoS One* **6**: e25710.

Minc, N. and Chang, F. (2010). "Electrical control of cell polarization in the fission yeast Schizosaccharomyces pombe." *Current Biology* **20**: 710–716.

Mosadegh, B., Saadi, W., Wang, S. J. and Jeon, N. L. (2008). "Epidermal growth factor promotes breast cancer cell chemotaxis in CXCL12 gradients." *Biotechnol Bioeng* **100**: 1205–1213.

Muller, A., Homey, B., Soto, H., Ge, N., Catron, D., Buchanan, M. E., Mcclanahan, T., et al. (2011). "Involvement of chemokine receptors in breast cancer metastasis." *Nature* **410**: 50–56.

Mycielska, M. and Djamgoz, M. (2004). "Cellular mechanisms of direct-current electric field effects: galvanotaxis and metastatic disease." *J Cell Sci* **117**: 1631–1639.

Nelson, R. D., Quie, P. G. and Simmons, R. L. (1975). "Chemotaxis under agarose: a new and simple method for measuring chemotaxis and spontaneous migration of human polymorphonuclear leukocytes and monocytes." *J Immunol* **115**: 1650–1656.

Pu, J., McCaig, C. D., Cao, L. Zhao, Z., Segall, J. E. and Zhao, M. (2007). "EGF receptor signalling is essential for electric-field-directed migration of breast cancer cells." *J Cell Sci* **120**: 3395–3403.

Rezai, P., Salam, S., Selvaganapathy, P. R. and Gupta, B. P. (2011). "Effect of pulse direct current signals on electrotactic movement of nematodes Caenorhabditis elegans and Caenorhabditis briggsae." *Biomicrofluidics* **5**: 44116–441169.

Rezai, P., Salam, S., Selvaganapathy, P. R. and Gupta, B. P. (2012). "Electrical sorting of Caenorhabditis elegans." *Lab on a Chip* **12**: 1831–1840.

Rezai, P., Siddiqui, A., Selvaganapathy, P. R. and Gupta, B. (2010a). "Electrotaxis of Caenorhabditis elegans in a microfluidic environment." *Lab Chip* **10**: 220–226.

Rezai, P., Siddiqui, A., Selvaganapathy, P. R. and Gupta, B. P. (2010b). "Behavior of Caenorhabditis elegans in alternating electric field and its application to their localization and control." *Applied Physics Letters* **96**: 153702–153703.

Saadi, W., Rhee, S. W., Lin, F., Vahidi, B., Chung, B. G. and Jeon, N. L. (2007). "Generation of stable concentration gradients in 2D and 3D environments using a microfluidic ladder chamber." *Biomed Microdevices* **9**: 627–635.

Sato, M. J., Ueda, M., Takagi, H., Watanabe, T. M. and Yanagida, T. (2007). "Input-output relationship in galvanotactic response of Dictyostelium cells." *Biosystems* **88**: 261–272.

Song, B., Gu, Y., Pu, J., Reid, B., Zhao, Z. and Zhao, M. (2007). "Application of direct current electric fields to cells and tissues in vitro and modulation of wound electric field in vivo." *Nat Protoc* **2**: 1479–1489.

Song, S., Han, H., Ko, U. H., Kim, J. and Shin, J. H. (2013). "Collaborative effects of electric field and fluid shear stress on fibroblast migration." *Lab on a Chip* **13**(8): 1602–1611.

Sudo, R., Chung, S., Zervantonakis, I. K., Vickerman, V., Toshimitsu, Y., Griffith, L. G. and Kamm, R. D. (2009). "Transport-mediated angiogenesis in 3D epithelial coculture." *The FASEB Journal* **23**: 2155–2164.

Sun, Y.–S., et al. (2012a). "In vitro electrical-stimulated wound-healing chip for studying electric field-assisted wound-healing process." *Biomicrofluidics* **6**: 034117.

Sun, Y. S., et al. (2012b). "Electrotaxis of lung cancer cells in ordered three-dimensional scaffolds." *Biomicrofluidics* **6**: 14102–14114.

Tai, G., Reid, B., Cao, L. and Zhao, M. (2009). "Electrotaxis and wound healing: experimental methods to study electric fields as a directional signal for cell migration." *Methods Mol Biol* **571**: 77–97.

Tsai, H.-F., Huang, C.-W., Chang, H.-F., Chen, J. J., Lee, C.-H. and Cheng, J.-Y. (2013). "Evaluation of EGFR and RTK signaling in the electrotaxis of lung adenocarcinoma cells under direct-current electric field stimulation." *PLoS One* **8**: e73418.

Wadhawan, N., Kalkat, H., Natarajan, K., Ma, X., Gajjeraman, S., Nandagopal, S., Hao, N., et al. (2012). "Growth and positioning of adipose-derived stem cells in microfluidic devices." *Lab on a Chip* **12**: 4829–4834.

Wang, C.-C., Kao, Y.-C., Chi, P.-Y., Huang, C.-W., Lin, J.-Y., Chou, C.-F., Cheng, J.-Y., et al. (2011). "Asymmetric cancer-cell filopodium growth induced by electric-fields in a microfluidic culture chip." *Lab on a Chip* **11**: 695–699.

Wang, S.-J., Saadi, W., Lin, F., Minh-Canh Nguyen, C. and Li Jeon, N. (2004). "Differential effects of EGF gradient profiles on MDA-MB-231 breast cancer cell chemotaxis." *Experimental Cell Research*, **300**, 180–189.

Whitesides, G. M. (2006). "The origins and the future of microfluidics." *Nature* **442**: 368–373.

Wu, D., et al. (2013a). "DC electric fields direct breast cancer cell migration, induce EGFR polarization, and increase the intracellular level of calcium ions." *Cell Biochemistry and Biophysics* **67**: 1115–1125.

Wu, D. and Lin, F. (2011). "A receptor-electromigration-based model for cellular electrotactic sensing and migration." *Biochemical and Biophysical Research Communications* **411**: 695–701.

Wu, J., et al. (2013b). "Recent developments in microfluidics-based chemotaxis studies." *Lab on a Chip* **13**: 2484–2499.

Wu, J. and Lin, F. (2014). "Recent developments in electrotaxis assays." *Advances in Wound Care* **3**: 149–155.

Zervantonakis, I., Chung, S., Sudo, R., Zhang, M., Charest, J. and Kamm, R. (2010). "Concentration gradients in microfluidic 3D matrix cell culture systems." *International Journal of Micro-Nano Scale Transport* **1**: 27–36.
Zhao, M. (2009). "Electrical fields in wound healing-An overriding signal that directs cell migration." *Semin Cell Dev Biol* **20**: 674–682.
Zhao, M., Agius-Fernandez, A., Forrester, J. and McCaig, C. (1996). "Directed migration of corneal epithelial sheets in physiological electric fields." *Invest Ophthalmol Vis Sci* **37**: 2548–2558.
Zhao, M., Dick, A., Forrester, J. and McCaig, C. (1999). "Electric field-directed cell motility involves up-regulated expression and asymmetric redistribution of the epidermal growth factor receptors and is enhanced by fibronectin and laminin." *Mol Biol Cell* **10**: 1259–1276.
Zhao, M., Song, B., Pu, J., Wada, T., Reid, B., Tai, G., Wang, F., et al. (2006). "Electrical signals control wound healing through phosphatidylinositol-3-OH kinase-gamma and PTEN." *Nature* **442**: 457–460.
Zhao, S., Gao, R., Devreotes, P. N., Mogilner, A. and Zhao, M. (2013). "3D arrays for high throughput assay of cell migration and electrotaxis." *Cell biology international* **37**: 995–1002.
Zigmond, S. 1977. "Ability of polymorphonuclear leukocytes to orient in gradients of chemotactic factors." *J Cell Biol* **75**: 606–616.

8 Laser ablation to investigate cell and tissue mechanics in vivo

Teresa Zulueta-Coarasa and Rodrigo Fernandez-Gonzalez

8.1 Introduction

An outstanding question in developmental biology is how cells acquire specific fates that determine their shape, function, and position within an animal. It is generally assumed that cell fates are specified according to intricate patterns of gene expression. But an embryo is also a dynamic system in which cells migrate, push other cells, and pull on their neighbors. Mechanical forces can regulate cell fate specification in tissue culture systems. In vivo, physical forces generated by the cytoskeleton regulate cell behaviors, molecular dynamics, and gene expression. However, the role of mechanical forces in determining cell fates in vivo is currently unclear, largely because it is difficult to measure and manipulate physical forces in living animals.

To investigate the impact of mechanical forces in cell behavior and dynamics, tools have been developed to quantify cell and tissue mechanics. However, some of these techniques are limited by the range of measurable forces, their invasiveness, or simply because they cannot be used in living organisms. Laser ablation has emerged as a tool to infer mechanical forces in vivo. In this approach, a laser is used to sever subcellular, cellular, or supracellular structures. The movements resulting from the ablation event are used to estimate the forces sustained by the ablated structures. Therefore, laser cutting allows measurements of cell and tissue mechanics with minimal disruption of the microenvironment and without establishing physical contact with the cells.

In this chapter, we provide an overview of laser ablation and its most recent applications to the investigation of physical forces in vivo. We begin with a description of the mechanisms by which lasers cut biological samples, and how the responses to laser ablation can be used to extract information about cell mechanics. We examine image analysis and modeling approaches used to quantify laser ablation data, and we discuss practical aspects and limitations that need to be considered when conducting these experiments. We conclude with a review of recent applications of laser ablation to investigate force generation and cell mechanics in living organisms. We propose that the use of laser ablation, in combination with other biophysical methods to build dynamic maps of cell and tissue mechanics during embryonic development, will have direct implications to understand how physical forces regulate cell fate in animal development.

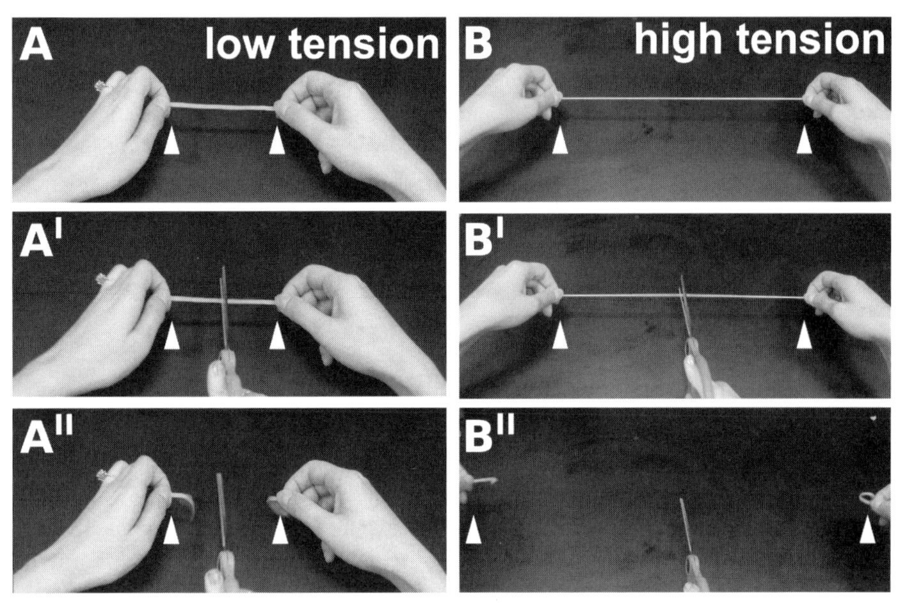

Figure 8.1 **A rubber band model of laser ablation.**
(A–B) A rubber band is held between two hands that exert low (A) or high (B) tension on the rubber band. Arrowheads indicate the position of the hands. (A'–B') The rubber band is cut using scissors. (A''–B'') Immediately after cutting, tension is released, causing a displacement of the hands that is proportional to the tension that the rubber band sustained.

8.2 Method

In recent years, laser-based ablation has been widely used to quantify mechanical properties in living organisms at the subcellular, cellular, and tissue scales. The principle behind laser ablation experiments can be illustrated with a simple analogy (Fig. 8.1). If a rubber band held under low tension is cut with scissors, the elements that hold the rubber band at its ends (the hands in Fig. 8.1) will experience little to no movement (Fig. 8.1A). In contrast, if the rubber band is stretched, cutting the rubber band will result in these same elements moving away from each other (Fig. 8.1B). The velocity of displacement of the elements that hold the rubber band, often referred to as "recoil velocity" or "retraction velocity," depends on the tension sustained by the rubber band. Therefore, filming the response to the cut of the rubber band, and measuring the velocity of displacement at its ends, it is possible to estimate the mechanical properties of the rubber band. In this section, we discuss the mechanisms by which lasers sever biological structures, the methods used to quantify velocities from time-lapse microscopy images, and the physical models necessary to interpret laser ablation data.

8.2.1 Principles of plasma-induced ablation

Biological tissues can reflect, refract, absorb, and scatter light (Niemz 2004). The interactions between light and tissue can result in tissue destruction. For instance, when a tissue is irradiated with a laser, the absorbed energy causes an increase in pressure and temperature in the sample that can lead to vaporization – microexplosions produced by the expansion of water in the tissue – or boiling. These mechanisms of tissue destruction provide little control over the size and morphology of the destroyed region, as the damage typically propagates outside of the targeted area (Vogel and Venugopalan 2003).

Plasma-induced laser ablation causes controlled destruction of biological samples. Plasma is a material containing charged particles that are not bound to specific atoms. The generation of plasma occurs when biological samples are irradiated with lasers at high power density (10^{10}–10^{13} W/cm^2). The high power densities induce an electric field that ionizes the medium (Vogel and Venugopalan 2003). During the ionization process, free electrons that can move in an unrestrained manner are generated. When a certain density of free electrons is achieved, plasma is formed. Three main mechanisms produce free electrons during plasma formation: multiphoton ionization, quantum tunneling, and impact ionization (Quinto-Su and Venugopalan 2007; Schaffer et al. 2001). During multiphoton ionization, several photons are absorbed simultaneously by a molecule, promoting it to an excited state in which electrons have high energy levels. Quantum tunneling refers to electrons that spontaneously appear in an excited state, which according to quantum mechanics, happens with a finite probability. Finally, impact ionization occurs when a free electron collides with an atom, generating new free electrons that can collide with other atoms, leading to an avalanche effect. Impact ionization depends on the pre-existence of free electrons in the sample, which often are generated by multiphoton ionization. These three mechanisms can act alone or simultaneously for different laser settings (Quinto-Su and Venugopalan 2007).

How does plasma destroy biological tissues? When the laser pulse duration is in the picosecond range, free electrons in the dense plasma absorb the energy provided by the laser. At the end of laser irradiation, the absorbed energy is deposited in the sample instantaneously, causing damage. Energy deposition occurs faster than thermal diffusion, and therefore the damage is confined to the irradiated region (Schaffer et al. 2001). Plasmas can sever subcellular structures, such as microtubules, with a resolution smaller than 1 μm (Heisterkamp et al. 2005). For longer laser pulses in the range of picoseconds to nanoseconds, the laser energy is transferred to the sample during the pulse and through thermal diffusion. Thermal diffusion causes a temperature increase in the sample with mechanical side effects, such as the formation of cavitation bubbles (Niemz 2004). Cavitation bubbles form as a by-product of tissue vaporization due to high temperatures. The mechanical stress generated by cavitation bubble formation and implosion breaks the tissue, causing ablation. Thermomechanical ablation is less precise than ablation caused purely by ionization, as the extent of the damage depends on the size of the cavitation bubbles (Hutson and Ma 2007).

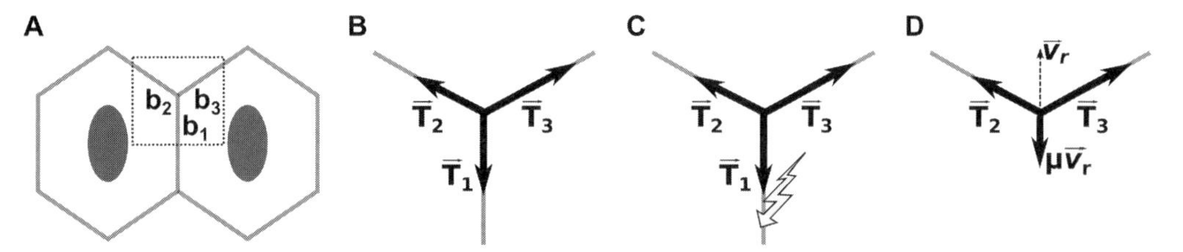

Figure 8.2 **Force balance at a tricellular vertex.**
(A) Schematic showing two adjacent cells. Dotted lines outline a vertex where the boundaries $b1$, $b2$, and $b3$ converge. (B) Force balance at the vertex before ablation of $b1$, when $\vec{T}_1$, $\vec{T}_2$ and $\vec{T}_3$ are in equilibrium. (C) The ablation of $b1$ releases $\vec{T}_1$ and drives the system out of equilibrium.
(D) Immediately after ablation, the vertex moves with a velocity of retraction $\vec{v}_r$, and a frictional force, $\mu\vec{v}_r$, opposes the movement.

However, even when long, nanosecond pulses are used, it is possible to locally disrupt single cell-cell interfaces in living animals due to the endogenous chromophores that act as a source of free electrons, lowering the energy density threshold necessary for plasma formation; and the extracellular matrix, which limits the expansion of cavitation bubbles (Hutson and Ma 2007).

8.2.2 Connecting laser cuts to cell mechanics

The effects of a laser ablation are related to the mechanical properties of cells. To quantify the mechanical forces sustained by subcellular, cellular, or tissue-level structures, it is helpful to establish a force balance around the structure in question (Hutson et al. 2003; Rauzi and Lenne 2011). As an example, we will describe the analysis of the forces sustained by individual cell-cell interfaces in epithelial tissues (Fig. 8.2) (Rauzi and Lenne 2011). Epithelial tissues serve as protective barriers in which cells are held in close apposition to prevent pathogen infiltration. In epithelia, cell interfaces are delimited by two vertices where three or more cells meet (Fig. 8.2A).

To establish a force balance, it is first necessary to consider the mechanical environment. If we consider an object moving through a fluid, the Reynolds number measures the ratio of inertial forces, which maintain an object in motion, to viscous forces, which slow down moving objects. Inertial forces depend on the mass and the acceleration of the object. Viscous forces are shear stresses generated by the friction between the object and the fluid. The viscosity of the cytoplasm is higher than that of water, and the size and velocity of subcellular structures is relatively small (del Alamo et al. 2008), leading to low Reynolds numbers inside cells (Purcell 1977). A low Reynolds number implies that viscous forces dominate, and that the movement of an object at a specific time point is determined by the forces acting on the object at that time point and not before (Purcell 1977). A low Reynolds number also implies that inertia is small and acceleration is negligible; therefore, the forces acting on an object at any given time, even outside equilibrium, must balance, as $m\vec{a} = 0$. In the case of tricellular vertices at the end of a cell-cell interface, the balance of forces at equilibrium is given by the tensile forces acting on each one of the three cell boundaries, $b1$, $b2$, and $b3$, that come together at the vertex (Fig. 8.2A):

$$\vec{T}_1 + \vec{T}_2 + \vec{T}_3 = 0, \tag{8.1}$$

where $\vec{T}_1$, $\vec{T}_2$, and $\vec{T}_3$ are vectors representing the tensile forces along *b1*, *b2*, and *b3*, respectively (Fig. 8.2B). Immediately after the ablation of *b1*, $\vec{T}_1$ disappears, but $\vec{T}_2$ and $\vec{T}_3$ remain unaffected. The net result is an instantaneous outward force on the vertex that will cause the vertex to move with velocity $\vec{v}_r$ (Fig. 8.2C, D). The movement of the vertex is opposed by a force $\mu\vec{v}_r$, where μ is the viscosity coefficient from the resistance of the fluid to retraction. Because there is no acceleration, the new force balance immediately after ablation is:

$$\mu\vec{v}_r + \vec{T}_2 + \vec{T}_3 = 0 \tag{8.2}$$

From Eqs. 8.1–8.2, and taking vector magnitudes:

$$v_r = T_1/\mu \tag{8.3}$$

Therefore, assuming that the viscosity coefficient is constant in time, the velocity of retraction of the vertices at the ends of an interface immediately after ablation can be used to estimate the tension sustained by the ablated interface.

8.2.3 Image analysis

An essential step in the analysis of laser ablation experiments is the quantification of retraction velocities from time-lapse microscopy sequences. Quantification can be done manually, by annotating corresponding structures in consecutive images and measuring their displacement. However, one of the advantages of laser ablation is that after the cut, a new force equilibrium is established rapidly (typically under one minute), and therefore a large number of experiments can be conducted in a short time. This advantage disappears if the data need to be manually annotated.

To automate the analysis of laser ablation data, particle image velocimetry, a technique that uses cross-correlation to calculate the displacements between two images, has been adopted (Mayer et al. 2010; Fernandez-Gonzalez and Zallen 2013). The cross-correlation between two real signals, $s1$ and $s2$, can be mathematically expressed as:

$$(s1 * s2)(r) = \int_{t=-\infty}^{t=+\infty} s1(t)s2(r+t)dt, \tag{8.4}$$

where r is a shift applied to $s2$ with respect to $s1$. In basic terms, $s2$ is overlaid on $s1$ ($r = 0$), and the similarity between the two signals is quantified by integrating the area under their product (Fig. 8.3A). $s2$ is then shifted by one position with respect to $s1$ ($r = 1$), and the similarity is recalculated. By iterating this process, it is possible to identify the shift that, when applied to $s2$, results in maximum overlap between $s1$ and $s2$ (Fig. 8.3A).

In particle image velocimetry, cross-correlation is applied to two images, I_s (the source image) and I_t (the target image). For the analysis of laser ablation experiments, I_s is the last image acquired before ablation, and I_t is the first image acquired after ablation (Fig. 8.3B). A window is defined around each pixel in I_s, and the cross-correlation

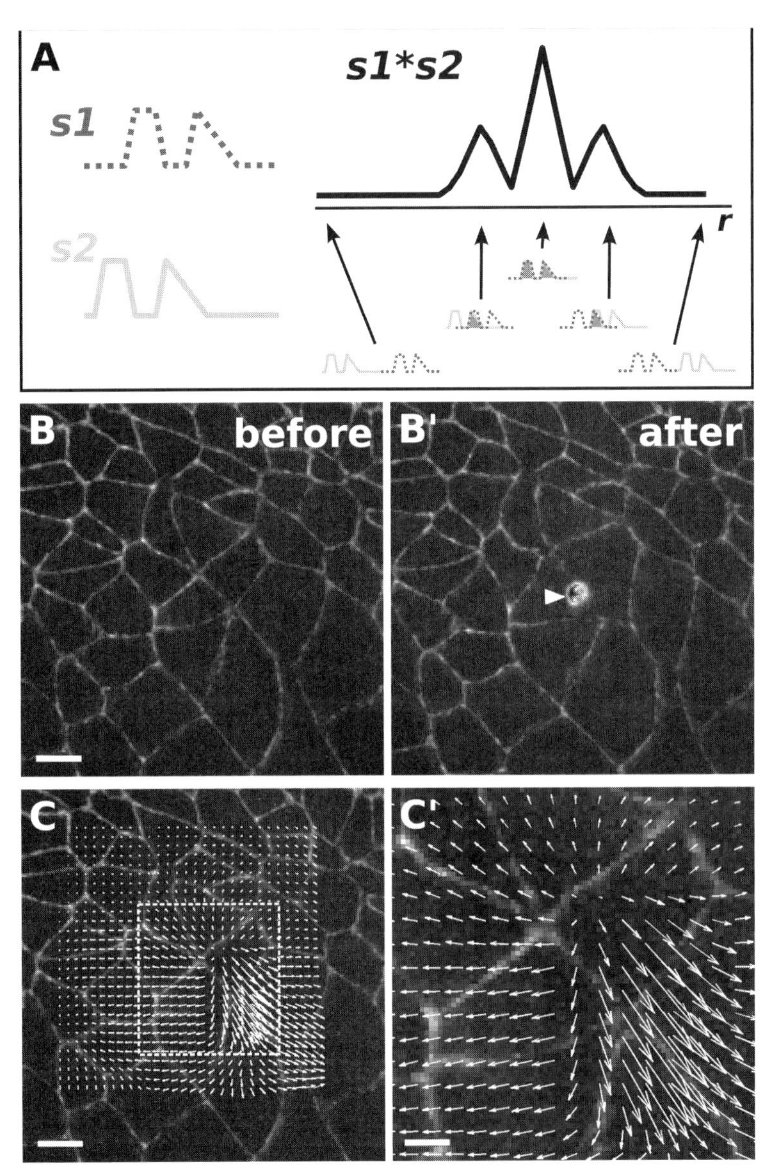

Figure 8.3 **Particle image velocimetry to quantify retraction velocities.**
(A) Schematic illustration of the cross-correlation between two one-dimensional signals, *s1* (dotted line) and *s2* (continuous gray line). The cross-correlation *s1**s2* is also shown (continuous black line). Arrows indicate the cross-correlation values for different degrees of overlap (shaded) between *s1* and *s2* resulting from different values of *r*. (B) Cells expressing E-cadherin:GFP in a *Drosophila* embryo to visualize cell outlines. The images were acquired immediately before (B) or after (B') ablation of a cell boundary. Arrowhead indicates an autofluorescent hole in the transparent membrane that encloses the embryo. Scale bar, 5 μm. (C) Vector field showing the magnitude and orientation of the local displacements between (B) and (B') overlaid on (B). (C) Overall view. (C') Magnified view of the area delineated by the dashed line in (C). Scale bars, 5 μm (C) and 2 μm (C').

between each of these windows, I_{si}, and the target image, I_t, is calculated. For cross-correlation between images, $\vec{r}_i$ is a vector that has as many dimensions as the images, and contains the shift in each dimension necessary to maximize the overlap between I_{si} and I_t. For each window, $\vec{r}_i$ indicates the displacement of the structures contained within that window. As a result, a vector field is obtained representing the directionality and magnitude of the movement of different structures in the image (Fig. 8.3C). The vector field can be filtered for smoothness and interpolated to obtain subpixel resolution (Raffel et al. 1998). By calculating the average magnitude of the radial components of the displacement vectors, it is possible to accurately estimate retraction velocities.

8.2.4 Models for data interpretation

While a simple force balance can connect laser cut experiments to mechanical forces, using this approach requires some knowledge of the viscoelastic properties of the sample (Hutson et al. 2003). However, the viscoelastic properties of a cell or a tissue are difficult to measure in vivo. To overcome this limitation, mechanical models have been used to fit laser ablation data and estimate the mechanical properties of living tissues. Most biological samples exhibit properties of solid (elastic) and liquid (viscous) materials simultaneously (Forgacs et al. 1998; Schotz et al. 2008). To capture these complex viscoelastic properties, elasticity can be modeled by springs, while viscosity can be represented by dashpots. Different mechanical circuits or combinations of springs and dashpots have been used to model the mechanical responses to laser ablation.

A mechanical circuit commonly used to model the results of laser ablation is the Kelvin-Voigt solid, where a spring and a dashpot are arranged in parallel (Fig. 8.4A). The force sustained by the spring, F_s, depends on its change in length:

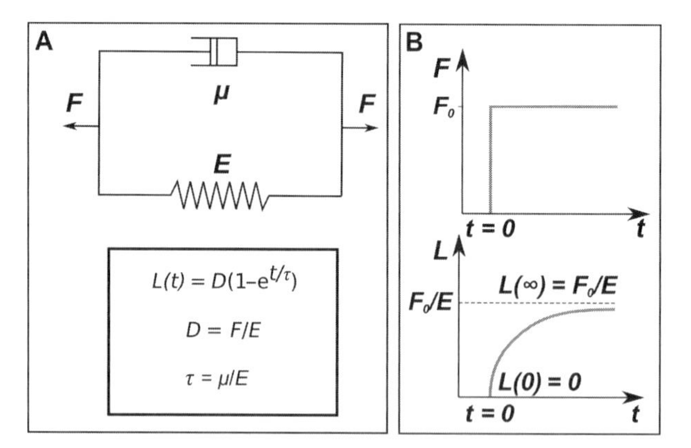

Figure 8.4 Kelvin-Voigt model for the viscoelastic response to laser ablation.
(A) A Kelvin-Voigt element consists of a dashpot and a spring connected in parallel. The equation that governs the model relates displacement, L, and force sustained, F, and can be used to fit the displacements caused by laser ablation. See text for details. (B) Creep solution. (Top) When a constant force is applied to a Kelvin-Voigt element (e.g., when a cell-cell interface is ablated and the tension that it sustained is released), (bottom) the initial displacement is zero, and the asymptotic displacement depends on the applied force, F_0, and the elastic modulus of the material, E.

$$F_s(t) = EL_s(t), \tag{8.5}$$

where t indicates time, $L_s(t)$ is the elongation of the spring at time t, and E is its elastic modulus. The force sustained by a dashpot, F_d, depends on the velocity of displacement of the dashpot:

$$F_d(t) = \mu \frac{d\left(L_d(t)\right)}{dt}, \tag{8.6}$$

where μ is the viscosity coefficient and L_d is the elongation of the dashpot. From Eqs. 8.5–8.6, the force F sustained by a Kelvin-Voigt element (in which $L_s(t) = L_d(t) = L(t)$), is:

$$F(t) = F_s(t) + F_d(t) = EL(t) + \mu \frac{d\left(L(t)\right)}{dt} \tag{8.7}$$

If we assume a constant force over time, F_0 (Fig. 8.4B, top), Eq. 8.7 can be rewritten as:

$$\frac{F_0}{\mu} = \frac{E}{\mu} L(t) + \frac{d\left(L(t)\right)}{dt} \tag{8.8}$$

The differential equation can be solved using the boundary condition $L(0) = 0$, where $t = 0$ is the time of ablation:

$$L(t) = \frac{F_0}{E}\left(1 - e^{-t\left(\frac{E}{\mu}\right)}\right) = D\left(1 - e^{-\frac{t}{\tau}}\right), \tag{8.9}$$

where D is the asymptotic value of the distance retracted after ablation ($L(\infty) = F_o/E$), and τ represents a relaxation time that depends on the viscoelastic properties of the material ($\tau = \mu/E$) (Fig. 8.4B, bottom). Eq. 8.9 can be used to fit the distance retracted after ablation and estimate the force-to-elasticity ratio (D) and the viscosity-to-elasticity ratio (τ). Kelvin-Voigt elements have been used to investigate the response of stress fibers (Kumar et al. 2006), cell junctions (Fernandez-Gonzalez et al. 2009) and the cell cortex (Mayer et al. 2010) to laser ablation. Kelvin-Voigt models have also been used to investigate the mechanical properties of epithelial tissues (Bonnet et al. 2012; Fischer et al. 2014).

More sophisticated models can be used to fit specific cell and tissue behaviors. For example, a Kelvin-Voigt element arranged in series with a dashpot (Mainardi and Spada 2011) has been used to model displacements that do not reach an asymptotic value (creep) after laser ablation (Fernandez-Gonzalez et al. 2009). Although complex models may fit the responses to ablation more accurately, the biological interpretation of the estimated parameters is less clear. Therefore, complex models have not been adopted to analyze laser ablation data.

8.2.5 Practical considerations

Several parameters and assumptions need to be taken into account when conducting laser ablation experiments to quantify cell mechanics:

- **Laser wavelength.** The choice of wavelength can dramatically affect the results of laser irradiation. Ultraviolet and near ultraviolet wavelengths display a significantly lower energy threshold for plasma formation than visible or infrared wavelengths (Vogel and Venugopalan 2003; Hutson and Ma 2007). The greater energy deposition required for plasma formation and ablation at longer wavelengths results in greater damage to the surrounding tissue (Hutson and Ma 2007). Therefore, shorter wavelengths are better suited for ablations that require higher spatial resolution.
- **Laser pulse duration.** Short laser pulses (<100 femtoseconds) cause ablation by plasma formation and optical breakdown of the sample in a spatially confined region. Short pulses result in sharp, clean cuts of reproducible morphology (Schaffer et al. 2001). Longer pulses (in the nanosecond range) induce plasma-mediated ablation, but also cause thermal ablation by local heating of the sample, and mechanical ablation due to the expansion of the plasma outside the irradiated region (Schaffer et al. 2001). Thermal and mechanical effects are not confined to the irradiated region and cause microexplosions, boiling, cracks, and fractures, leading to less precise cuts of irregular morphology.
- **Speed of ablation and image acquisition.** The relationship between recoil velocity and mechanical tension is based on a force balance immediately after ablation. The recoil velocity decays exponentially with time after ablation (Mayer et al. 2010). To estimate forces it is important to measure recoil velocity as close as possible to the time of ablation, which requires a reduced number of pulses of short duration for ablation and short exposure times for imaging.
- **PIV window size.** The window size used to subdivide the source image and calculate the cross-correlation with the target image must be carefully considered. If the window size is too small, it may not contain enough information to find the matching position in the target image, creating errors in the velocity field. On the contrary, if the window is too large, the resulting vector field will not have sufficient resolution to capture local differences in the pattern of displacement after ablation. A good rule of thumb is to choose a window size large enough to fully contain one of the basic elements being tracked (i.e., one cell, one node in a network, etc.).
- **Pixels used for PIV analysis.** Laser ablation destroys material in the irradiated site and, depending of the physical mechanisms of ablation, possibly also around it. The destruction of material typically results in a loss of signal (e.g., fluorescence), making it difficult to estimate displacements around the site of ablation. The displacements caused by laser ablation decay with distance from the ablation site (Ma et al. 2009). Therefore, including pixels that are too far from the ablation site in the analysis could underestimate the recoil velocities. In general, it is best to

restrict PIV analysis to pixels within a box (Mayer et al. 2010) or a ring (Fernandez-Gonzalez and Zallen 2013) that exclude the ablated region and include only structures directly connected to the ablated object.

- **Homogeneous material properties.** In the analysis of laser ablation experiments, the material properties of the sample, and particularly the viscosity coefficient that modulates the proportionality between tension and recoil velocity, are assumed to be homogeneous throughout the sample. It is important to keep this assumption in mind, as microrheology studies have found that some mechanical properties can change in different subcellular domains, and for instance, the region around the nucleus in fibroblasts is more compliant than regions at the periphery of the cell (Tseng et al. 2002).

8.3 Applications

Light-induced ablation has been used since the 1960s to investigate contractile behaviors and mechanical properties in living cells and animals. In pioneering studies, ultraviolet light was used to sever myofibrils, the contractile units in muscle cells formed by the cytoskeletal proteins actin and myosin, and investigate their mechanism of contraction (Stephens 1965). Fibrils were optically severed at different anatomical locations, and contraction was chemically induced. Based on the observed movements when the severed fibrils contracted, this study suggested that actin filament sliding was responsible for the contraction of striated muscles.

Subsequent advances, including the use of lasers as light sources for cutting, the development of the green fluorescent protein and transgenic animal technologies, the advent of confocal microscopy, and the increased sensitivity of detectors, have enabled the use of optical ablation in living animals. In vivo laser ablation studies have used worm (*C. elegans*), fruit fly (*Drosophila melanogaster*) and zebrafish embryos as model organisms, taking advantage of their transparency; or frog (*Xenopus*) tissue explants, which can be easily manipulated surgically. In this section, we review recent contributions of laser ablation experiments to our understanding of cytoskeletal organization and dynamics, cell behavior, and tissue architecture in vivo.

8.3.1 Cell division

One of the first applications of laser ablation in living cells was to investigate the mechanisms of chromosome segregation during cell division (Aist and Berns 1981). The initial studies relied on phase-contrast and electron microscopies to visualize the effects of laser cuts at different positions in the mitotic spindle, the subcellular structure responsible for chromosome segregation (Fig. 8.5A), in fungal cells. Using this approach, the authors found that the speed of nuclear separation was greatly increased when the spindle was ablated, suggesting that the spindle resisted chromosome segregation. Ablation of the asters, microtubule-based structures that anchor the spindle to

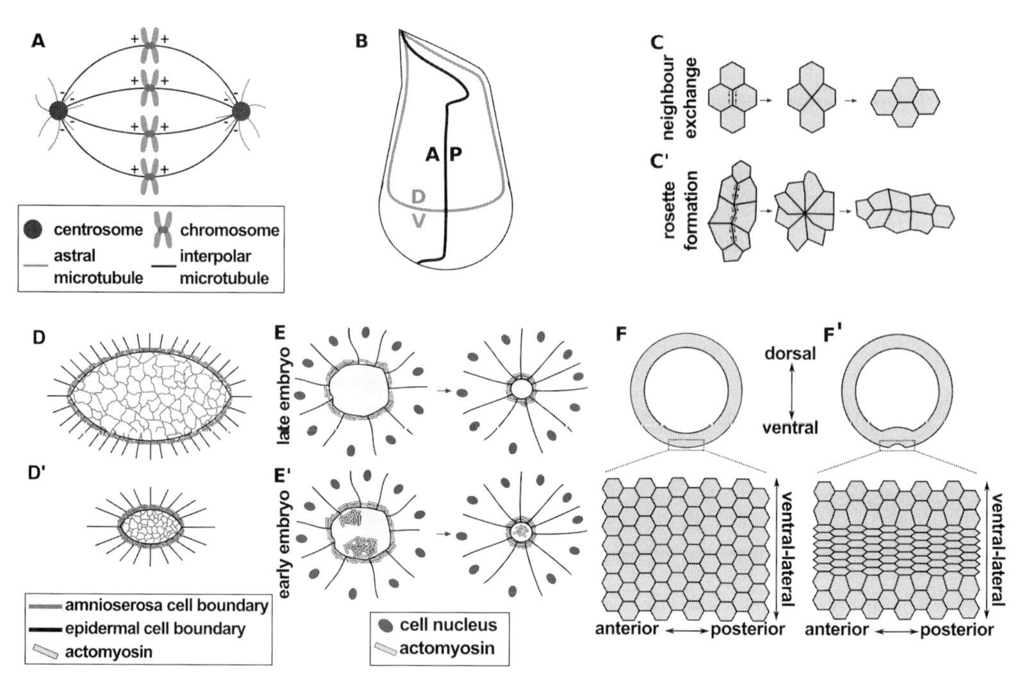

Figure 8.5 **Biological processes investigated with laser ablation.**
(A) Spindle apparatus during metaphase. The chromosomes are located in the center of the spindle and are connected to the spindle poles through microtubules. The centrosomes are positioned in the spindle poles and are surrounded by astral microtubules that connect the poles to the cell cortex. (B) Fly larval wing disk schematic showing the anterior-posterior (black) and the dorsal-ventral (gray) compartment boundaries. (C–C') Cartoons depicting the two forms of cell intercalation that drive axis elongation in *Drosophila*. In neighbor exchange (C) and rosette formation (C'), myosin-enriched interfaces (vertical) contract (leftmost panel), forming a vertex where four (neighbor exchange) or more than four (rosette) cells converge (middle panel). The cells form new interfaces perpendicular to the ones that contracted (rightmost panel), driving axis elongation. (D) Actomyosin-based contractility mediates dorsal closure in *Drosophila* embryos. During dorsal closure, an actomyosin cable forms around the dorsal discontinuity in the epidermis (D). The cable contracts, pulling the epidermal cells (black lines) (D'). Amnioserosa cells (gray lines) constrict apically (D'), pulling the epidermis over them and driving dorsal closure. (E) Wound repair in late (E) and early (E') *Drosophila* embryos. After wounding, actin and myosin accumulate at the interface between wounded and adjacent cells, forming a supracellular cable around the wound. The cable contracts, pulling the adjacent cells toward the center of the wound and driving wound closure. In the early embryo (E') actin and myosin also accumulate on the medial-apical surface of the wounded cells. The contraction of these medial actomyosin networks drives apical constriction of the wounded cells and rapid wound closure. (F) Apical constriction during mesoderm internalization in *Drosophila* embryos. (Top panels) Cross-section of a *Drosophila* embryo before (F) and during (F') mesoderm invagination. (Bottom panels) Apical view of mesoderm cells demonstrating the anisotropic constriction of their apical surface (F').

the cell cortex, demonstrated that asters pull on the spindle (Aist et al. 1991; Aist et al. 1993). Years later, a similar approach was used to investigate the role of mechanical forces in asymmetric cell division in the *C. elegans* embryo (Grill et al. 2001). These studies suggested that during asymmetric cell division, the two asters exert different forces on the spindle, as indicated by differences in the speed of movement of the centrosomes, the structures that organize microtubules during cell division, after spindle ablation. These differences result in the asymmetric positioning of the spindle inside the cell and asymmetric cell division. Ablation of a centrosome, led to the fragmentation of the associated aster (Grill et al. 2003). Aster fragments moved at a range of velocities, but interestingly, the velocity variance was very small at both low and high velocities, suggesting that the molecules responsible for force generation at the asters are limited and can be saturated.

More recently, laser ablation has been used to dissect the spindle architecture in *Xenopus* egg extracts (Brugues et al. 2012). Laser ablation of microtubules induces depolymerization at the plus but not the minus end. The authors ablated the spindle at different locations and used fluorescence microscopy to track microtubule depolymerization dynamics. The results demonstrated that spindle microtubules are shorter close to the spindle poles. Combining laser ablation with visualization and quantification of microtubule polymerization (using fluorescent tubulin, the main constituent of microtubules), mathematical modeling, and drug treatments to block microtubule transport, the authors found that microtubule nucleation is faster at the center of the spindle, and that motor-mediated microtubule transport is necessary for the observed organization of microtubule lengths in the spindle. Together, these studies demonstrate that laser ablation can be used as a tool to study the mechanics of cell division, and also the organization and dynamics of cytoskeletal networks.

8.3.2 Compartment boundary formation

The application of light-induced ablation to living animals has allowed investigation of how mechanical tension affects tissue formation. A well-studied example is the development of the wing discs in *Drosophila* larvae. Wing discs, which become the wings in the adult, can be surgically removed, cultured, and imaged live throughout their development at cellular and subcellular scales, and therefore constitute an ideal system to investigate tissue morphogenesis. During wing disc development, cells dramatically change their geometry, from irregular packing to a hexagonal array of cells (Classen et al. 2005). To understand the mechanical forces driving this dramatic change in tissue architecture, a theoretical model was proposed to determine the interplay between cytoskeletal tension and cell adhesion (Farhadifar et al. 2007). The model was used to simulate laser ablation of individual cell-cell interfaces, and the results of the simulations were compared to actual laser ablation experiments to estimate the magnitudes of cell adhesion, contractility, and elasticity in the disc. The authors found that contractility dominates over cell adhesion, and that perimeter elasticity was necessary to account for their laser ablation results.

The wing disc is divided into regions called compartments, in which cells produce different signals that are essential for patterning the wing and establishing different cell fates (McNeill 2000; Tepass et al. 2002). Wing disc compartments have well-defined boundaries that prevent cell mixing (Fig. 8.5B). Laser severing has been used to investigate the mechanisms by which compartment boundaries are formed and maintain the organization of the wing disc (Landsberg et al. 2009; Aliee et al. 2012). Ablation of single cell-cell interfaces on the boundaries demonstrated that they sustain greater tension than cell interfaces in other regions of the disc (Landsberg et al. 2009; Aliee et al. 2012), coinciding with the presence of strong cables of actin and myosin spanning the length of the boundaries (Major and Irvine 2005; Major and Irvine 2006; Landsberg et al. 2009). Ablation of the dorsal-ventral boundary at different stages of development showed that increased tension is correlated with a straightening of the boundary (Aliee et al. 2012). Increased mechanical tension seems to be a common feature of compartment boundaries (Umetsu et al. 2014), where tension could prevent the cellular rearrangements that promote mixing of cells from different compartments (Umetsu et al. 2014).

8.3.3 Axis elongation

To investigate how mechanical tension affects cell rearrangements, laser ablation has been applied to the study of axis elongation in *Drosophila*. Axis elongation is a conserved process in which the head-to-tail (anterior-posterior) axis of the animal is established. In *Drosophila*, axis elongation occurs in an epithelial monolayer referred to as the germband, and is driven by polarized cell rearrangements that promote the extension of the tissue along the anterior-posterior axis and its shortening along the perpendicular, dorsal-ventral axis (Irvine and Wieschaus 1994). Two types of cell rearrangements drive *Drosophila* axis elongation. In neighbor exchange, a cell interface between anterior and posterior neighbors (an AP interface, Fig. 8.5C) contracts, bringing four cells into contact at a vertex (Bertet et al. 2004). The vertex resolves through the assembly of a new interface between dorsal and ventral neighbors (a DV interface, Fig. 8.5C). In rosette formation, the coordinated contraction of several AP interfaces linked together by a supracellular actomyosin cable results in the formation of a rosette vertex where five or more cells meet (Blankenship et al. 2006) (Fig. 8.5C'). The rosette vertex resolves through the assembly of new DV interfaces (Fig. 8.5C'). Laser ablation of AP and DV interfaces demonstrated that AP interfaces, which are enriched with actin and the molecular motor nonmuscle myosin II (Bertet et al. 2004; Zallen and Wieschaus 2004; Blankenship et al. 2006), sustain increased mechanical tension (Rauzi et al. 2008; Fernandez-Gonzalez et al. 2009). Notably, laser ablation of AP interfaces that were part of supracellular myosin cables resulted in greater recoil velocities than laser ablation of AP interfaces that were not part of supracellular cables (Fernandez-Gonzalez et al. 2009), suggesting that the coordinated contraction of multiple cell interfaces during rosette formation requires greater mechanical tension than neighbor exchange events. The asymmetric distribution of mechanical forces is conserved during axis elongation in *Xenopus* frogs (Shindo and Wallingford 2014) and during limb bud elongation in mice (Lau et al. 2015). By severing similar cellular structures (single cell-cell interfaces) with different orientations, the use of laser ablation enables a quantitative assessment of the distribution of mechanical forces in tissues, thus bridging cell and tissue mechanics.

Laser ablation has been used to investigate the molecular mechanisms that sustain specific force distributions during axis elongation. In *Drosophila*, laser cutting assays demonstrated that the mechanical anisotropy of the germband is at least in part dependent on the actin-associated protein Shroom, which contributes to the enrichment of myosin on AP edges (Simoes Sde et al. 2014); and on the phosphorylation of the myosin light chain (Kasza et al. 2014), which regulates myosin motor activity. In *Xenopus*, the septin cytoskeleton, which regulates actomyosin organization, is essential to establish mechanical asymmetries (Shindo and Wallingford 2014). And in mouse embryos, β-catenin is required for the directional bias in tissue stress in the limb bud ectoderm (Lau et al. 2015). These studies demonstrate the power of combining laser ablation with the genetic and pharmacological tools available for in vivo animal models. Automation of the microscopy and image analysis involved in laser ablation experiments will enable screening strategies for the molecular factors that control force distribution during developmental processes.

8.3.4 Cell sheet fusion

An alternative approach to quantify tissue-level forces is to ablate multiple cells, thus releasing tension in small patches of tissue. Multicellular laser ablation has been used to investigate the fusion of epithelial cell sheets, a process essential for eyelid closure during embryonic development in mammals (Heller et al. 2014). One of the best-characterized examples of cell sheet fusion is dorsal closure in *Drosophila* embryos. During dorsal closure, a discontinuity on the epidermis at the back of the embryo is sealed (Fig. 8.5D). The epidermal discontinuity arises as part of normal embryonic development, and several forces contribute to its closure, including contraction of cytoskeletal structures formed by actin and myosin; and the formation and interdigitation of actin-based protrusions from opposite sites of the discontinuity (Kiehart et al. 2000; Hutson et al. 2003). Laser ablation combined with quantitative measurements of retraction velocity, wound morphology, and closure speed allowed dissection of the tissues and forces implicated in dorsal closure, and of the cellular basis of force generation in these tissues.

Actomyosin contractility is essential for dorsal closure and occurs in two different forms. A supracellular cable formed by the cytoskeletal proteins actin and myosin is assembled around the perimeter of the discontinuity, at the leading edge of the epidermis (Fig. 8.5D). Laser ablation of the actomyosin cable resulted in retraction of the severed ends, indicating that the cable was contractile (Kiehart et al. 2000). These data suggest that contraction of the actomyosin cable pulls the epidermis over the amnioserosa, a population of epithelial cells that cover the discontinuity (Fig. 8.5D). Furthermore, ablations of the epidermis created wounds that were elongated toward the discontinuity (Kiehart et al. 2000), suggesting that the epidermis was under tension, likely due to the contraction of the actomyosin cable. In addition to the actomyosin cable, amnioserosa cells also generate forces during dorsal closure. The amnioserosa is attached to the epidermis. Laser ablation experiments showed that amnioserosa cells are under tension (Kiehart et al. 2000). Removing the amnioserosa by using laser incisions slowed down dorsal closure (Hutson et al. 2003). Further studies have demonstrated that amnioserosa cells undergo actomyosin-based, pulsed constriction of their external (apical) surface (Solon et al. 2009; Blanchard et al. 2010; David et al. 2010; Sokolow et al. 2012) (Fig. 8.5D). As they constrict apically, amnioserosa cells pull on the epidermis, thus contributing to dorsal closure. Notably, laser ablation of one amnioserosa cell disrupts the pulsatile behavior of adjacent cells (Solon et al. 2009), suggesting that the contractile behaviors of adjacent amnioserosa cells are mechanically coupled. Pharmacological inhibition of the channels that promote ion flux between cells resulted in reduced retraction velocities when ablations were conducted in the amnioserosa (Hunter et al. 2014), suggesting that ion channels are required for the efficient generation of contractile forces by amnioserosa cells during dorsal closure. Present efforts (Fischer et al. 2014) are combining laser ablation with the power of *Drosophila* genetics to elucidate the molecular basis of the contractile behaviors of amnioserosa cells, and how these behaviors are coupled to generate tissue-level forces.

Similar to dorsal closure, cell sheet fusion mediates wound repair in the embryonic epidermis. Upon wounding, actin and myosin become polarized in the cells adjacent to the wound, forming a supracellular cable around the wound (Martin and Lewis 1992) (Fig. 8.5E). Laser ablation studies showed that the actomyosin cable around the wound is contractile (Fernandez-Gonzalez and Zallen 2013), acting as a purse string to drive rapid wound closure. In addition to the purse string, and specifically during early embryonic development in *Drosophila*, the wounded cells can also participate in the wound response by assembling an actomyosin network on their medial-apical surface that is associated with apical constriction of the wounded cells and wound closure (Fernandez-Gonzalez and Zallen 2013) (Fig. 8.5E'). Laser severing of medial-apical actomyosin networks resulted in retraction of the severed ends of the network and slowed down wound closure, indicating that the medial-apical networks in the wounded cells are contractile and contribute to wound repair. Furthermore, comparison of the retraction velocities after ablation of medial-apical networks and purse strings suggested that the contractile forces generated by the wounded cells were the main contribution to wound repair during the early stages of embryonic development in *Drosophila*.

8.3.5 Tissue internalization

Laser ablation can be used to establish if forces are isotropically distributed in a tissue. During the internalization of the mesoderm in *Drosophila*, mesodermal precursors undergo pulsed apical constriction, eventually leading to buckling and internalization of the tissue (Martin et al. 2009). Mesoderm cells predominantly contract along the ventral-lateral axis (Martin et al. 2010) (Fig. 8.5F), increasing their shape anisotropy as they constrict their surface. The increase in anisotropy is not strongly correlated with pulses of constriction, suggesting that additional mechanisms may be responsible for cell shape anisotropy. Single-spot ablations in the apical surface of mesoderm cells resulted in greater retractions along the anterior-posterior axis than along the ventral-lateral axis, suggesting that tension is anisotropically distributed in the mesoderm

(Martin et al. 2010). These findings were further confirmed by multicellular laser cuts. Cuts to the mesoderm across the anterior-posterior axis resulted in greater retractions than cuts across the dorsal-ventral axis, suggesting that tension is highest along the anterior-posterior axis of the tissue. The authors proposed that tissue-wide tension along the anterior-posterior axis of the mesoderm resists contraction of the cells along this axis, resulting in anisotropic constriction.

Alternative methods to measure mechanical anisotropies in space include ring ablations around tissue patches to quantify differences in the displacements at different positions around the patch (Bonnet et al. 2012). Regardless of the method, the use of laser ablation allows investigation of mechanical properties at supracellular scales that are not accessible for other methods used to quantify subcellular mechanics, such as particle-tracking microrheology or optical tweezers.

8.4 Conclusions

Laser ablation has emerged as an ideal method to quantify tissue mechanics in vivo. Unlike other techniques, such as atomic force microscopy or microrheology, laser ablation can be used to probe cells that may not be physically accessible. In addition, laser ablation experiments are short and can be automated, thus enabling genome-wide in vivo screens to identify the factors that regulate mechanical force generation and transmission during animal development. It should be noted that the applications discussed here do not constitute a comprehensive list; laser ablation has also been used to measure the interplay between cell density and cell tension (Marinari et al. 2012); to investigate the mechanical forces implicated in vertebrate development (Behrndt et al. 2012; Heller et al. 2014); or to increase, orient, or reduce the mechanical forces sustained by cells in vivo (Campinho et al. 2013; Herszterg et al. 2013).

The main limitation of laser ablation is that it is only a semiquantitative approach: laser ablation experiments cannot provide absolute measurements of force, only relative comparisons. The development of methods to estimate tissue viscosity (Bambardekar et al. 2015; Daniels et al. 2006) or adhesion force (Campas et al. 2014) in vivo will complement laser ablation and allow the quantification of absolute force values from laser cut experiments. The advent of these new approaches, together with advances in laser technology and increased automation of data acquisition and analysis, will allow us to address the challenge of mapping and tracking mechanical forces throughout embryonic development, and to establish the relationship between physical forces and cell fate determination in vivo.

Acknowledgments

We thank Miranda Hunter and Jessica Yu for assistance with Fig. 8.1, and Anna Kobb, Miranda Hunter, and Jessica Yu for comments of this manuscript. Teresa

Zulueta-Coarasa is supported by an Ontario Trillium Scholarship. Work in our laboratory is supported by a Connaught Fund New Investigator Award to Rodrigo Fernandez-Gonzalez, and by grants from the University of Toronto Faculty of Medicine Dean's New Staff Fund, the Canada Foundation for Innovation [#30279], the Ontario Research Fund, and the Natural Sciences and Engineering Research Council of Canada Discovery Grant program [#418438–13 to R.F.-G.].

References

Aist, J. R. and Berns, M. W. (1981). "Mechanics of chromosome separation during mitosis in Fusarium (Fungi imperfecti): new evidence from ultrastructural and laser microbeam experiments." *J Cell Biol* **91**: 446–458.

Aist, J. R., Liang, H. and Berns, M. W. (1993). "Astral and spindle forces in PtK2 cells during anaphase B: a laser microbeam study." *J Cell Sci* **104(4)**: 1207–1216.

Aliee, M., Roper, J. C., Landsberg, K. P., Pentzold, C., Widmann, T. J., Julicher, F. and Dahmann, C. (2012). "Physical mechanisms shaping the Drosophila dorsoventral compartment boundary." *Curr Biol* **22**: 967–976.

Bambardekar, K., Clement, R., Blanc, O., Chardes, C. and Lenne, P. F. (2015). "Direct laser manipulation reveals the mechanics of cell contacts in vivo." *Proc Natl Acad Sci USA* **112(5)**: 1416–1421.

Behrndt, M., Salbreux, G., Campinho, P., Hauschild, R., Oswald, F., Roensch, J., Grill, S. W., et al. (2012). "Forces driving epithelial spreading in zebrafish gastrulation." *Science* **338**: 257–260.

Bertet, C., Sulak, L. and Lecuit, T. (2004). "Myosin-dependent junction remodelling controls planar cell intercalation and axis elongation." *Nature* **10**: 667–671.

Blanchard, G. B., Murugesu, S., Adams, R. J., Martinez-Arias, A. and Gorfinkiel, N. (2010). "Cytoskeletal dynamics and supracellular organisation of cell shape fluctuations during dorsal closure." *Development* **137**: 2743–2752.

Blankenship, J. T., Backovic, S. T., Sanny, J. S. P., Weitz, O. and Zallen, J. A. (2006). "Multicellular rosette formation links planar cell polarity to tissue morphogenesis." *Dev Cell* **11**: 459–470.

Bonnet, I., Marcq, P., Bosveld, F., Fetler, L., Bellaiche, Y. and Graner, F. (2012). "Mechanical state, material properties and continuous description of an epithelial tissue." *J R Soc Interface* **9**: 2614–2623.

Brugues, J., Nuzzo, V., Mazur, E. and Needleman, D. J. (2012). "Nucleation and transport organize microtubules in metaphase spindles." *Cell* **149**: 554–564.

Campas, O., Mammoto, T., Hasso, S., Sperling, R. A., O'Connell, D., Bischof, A. G., Maas, R., et al. (2014). "Quantifying cell-generated mechanical forces within living embryonic tissues." *Nat Methods* **11**: 183–189.

Campinho, P., Behrndt, M., Ranft, J., Risler, T., Minc, N. and Heisenberg, C. P. (2013). "Tension-oriented cell divisions limit anisotropic tissue tension in epithelial spreading during zebrafish epiboly." *Nat Cell Biol* **15**: 1405–1414.

Classen, A. K., Anderson, K. I., Marois, E. and Eaton, S. (2005). "Hexagonal packing of Drosophila wing epithelial cells by the planar cell polarity pathway." *Dev Cell* **9**: 805–817.

Daniels, B. R., Masi, B. C. and Wirtz, D. (2006). "Probing single-cell micromechanics in vivo: the microrheology of C. elegans developing embryos." *Biophys J* **90**: 4712–4719.

David, D. J., Tishkina, A. and Harris, T. J. (2010). "The PAR complex regulates pulsed actomyosin contractions during amnioserosa apical constriction in Drosophila." *Development* **137**: 1645–1655.

del Alamo, J. C., Norwich, G. N., Li, Y. S., Lasheras, J. C. and Chien, S. (2008). "Anisotropic rheology and directional mechanotransduction in vascular endothelial cells." *Proc Natl Acad Sci USA* **105**: 15411–15416.

Farhadifar, R., Röper, J.-C., Aigouy, B., Eaton, S. and Jülicher, F. (2007). "The influence of cell mechanics, cell-cell interactions, and proliferation on epithelial packing." *Curr Biol* **17**(24): 2095–2104.

Fernandez-Gonzalez, R., Simoes Sde, M., Roper, J. C., Eaton, S. and Zallen, J. A. (2009). "Myosin II dynamics are regulated by tension in intercalating cells." *Dev Cell* **17**: 736–743.

Fernandez-Gonzalez, R. and Zallen, J. A. (2013). "Wounded cells drive rapid epidermal repair in the early Drosophila embryo." *Mol Biol Cell* **24**: 3227–3237.

Fischer, S. C., Blanchard, G. B., Duque, J., Adams, R. J., Arias, A. M., Guest, S. D. and Gorfinkiel, N. (2014). "Contractile and mechanical properties of epithelia with perturbed actomyosin dynamics." *PLoS One* **9**: e95695.

Forgacs, G., Foty, R. A., Shafrir, Y. and Steinberg, M. S. (1998). "Viscoelastic properties of living embryonic tissues: a quantitative study." *Biophys J* **74**: 2227–2234.

Grill, S. W., Gonczy, P., Stelzer, E. H. and Hyman, A. A. (2001). "Polarity controls forces governing asymmetric spindle positioning in the Caenorhabditis elegans embryo." *Nature* **409**: 630–633.

Grill, S. W., Howard, J., Schaffer, E., Stelzer, E. H. and Hyman, A. A. (2003). "The distribution of active force generators controls mitotic spindle position." *Science* **301**: 518–521.

Heisterkamp, A., Maxwell, I. Z., Mazur, E., Underwood, J. M., Nickerson, J. A., Kumar, S. and Ingber, D. E. (2005). "Pulse energy dependence of subcellular dissection by femtosecond laser pulses." *Opt Express* **13**: 3690–3696.

Heller, E., Kumar, K. V., Grill, S. W. and Fuchs, E. (2014). "Forces generated by cell intercalation tow epidermal sheets in mammalian tissue morphogenesis." *Dev Cell* **28**: 617–632.

Herszterg, S., Leibfried, A., Bosveld, F., Martin, C. and Bellaiche, Y. (2013). "Interplay between the dividing cell and its neighbors regulates adherens junction formation during cytokinesis in epithelial tissue." *Dev Cell* **24**: 256–270.

Hunter, G. L., Crawford, J. M., Genkins, J. Z. and Kiehart, D. P. (2014). "Ion channels contribute to the regulation of cell sheet forces during Drosophila dorsal closure." *Development* **141**: 325–334.

Hutson, M. S. and Ma, X. (2007). "Plasma and cavitation dynamics during pulsed laser microsurgery in vivo." *Phys Rev Lett* **99**: 158104.

Hutson, M. S., Tokutake, Y., Chang, M.-S., Bloor, J. W., Venakides, S., Kiehart, D. P. and Edwards, G. S. (2003). "Forces for morphogenesis investigated with laser microsurgery and quantitative modeling." *Science* **300**: 145–149.

Irvine, K. D. and Wieschaus, E. (1994). "Cell intercalation during Drosophila germband extension and its regulation by pair-rule segmentation genes." *Development* **120**: 827–841.

Kasza, K. E., Farrell, D. L. and Zallen, J. A. (2014). "Spatiotemporal control of epithelial remodeling by regulated myosin phosphorylation." *Proc Natl Acad Sci USA* **111**: 11732–11737.

Kiehart, D. P., Galbraith, C. G., Edwards, K. A., Rickoll, W. L. and Montague, R. A. (2000). "Multiple forces contribute to cell sheet morphogenesisfor dorsal closure in Drosophila." *J Cell Biol* **149**: 471–490.

Kumar, S., Maxwell, I. Z., Heisterkamp, A., Polte, T. R., Lele, T. P., Salanga, M., Mazur, E., et al. (2006). "Viscoelastic retraction of single living stress fibers and its impact on cell shape, cytoskeletal organization, and extracellular matrix mechanics." *Biophys J* **90**: 3762–3773.

Landsberg, K. P., Farhadifar, R., Ranft, J., Umetsu, D., Widmann, T. J., Bittig, T., Said, A., et al. (2009). "Increased cell bond tension governs cell sorting at the Drosophila anteroposterior compartment boundary." *Curr Biol* **19**: 1950–1955.

Lau, K., Tao, H., Liu, H., Wen, J., Sturgeon, K., Sorfazlian, N., Lazic, S., et al. (2015). "Anisotropic stress orients remodelling of mammalian limb bud ectoderm." *Nat Cell Biol* **17**: 569–579.

Ma, X., Lynch, H. E., Scully, P. C. and Hutson, M. S. (2009). "Probing embryonic tissue mechanics with laser hole drilling." *Phys Biol* **6**: 036004.

Mainardi, F. and Spada, G. (2011). "Creep, relaxation and viscosity properties for basic fractional models in rheology." *EPJ-Special Topics* **193**: 133–160.

Major, R. J. and Irvine, K. D. (2005). "Influence of Notch on dorsoventral compartmentalization and actin organization in the Drosophila wing." *Development* **132**: 3823–3833.

Major, R. J. and Irvine, K. D. (2006). "Localization and requirement for Myosin II at the dorsal-ventral compartment boundary of the Drosophila wing." *Dev Dyn* **235**: 3051–3058.

Marinari, E., Mehonic, A., Curran, S., Gale, J., Duke, T. and Baum, B. (2012). "Live-cell delamination counterbalances epithelial growth to limit tissue overcrowding." *Nature* **484**(7395): 542–545.

Martin, A. C., Gelbart, M., Fernandez-Gonzalez, R., Kaschube, M. and Wieschaus, E. F. (2010). "Integration of contractile forces during tissue invagination." *J Cell Biol* **188**: 735–749.

Martin, A. C., Kaschube, M. and Wieschaus, E. F. (2009). "Pulsed contractions of an actin-myosin network drive apical constriction." *Nature* **457**: 495–499.

Martin, P. and Lewis, J. (1992). "Actin cables and epidermal movement in embryonic wound healing." *Nature* **360**: 179–183.

Mayer, M., Depken, M., Bois, J. S., Julicher, F. and Grill, S. W. (2010). "Anisotropies in cortical tension reveal the physical basis of polarizing cortical flows." *Nature* **467**: 617–621.

McNeill, H. (2000). "Sticking together and sorting things out: adhesion as a force in development." *Nat Rev Genet* **1**: 100–108.

Niemz, M. H. (2004). *Laser-Tissue Interactions: Fundamentals of Microscopy*. Berlin: Springer.

Purcell, E. M. (1977). "Life at low Reynolds number." *Am J Phys* **45**: 3–11.

Quinto-Su, P. A. and Venugopalan, V. (2007). "Mechanisms of laser cellular micro-surgery." *Methods Cell Biol* **82**: 113–151.

Raffel, M., Willert, C. E. and Kompenhans, J. (1998). *Particle Image Velocimetry: a Practical Guide*. Berlin and New York: Springer.

Rauzi, M. and Lenne, P. F. (2011). "Cortical forces in cell shape changes and tissue morphogenesis." *Curr Top Dev Biol* **95**: 93–144.

Rauzi, M., Verant, P., Lecuit, T. and Lenne, P. F. (2008). "Nature and anisotropy of cortical forces orienting Drosophila tissue morphogenesis." *Nat Cell Biol* **10**: 1401–1410.

Schaffer, C. B., Brodeur, A. and Mazur, E. (2001). "Laser-induced breakdown and damage in bulk transparent materials induced by tightly focused femtosecond laser pulses." *Meas Sci Technol* **12**: 1784–1794.

Schotz, E. M., Burdine, R. D., Julicher, F., Steinberg, M. S., Heisenberg, C. P. and Foty, R. A. (2008). "Quantitative differences in tissue surface tension influence zebrafish germ layer positioning." *HFSP J* **2**: 42–56.

Shindo, A. and Wallingford, J. B. (2014). "PCP and septins compartmentalize cortical actomyosin to direct collective cell movement." *Science* **343**: 649–652.

Simoes, S. de M., Mainieri, A. and Zallen, J. A. (2014). "Rho GTPase and Shroom direct planar polarized actomyosin contractility during convergent extension." *J Cell Biol* **204**: 575–589.

Sokolow, A., Toyama, Y., Kiehart, D. P. and Edwards, G. S. (2012). "Cell ingression and apical shape oscillations during dorsal closure in Drosophila." *Biophys J* **102**: 969–979.

Solon, J., Kaya-Copur, A., Colombelli, J. and Brunner, D. (2009). "Pulsed forces timed by a ratchet-like mechanism drive directed tissue movement during dorsal closure." *Cell* **137**: 1331–1342.

Stephens, R. E. (1965). "Analysis of muscle contraction by ultraviolet microbeam disruption of sarcomere structure." *J Cell Biol* **25**: 129–139.

Tepass, U., Godt, D. and Winklbauer, R. (2002). "Cell sorting in animal development: signalling and adhesive mechanisms in the formation of tissue boundaries." *Curr Opin Genet Dev* **12**: 572–582.

Tseng, Y., Kole, T. P. and Wirtz, D. (2002). "Micromechanical mapping of live cells by multiple-particle-tracking microrheology." *Biophys J* **83**(6): 3162–3176:

Umetsu, D., Aigouy, B., Aliee, M., Sui, L., Eaton, S., Julicher, F. and Dahmann, C. (2014). "Local increases in mechanical tension shape compartment boundaries by biasing cell intercalations." *Curr Biol* **24**: 1798–1805.

Vogel, A. and Venugopalan, V. (2003). "Mechanisms of pulsed laser ablation of biological tissues." *Chem Rev* **103**: 577–644.

Zallen, J. A. and Wieschaus, E. (2004). "Patterned gene expression directs bipolar planar polarity in Drosophila." *Dev Cell* **6**: 343–355.

9 Computational image analysis techniques for cell mechanobiology

Ge Yang and Hao-Chih Lee

Light microscopy techniques are essential tools for visualizing the mechanobiology of cells. Computational image analysis transforms light microscopy techniques beyond tools of visualization by making it possible to extract from collected images quantitative measurements of cellular mechanical processes and to understand their behavior and mechanisms. The main goal of this chapter is to provide an up-to-date and selective review of computational image analysis techniques for cell mechanobiology applications. We aim to provide practical information to cell mechanobiology practitioners looking for image analysis techniques as well as to image analysis practitioners looking for cell mechanobiology applications. The focus of the chapter is exclusively on computational analysis techniques for dynamic fluorescence microscopy images. We first classify the images into two different categories: *singe particle images* and *continuous region images*. We then review computational analysis techniques for each category, respectively. For single particle images, we review related particle detection and particle tracking techniques and their cell mechanobiology applications. Similarly, for continuous region images, we review related region detection and region tracking techniques and their cell mechanobiology applications. We conclude with an outlook on future development of computational image analysis techniques for cell mechanobiology.

9.1 Overview

Light microscopy, especially fluorescence light microscopy, provides an essential tool for studying the biology of cells by visualizing their structure and function. Because of its unique properties, light microscopy is particularly suitable for studying the mechanobiology of cells. The image formation process in light microscopy is based on collection of photons within the visible spectrum and is noncontact and noninvasive in nature. The use of visible light allows imaging of cellular mechanical processes such as force transduction under live and physiologically relevant conditions. This, in particular, makes it possible to understand their spatiotemporal dynamics. Advances in fluorescence microscopy techniques in areas such as single-molecule imaging (Selvin and Ha 2008), super-resolution imaging (Huang et al. 2009), and fluorescence probe development (Giepmans et al. 2006) have enabled significant improvements in imaging resolution, sensitivity, specificity, and multiplicity. Such improvements have provided new and exciting opportunities for cell mechanobiology studies. Computational image analysis, through its integration with different light microscopy techniques, transforms them beyond tools of visualization by making it possible not only to extract from collected images quantitative measurements of cellular processes but also to understand their behavior and mechanisms (Yang 2013).

Computational image analysis techniques have found many applications in cell mechanobiology. For example, image correlation and single particle tracking techniques are essential tools for quantifying substrate displacement in traction force microscopy (Kraning-Rush et al. 2012; Style et al. 2014). Single particle tracking techniques are also essential tools for quantifying movement of intracellular markers in cell microrheology (Crocker and Hoffman 2007; Wirtz 2009). Driven by the creation of new imaging techniques, new computational image analysis techniques are constantly being developed for cell mechanobiology applications. For example, development of fluorescence speckle microscopy techniques has driven the development of new single particle tracking techniques for following complex movement of hundreds of thousands of fluorescently labeled subunits of cytoskeletal filaments (Danuser and Waterman-Storer 2006).

Development of computational image analysis techniques is also the focus of several related areas, especially computer vision (Szeliski 2010), medical image analysis (Dhawan 2011), and bioimage informatics (Peng 2008). Some of the techniques developed in these areas can be adapted for cell mechanobiology applications. However, despite the similarities shared by image analysis in different areas, a unique requirement of cell mechanobiology is the integration of physical modeling with computational image analysis. This is because of the focus of cell mechanobiology on mechanical readouts. For example, particle displacement and object deformation are common readouts from computational image analysis of cellular mechanical processes. To convert these image analysis readouts into cell mechanical readouts such as force, stress, or elasticity, physical modeling of the cellular structures involved is required.

This chapter aims to provide an up-to-date and selective review of techniques and applications of computational image analysis for cell mechanobiology, with an emphasis on providing practical information to cell mechanobiology practitioners searching for image analysis techniques as well as to image analysis practitioners searching for cell mechanobiology application. The focus is exclusively on computational analysis techniques for dynamic fluorescence light microscopy images. The rest of the chapter is organized as follows. We start with a classification of cell biological images into two different categories: *single particle images* and *continuous region images*. We then review computational analysis techniques for each category respectively. For single particle images, we start with an introduction to particle detection and particle tracking techniques and then review their cell mechanobiology applications. Similarly, for continuous region images, we start with an overview of region detection and region tracking techniques and then summarize their cell mechanobiology applications. We conclude with an outlook on future development of computational image analysis techniques for cell mechanobiology.

9.2 A classification of cell biological images

The image of a point light source in perfect focus under a microscope is an Airy disk, which also defines the point-spread function of the microscope (Born and Wolf 1999). The radius of the Airy disk is $\frac{0.61 \cdot \lambda}{NA}$, where λ is the light wavelength and NA is the numerical aperture of the microscope objective lens used. This radius is also referred to as the Rayleigh limit, which is used as a definition of the resolution of a conventional light microscope (Born and Wolf 1999). The Airy disk defines the smallest image object that can be generated by a conventional microscope. We identify an image object as a particle if it cannot be distinguished from an Airy disk.

Based on the definition of a particle, we propose a classification of cell biological images according to the spatial and geometrical properties of their image objects. Although different cellular processes and imaging techniques can produce images with very different properties, we can generally classify them into two categories (Fig. 9.1). We refer to images under the first category as *single particle images*. Objects within these images are spatially separated particles. We refer to images under the second category as *continuous region images*. Objects within these images are regions or blobs (Dorn et al. 2008), which are larger than the Airy disk and thus cannot be considered as particles. Analysis of images of different categories may require different computational techniques. Some cell biological images contain both single particles as well as continuous regions and can be analyzed using techniques of both categories.

Here we use a few examples to illustrate these two types of images (Fig. 9.1). We start with images of microtubules collected using fluorescent speckle microscopy (FSM), a technique for generating single particle images of cytoskeletal filament networks (Waterman-Storer and Salmon 1998; Danuser and Waterman-Storer 2006). Individual microtubules are visualized by fluorescently labeled tubulins randomly incorporated during polymerization (Fig. 9.1A). When the fraction of labeled tubulins is low (e.g., 1.25–2.5%), incorporated fluorescent tubulins appear as spatially separated particles (Waterman-Storer and Salmon 1998) and produce single particle images (Fig. 9.1B, upper two panels). However, when the fraction of labeled tubulins is high (e.g., 10–50%), labeled tubulins are no longer spatially separated after incorporation and produce continuous region images (Fig. 9.1B, lower three panels). The case for microtubule networks is similar, although much lower fractions of labeled tubulins than for single microtubules are required for generating single particle images because labeled tubulins from different microtubules can get close to each other and become unresolvable (Fig. 9.1C). This has been confirmed for example in the microtubule networks of *Xenopus* egg extract spindles (Fig. 9.1D). Another example of continuous region images (Fig. 9.1E) shows fluorescently labeled mitochondria within a single axon of a *Drosophila* third instar larva.

Single particle images such as FSM images generally provide higher spatial and temporal resolutions than continuous region images. This is because individual particles

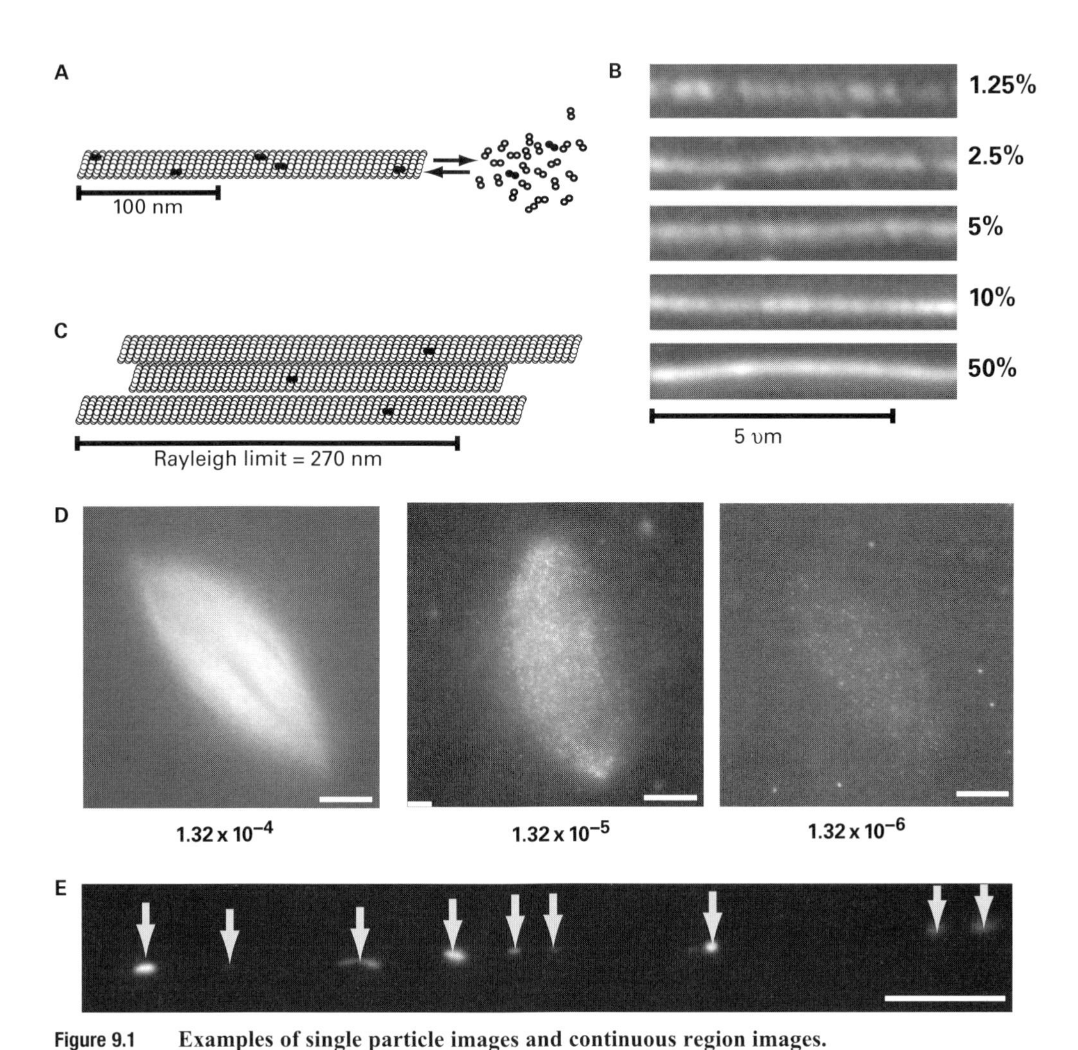

Figure 9.1 **Examples of single particle images and continuous region images.**
(A) Random incorporation of labeled tubulins (solid black) into a single microtubule through polymerization. (B) Single microtubules under different fractions of labeled tubulins (shown to the right of each panel). (C) Incorporated fluorescent tubulins in a microtubule network. The scale bar shows the Rayleigh limit. (D) Microtubule networks in *Xenopus* extract spindles under different fractions of labeled tubulins (shown at the bottom of each panel). The left panel shows a continuous region image; the middle and right panels show single particle images. Scale bars: 10 μm. (E) Another continuous region image, which shows mitochondria (yellow arrows) within a single axon of a Drosophila third instar larva. Scale bar: 10 μm.

(A) and (B) reprinted with permission from Danuser and Waterman-Storer (2006). Copyright © 2006, *Annual Reviews*; (D) reprinted with permission from Yang et al. (2007).

serve as sensitive local reporters of cellular processes. Their activities can be followed with high spatial and temporal resolutions using single particle tracking techniques. However, a potential limitation of single particle images is that if the particles are too sparse, they can no longer fully represent the dynamics of cellular processes. Another potential limitation of single particle images is that they can be technically more challenging to produce due to stricter requirements on imaging conditions. For example, to produce FSM images of the spindle microtubule network, the fraction of labeled tubulins must be kept at a very low level, which in turn requires stable long exposure for image formation. Computational analysis techniques for single particle images are reviewed in Sections 9.3 and 9.4.

Continuous region images generally provide lower spatial and temporal resolutions because of their lower contrast, although specialized techniques such as photoactivation (PA) and fluorescence recovery after photobleaching (FRAP) (Lippincott-Schwartz et al. 2001) can be used to improve image contrast. However, continuous region images are often more convenient to produce than single particle images. Computational analysis techniques for continuous region images are reviewed in Section 9.5.

It is worth noting that because of their significantly higher resolutions, super-resolution imaging techniques (Huang et al. 2009) can produce single particle images of cellular processes that would otherwise render continuous region images under conventional fluorescence microscopy. For example, connected cellular structures may appear as clusters of particles under STORM (stochastic optical reconstruction microscopy) (Rust et al. 2006) or PALM (photoactivation localization microscopy) (Betzig et al. 2006). Such clusters can be converted into regions through image segmentation (Chen et al. 2014).

9.3 Computational analysis techniques for single particle images in cell mechanobiology

Single particle images are used widely in cell mechanobiology applications such as traction force microscopy (Kraning-Rush et al. 2012; Style et al. 2014) and cell micro-rheology (Crocker and Hoffman 2007; Wirtz 2009). Activities of individual particles can be followed and quantified with high spatial and temporal resolutions using single particle detection and tracking techniques. Through analyzing activities of particles individually and collectively, we can obtain high-resolution quantitative readouts of the local and global mechanical properties of cells.

9.3.1 Computational image analysis techniques for particle detection

The first step in computational analysis of single particle images is usually particle detection, which identifies the particles and determines their positions and intensities. For cell mechanobiology applications, to achieve the best localization accuracy is often essential for subsequent single particle tracking and particle trajectory analysis. For example, in calculating diffusion coefficients of particles undergoing random walk, insufficient localization accuracy can lead to large errors (Qian et al. 1991).

A wide variety of computational techniques have been developed for single particle detection. A critical determinant of which technique should be adopted is the imaging configuration, specifically the effective pixel size, which is defined as the physical size of the camera pixels divided by the optical magnification. This parameter determines the spatial sampling rate of imaging. If it is greater than one third of the Rayleigh limit, individual particles can only be localized with pixel-level resolution (Cheezum et al. 2001). That is, positions of particles are determined to integer pixel coordinates in the image plane. In this case, localization accuracy is limited by the physical sizes of camera pixels. However, if the effective pixel size is less than one third of the Rayleigh limit, individual particles can be localized with subpixel-level resolution, under which localization accuracy is no longer limited by the physical sizes of camera pixels. Since the Rayleigh limit is ~200 nm for conventional fluorescence microscopy, the effective pixel size should be ~70 nm or smaller. The criterion of one third of the Rayleigh limit is empirically based on the Nyquist criterion on spatial sampling. When the image signal-to-noise ratio is high, the effective pixel size allowed can be larger. Although subpixel resolution detection is usually preferred, there are cases when the effective pixel size requirement cannot be met. For example, when particle signals are very weak, cameras with large pixel sizes may have to be used for effective photon collection.

A variety of algorithms are available for pixel-level resolution particle detection (Nixon and Aguado 2012). A representative algorithm is to first identify local intensity maxima as potential candidates (Ponti et al. 2003). Those candidates whose intensities are statistically significantly higher than their local background intensities are classified as true particles. That is, given a local intensity maximum $I_{max}(x,y)$ whose local background intensity is $I_{BG}(x,y)$, it is classified as a true particle if the condition $(I_{max}(x,y) - I_{BG}(x,y)) > Q_\alpha\sqrt{\sigma_I^2(x,y) + \sigma_{BG}^2(x,y)}$ is satisfied. Here $\sigma_I(x,y)$ and $\sigma_{BG}(x,y)$ denote standard deviations of the signal and the background, respectively, subscript α is a user specified detection confidence level, and Q_α is the student's t-distribution quantile corresponding to α. The level of detection selectivity can be tuned by adjusting Q_α.

To allow users to adjust the selectivity level of particle detection is desirable in certain cases. For example, if the goal is to ensure weak particles are detected, a low selectivity level can be applied, although this may result in more false positive detections. Conversely, if the goal is to minimize the number of false positive detections, a high selectivity level can be applied, although this may result in loss of weak particles. There are other cases when user adjustment of selectivity level is not preferred. For example, to process large volume of image data from high-throughput experiments, automated setting of selectivity levels is required. Several studies proposed automated detection algorithms, especially for low signal to noise ratio (SNR) (e.g., <5) images (Smal et al. 2010; Lei et al. 2012). However, the proposed methods were designed for detection of small pixel patches (regions) rather than particles. A strategy to overcome this limitation is to combine them with a search for intensity maxima within the segmented pixel patches.

Similar to the case of pixel-level particle detection, a variety of algorithms for subpixel resolution detection have been developed, such as maximum cross-correlation and kernel fitting (Cheezum et al. 2001; Yildiz and Selvin 2005). Many studies have shown that image fitting using a Gaussian kernel as an approximation of the point-spread function of the microscope used provides overall the best accuracy (Cheezum et al. 2001; Yildiz and Selvin 2005). Given an initial position (x, y), the kernel fitting can be formulated as an optimization search for a nearby particle position (x_0, y_0) as well as intensity I_0 and background noise B that minimize the following fitting error function:

$$E(x,y) = \int_{x-h}^{x+h} \int_{y-h}^{y+h} \left(I(x,y) - B - I_0 e^{-\frac{(x-x_0)^2 + (y-y_0)^2}{\sigma^2}} \right)^2 dxdy \tag{9.1}$$

Here σ is the standard deviation of the Gaussian kernel and is often chosen as $0.21 \cdot \lambda / NA$, and h is half of the fitting window size. The initial position for kernel fitting may be provided by algorithms for pixel-level resolution detection. The best achievable localization uncertainty σ_L can be estimated using the following formula (Thompson et al. 2002):

$$\sigma_L^2 = \frac{s^2}{N} + \frac{a^2/12}{N} + \frac{8\pi s^4 b^2}{a^2 N^2} \tag{9.2}$$

Here s is the standard deviation of the microscope point-spread function and is often estimated as $0.21 \cdot \lambda / NA$, N is the total number of photon collected, a is the pixel size, and b is background noise. This formula also reveals another critical determinant of localization accuracy in addition to the effective pixel size, namely the total number of photons, or equivalently, the SNR. Localization accuracy of 1 nm has been demonstrated when the number of photons or the SNR was sufficiently high (Yildiz and Selvin 2005). It is worth noting that when the spatial sampling criterion is not satisfied, kernel fitting techniques may actually give worse localization accuracy than techniques for pixel-level resolution particle detection.

As an essential image analysis technique, single particle detection is provided as a function of several general-purpose software packages for biological image analysis (Eliceiri et al. 2012). Kernel fitting based subpixel resolution particle detection is also a standard technique for localization-based super-resolution imaging techniques such as STORM and PALM (Huang et al. 2009; Patterson et al. 2010). Related techniques and software tools developed for super-resolution imaging often can be adopted for cell mechanobiology applications. A list of these software tools can be found at bigwww.epfl.ch/smlm/software/index.html.

9.3.2 Computational image analysis techniques for particle tracking

After detection of particles within each frame of an image sequence, correspondence between detected particles must be established to recover complete particle trajectories. This is the goal of single particle tracking. In the example of a particle within a 2D image sequence of N frames, its trajectory is typically represented by a time series $\{(x_1^k, y_1^k; I_1^k), (x_2^k, y_2^k; I_2^k), \ldots (x_j^k, y_j^k; I_j^k), \ldots (x_N^k, y_N^k; I_N^k)\}$, where (x_j^k, y_j^k) and I_j^k denote the position and intensity, respectively, of this k-th numbered particle in the j-th image frame. A basic strategy for recovering such a trajectory is to establish correspondence between detected particles within each pair of consecutive frames. Single particle tracking is an essential technique for characterizing spatiotemporal dynamics of cellular processes, and a wide variety of algorithms have been developed (Saxton and Jacobson 1997; Meijering et al. 2012; Chenouard et al. 2014).

Local nearest neighbor (LNN) tracking is perhaps the simplest single particle tracking algorithm (Blackman and Popoli 1999). Its basic idea is to associate each detected particle in a given frame to its nearest neighbor in the next frame, and then repeat this procedure for all particles through the entire image sequence. LNN is straightforward to implement and generally performs well when displacements of particles within each frame are substantially smaller than distances between them (Gao and Kilfoil 2009). However, in many cell biological applications, this assumption is not satisfied because of the high spatial density of particles. Consequently, correspondence conflicts arise under LNN tracking. For example, multiple particles may compete for correspondence with one particle, or one particle may have to choose among multiple particles for correspondence. To resolve such conflicts, an effective solution is to use the global nearest neighbor (GNN) algorithm, which formulates the tracking as the following linear assignment problem (Burkard et al. 2009):

$$\begin{aligned} & min \sum_{i \in G_k} \sum_{j \in G_{k+1}} a^k(i,j) w^k(i,j) \\ & st. \quad \sum_i a(i,j) = 1, \quad \sum_i a(i,j) = 1, \quad a(i,j) \in \{0,1\} \end{aligned} \tag{9.3}$$

where G_k and G_{k+1} denote the sets of detected particles in frames k and $k+1$, respectively, and $a(i,j)$ is a binary variable that equals 1 if particle i in frame k is assigned to particle j in frame $k+1$ at the assignment cost denoted by $w^k(i,j)$. The constraint equations $\sum_i a(i,j) = 1$ and $\sum_j a(i,j) = 1$ ensure a one-to-one correspondence between particles in frame k and frame $k+1$.

The basic idea of GNN is to identify the assignments between the two sets of particles that give the lowest *total* assignment cost. The assignment cost between particle i and particle j can take different forms, such as their Euclidean distance, the Mahalanobis distance between the predicted position of particle i in frame $k+1$ and particle j (Cox 1993), or the smoothness of movement (Veenman et al. 2001). Because highly efficient algorithms are available to solve large scale linear assignment problems (Burkard et al. 2009), an important advantage of the GNN algorithm is that it can be used to track hundreds of thousands of particles for applications such as fluorescent speckle microscopy (Yang et al. 2008). However, a basic limitation of the GNN algorithm is that assignment decisions are made based on information from two consecutive frames and are not necessarily optimal over the entire image sequence. This limitation is alleviated partially by using various models of particle movement. For example, the known history of particle movement can be used to train such models to predict the next position of each particle, which can then be used to calculate, for example, the Mahalanobis distance (Cox 1993). Simple particle movement is often modeled using Kalman filters (Bar-Shalom et al. 2001). When particle movement switches between different modes, interacting multiple model (IMM) filters can be used (Bar-Shalom et al. 2001; Genovesio et al. 2006; Lei et al. 2012).

As an alternative to the deterministic GNN algorithm, the particle-tracking problem can also be formulated in a statistical framework and solved using for example the joint-probabilistic data-association filtering algorithm (JPDAF) (Cox 1993; Bar-Shalom et al. 2001; Smal et al. 2008). Performance of such statistical algorithms depends critically on how accurately the probabilistic models represent particle behavior. Some of these algorithms, including JPDAF, have high computational cost and therefore cannot be used for tracking large numbers of particles.

Deterministic and statistical single particle-tracking algorithms such as GNN and JPDAF have been successfully applied in a variety of cell biological studies (Meijering et al. 2009; Meijering et al. 2012). However, a common limitation of them is that they cannot effective handle complex events such as particle appearance and disappearance as well as merging and splitting. For example, the basic formulation of GNN (equation 9.3) stipulates a one-to-one correspondence between particles within two consecutive frames. This requires that the number of particles within each frame to remain the same, which is unrealistic for most cell biological images. Virtual points must be introduced to handle particle appearance and disappearance within GNN, a cumbersome process (Veenman et al. 2001). An effective solution to overcome this limitation is multiple hypothesis tracking (MHT), in which particle trajectories are represented using data structures such as trees (Blackman and Popoli 1999; Padfield et al. 2011). The tree representation has the flexibility to handle complex particle events. However, these algorithms may not be suitable for tracking large numbers (e.g., >1000) of particles because of their high computational cost of growing and maintaining the trees. Another important class of algorithms first recover segments of trajectories, referred to as tracklets, and then link them into full trajectories in postprocessing (Jaqaman et al. 2008; Li et al. 2008). These algorithms have lower computational cost than MHT and can effectively track large numbers of particles but are less effective in handling complex particle events.

Particle movement in cellular processes is often constrained by the underlying cellular structures. Tracking performance can be improved substantially by incorporating knowledge of such constraints in forms such as prior probability distributions into single particle-tracking algorithms. For example, performance of tracking tubulin speckles in *Xenopus* egg extract spindles was significantly improved by using the local vector field of tubulin movement (Yang et al. 2008). This was also shown by a recent study that compared the performance of fourteen representative single particle-tracking algorithms (Chenouard et al. 2014), which found that using application specific knowledge and models can substantially improve tracking performance. It also found that there were no universally optimal algorithms for different cell biological applications and that performance of all fourteen algorithms deteriorated substantially under low SNR and high particle density (Chenouard et al. 2014).

So far, we have been focusing on the question of recovering complete particles trajectories using single particle tracking algorithms. However, as mentioned previously, tracking performance often deteriorates substantially when the spatial density of particles is high (Chenouard et al. 2014). In such cases, to recover complete particle trajectories may no longer be feasible. An alternative is to recover the vector field of particle movement, which defines the velocity and direction of particle movement at different locations using algorithms such as graph optimization (Matov et al. 2011) or image correlation (Ji and Danuser 2005).

From recovered particle trajectories, simple quantitative descriptors can be computed to characterize particle behavior. For a particle undergoes directed motion, whose trajectory is represented by $\{(x_1^k, y_1^k; I_1^k), (x_2^k, y_2^k; I_2^k), \ldots (x_j^k, y_j^k; I_j^k), \ldots (x_N^k, y_N^k; I_N^k)\}$, its behavior can be characterized using descriptors such as its average velocity, defined as $v^k = \frac{1}{(N-1)\cdot T}\sum_{i=2}^{N}\sqrt{\left(x_i^k - x_{i-1}^k\right)^2 + \left(y_i^k - y_{i-1}^k\right)^2}$ where T is the time interval between two consecutive frames. For a particle undergoing random walk, its behavior can be characterized using the relation of its mean-square displacement (MSD) with respect to time (Qian et al. 1991; Saxton 2007). A particle undergoing one-dimensional pure diffusion satisfies the linear relation of $MSD(t) = 2Dt$ where D is its diffusion coefficient. If the particle undergoes diffusion superimposed by directed movement or local flow, its MSD follows $MSD(t) = 2Dt + v^2t^2$, where v is the movement or flow velocity. If the particle undergoing locally confined diffusion, its MSD follows $MSD(t) = MSD(\infty)\left[1 - e^{-t/\tau}\right]$ where τ is a constant reflecting the local confinement.

Because of its many important applications in biological image analysis, single particle tracking is provided as a function of several general-purpose biological image analysis software packages such as ICY (de Chaumont et al. 2012) and Fiji (Schindelin et al. 2012).

9.4 Applications of single particle detection and tracking techniques in cell mechanobiology

Single particle detection and tracking techniques have been used to study cell mechanobiology at both the single particle level and over the whole cell. In the first case, dynamic behavior of particles is analyzed individually to study local cell mechanobiology. In the second case, dynamic behavior of individual particles is analyzed collectively, often in combination with physical modeling, to study global cell mechanobiology. We review these two cases separately.

9.4.1 Applications in studying cell mechanobiology at the single particle scale

A representative application is to combine single particle detection and tracking with optical or magnetic micromanipulation for characterizing mechanical properties of molecules such as motor proteins, RNA polymerase, or DNA (Svoboda and Block 1994; Herbert et al. 2008; De Vlaminck and Dekker 2012). Subpixel resolution particle detection was used to measure step sizes of kinesin and dynein (Gelles et al. 1988; Svoboda and Block 1994; Moffitt et al. 2008). Based on calibration of the stiffness of optical tweezers and modeling its physical properties, the stall forces of kinesin and dynein were determined from their movement quantified by single particle tracking (Svoboda and Block 1994). Another representative application is to combine single particle tracking with physical modeling for characterizing cellular mechanical properties in cell microrheology (Crocker and Hoffman 2007; Wirtz 2009). For example, the relation between cell viscosity and particle movement is described by the Einstein-Stokes relation (Crocker and Hoffman 2007):

$$\eta = \frac{k_B T}{\langle \Delta r(\tau)^2 \rangle \ \pi a}\tau \tag{9.4}$$

Where η is the local viscosity of the cell cytoplasm, k_B is the Boltzman's constant, T is the absolute temperature, $\langle \Delta(\tau)^2 \rangle$ is the *MSD* over time interval τ, and a is the physical radius of the particle. *MSD* of individual particles are calculated from their trajectories recovered by single particle tracking. The local cell cytoplasm viscosity can then be determined from the *MSD* of individual particles.

An implicit assumption often made in single particle tracking is that particles remain in the same mode of dynamic behavior. However, this assumption may not hold within the heterogeneous cellular environment. Recent development in single particle tracking techniques has provided tools to detect and characterize changes in particle behavior (Das et al. 2009). Here we introduce these techniques using analysis of vesicle transport in the axon of Drosophila as an example (Fig. 9.2A). Complete trajectories of individual vesicles were recovered using single particle tracking (Qiu et al. 2012) (Fig. 9.2B–C). Individual vesicles can switch between three states: movement toward the synaptic terminal (anterograde movement), movement toward the cell body (retrograde movement), and pause. Such switches were described using a hidden Markov model (Fig. 9.2D). Movement of vesicle in each direction also can switch between three velocity modes (Fig. 9.2E) (Reis et al. 2012), which were also described using a hidden Markov model (Fig. 9.2F). Overall, this example demonstrates the application of computational models in characterizing complex behavior of single particles, which is essential for understanding the underlying molecular mechanisms.

9.4.2 Applications in studying cell mechanobiology at the whole cell scale

In Section 9.4.1 we showed examples of using single particle detection and tracking techniques for studying local cell mechanobiology. Here we survey applications in which activities of particles are analyzed collectively for studying cell mechanobiology at the whole cell scale.

Traction force microscopy (Kraning-Rush et al. 2012; Style et al. 2014) provides a representative application of single particle tracking in studying global cell

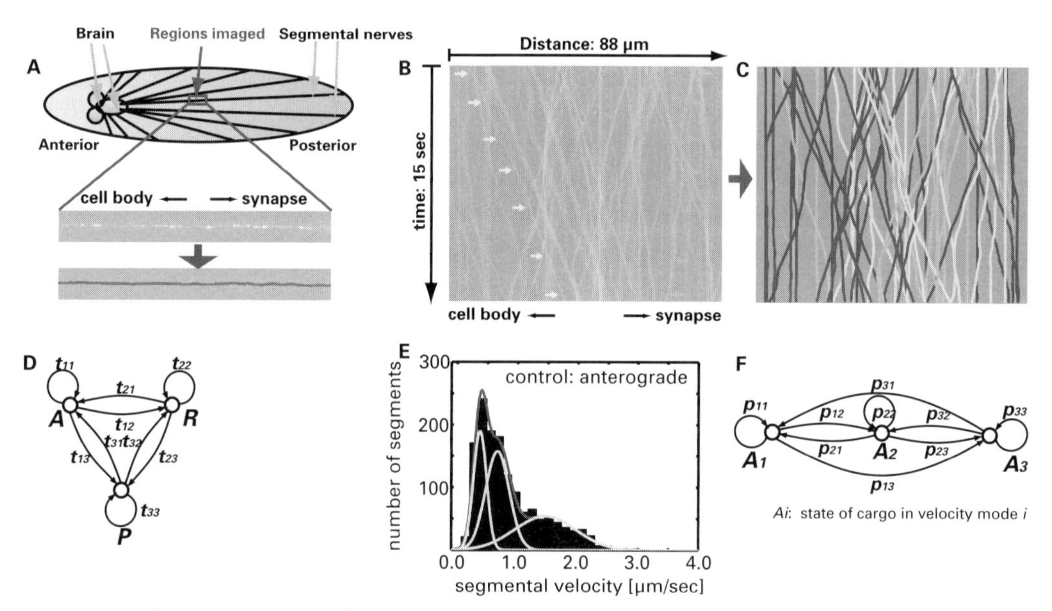

Figure 9.2 **Tracking and modeling complex behavior of vesicles in axonal transport.** (A) Upper panel: regions selected for imaging. Middle panel: one frame from a time lapse video of APP vesicle transport. Lower panel: a band (5 pixels between green curves) following the axon (marked in cyan) is taken from each frame. (B) Bands from all frames were placed sequentially to generate a kymograph, a map of vesicle movement along the axon over time. Yellow and green arrows point to trajectories of two vesicles moving towards the synapse and the cell body, respectively. (C) Trajectories recovered by single particle tracking software were randomly colored and overlaid onto the kymograph for visual inspection. (D) A hidden Markov model (HMM) of cargo behavior, which characterizes transitions between anterograde movement A, retrograde movement R, and pause P. (E) Velocities of anterograde cargoes follow three modes (red: total distribution. cyan: individual modes). (F) Another hidden Markov model (HMM) of cargo behavior, which characterizes transitions between different anterograde velocity modes.

(A–C, E) reproduced with permission from Reis et al. (2012).

mechanobiology. After displacements of individual particles are determined from single particle tracking, the field of substrate displacement is computed through interpolation and used to calculate mechanical properties of the whole cell. For example, the traction stress field of the cell can be calculated from the following equation (Legant et al. 2010; Style et al. 2014)

$$\sigma(x,y,z) = \frac{E}{1-v}\left(\frac{1}{2}\left(\nabla u(x,y,z) + \nabla u^T(x,y,z)\right) + \frac{1-2v}{v}\nabla \cdot u(x,y,z)\cdot I\right) \quad (9.5)$$

where E is the Young's modulus, v is the Poisson's ratio, and $u(x,y,z)$ denotes substrate displacement at position (x,y,z).

Traction force microscopy measures *extracellular* forces applied to the cells. Recently, a new technique was developed to map *intracellular* forces from the vector field of actin flow that was visualized and mapped using fluorescent actin speckles as tracers (Ji et al. 2007; Ji et al. 2008). The technique assumed a continuum mechanical model of the actin cytoskeleton. The basic idea was to calculate the deformation of the actin network based on single particle tracking of actin speckles. Then, based on the continuum mechanical model of the actin network, the intracellular force was inversely estimated through regularized optimization (Ji et al. 2008).

9.5 Computational image analysis techniques for continuous region images and their applications in cell mechanobiology

Individual particles in single particle images have the same appearance of an Airy disk. Although their corresponding cellular structures may have different sizes, these structures cannot be differentiated because of the microscope resolution limit. Consequently, shapes of particles are generally not considered in computational analysis of single particle images. In contrast, shapes and intensity distributions of region objects in continuous region images carry essential information about their corresponding cellular structures and must be characterized. For cell mechanobiology applications, to detect and track regions objects is one of the main goals for computation analysis of continuous regions images. We review related computational technique and their applications.

9.5.1 Computational image analysis techniques for region detection

The first step in computational analysis of continuous region images is to detect their region objects, a procedure referred to as image segmentation. Because of the critical importance of image segmentation to many image analysis applications, a very large variety of image segmentation techniques have been developed, such as intensity thresholding, region growth, pixel clustering, and deformable model fitting (Pham et al. 2000; Dima et al. 2011). Detailed introduction to these techniques can be found in references such as (Sonka et al. 2007; Szeliski 2010; Nixon and Aguado 2012). Here we address the question of how to choose image segmentation techniques for cell mechanobiology applications.

Cell biological images often have low signal-to-noise ratios for reasons such as the need to minimize phototoxicity and photodamage. In addition, cell biological images often lack sharp edges because of the limited depth-of-field of high resolution microscopy (Fig. 9.3). These properties can have significant impact on the performance of image segmentation techniques (Dima et al. 2011). For example, image segmentation techniques that do not rely on sharp edges have been shown to give better performance than conventional techniques (Srinivasa et al. 2009; Chen et al. 2012) (Fig. 9.3). Overall, there are no universally optimal segmentation algorithms. Instead, to test and compare different segmentation techniques is essential. Importantly, segmentation performance can be improved by utilizing application-specific knowledge and models (Maska et al. 2014) as well as interactive user inputs (Heimann et al. 2009).

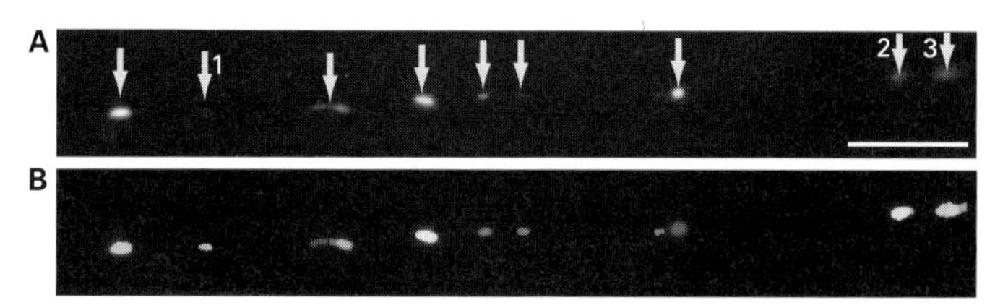

Figure 9.3 **Identification of region objects through image segmentation.**
(A) Mitochondria (yellow arrows) distributed along a single axon of a Drosophila third instar larva (same as Fig. 9.1E). Many of them have weak edges because they are partially out of focus, especially those marked by arrows 1, 2, and 3. Scale bar: 10 μm. (B) Segmented mitochondria shown in random colors.

9.5.2 Computational image analysis techniques for region tracking

A particle can be completely characterized by its position and intensity. Tracking of the particle is therefore equivalent to determining its positions and intensities over time. In contrast, complete characterization of a region requires its position, shape, as well as internal intensity distribution, all of which must be determined through image segmentation. Region tracking can take different forms depending on specific applications. Cell tracking is an important and representative application of region tracking. We review related computational image analysis techniques using cell tracking as an example. Such techniques are usually applicable to tracking other region objects.

Computational techniques for tracking positions and shapes of regions
The overall goal of cell tracking is to follow positions and shapes of individual cells over time. A variety of tracking algorithms have been developed (Meijering et al. 2012; Maska et al. 2014). The algorithms can be classified into two categories (Maska et al. 2014). Algorithms in the first category first detect cells within each image frame through segmentation and then establish correspondence between the detected cells over time. In this case, the segmentation and tracking are conducted separately. Positions of the cells are often represented by positions of their centroids, whereas shapes of the cells are used as a cue in establishing correspondence. A main advantage of these algorithms is their flexibility. Different segmentation algorithms and tracking algorithms can be combined, and segmentation performance and tracking performance are largely decoupled. On the other hand, a main disadvantage of these algorithms is that shape and correspondence information cannot be integrated effectively. For example, segmentation of region objects over time cannot utilize their correspondence information. Algorithms in the second category use various models of cell boundary evolution to effectively integrate shape and correspondence information (Meijering et al. 2009; Maska et al. 2014). However, these models are often rather complex, and their performance depends heavily on whether they accurately reflect cell behavior. Overall, techniques for tracking cells can be used to track other region objects.

Computational techniques for tracking deformations of regions
For cell tracking, the main goal is to follow individual cells over time. Tracking cell shape changes may not always be necessary. However, certain cell mechanobiology applications such as studies of cell protrusion (Machacek and Danuser 2006) and cell aspiration (Vaziri and Mofrad 2007) require high-resolution characterization of shape changes. Here the main goal is to determine deformation of a region object over time at different locations along its boundary. This usually requires placing sampling points along the region boundary and tracking their positions over time in a way that is consistent with the physical properties of the underlying cellular structures and processes without violating topological constraints (Machacek and Danuser 2006). Such applications are closely connected with non-rigid image object registration (Crum et al. 2004). The main challenge here is to develop models that accurately represent physical properties of the underlying cellular structures and processes. An example of such models for tracking protruding cell edges can be found in (Machacek and Danuser 2006).

Computational techniques for tracking intensity changes within regions
Certain cellular mechanical processes are characterized by changes of image intensity distributions with regions. To track intensity changes within regions gives another form of region tracking. The translocation of spindle microtubule networks (Fig. 9.1D) provides an example. Tubulins within the spindle microtubule network undergoes poleward movement in a process called microtubule flux, which is thought to be essential for the faithful separation of replicated chromosomes during mitosis (Mitchison 2005). The movement can be directly visualized using single particle images of FSM (Danuser and Waterman-Storer 2006) (Fig. 9.1D). However, in cases such as when tubulins are labelled by genetically encoded fluorescent proteins, only continuous region images can be produced. In such cases, poleward movement of the tubulins can no longer be followed by single particle tracking. Instead, it can be estimated from the changes of image intensity using a computational image analysis technique called optical flow estimation (Fleet and Weiss 2005; Szeliski 2010).

Optical flow estimation requires stable or quasi-stable intensity signals. Specifically, if $I(x,t)$, the image intensity signal at position x and time t, remains stable after undergoing a displacement at a velocity of u for a short interval of τ, we have $I(x,t) = I(x+u\tau, t+\tau)$. If we assume that the movement can be well approximated by a first-order Taylor expansion:

$$I(x+u\tau, t+\tau) \approx I(x,t) + \nabla I(x,t)u\tau + \tau\frac{\partial I(x,t)}{\partial t} \tag{9.6}$$

we have the following basic equation of optical flow estimation:

$$\nabla I(x,t)u + \frac{\partial I(x,t)}{\partial t} = 0 \tag{9.7}$$

Here $\frac{\partial I(x,t)}{\partial t}$ denotes changes of intensity over time, and $\nabla I(x,t)$ denotes changes of intensity over space. When the image contrast is sufficiently high, $\nabla I(x,t)$ is nonsingular, and u can be estimated using Eq. (9.7). A wide variety of optical flow estimation techniques have been developed. Performance comparison of these techniques can be found in references such as (Barron et al. 1994; Sun et al. 2014).

9.5.3 Applications of region detection and tracking techniques in cell mechanobiology

Region tracking techniques have been used extensively in cell mechanobiology applications. For example, cell-tracking techniques have been used to quantify cell migration in many cell and developmental biology studies (Meijering et al. 2009). Deformation tracking has been used for characterizing cell mechanical properties through pipette aspiration (Hochmuth 2000; Vaziri and Mofrad 2007) and for studying spatiotemporal integration of mechanochemical signals in cell protrusion (Machacek and Danuser 2006; Machacek et al. 2009). Customized optical flow estimation techniques have been developed to quantify dynamic turnover of cytoskeletal filament networks (Ji and Danuser, 2005; Wilson et al. 2010) as well as stress propagation in living cells (Wang et al. 2007).

9.6 Outlook

We conclude by presenting some personal views on the future development of computational image analysis techniques for cell mechanobiology. Overall, we think that the development of experimental techniques, especially image formation and fluorescence sensor techniques, will continue to be a major driving force of the development of new image analysis techniques. Currently, the majority of image analysis techniques in cell mechanobiology are for 2-D images. This is due to the fact that most of the image formation techniques as well as experimental assays remain two-dimensional. However, new assays for studying cell mechanobiology in 3-D are increasingly being developed (Legant et al. 2010; Ulrich et al. 2010). New imaging techniques such as plane illumination microscopy have made it possible to image cells in 3-D under live conditions with high temporal resolution and minimal photobleaching (Planchon et al. 2011). These techniques will increasingly shift the focus of computational image analysis to 3-D images. In the meantime, super-resolution imaging techniques will continue to push the boundaries of spatial resolutions in cell mechanobiology. This will drive the development of new image analysis techniques specifically for super-resolution images. In addition, novel fluorescence biosensors will allow direct visualization of intracellular forces (Grashoff et al. 2010). This will enable us to follow how cellular mechanical processes such as force transduction are integrated with other cellular processes such as reorganization of cytoskeletal filament networks and propagation of intracellular signals. Overall, we envision that computational image analysis techniques will continue to play an essential role in helping investigators to understand complex behavior and mechanisms of cellular mechanical processes quantitatively and at the systems level.

Acknowledgments

We apologize to colleagues whose work could not be cited due to space limitations. We acknowledge support of NSF grants MCB-1052660 and DBI-1052925 as well as NSF Faculty Early Career Award DBI-1149494.

References

Bar-Shalom, Y., Li, X. R. and Kirubarajan, T. (2001). *Estimation with Applications to Tracking and Navigation*. New York: Wiley-Interscience.

Barron, J. L., Fleet, D. J. and Beauchemin, S. S. (1994). "Performance of optical flow techniques." *Int J Comp Vis* **12**: 43–77.

Betzig, E., Patterson, G. H., Sougrat, R., Lindwasser, O. W., Olenych, S., Bonifacino, J. S., Davidson, M. W., et al. (2006). "Imaging intracellular fluorescent proteins at nanometer resolution." *Science* **313**: 1642–1645.

Blackman, S. and Popoli, R. (1999). *Design and Analysis of Modern Tracking Systems*. Norwood, MA: Artech House.

Born, M. and Wolf, E. (1999). *Principles of Optics*. Cambridge University Press.

Burkard, R., Dell'amico, M. and Martello, S. (2009). *Assignment Problems*. Philadelphia: Society for Industrial and Applied Mathematics.

Cheezum, M. K., Walker, W. F. and Guilford, W. H. (2001). "Quantitative comparison of algorithms for tracking single fluorescent particles." *Biophys J* **81**: 2378–2388.

Chen, K. C., Yang, G. and Kovacevic, J. (2014). "Spatial density estimation based segmentation of super-resolution localization microscopy images." *Proc 2014 IEEE Int Conf Image Proc (ICIP)*: 867–871.

Chen, K. C. J., Yiyi, Y., Ruiqin, L., Hao-Chih, L., Ge, Y. and Kovacevic, J. (2012). "Adaptive active-mask image segmentation for quantitative characterization of mitochondrial morphology." *Proc 2012 IEEE Int Conf Image Proc (ICIP)*: 2033–2036.

Chenouard, N., Smal, I., de Chaumont, F., Maska, M., Sbalzarini, I. F., Gong, Y., Cardinale, J., et al. (2014). "Objective comparison of particle tracking methods." *Nat Meth* **11**: 281–289.

Cox, I. (1993). "A review of statistical data association techniques for motion correspondence." *Int J Comp Vis* **10**: 53–66.

Crocker, J. C. and Hoffman, B. D. (2007). "Multiple-particle tracking and two-point microrheology in cells." *Meth in Cell Biol* **83**: 141–178.

Crum, W. R., Hartkens, T. and Hill, D. L. G. (2004). "Non-rigid image registration: theory and practice." *British J Radiology* **77**: S140–S153.

Danuser, G. and Waterman-Storer, C. M. (2006). "Quantitative fluorescent speckle microscopy of cytoskeleton dynamics." *Annu Rev Biophys Biomol Struct* **35**: 361–387.
Das, R., Cairo, C. W. and Coombs, D. (2009). "A hidden Markov model for single particle tracks quantifies dynamic interactions between LFA-1 and the actin cytoskeleton." *PLoS Comp Biol* **5**: e1000556.
De Chaumont, F., Dallongeville, S., Chenouard, N., Herve, N., Pop, S., Provoost, T., Meas-Yedid, V., et al. (2012). "Icy: an open bioimage informatics platform for extended reproducible research." *Nat Meth* **9**: 690–696.
De Vlaminck, I. and Dekker, C. (2012). "Recent advances in magnetic tweezers." *Annu Rev Biophys* **41**: 453–472.
Dhawan, A. P. (2011). *Medical Image Analysis*. New York: Wiley-IEEE Press.
Dima, A. A., Elliott, J. T., Filliben, J. J., Halter, M., Peskin, A., Bernal, J., Kociolek, M., et al. (2011). "Comparison of segmentation algorithms for fluorescence microscopy images of cells." *Cytometry Part A* **79A**: 545–559.
Dorn, J. F., Danuser, G., and Yang, G. (2008). "Computational processing and analysis of dynamic fluorescence image data." *Meth Cell Biol* **85**: 497–538.
Eliceiri, K. W., Berthold, M. R., Goldberg, I. G., Ibanez, L., Manjunath, B. S., Martone, M. E., Murphy, R. F., et al. (2012). "Biological imaging software tools." *Nat Meth* **9**: 697–710.
Fleet, D. J. and Weiss, Y. (2005). Optical flow estimation. In Paragios, N., Chen, Y. and Faugeras, O. (eds.), *Mathematical Models in Computer Vision*. New York: Springer.
Gao, Y. and Kilfoil, M. L. (2009). "Accurate detection and complete tracking of large populations of features in three dimensions." *Opt Exp* **17**: 4685–4704.
Gelles, J., Schnapp, B. J. and Sheetz, M. P. (1988). "Tracking kinesin-driven movements with nanometre-scale precision." *Nature* **331**: 450–453.
Genovesio, A., Liedl, T., Emiliani, V., Parak, W. J., Coppey-Moisan, M. and Olivo-Marin, J. C. (2006). "Multiple particle tracking in 3-D+t microscopy: method and application to the tracking of endocytosed quantum dots." *IEEE Trans Image Proc* **15**: 1062–1070.
Giepmans, B. N. G., Adams, S. R., Ellisman, M. H. and Tsien, R. Y. (2006). "The fluorescent toolbox for assessing protein location and function." *Science* **312**: 217–224.
Grashoff, C., Hoffman, B. D., Brenner, M. D., Zhou, R., Parsons, M., Yang, M. T., Mclean, M. A., et al. (2010). "Measuring mechanical tension across vinculin reveals regulation of focal adhesion dynamics." *Nature* **466**: 263–266.
Heimann, T., van Ginneken, B., Styner, M. A., Arzhaeva, Y., Aurich, V., Bauer, C., Beck, A., et al. (2009). "Comparison and evaluation of methods for liver segmentation from CT datasets." *IEEE Trans Med Imaging* **28**: 1251–1265.
Herbert, K. M., Greenleaf, W. J. and Block, S. M. (2008). "Single-molecule studies of RNA polymerase: motoring along." *Annu Rev Biochem* **77**: 149–176.
Hochmuth, R. M. (2000). "Micropipette aspiration of living cells." *J Biomechanics* **33**: 15–22.
Huang, B., Bates, M. and Zhuang, X. (2009). "Super-resolution fluorescence microscopy." *Annu Rev Biochem* **78**: 993–1016.
Jaqaman, K., Loerke, D., Mettlen, M., Kuwata, H., Grinstein, S., Schmid, S. L. and Danuser, G. (2008). "Robust single-particle tracking in live-cell time-lapse sequences." *Nat Meth* **5**: 695–702.

Ji, L. and Danuser, G. (2005). "Tracking quasi-stationary flow of weak fluorescent signals by adaptive multi-frame correlation." *J Microscopy* **220**: 150–167.

Ji, L., Lim, J. and Danuser, G. (2008). "Fluctuations of intracellular forces during cell protrusion." *Nat Cell Biol* **10**: 1393–1400.

Ji, L., Loerke, D., Gardel, M., Danuser, G. (2007). "Probing intracellular force distributions by high-resolution live cell imaging and inverse dynamics." *Meth Cell Biol* **83**: 199–235.

Kraning-Rush, C. M., Carey, S. P., Califano, J. P. and Reinhart-King, C. A. (2012). "Quantifying traction stresses in adherent cells." *Methods in Cell Biology* **110**: 139–178.

Legant, W. R., Miller, J. S., Blakely, B. L., Cohen, D. M., Genin, G. M. and Chen, C. S. (2010). "Measurement of mechanical tractions exerted by cells in three-dimensional matrices." *Nat Meth* **7**: 969–971.

Lei, Y., Zhen, Q., Greenaway, A. H. and Weiping, L. (2012). "A new framework for particle detection in low-SNR fluorescence live-cell images and its application for improved particle tracking." *IEEE Trans Biomed Eng* **59**: 2040–2050.

Li, K., Miller, E. D., Chen, M., Kanade, T., Weiss, L. E. and Campbell, P. G. (2008). "Cell population tracking and lineage construction with spatiotemporal context." *Med Image Analy* **12**: 546–566.

Lippincott-Schwartz, J., Snapp, E. and Kenworthy, A. (2001). "Studying protein dynamics in living cells." *Nat Rev Mol Cell Biol* **2**: 444–456.

Machacek, M. and Danuser, G. (2006). "Morphodynamic profiling of protrusion phenotypes." *Biophys J* **90**: 1439–1452.

Machacek, M., Hodgson, L., Welch, C., Elliott, H., Pertz, O., Nalbant, P., Abell, A., et al. (2009). "Coordination of Rho GTPase activities during cell protrusion." *Nature* **461**: 99–103.

Maska, M., Ulman, V., Svoboda, D., Matula, P., Matula, P., Ederra, C., Urbiola, A., et al. (2014). "A benchmark for comparison of cell tracking algorithms." *Bioinformatics* **30**: 1609–1617.

Matov, A., Edvall, M. M., Yang, G. and Danuser, G. (2011). "Optimal-flow minimum-cost correspondence assignment in particle flow tracking." *Comput Vis Image Underst* **115**: 531–540.

Meijering, E., Dzyubachyk, O. and Smal, I. (2012). "Methods for cell and particle tracking." *Meth Enzymol* **504**: 183–200.

Meijering, E., Dzyubachyk, O., Smal, I. and van Cappellen, W. A. (2009). "Tracking in cell and developmental biology." *Semi Cell Dev Biol* **20**: 894–902.

Mitchison, T. J. (2005). "Mechanism and function of poleward flux in Xenopus extract meiotic spindles." *PhiloTrans Royal Soc B: BiolSci* **360**: 623–629.

Moffitt, J. R., Chemla, Y. R., Smith, S. B. and Bustamante, C. (2008). "Recent advances in optical tweezers." *Annu Rev Biochem* **77**: 205–228.

Nixon, M. and Aguado, A. (2012). *Feature Extraction and Image Processing for Computer Vision*. Waltham, MA: Academic Press.

Padfield, D., Rittscher, J. and Roysam, B. (2011). "Coupled minimum-cost flow cell tracking for high-throughput quantitative analysis." *Med Image Analy* **15**: 650–668.

Patterson, G., Davidson, M., Manley, S. and Lippincott-Schwartz, J. (2010). "Superresolution imaging using single-molecule localization." *Annu Rev Phys Chem* **61**: 345–367.

Peng, H. (2008). "Bioimage informatics: a new area of engineering biology." *Bioinformatics* **24**: 1827–1836.
Pham, D. L., Xu, C. and Prince, J. L. (2000). "Current methods in medical image segmentation." *Annu Rev Biomed Eng* **2**: 315–337.
Planchon, T. A., Gao, L., Milkie, D. E., Davidson, M. W., Galbraith, J. A., Galbraith, C. G. and Betzig, E. (2011). "Rapid three-dimensional isotropic imaging of living cells using Bessel beam plane illumination." *Nat Meth* **8**: 417–423.
Ponti, A., Vallotton, P., Salmon, W. C., Waterman-Storer, C. M. and Danuser, G. (2003). "Computational analysis of F-actin turnover in cortical actin meshworks using fluorescent speckle microscopy." *Biophys J* **84**: 3336–3352.
Qian, H., Sheetz, M. P. and Elson, E. L. (1991). "Single particle tracking: analysis of diffusion and flow in two-dimensional systems." *Biophys J* **60**: 910–921.
Qiu, M., Lee, H.-C. and Yang, G. (2012). "Nanometer resolution tracking and modeling of bidirectional axonal cargo transport." *Proc 2012 IEEE Int Symp Biomed Imaging (ISBI)*: 992–995.
Reis, G. F., Yang, G., Szpankowski, L., Weaver, C., Shah, S. B., Robinson, J. T., Hays, T. S., et al. (2012). "Molecular motor function in axonal transport in vivo probed by genetic and computational analysis in Drosophila." *Mol Biol Cell* **23**: 1700–1714.
Rust, M. J., Bates, M. and Zhuang, X. (2006). "Sub-diffraction-limit imaging by stochastic optical reconstruction microscopy (STORM)." *Nat Meth* **3**: 793–795.
Saxton, M. J. (2007). "Modeling 2D and 3D diffusion." *Meth Mol Biol* **400**: 295–321.
Saxton, M. J. and Jacobson, K. (1997). "Single-particle tracking: applications to membrane dynamics." *Annu Rev Biophys Biomol Struct* **26**: 373–399.
Schindelin, J., Arganda-Carreras, I., Frise, E., Kaynig, V., Longair, M., Pietzsch, T., Preibisch, S., et al. (2012). "Fiji: an open-source platform for biological-image analysis." *Nat Meth* **9**: 676–682.
Selvin, P. and Ha, T. (2008). *Single-Molecule Techniques*. Cold Spring Harbor, NY: Cold Spring Harbor Laboratory Press.
Smal, I., Loog, M., Niessen, W. and Meijering, E. (2010). "Quantitative comparison of spot detection methods in fluorescence microscopy." *IEEE Trans Med Img* **29**: 282–301.
Smal, I., Niessen, W. and Meijering, E. (2008). "A new detection scheme for multiple object tracking in fluorescence microscopy by joint probabilistic data association filtering." *Proc 2008 IEEE Int Symp Biomed Imaging (ISBI)*: 264–267.
Sonka, M., Hlavac, V. and Boyle, R. (2007). *Image Processing, Analysis, and Machine Vision*. Toronto: Thomson Learning.
Srinivasa, G., Fickus, M. C., Yusong, G., Linstedt, A. D. and Kovacevic, J. (2009). "Active mask segmentation of fluorescence microscope images." *IEEE Trans Image Proc* **18**: 1817–1829.
Style, R. W., Boltyanskiy, R., German, G. K., Hyland, C., Macminn, C. W., Mertz, A. F., Wilen, et al. (2014). "Traction force microscopy in physics and biology." *Soft Mat* **10**: 4047–4055.
Sun, D., Roth, S. and Black, M. J. (2014). "A quantitative analysis of current practices in optical flow estimation and the principles behind them." *Int J Comp Vis* **106**: 115–137.
Svoboda, K. and Block, S. M. (1994). "Biological applications of optical forces." *Annu Rev Biophy Biomol Struc* **23**: 247–285.

Szeliski, R. (2010). *Computer Vision: Algorithms and Applications*. New York: Springer.

Thompson, R. E., Larson, D. R. and Webb, W. W. (2002). "Precise nanometer localization analysis for individual fluorescent probes." *Biophys J* **82**: 2775–2783.

Ulrich, T. A., Jain, A., Tanner, K., Mackay, J. L. and Kumar, S. (2010). "Probing cellular mechanobiology in three-dimensional culture with collagen-agarose matrices." *Biomaterials* **31**: 1875–1884.

Vaziri, A. and Mofrad, M. R. K. (2007). "Mechanics and deformation of the nucleus in micropipette aspiration experiment." *J Biomechanics* **40**: 2053–2062.

Veenman, C. J., Reinders, M. J. T. and Backer, E. (2001). "Resolving motion correspondence for densely moving points." *IEEE Trans Patt Analy Mach Intel* **23**: 54–72.

Wang, N., Hu, S., and Butler, J. P. (2007). "Imaging stress propagation in the cytoplasm of a living cell." *Methods in Cell Biology* **83**: 179–198.

Waterman-Storer, C. M. and Salmon, E. D. (1998). "How microtubules get fluorescent speckles." *Biophys J* **75**: 2059–2069.

Wilson, C. A., Tsuchida, M. A., Allen, G. M., Barnhart, E. L., Applegate, K. T., Yam, P. T., Ji, L., et al. (2010). "Myosin II contributes to cell-scale actin network treadmilling through network disassembly." *Nature* **465**: 373–377.

Wirtz, D. (2009). "Particle-tracking microrheology of living cells: principles and applications." *Annu Rev Biophys* **38**: 301–326.

Yang, G. (2013). "Bioimage informatics for understanding spatiotemporal dynamics of cellular processes." *Wiley Inter Rev Sys Biol Med* **5**: 367–380.

Yang, G., Cameron, L. A., Maddox, P. S., Salmon, E. D. and Danuser, G. (2008). "Regional variation of microtubule flux reveals microtubule organization in the metaphase meiotic spindle." *J Cell Biol* **182**: 631–639.

Yang, G., Houghtaling, B. R., Gaetz, J., Liu, J. Z., Danuser, G. and Kapoor, T. M. (2007). "Architectural dynamics of the meiotic spindle revealed by single-fluorophore imaging." *Nat Cell Biol* **9**: 1233–1242.

Yildiz, A. and Selvin, P. R. (2005). "Fluorescence imaging with one nanometer accuracy: application to molecular motors." *Acc Chem Res* **38**: 574–582.

10 Micro- and nanotools to probe cancer cell mechanics and mechanobiology

Yasaman Nematbakhsh and Chwee Teck Lim

10.1 Introduction

When cells are exposed to mechanical stimulation, they activate several mechanisms to sense and respond to these mechanical cues. Examples of such processes can be seen in the alignment of endothelial cells on the blood vessel wall when subjected to shear stress arising from blood flow or in hypertrophy of muscle cells in response to weight lifting. While external mechanical stimulations can lead to changes at the cellular and molecular level, internal biological and structural alterations can also elicit mechanical property changes of the cells. One example is the biological and structural changes induced by the progression of certain diseases such as sickle cell anemia, malaria, and cancer (Lim 2006; Suresh et al. 2005). As such, changes in cellular mechanical properties can serve as an indicator not only of cell regulation, but also of the presence and state of a disease (Dufrene and Pelling 2013). This chapter will focus on the disease cancer and review available micro- and nanomechanical tools that have been used to probe the roles cell mechanics and mechanobiology play in pathophysiological processes such as cancer initiation and progression.

Cancer is a disease of uncontrolled cell growth and proliferation (Prabhune et al. 2012). Its initiation and progression is usually accompanied by mechanical alteration at the cellular or tissue level (Cross et al. 2007; Scianna and Preziosi 2013) and can be related to enzymatic, nuclear, and cytoskeletal changes as well as adhesion and migration. During enzymatic changes, different enzymes are secreted by the cells to facilitate their invasion and metastasis. For instance, excessive activation of telomerase (an enzyme responsible for maintenance of the length of telomeres) in cancer cells plays a role in their uncontrolled proliferation, or their movement is facilitated by secretion of collagenase (an enzyme that breaks down the extracellular matrix) (Abdolahad et al. 2012; Cree 2011). At the nuclear level, cancer cells usually have disrupted chromatin arrangements that affect nuclear morphology and stiffness. For instance, cancer cells have altered expression of lamins, a major component of nuclear protein meshwork, or their nuclei are usually more deformable than that of normal cells (Davidson and Lammerding 2014). This change in nuclear stiffness can facilitate cell movement through extracellular matrix (ECM) and microvessels (Giverso et al. 2014; Leong et al. 2011). In terms of cytoskeletal changes, redistribution and activity of microfilaments and microtubules may occur to affect cell deformability (Abdolahad et al. 2012; Scianna and Preziosi 2013). As such, tumor cells tend to lose their original mechanical integrity such as cell stiffness, polarity, and adhesion when they are in the organ, and these can consequently contribute to their metastasis or migration to distant sites. Degree of invasiveness is usually reversely associated with cell stiffness; the higher the invasiveness, the lower the stiffness (Giverso et al. 2014). All these changes will then affect the ability of cancer cells to detach, migrate, invade, or metastasize.

The ECM around tumor cells is another important factor that may initiate and drive cancer (Cheng et al. 2009). The best example will be that of breast cancer, where a tumor is clearly felt as a denser tissue but the individual cancerous cells are usually softer than their healthy counterparts. Increased stiffness of a tumor is usually caused by increased stromal rigidity due to fibrosis, presence of other cell types, and ECM remodeling (Paszek et al. 2005; Plodinec et al. 2012; Ramos et al. 2014). Thus, lowering of elastic modulus of cells and alterations in stiffness of surrounding ECM might aid cancer cells in their migration and survival. (Ramos et al. 2014; Tan et al. 2014).

As mentioned, some of the malignant cells are capable of shedding from the primary tumor and intravasating into the blood stream to become circulating tumor cells (CTCs). However, these cells are very rare, as few as one in a billion blood cells (Lim and Hoon 2014). In the course of metastasis, some of these CTCs can squeeze back into another tissue or organ to form secondary tumors. Due to the metastatic nature of these CTCs, many studies have been done to characterize these cells (Chen et al. 2013). CTCs are usually very heterogeneous and can possess different mechanical properties, nuclear to cytoplasm ratio, nuclear morphology and surface antigen expression (Park et al. 2014). For example, some CTCs may down-regulate their surface receptors such as epithelial cell adhesion molecules (EpCAM) as well as E-cadherins to alter cells from epithelial-like to more mesenchymal-like – a process known as epithelial to mesenchymal transition (EMT) (Chen et al. 2013; Thiery and Lim 2013).

Changes in mechanical properties of a cancer cell from the point of initiation of a tumor to its spread to other parts of a body have been considered as a promising biomarker in detecting and investigating the metastatic potential of these cancer cells. Mechanical transformations can lead to changes in the cell's ability to migrate and invade (Fig. 10.1). Aside from mechanical changes taking place while a cell evolves from the normal to a diseased state, there are also other markers involving genetic mutations and chromosomal changes that have already been used as clinical diagnostic markers, but this is outside the scope of this chapter.

So far, we have explained the importance of cell mechanics and mechanobiology in enabling us to better understand cancer. One practical application of such studies is to evaluate the effectiveness of a drug. Paclitaxel, an antineoplastic agent commonly used in treatment of human carcinomas, inhibits cell replication and induces apoptosis (programmed cell death) by altering biophysical properties and stiffness of cancer

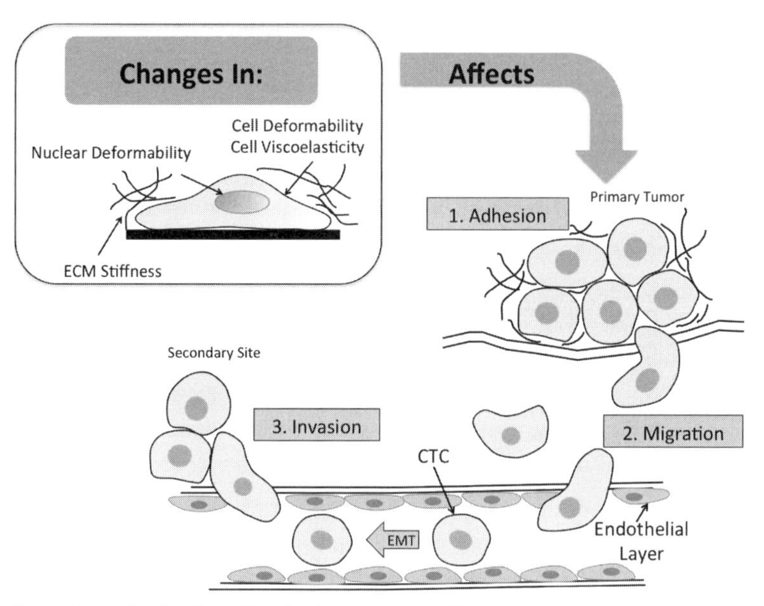

Figure 10.1 **Overview of role of mechanics in cancer studies.**

cells. Studying the mechanical property changes induced in response to anticancer drugs can be crucial in understanding the efficacy of a drug (Muller et al. 2012; Wuang et al. 2011).

In recent years, the advancement of micro- and nanotechnologies has enabled researchers to use available nano- and micromechanical tools, as shown in Fig. 10.2, to probe cancer cells. For example, in probing the mechanics or mechanical properties of the cell, information is obtained by mechanically perturbing the cell with little or no regard to any activity that may be occurring within the cell such as cell signaling. On the other hand, in probing the mechanobiology of the cell, cells are seeded in an engineered environment or mechanically perturbed; then observations are made to see how they respond and interact. Such approaches include putting cells on a hard or soft substrate (Liu et al. 2012; Rabineau et al. 2013; Tilghman et al. 2010), micropillars of varying lengths or cross sections (Wuang et al. 2011; Jaeger et al. 2013; Tzvetkova-Chevolleau et al. 2008), or in a stiffness tunable 3-D collagen matrix microenvironment (Sun, Kurniawan et al. 2014; Sun, Lim et al. 2014; Mak et al. 2013). Each of the available technologies varies in terms of ease of use, throughputness, and precision. Each may also measure mechanical properties or responses of different components such as cytoskeletal organization or gene expression in the cell. Therefore, each technique can provide useful information depending on specific application or need (Dudani et al. 2013). This chapter highlights available micro- and nanomechanical tools to study cell mechanics and mechanobiology (summarized in Fig. 10.2) and provides recent examples of how these tools have been used to better understand mechanical and mechanobiological changes due to development of cancer.

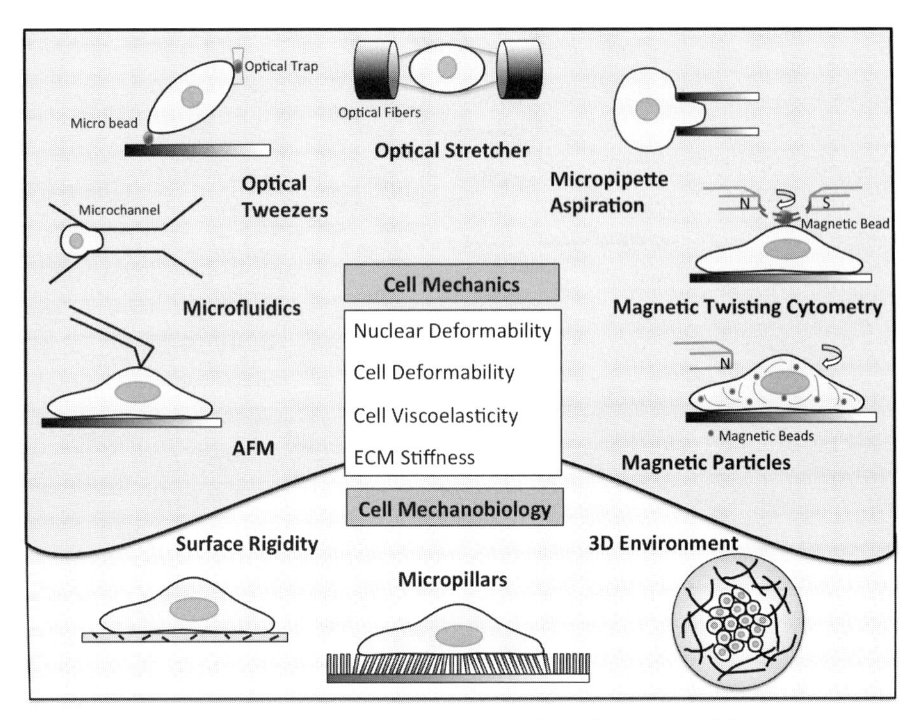

Figure 10.2 Overview of the available tools to study cell mechanics and mechanobiology.

10.2 Nano- and micromechanical tools

10.2.1 Atomic force microscopy (AFM)

One of the most common technologies to probe mechanical properties of cells is atomic force microscopy (AFM) invented by Binnig, Quate and Gerber in 1986 (Binnig et al. 1986). AFM is a powerful tool that can help us investigate how cells respond to locally applied pico- and nanoscale forces when an AFM tip comes into contact with the cell surface. We can also study how a cell responds and interacts with specific proteins when they are functionalized on the AFM tip (Dufrene and Pelling 2013). AFM can provide measurements on elasticity, adhesion, and topography (Kirmizis and Logothetidis 2010).

One of the major applications of AFM in cell mechanics is in measuring cell elasticity (Fig. 10.3A). Using AFM, it has been shown that cancerous thyroid cells are almost three to five times softer than normal types; or ovarian cancer cells exhibit lower stiffness but increased migration and invasion compared to healthy ovarian cells (Prabhune et al. 2012; Xu et al. 2012). Overall, using AFM, researchers have observed that independent of cancer type (breast, colon, prostate, melanoma, and bladder), cancer cells are in general softer than their normal counterparts (Wu et al. 2010; Lekka et al. 1999; Faria et al. 2008; Li et al. 2008; Lekka et al. 2012). AFM also provides a method to measure the viscoelastic response of cells by varying the rate of

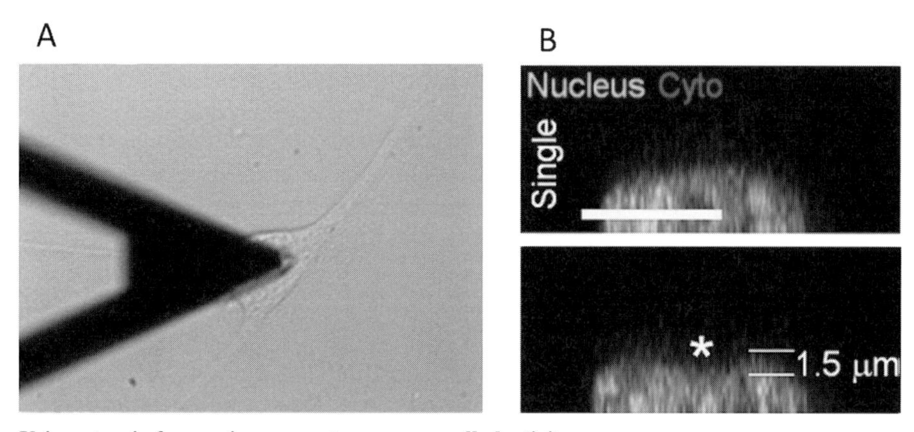

Figure 10.3 **Using atomic force microscopy to measure cell elasticity.**
(A) AFM indentation of an ovarian cancer cell. (B) Nuclear deformation of a fibrosarcoma cell under AFM probe.
(A) reproduced from Xu et al. (2012); (B) reproduced from Krause et al. (2013).

indentation. Rebelo et al. (2013) measured the viscoelasticity of kidney cancer cells and found out that cancer cells are more deformable but less viscous. Few other groups have reached similar conclusions (Darling et al. 2007; Ketene et al. 2012; Li et al. 2008).

Apart from comparing normal with cancerous cells, researchers are also interested in understanding mechanical property differences at the different stages of cancer. AFM studies have shown that highly malignant and more invasive breast cancer cells are more deformable than the less malignant cells, thus suggesting that deformability is one factor that can contribute towards cancer cells' ability to invade (Lee et al. 2012). AFM has also been used to illustrate that bladder cancer cells of different grades vary in stiffness with well differentiated types being stiffer than poorly differentiated ones (Liu et al. 2014). Weder et al. (2014) studied how stiffness of melanoma cells decreased during progression from noninvasive radial growth phase (RGP) – an early phase of the disease – to invasive vertical growth phase (VGP), to metastatic tumors. As mentioned earlier, mechanical changes in a cell induced by anticancer drug can aid in better investigation of drug efficacy. Sharma et al. (2012) has used AFM to study the effect of Cisplatin treatment, a chemotherapy drug, on ovarian cancer cells at nanoscale level and observed that Cisplatin affects cell stiffness by modulating changes in cytoskeleton remodeling.

Changes in mechanical properties of cancer cells are apparent not only at the cellular level but also at the nuclear level. The nucleus is the largest and stiffest compartment of a cell and a major limiting factor in cancer cell invasion. AFM has been used to measure nuclear deformability of cells such as tumorigenic breast cancer cells and adherent fibrosarcoma cells, and results showed that cancer cell nuclei are more deformable than normal cells (Lee et al. 2012; Krause et al. 2013). Figure 10.3B shows nuclear deformation of a fibrosarcoma cell under an AFM probe (Krause et al. 2013).

AFM also provides a great tool to study cell mechanobiology. Nikkhah et al. (2010) developed 3-D silicon microstructures and used AFM to measure elasticity of breast cancer cells with different invasiveness. They observed that more invasive breast cancer cells exhibited lower stiffness on these 3-D microstructures. Another group used AFM to investigate the effect of surface adhesion on the stiffness of melanoma cells and observed that these cells modulate their stiffness in response to different ECM stimuli (Weder et al. 2014).

Overall, AFM has many advantages. It can provide information on not only the structural but also mechanical properties of cells at the micro- and nanometer and the nano and piconewton scales. It involves a simple preparation of mounting cells on a surface and thus provides the option of probing these cells in their physiologically wet or fixed dry conditions (Kirmizis and Logothetidis 2010; Rebelo et al. 2013). However, AFM is not a high throughput tool as it requires the sample to be adhered to a surface, and it demands some amount of training or technical expertise to operate the equipment.

10.2.2 Optical manipulation

Another tool to study cell mechanics is optical tweezers. Arthur Ashkin was one of the pioneers of optical tweezers who conducted a series of experiments on trapping and manipulating neutral atoms, viruses, and E.coli bacteria (Ashkin and Dziedzic 1987). Since then, optical tweezers have been used for various applications to obtain information on cell membrane, protein-protein interactions, and elasticity of DNA molecules (Lim et al. 2006). Optical tweezers use the effect of light on matter and can manipulate objects from atoms to cells (Guck et al. 2001). The working concept is that when light hits a cell, due to the difference in refractive index of the two media, some of the momentum is transferred from light to the cell, which results in exerting a force on it (Adamo et al. 2012; Guck et al. 2001).

One technique that uses this principle is the optical stretcher. This optical stretcher measures the bulk mechanical properties of a cell from observing the deformation of a cell as a laser induced force is exerted over the entire cell surface. Optical stretcher is a convenient and relatively high throughput technique as compared to that of the AFM. This tool also avoids direct contact with the cells that may otherwise result in cell damage and loss of viability (Preira et al. 2013). First developed by Guck et al. in 2002, this tool provides a versatile tool for noncontact and sterile manipulation of biological objects (Sawetzki et al. 2013). An optical stretcher is composed of two optical fibers placed in a way that their ends face each other to create a dual beam optical trap (Fig. 10.4A). The cells are then stretched by increasing the laser power. Relating the cell elongation to the applied force can lead to useful information about the elasticity of a cell. This technique was used to compare deformability of breast cancer cells with different metastatic ability. Nonmalignant mammary epithelial cells show less deformability under the power of laser comparing to their invasive counterparts (Fig. 10.4B–G) (Guck et al. 2005). Another group has also used the optical stretcher to study

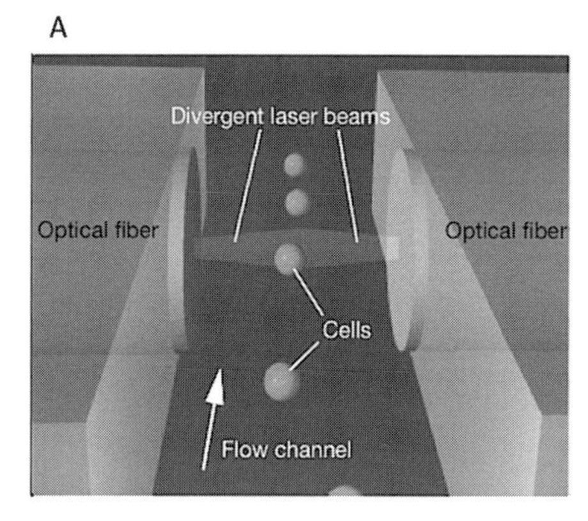

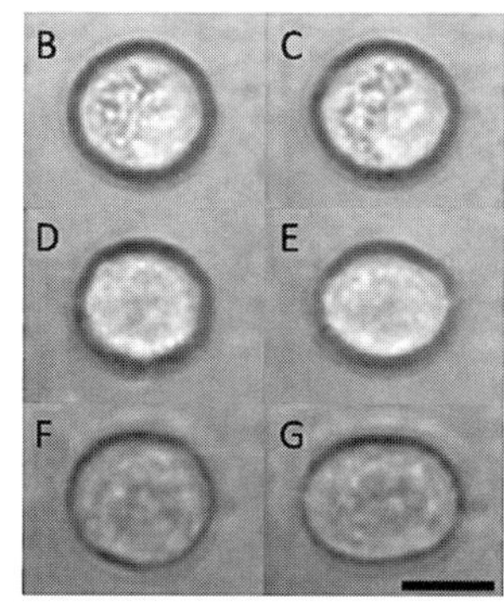

Figure 10.4 **Deformability comparison of nonmalignant and invasive mammary epithelial cells.** (A) Schematic of an optical stretcher. (B–G) Trapped (left column) and stretched (right column) breast cancer cells with different malignancy under an optical stretcher. Malignancy increases going from first row to last (B/C–F/G). Scale bar is 10 μm (Guck et al. 2005).
Reprinted with permission from Elsevier.

deformation of leukemia cells under their custom-made optical stretcher (Bellini et al. 2012).

One of the main advantages of the optical stretcher is the lack of physical contact due to the use of a laser. Another advantage is that it can sense forces up to sub-piconewtons. However, this technique is limited to characterizing suspended cells but not adherent cells.

10.2.3 Microfluidics

Each of the technologies mentioned above has its own limitations. Most have also limited throughputness. Statistically though, it is better to have data on thousands of cells. Microfluidics has the potential to significantly change the way we perform cell mechanics studies. It can offer a high throughput tool with precise spatial and temporal control (Liu et al. 2012). Microfluidic devices can also offer superior observation capability (Cheung et al. 2009). In general, microfluidics offers a reliable, efficient, and flexible tool with possibility of parallel operation for high throughput testing (Beebe et al. 2002).

One of the major applications of microfluidics in studying cell mechanics is the use of microchannels to determine cell deformability. In a microchannel system, a cell is deformed as it passes through a channel with a size smaller than its diameter (Tsai et al. 2014). Cell size and the time it takes for a cell to pass through the channel can be used to determine cell deformability. For instance, one group designed a microchannel (Fig. 10.5A) to relate transit time through a narrow channel to probe the elasticity of breast cancer cells and concluded that as a cell becomes tumorigenic, it takes less time for it to pass through a narrow channel (Babahosseini et al. 2012; Hou et al. 2009). High throughput microfluidic devices can be designed using the same concept. Mak and Erickson (2013) designed a microfluidic device composed of a series of channels

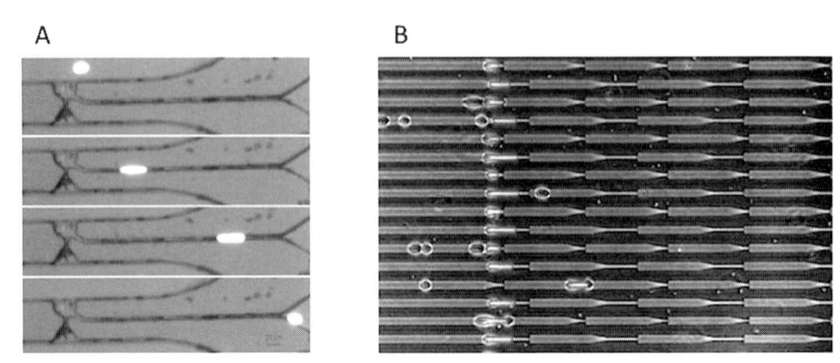

Figure 10.5 **Microchannels as a tool to determine cell deformability.**
(A) A cancer cell traversing through a constriction channel (Babahosseini et al. 2012). (B) Serial microchannel system to measure deformability of metastatic breast cancer cells (Mak and Erickson 2013).
(A) reprinted with permission from the IEEE; (B) reprinted with permission from the Royal Society of Chemistry.

where each channel is a series of micro-constrictions to measure deformability of metastatic breast cancer cells. By letting the cells pass through each set of constrictions using pressure difference, they observed that the initial deformation facilitates the subsequent deformations; this shows that the cells that undergo perpetual deformation have less difficulty squeezing through subnuclear scale constrictions. Such observation may suggest that it is easier for cancer cells to invade as they constantly deform to travel through the ECM (Fig. 10.5B) (Mak and Erickson 2013). Alternatively, transit time can be recorded using electronic concepts such as measuring the resistance difference between the two electrodes on each side of a funnel shaped channel as cell passes through (Adamo et al. 2012). Microchannels can also be used to obtain information on nuclear deformation. One group used arrays of parallel microchannels to study nuclear deformation as cells passed through each channel. Using their device, they observed that nuclear deformation is the primary rate-limiting step during transmigration. They also investigated the effect of chromatin-condensing drugs and found that such drugs can play a role in reducing nuclear deformability and consequently decreasing cell transmigration probability (Fu et al. 2012).

There are few arguments about using transit time to measure cell stiffness; this method is confounded by precise knowledge of cell size and its friction with the channel wall. However, knowledge of entry velocity as well as transit velocity plays a role in predicting cell deformability. Entry velocity relates to cytoskeleton rearrangements of a cell to pass through the narrow channel whereas transit time relates to alteration in surface friction properties. To accomplish a more accurate value for deformability, this group has integrated a constriction near a suspended microchannel resonator (SMR), a hollow channel embedded in a silicon cantilever where its frequency is measured using a laser beam. As the cell passes through the constriction, the resonant frequency is decreased depending on buoyant mass and the position of center of mass. Based on this frequency change, one can determine transit time, entry velocity, and transit velocity with high throughputness of a few thousand cells per hour (Byun et al. 2013). Traditional evaluation of transit time also includes information on not only cell stiffness, but also cell viscosity. Therefore, the conventional method might misinterpret the stiffness of two characteristically distinct cells as similar because same transit time is detected. Tsai et al. (2014) proposes that by using equilibrium velocity during the transit time instead of the entering velocity, one will eliminate the viscosity effect of a cell.

Microfluidic devices can also be designed to measure deformability using hydrodynamic forces. Some ways of creating hydrodynamic forces are by exposing cells to a pinch flow or using a microchannel with hyperbolic shape in which fluid experiences a strong extensional flow (Faustino et al. 2014). Deformability of prostate carcinoma cells was measured by exposing them to hydrodynamic forces and relating flow rate and flow acceleration (flow rate increase per time) to the amount the cell deforms (Cheung et al. 2009). It was observed that for the same flow rate, and as flow acceleration increases, cell deformability decreases due to cell's viscoelastic behavior. For a more high throughput setup, Gossette et al. (2012) developed a microfluidic device that utilizes hydrodynamic forces to align the cells at the center of a channel and uses perpendicular pinch flow to deform cervical and breast cancer cells. A high-speed camera is then used to record the degree of each cell deformation and relates it to cell initial diameter (Fig. 10.6) (Gossett et al. 2012). Using this device, they could identify malignancy in clinical samples of pleural fluid, the excess fluid that accumulates between lungs and the chest cavity, as a way to detect metastasis in cancer patients. From deformability scatter plots, they could observe that samples from patients with malignancy contain larger but more deformable cells. One of the main drawbacks of this technique is the lack of knowledge on the amount of force exerted on each cell. For instance, in the example shown (Fig. 10.6), one can relate diameter change of a cell as it undergoes deformation, but to quantify the elastic modulus of the cell, knowledge of the force on each cell is necessary. This limitation, however, should not undermine the fact that such devices can provide great opportunities to perform comparative studies on thousands of cells in a relatively short amount of time.

Other ways that microfluidics can be used to study cancer cell mechanics is the use of a microchannel to find the threshold pressure required to deform cells through a

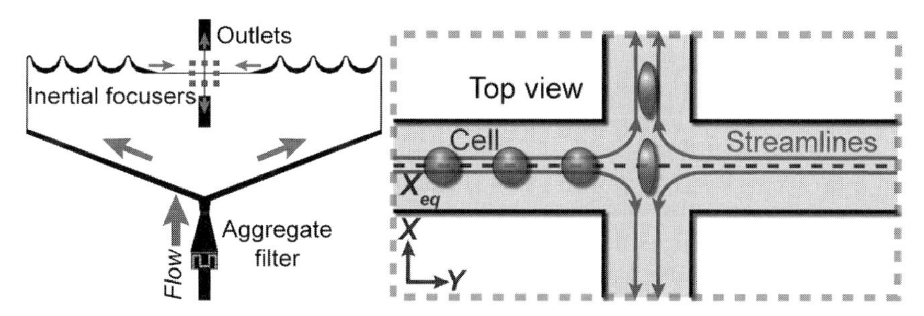

Figure 10.6 **Measuring cell deformability.**
Example of a microfluidic device that uses hydrodynamic forces to deform cells.
Reproduced from Gossett et al. (2012).

funnel-shaped constriction. Using such a setup, it was observed that bladder cancer cells are more rigid than leukocytes due to having larger nuclei and more defined cytoskeleton (Guo et al. 2012).

To study cancer cells in a more dynamic way, one needs to consider the surrounding ECM. As mentioned, tumors are usually stiffer than normal tissue and their growth is strongly affected by their microenvironment (Cheng et al. 2009). It has been shown that breast cancer progression is accompanied by stiffening of surrounding ECM (Werfel et al. 2013). Therefore, it is of great interest to understand the stress-field around growing tumor spheroids and how the surrounding mechanical stress can affect the tumor growth (Cheng et al. 2009). One way to study this is to mimic cell confinement; a factor that has an impact on the microenvironment of a tumor both biologically and mechanically. Microfluidic devices have been designed to study the effect of confinement on mitotic process of cancer cells and observed delayed mitosis, unevenly sized daughter cells, and induction of cell death due to not having proper space for cell rounding (Kittur et al. 2014; Tse et al. 2012). Another group investigated the effect of compression on tumor growth and observed an increase in growth in low stress region but increased apoptosis in the high stress region (Cheng et al. 2009).

Cell confinement can also be used to study dynamics of cell invasion during metastasis. One group developed a microfluidic device composed of subnuclear-size channels to mimic cell penetration into small pores in ECM of tumor stroma or endothelial junctions during intravasation (Mak et al. 2013). Microfluidics has been used to observe the interaction between two breast cancer cell lines with different metastatic potential to investigate the dynamics of metastasis from self-seeding, to invasion, and to coexistence (Liu et al. 2012). One group integrated electrical sensors with microfluidics to design a device that tracks and compares migration of different breast cancer cell types in a 3-D matrix. The existence of a 3-D matrix changes cell behaviors such as migration speed and can present a closer resemblance of an ECM surrounding a tumor (Nguyen et al. 2013).

Overall, microfluidics can offer an excellent tool to study cancer cell mechanics and mechanobiology. Single cell study is especially useful in cancer studies due to the heterogeneity of cells in a tumor. The ability to customize microfluidics provides many opportunities to design novel microfluidics that help us better understand how transformation of a cell from normal to disease state affects its mechanical properties and vice versa.

10.2.4 Other technologies

Aside from the three most common technologies described above, there are other micro- and nanotools that researchers have used to study cell mechanics and mechanobiology. For instance, micropipette aspiration is a technique that can be used to study mechanics of suspended cells to obtain information on cell elasticity, membrane elasticity, or cell viscoelasticity. This is done by aspirating a cell into a pipette using a negative pressure and quantifying the relationship between the changes in cell geometry inside the pipette (strain) and suction pressure (stress) (Adamo et al. 2012; Shojaei-Baghini et al. 2013). Magnetic twisting cytometry is another technique where magnetic beads are attached to the cell membrane and then an external magnetic field is activated to induce a twisting moment to deform the cells. Both elastic and viscoelastic properties of a cell can be obtained using this method (Shojaei-Baghini et al. 2013). There are also other techniques that might not give macroscopic information on cell mechanics or mechanobiology in a sense that previously described techniques did; but they do provide useful information on how external factors can induce microscopic changes that may eventually affect the ability of a cancer cell to migrate and proliferate.

For instance, cues from ECM environment such as substrate topography and rigidity have strong effect on cancer cell behavior. One way to alter surface rigidity is by using micropillars or gels of different stiffness. Badique et al. (2013) used micropillars of different size and spacing to investigate nuclear deformability of three different osteosarcoma cell lines and observed that stiffer cell line exhibited higher nuclear deformability. Micropillars also provide a powerful tool to study migration and traction force of cancer cells. Cancer fibroblastic cells showed an increase in migration speed as substrate rigidity was increased using shorter micropillars (Tzvetkova-Chevolleau et al. 2008). Breast cancer cells seeded on micropillars showed higher disruption in their traction forces in response to emodin treatment – a chemotherapy drug (Wuang et al. 2011). Finally, breast, lung, and pancreatic cancer cells have shown an increase in their proliferation on stiffer substrates (Tilghman et al. 2010).

Another tool to study how cells respond to external cues is the use of nano- and micromagnetic particles (Kozissnik and Dobson 2013). The concept involves using an external magnetic field to manipulate magnetic-labeled particles, either attached to the cell surface or within the cell, to exert a force (Fu et al. 2012). Among the pioneers using magnetic particles to study cell mechanics were Crick and Hughes in 1950 (Crick 1950). Since then, magnetic beads have been used by many groups to study cell mechanics. For a high throughput setup, Tseng et al. (2012) embedded nanomagnetic particles in cells to perform simultaneous mechanical stimulation of thousands of HeLa cells. They obtained quantitative data on cellular responses to mechanical stimulation over a range of forces. They fabricated a series of micro magnets and seeded single cells to align with each one (Fig. 10.7) Once an external magnetic field was activated, it exerted a force on each cell through the embedded nanomagnetic particles. Amount of force could be adjusted by varying the range of the external magnet field. They then investigated the effect of force on cell actin polymerization and mitosis; they observed that by increasing the force, larger fraction of cells exhibited protrusion asymmetry and that applying force will bias mitotic spindle orientation.

The micropillars and magnetic bead manipulation both offer great opportunities to investigate the intracellular response of cancer cells to external cues. For more cell mechanics studies, these tools can be combined with other technologies such as AFM to offer an opportunity to compare cell elasticity with and without such external stimulations.

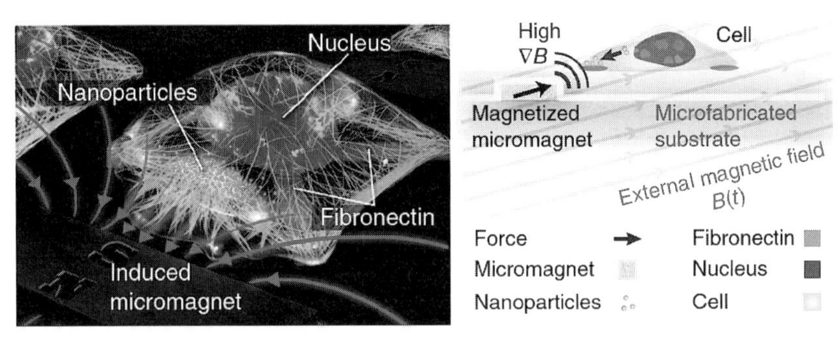

Figure 10.7 **Schematic of force generating platform used to exert mechanical stimulations on cells in a high throughput manner.**
Tseng et al. 2012. Reprinted with permission from Macmillan Publishers Ltd.

10.3 Conclusions

This chapter reviews some of the most popular techniques that have been used to study cancer cell mechanics and mechanobiology. Using these technologies, we can now have better insights into the biomechanical changes that occur between a healthy cell and a cancer cell and the effects these changes have on the ability of the cancer cell to invade and metastasize. However, the use of cell mechanics or mechanobiology to understand a disease like cancer is still in its infancy. There are still many questions that are unknown and need to be addressed. We still have limited knowledge of how cancer induces the cellular and molecular changes that eventually lead to structural and biomechanical property changes in the cells; or how structural and biomechanical changes at the cellular and tissue level can lead to formation of a tumor. It is also still not fully known how mechanical properties of CTCs can enable them to better traverse through the blood circulatory or lymphatic systems to reach their final destinations; or how chemokines in the microenvironment of a target organ affect CTC metastasis. Nevertheless, the currently available nano- and micromechanical technologies are certainly powerful state-of-the-art tools that can help to advance this important interdisciplinary field. While they may have their own sets of limitations, they can still provide us with the capability and opportunity to address some of these important questions. Thus, it is hoped that arising from this, we can not only better understand cancer from a different perspective, but we can also translate what we eventually learn to better detect, diagnose, and treat this disease.

References

Abdolahad, M., Z. Sanaee, M. Janmaleki, et al. (2012). “Vertically aligned multiwall-carbon nanotubes to preferentially entrap highly metastatic cancerous cells.” *Carbon* **50**(5): 2010–2017.

Adamo, A., A. Sharei, L. Adamo, et al. (2012). "Microfluidics-based assessment of cell deformability." *Anal Chem* **84**(15): 6438–6443.

Ashkin, A. and J. M. Dziedzic. (1987). "Optical trapping and manipulation of viruses and bacteria." *Science* **235**(4795): 1517–1520.

Babahosseini, H., V. Srinivasaraghavan and M. Agah. (2012). "Microfluidic chip bio-sensor for detection of cancer cells." *Sensors*. Taipei, IEEE 1–4, 28–31 Oct. 2012.

Badique, F., D. R. Stamov, P. M. Davidson, et al. (2013). "Directing nuclear deformation on micropillared surfaces by substrate geometry and cytoskeleton organization." *Biomaterials* **34**(12): 2991–3001.

Beebe, D. J., G. A. Mensing and G. M. Walker. (2002). "Physics and applications of microfluidics in biology." *Annu Rev Biomed Eng* **4**: 261–286.

Bellini, N., F. Bragheri, I. Cristiani, et al. (2012). "Validation and perspectives of a femtosecond laser fabricated monolithic optical stretcher." *Biomed Opt Express* **3**(10): 2658–2668.

Binnig, G., C. F. Quate and C. Gerber. (1986). "Atomic force microscope." *Phys Rev Lett* **56**(9): 930–933.

Byun, S., S. Son, D. Amodei, et al. (2013). "Characterizing deformability and surface friction of cancer cells." *Proc Natl Acad Sci USA* **110**(19): 6.

Chen, C. L., D. Mahalingam, P. Osmulski, et al. (2013). "Single-cell analysis of circulating tumor cells identifies cumulative expression patterns of EMT-related genes in metastatic prostate cancer." *Prostate* **73**(8): 813–826.

Cheng, G., J. Tse, R. K. Jain, et al. (2009). "Micro-environmental mechanical stress controls tumor spheroid size and morphology by suppressing proliferation and inducing apoptosis in cancer cells." *PLoS One* **4**(2): e4632.

Cheung, L. S., X. Zheng, A. Stopa, et al. (2009). "Detachment of captured cancer cells under flow acceleration in a bio-functionalized microchannel." *Lab Chip* **9**(12): 1721–1731.

Cree, I. A. (2011). "Principles of cancer cell culture." *Methods Mol Biol* **731**: 13–26.

Crick, F. H. C., and A. F. W Hughes. (1950). "The physical properties of cytoplasm: a study by means of the magnetic particle method." *Experimental Cell Research* **1**(1): 44.

Cross, S. E., Y. S. Jin, J. Rao, et al. (2007). "Nanomechanical analysis of cells from cancer patients." *Nat Nanotechnol* **2**(12): 780–783.

Darling, E. M., S. Zauscher, J. A. Block, et al. (2007). "A thin-layer model for viscoelastic, stress-relaxation testing of cells using atomic force microscopy: do cell properties reflect metastatic potential?" *Biophys J* **92**(5): 1784–1791.

Davidson, P. M. and J. Lammerding. (2014). "Broken nuclei–lamins, nuclear mechanics, and disease." *Trends Cell Biol* **24**(4): 247–256.

Dudani, J. S., D. R. Gossett, H. T. Tse, et al. (2013). "Pinched-flow hydrodynamic stretching of single-cells." *Lab Chip* **13**(18): 3728–3734.

Dufrene, Y. F. and A. E. Pelling. (2013). "Force nanoscopy of cell mechanics and cell adhesion." *Nanoscale* **5**(10): 4094–4104.

Faria, E. C., N. Ma, E. Gazi, et al. (2008). "Measurement of elastic properties of prostate cancer cells using AFM." *Analyst* **133**(11): 1498–1500.

Faustino, V., D. Pinho, T. Yaginuma, et al. (2014). "Extensional flow-based microfluidic device: deformability assessment of red blood cells in contact with tumor cells." *BioChip Journal* **8**(1): 42–47.

Fu, C., C. Han, C. Cheng, et al. (2012). "Bio-mechanical properties of human renal cancer cells probed by magneto-optical tweezers." *Sensors*. Taipei, IEEE, 1–4, 28–31 Oct. 2012.

Fu, Y., A. M. J. Vandongen, T. Bourouina, et al. (2012). "A study of cancer cell metastasis using microfluidic transmigration device." *MEMS*. Paris, IEEE: 773–776, 29 Jan. 2012–2 Feb. 2012.

Giverso, C., A. Grillo and L. Preziosi. (2014). "Influence of nucleus deformability on cell entry into cylindrical structures." *Biomech Model Mechanobiol* **13**(3): 481–502.

Gossett, D. R., H. T. Tse, S. A. Lee, et al. (2012). "Hydrodynamic stretching of single cells for large population mechanical phenotyping." *Proc Natl Acad Sci USA* **109**(20): 7630–7635.

Guck, J., R. Ananthakrishnan, H. Mahmood, et al. (2001). "The optical stretcher: a novel laser tool to micromanipulate cells." *Biophys J* **81**(2): 767–784.

Guck, J., S. Schinkinger, B. Lincoln, et al. (2005). "Optical deformability as an inherent cell marker for testing malignant transformation and metastatic competence." *Biophys J* **88**(5): 3689–3698.

Guo, Q., S. Park and H. Ma. (2012). "Microfluidic micropipette aspiration for measuring the deformability of single cells." *Lab Chip* **12**(15): 2687–2695.

Hou, H. W., Q. S. Li, G. Y. Lee, et al. (2009). "Deformability study of breast cancer cells using microfluidics." *Biomed Microdevices* **11**(3): 557–564.

Jaeger, A. A., C. K. Das, N. Y. Morgan, et al. (2013). "Microfabricated polymeric vessel mimetics for 3-D cancer cell culture." *Biomaterials* **34**(33): 8301–8313.

Ketene, A. N., E. M. Schmelz, P. C. Roberts, et al. (2012). "The effects of cancer progression on the viscoelasticity of ovarian cell cytoskeleton structures." *Nanomedicine* **8**(1): 93–102.

Kirmizis, D. and S. Logothetidis. (2010). "Atomic force microscopy probing in the measurement of cell mechanics." *Int J Nanomedicine* **5**: 137–145.

Kittur, H., W. Weaver and D. Di Carlo. (2014). "Well-plate mechanical confinement platform for studies of mechanical mutagenesis." *Biomed Microdevices* **16**(3): 439–447.

Kozissnik, B. and J. Dobson. (2013). "Biomedical applications of mesoscale magnetic particles." *MRS Bulletin* **38**(11): 927–932.

Krause, M., J. Te Riet and K. Wolf. (2013). "Probing the compressibility of tumor cell nuclei by combined atomic force-confocal microscopy." *Phys Biol* **10**(6): 065002.

Lee, M. H., P. H. Wu, J. R. Staunton, et al. (2012). "Mismatch in mechanical and adhesive properties induces pulsating cancer cell migration in epithelial monolayer." *Biophys J* **102**(12): 2731–2741.

Lekka, M., P. Laidler, D. Gil, et al. (1999). "Elasticity of normal and cancerous human bladder cells studied by scanning force microscopy." *Eur Biophys J* **28**(4): 312–316.

Lekka, M., K. Pogoda, J. Gostek, et al. (2012). "Cancer cell recognition–mechanical phenotype." *Micron* **43**(12): 1259–1266.

Leong, F. Y., Q. Li, C. T. Lim, et al. (2011). "Modeling cell entry into a micro-channel." *Biomech Model Mechanobiol* **10**(5): 755–766.

Li, Q. S., G. Y. Lee, C. N. Ong, et al. (2008). "AFM indentation study of breast cancer cells." *Biochem Biophys Res Commun* **374**(4): 609–613.

Lim, C. T. (2006). "Single Cell Mechanics Study of the Human Disease Malaria." *Journal of Biomechanical Science and Engineering* **1**(1): 82–92.

Lim, C. T. and S. B. Hoon (2014). "Circulating tumor cells: Cancer's deadly couriers." *Physics Today* **67**(2): 5.

Lim, C. T., E. H. Zhou, A. Li, et al. (2006). "Experimental techniques for single cell and single molecule biomechanics." *Materials Science and Engineering C* **26**(8): 1278–1288.

Liu, A., W. Liu, Y. Wang, et al. (2012). "Microvalve and liquid membrane double-controlled integrated microfluidics for observing the interaction of breast cancer cells." *Microfluidics and Nanofluidics* **14**(3–4): 515–526.

Liu, H., Q. Tan, W. R. Geddie, et al. (2014). "Biophysical characterization of bladder cancer cells with different metastatic potential." *Cell Biochem Biophys* **68**(2): 241–246.

Liu, J., Y. Tan, H. Zhang, et al. (2012). "Soft fibrin gels promote selection and growth of tumorigenic cells." *Nat Mater* **11**(8): 734–741.

Mak, M. and D. Erickson. (2013). "A serial micropipette microfluidic device with applications to cancer cell repeated deformation studies." *Integr Biol (Camb)* **5**(11): 1374–1384.

Mak, M., C. A. Reinhart-King and D. Erickson. (2013). "Elucidating mechanical transition effects of invading cancer cells with a subnucleus-scaled microfluidic serial dimensional modulation device." *Lab Chip* **13**(3): 340–348.

Muller, D. J., K. S. Kim, C. H. Cho, et al. (2012). "AFM-detected apoptotic changes in morphology and biophysical property caused by paclitaxel in Ishikawa and HeLa cells." *PLoS One* **7**(1): e30066.

Nguyen, T. A., T. I. Yin, D. Reyes, et al. (2013). "Microfluidic chip with integrated electrical cell-impedance sensing for monitoring single cancer cell migration in three-dimensional matrixes." *Anal Chem* **85**(22): 11068–11076.

Nikkhah, M., J. S. Strobl, R. De Vita, et al. (2010). "The cytoskeletal organization of breast carcinoma and fibroblast cells inside three dimensional. (3-D) isotropic silicon microstructures." *Biomaterials* **31**(16): 4552–4561.

Park, S., R. R. Ang, S. P. Duffy, et al. (2014). "Morphological differences between circulating tumor cells from prostate cancer patients and cultured prostate cancer cells." *PLoS One* **9**(1): e85264.

Paszek, M. J., N. Zahir, K. R. Johnson, et al. (2005). "Tensional homeostasis and the malignant phenotype." *Cancer Cell* **8**(3): 241–254.

Plodinec, M., M. Loparic, C. A. Monnier, et al. (2012). "The nanomechanical signature of breast cancer." *Nat Nanotechnol* **7**(11): 757–765.

Prabhune, M., G. Belge, A. Dotzauer, et al. (2012). "Comparison of mechanical properties of normal and malignant thyroid cells." *Micron* **43**(12): 1267–1272.

Preira, P., V. Grandne, J. M. Forel, et al. (2013). "Passive circulating cell sorting by deformability using a microfluidic gradual filter." *Lab Chip* **13**(1): 161–170.

Rabineau, M., L. Kocgozlu, D. Dujardin, et al. (2013). "Contribution of soft substrates to malignancy and tumor suppression during colon cancer cell division." *PLoS One* **8**(10): e78468.

Ramos, J. R., J. Pabijan, R. Garcia, et al. (2014). "The softening of human bladder cancer cells happens at an early stage of the malignancy process." *Beilstein J Nanotechnol* **5**: 447–457.

Rebelo, L. M., J. S. de Sousa, J. Mendes Filho, et al. (2013). "Comparison of the viscoelastic properties of cells from different kidney cancer phenotypes measured with atomic force microscopy." *Nanotechnology* **24**(5): 055102.

Sawetzki, T., C. D. Eggleton, S. A. Desai, et al. (2013). "Viscoelasticity as a biomarker for high-throughput flow cytometry." *Biophys J* **105**(10): 2281–2288.

Scianna, M. and L. Preziosi. (2013). "Modeling the influence of nucleus elasticity on cell invasion in fiber networks and microchannels." *J Theor Biol* **317**: 394–406.

Sharma, S., C. Santiskulvong, L. A. Bentolila, et al. (2012). "Correlative nanomechanical profiling with super-resolution F-actin imaging reveals novel insights into mechanisms of cisplatin resistance in ovarian cancer cells." *Nanomedicine* **8**(5): 757–766.

Shojaei-Baghini, E., Y. Zheng and Y. Sun. (2013). "Automated micropipette aspiration of single cells." *Ann Biomed Eng* **41**(6): 1208–1216.

Sun, W., N. A. Kurniawan, A. P. Kumar, et al. (2014). "Effects of migrating cell-induced matrix reorganization on 3d cancer cell migration." *Cellular and Molecular Bioengineering* **7**(2): 205–217.

Sun, W., C. T. Lim and N. A. Kurniawan. (2014). "Mechanistic adaptability of cancer cells strongly affects anti-migratory drug efficacy." *J R Soc Interface* **11**(99).

Suresh, S., J. Spatz, J. P. Mills, et al. (2005). "Connections between single-cell biomechanics and human disease states: gastrointestinal cancer and malaria." *Acta Biomater* **1**(1): 15–30.

Tan, Y., A. Tajik, J. Chen, et al. (2014). "Matrix softness regulates plasticity of tumour-repopulating cells via H3K9 demethylation and Sox2 expression." *Nat Commun* **5**: 4619.

Thiery, J. P. and C. T. Lim. (2013). "Tumor dissemination: an EMT affair." *Cancer Cell* **23**(3): 272–273.

Tilghman, R. W., C. R. Cowan, J. D. Mih, et al. (2010). "Matrix rigidity regulates cancer cell growth and cellular phenotype." *PLoS One* **5**(9): e12905.

Tsai, C. H., S. Sakuma, F. Arai, et al. (2014). "A new dimensionless index for evaluating cell stiffness-based deformability in microchannel." *IEEE Trans Biomed Eng* **61**(4): 1187–1195.

Tse, H. T., W. M. Weaver and D. Di Carlo. (2012). "Increased asymmetric and multi-daughter cell division in mechanically confined microenvironments." *PLoS One* **7**(6): e38986.

Tseng, P., J. W. Judy and D. Di Carlo. (2012). "Magnetic nanoparticle–mediated massively parallel mechanical modulation of single-cell behavior." *Nat Methods* **09**(11): 9.

Tzvetkova-Chevolleau, T., A. Stephanou, D. Fuard, et al. (2008). "The motility of normal and cancer cells in response to the combined influence of the substrate rigidity and anisotropic microstructure." *Biomaterials* **29**(10): 1541–1551.

Weder, G., M. C. Hendriks-Balk, R. Smajda, et al. (2014). "Increased plasticity of the stiffness of melanoma cells correlates with their acquisition of metastatic properties." *Nanomedicine* **10**(1): 141–148.

Werfel, J., S. Krause, A. G. Bischof, et al. (2013). "How changes in extracellular matrix mechanics and gene expression variability might combine to drive cancer progression." *PLoS One* **8**(10): e76122.

Wu, Y., G. D. McEwen, S. Harihar, et al. (2010). "BRMS1 expression alters the ultrastructural, biomechanical and biochemical properties of MDA-MB-435 human breast carcinoma cells: an AFM and Raman microspectroscopy study." *Cancer Lett* **293**(1): 82–91.

Wuang, S. C., B. Ladoux and C. T. Lim. (2011). "Probing the chemo-mechanical effects of an anti-cancer drug Emodin on breast cancer cells." *Cellular and Molecular Bioengineering* **4**(3): 466–475.
Xu, W., R. Mezencev, B. Kim, et al. (2012). "Cell stiffness is a biomarker of the metastatic potential of ovarian cancer cells." *PLoS One* **7**(10): e46609.

11 Stimuli-responsive polymeric substrates for cell-matrix mechanobiology

Mitsuhiro Ebara and Koichiro Uto

11.1 Introduction

Cell responsiveness to surrounding mechanical cues has been the subject of a number of researchers. Especially, mechanical and topographical properties of the substrate are critical parameters in stem cell maintenance and differentiation. In spite of a considerable amount of ongoing research, however, current efforts are centered on rather static patterns. Due to the dynamic nature of the regeneration processes, static substrates seem to be deficient in mimicking changing physiological conditions, such as would be expected during development, wound healing, and disease. Therefore, the scientific community has recently shown increased interest in developing surfaces with tunable abilities. In this context, "smart" or "stimuli-responsive" materials have emerged as powerful tools for basic cell studies as well as promising biomedical applications. These materials can change their properties with time and in response to user-defined triggers or cellular behavior. Recent examples of smart materials include temperature-responsive polymer surfaces where the surface energy can be controlled with temperature. Photochemical reactions are also used to induce switching the surface physicochemical properties involving either irreversible photocleavage or reversible photoisomerization, because light is able to present precise control over temporal and spatial signals. As an alternative approach, shape-memory materials have recently been developed to direct cell behavior. In the following sections of this chapter, we review different types of stimuli-responsive polymeric substrates that have been developed for cell-matrix mechanobiology in the last decades. In Section 11.2, we describe the designing of dynamic cell culture substrates with tunable stiffness. Softening and stiffening materials are both introduced. In addition to sensing stiffness, topographical cues also play an integral role in influencing cell fate. Therefore, Section 11.3 focuses on the developments of surfaces with tunable topography. In Section 11.4, recent examples of dynamically switchable surfaces to control cell-matrix interactions are discussed. The chapter ends with an overview of some of the future trends in applications in biotechnology and biomedicine.

11.2 Dynamic stiffness

Recent reports have revealed that many types of cells have the capability of sensing and reacting to mechanical stiffness. For example, stiff gels promote spreading and scattering of adherent cells, while soft gels promote soft tissue differentiation and tissue-like cell-cell associations. In addition, adherent cells migrate preferentially toward stiffer regions. Although many studies have shown that the stiffness of the synthetic substrate can influence cell fate, current efforts are centered on rather static effects because properties of synthetic substrates are usually constant in time. From these perspectives, much attention has been focused on the designing of dynamic cell culture substrates or matrices with tunable abilities. To allow dynamic control of substrate stiffness, Frey and Wang (2009) have developed a modulatable hydrogel with UV-mediated control of rigidity. This gel is formed by crosslinking polyacrylamide (PA) with a UV-cleavable agent. They were able to soften the gel from 7.2 kPa to 5.5 kPa by UV irradiation. The spread area of NIH fibroblast (NIH3T3) cells decreased substantially upon UV irradiation. Furthermore, localized softening of the posterior substratum of polarized cells causes no apparent effect, while softening of the anterior substratum elicits pronounced retraction, indicating that rigidity sensing is localized to the frontal region. This result indicates that mechanosensing is largely localized to the anterior of polarized cells. Kloxin et al. (2009) demonstrated that cell-material interactions are dynamically and externally directed by photodegradable hydrogels. They have synthesized monomers capable of polymerizing in the presence of cells to produce photolytically degradable hydrogels whose physical or chemical properties are tunable temporally and spatially with light. They successfully generated 3-D features within a hydrogel to allow cell migration. DeForest and Anseth (2011) designed three-dimensional hydrogels that allow orthogonal and dynamic control of photocleavage of crosslinks and photoconjugation of pendant functionalities. Photocleavage and photoconjugation reactions are both used to independently regulate physical and chemical properties of hydrogels. The hydrogel is formed by means of a copper-free alkyne-azide reaction. In the presence of visible light, thiol-containing biomolecules are covalently affixed to pendant vinyl functionalities throughout the hydrogel network via the thiol-ene reaction. In the presence of UV light, on the other hand, a nitrobenzyl ether moiety results in photodegradation of the network. This system allows for real-time manipulation of cell function within a simplified synthetic microenvironment. Yang et al. (2014) have recently investigated whether or not stem cell fate is influenced by culture history, that is, past mechanical memory, using phototunable PEG hydrogels with initially stiff (~10 kPa) and then soft (~2 kPa) substrates (Fig. 11.1A). The activation of the Yes-associated protein (YAP) and transcriptional coactivator with PDZ binding domain (TAZ), as well as the pre-osteogenic transcription factor RUNX2 in human mesenchymal stem cells (hMSCs), depended on previous culture time on stiff substrate (Fig. 11.1B). This result indicates that the YAP/TAZ transcriptional coactivators may act as intracellular mechanical rheostats mediating the effectors of mechanical dosing on stem cell plasticity by a persistent presence in the nucleus.

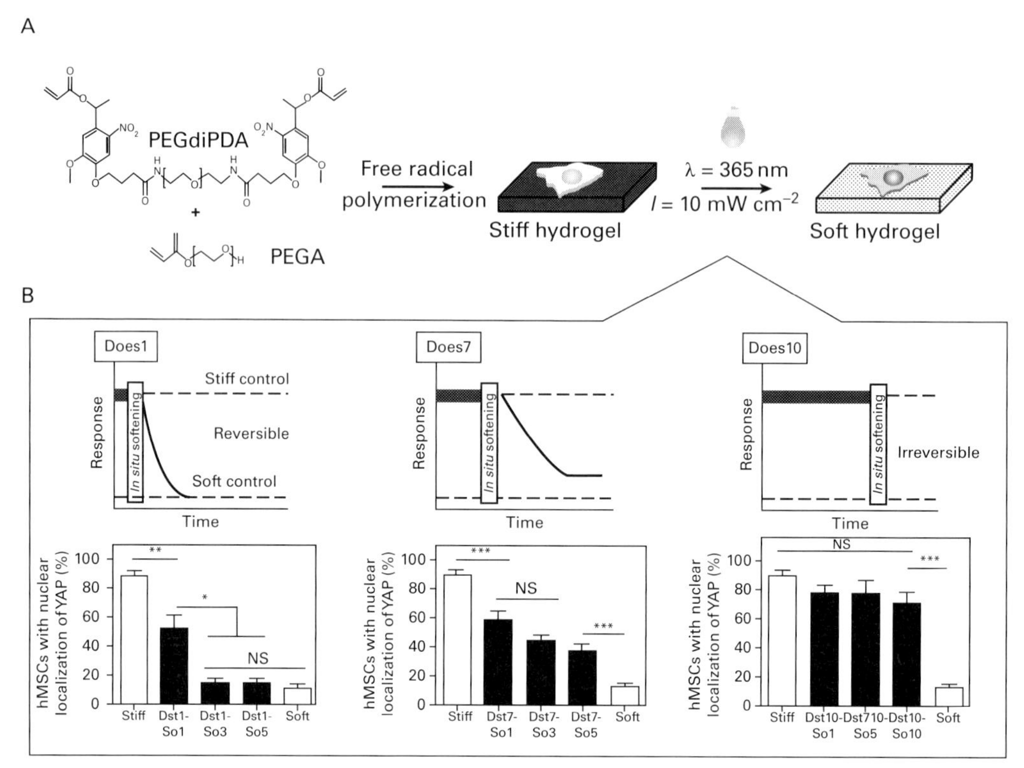

Figure 11.1 **Determining cell fate by dynamic stiffness.**
(A) Photodegradable hydrogels are fabricated from the free-radical polymerization of a photodegradable crosslinker, a poly(ethylene glycol) di-photodegradable acrylate (PEGdiPDA), with a monoacrylated PEG (PEGA). Polymerization results in a stiff hydrogel (~10 kPa). Light exposure softens the substrate to a soft hydrogel (~2 kPa). hMSCs are cultured on stiff hydrogels in growth media for 1 to 10 days before in situ softening the underlying culture substrata to soft hydrogels. The hMSCs are cultured subsequently on the softened hydrogels for 1 to 10 days in growth media before collection and analysis. (B) YAP response to in situ softening after 1, 7, and 10 days (DSt1, 7 and 10) of mechanical dosing on stiff hydrogels. Stiff control is the average hMSC expression of YAP over 3, 5, 7, and 10 days (So1, 3, 7 and 10) on stiff hydrogels. Soft control is the average hMSC expression of YAP over 3, 5, 7, and 10 days of soft hydrogels (Yang et al. 2014).
Reprinted with permission from Macmillan Publishers Ltd.

Alternatively, temperature has been also used to control substrate stiffness. We have recently developed a dynamic cell culture platform with dynamically tunable elasticity using temperature-responsive poly(ε-caprolactone) (PCL) films (Uto et al. 2014). While the crosslinked films are relatively stiff (20 MPa) below the melting temperature (T_m), they suddenly become soft (1 MPa) above the T_m. Correspondingly, spread myoblasts on the film became rounded when temperature was suddenly increased to the temperature above the T_m, while significant changes in cell morphology were not observed for fibroblasts. These results indicate that cells can sense dynamic changes in the surrounding environment but the sensitivity depends on cell types. PCL films also demonstrated

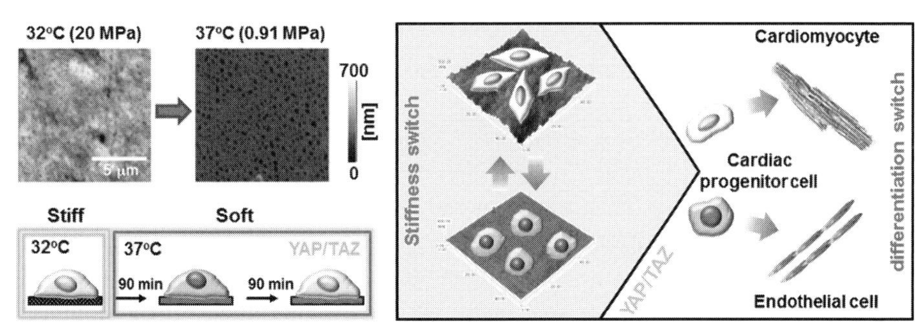

Figure 11.2 **Stiffness switches in semi-crystalline substrates.**
YAP/TAZ intracellular localization is dynamically regulated in cardiac progenitor cells as a function of substrate stiffness (Mosqueira et al. 2014). While the crosslinked films are relatively stiff (20 MPa) below the melting temperature (T_m), they suddenly become soft (1 MPa) above the T_m. Following the stiffness switch, a significant decrease in nuclear expression of YAP/TAZ (green) could be detected after 90 minutes. However, the percentage of YAP nuclear positive cells returned to the original values after 180 min.

that intracellular localization of YAP/TAZ is dynamically regulated in cardiac progenitor cells as a function of substrate stiffness (Mosqueira et al. 2014). Following the stiffness switch, a significant decrease in nuclear expression of YAP/TAZ could be detected after 90 minutes. However, the percentage of YAP nuclear positive cells returned to the original values after 180 min (Fig. 11.2).

Although the majority of studies to date have focused on hydrogels that respond to physical stimuli such as light or heat, these approaches may also lead to other undesirable changes in properties. Emerging concepts, therefore, are focusing on hydrogels that respond to specific biological stimuli. Jiang et al. (2010a) have investigated the effects of the dynamic stiffness of the microenvironment on neuronal responses using DNA-crosslinked PA hydrogels. In this gel system, DNA strands are covalently attached to polymer chains, forming zips. Reverse of gelation occurs when the complement strand is introduced to displace the crosslinker DNA out of the gel network. When the substrates become softer, the primary dendrite number decreased and the axonal length markedly increased. This result indicates that neurons are capable of detecting alterations in the mechanical stiffness in the local microenvironment and responding to the alterations as manifested by changes in neurite outgrowth. This offers a unique opportunity for neural tissue engineering in cases where promoting axonal regeneration and growth are desired. The differentiation of hMSCs was also directed by degradation-mediated cellular traction in HA hydrogels through the proteolytically cleavable crosslinks (Khetan et al. 2013). HA was first functionalized with both methacrylate and maleimide groups. The gel was then formed using Michael-type reactions between maleimides and thiols on bifunctional matrix metalloproteinase (MMP)-degradable peptides. Finally, the gel was incubated with photoinitiator and exposed to UV light to initiate free-radical photpolymerization of methacrylates. For gel without UV, hMSCs spread and deformed the surrounding matrix and underwent primarily osteogenesis. On the other hand, cells within UV-irradiated gel underwent adipogenesis. These results suggest that upregulating tension induced osteogenesis even in the restrictive environment.

In addition to softening materials, current efforts are centered on stiffening materials because recent researches have also suggested that changes in tissue stiffness are related to specific disease characteristics. For example liver fibrosis, which results in cirrhosis, portal hypertension, and hepatocellular carcinoma are initiated by stiffening of liver tissue (Georges et al. 2007). Healthy lung tissue has been shown to have an elastic modulus in the range of 5–30 kPa when deformed at physiologically relevant rates, whereas tissues treated with proteases to mimic progression of alveolar disease showed a loss in mechanical rigidity (Yuan et al. 2000). For an embryonic chicken heart, maturing from mesoderm to adult myocardium results in a 9-fold stiffening originating in part from a change in collagen expression and localization. To mimic this temporal stiffness change in vitro, Young and Engler (2011) synthesized thiolated-hyaluronic acid (HA) hydrogels crosslinked with poly(ethylene glycol) diacrylate. The HA hydrogels were found to stiffen from 1.9 kPa to 8.2 kPa over days due to the Michael-type addition reaction. When precardiac cells were cultured on collagen-coated HA hydrogels, they exhibited a 3-fold increase in mature cardiac specific markers and formed up to 60% more maturing muscle fibers than they did when grown on compliant but static polyacrylamide hydrogels over 2 weeks. Jiang et al. (2010b) also utilized the aforementioned DNA-crosslinked PA hydrogel system to investigate the effects of dynamic stiffening on fibroblast growth. Two types of fibroblasts, L929 and GFP, were subject to the alterations in substrate rigidity. The stiffness was increased by incorporating DNA strands into the DNA gel network, which results in the changes in the crosslinking density. The range of mechanical stiffness of the resulting gels is 5.9–22.9 kPa. Both cells shared specific responses in common, but the projection area and polarity responded differently. Guvendiren and Burdick (2012) developed a stiffening hydrogel system that provides fast dynamic changes, long-term stability and structural uniformity. By using a model system that permits osteoblast and adipocyte differentiation in a bipotential media, they investigate the short-term (minutes-to-hours) and long-term (days-to-weeks) hMSC response to substrate stiffening from soft (~3 kPa) to stiff (~30 kPa). First, methacrylated HA was crosslinked with dithiothreitol via a Michael-type reaction. Then, the hydrogel was stiffened within minutes by radical polymerization of the remaining methacrylate groups when the hydrogel is swollen with a photoinitiator and exposed to ultraviolet light. The hMSCs selectively differentiated based on the period of culture (Fig. 11.3). For example, adipogenic differentiation is favored for later stiffening, whereas osteogenic differentiation is favored for earlier stiffening.

11.3 Dynamic topographies

In addition to sensing stiffness, topographical cues also play an integral role in influencing cell fate. However, current efforts are centered on rather static patterns. Therefore, much

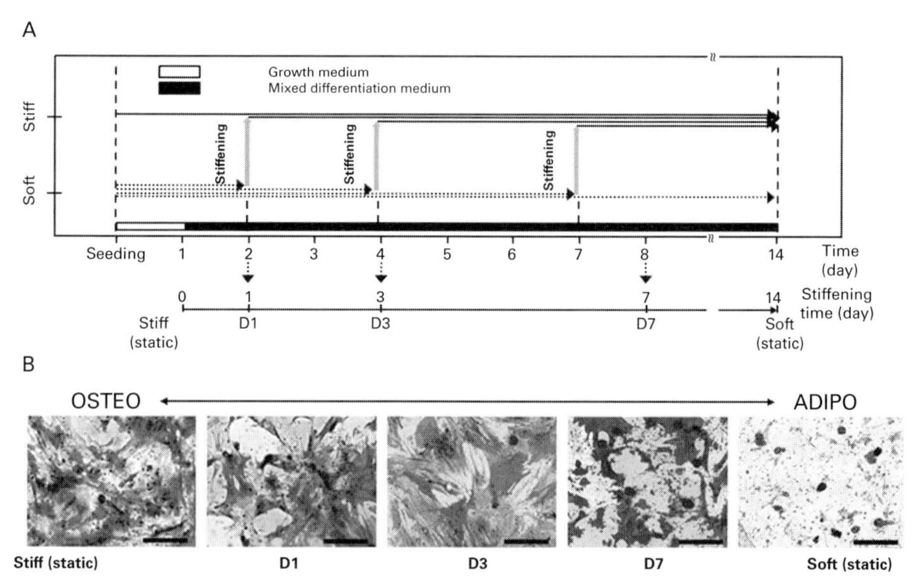

Figure 11.3 **Effects of temporal stiffening of hydrogel substrates on hMSC differentiation.** (A) Plot indicating the time line for temporal stiffening during differentiation experiments. Cells were cultured 1 day in growth media followed by 14 days in bipotential (adipo/osteo) differentiation medium (mixed media). Substrate condition is defined as static or dynamic, and for dynamic gels stiffening time (D1, D3, D7 for 1, 3, and 7 days, respectively, of culture in mixed media before stiffening) is reported. (B) Bright-field images of hMSCs stained for ALP (osteo) and lipid droplets with oil red O (adipo) after 14 days of culture in mixed media on MeHA hydrogels stiffened at indicated days. Scale bars are 50 μm (Guvendiren and Burdick 2012).
Reprinted with permission from Macmillan Publishers Ltd.

attention has been focused on development of surfaces with tunable topography. Lam et al. (2008) first developed a reconfigurable microtopographical system customized for cell culture that consists of reversible wavy microfeatures on poly(dimethylsiloxane) (PDMS). The wavy features are created by first plasma oxidizing the PDMS substrate to create a thin, brittle film on the surface and then applying and releasing compressive strain, to introduce and remove the microfeatures, respectively (Fig. 11.4A). Average amplitude of waves was around 670 nm. The reversible wavy microfeatures were able to align, unalign, and realign C2C12 myogenic cells repeatedly on the same substrate (Fig. 11.4B). Kischner and Anseth (2013), on the other hand, demonstrated real-time cell morphology control by controlled erosion of a photolabile, PEG-based hydrogel system. Photolithographic techniques were employed to present dynamic topographies with features of subcellular dimensions (~5–40 μm) and with various aspect ratios increasing from 1:1 to infinity. The hMSCs were initially seeded onto smooth surfaces, which were in turn patterned sequentially using photolithography. After photo-irradiation, cell morphology and alignment had begun to respond and change based on the new underlying pattern. The presentation of new pattern led to an increase in the average cellular aspect ratio and an increase in cellular alignment along the pattern. These changes in cell morphology were reversed with the next patterning step, which returned

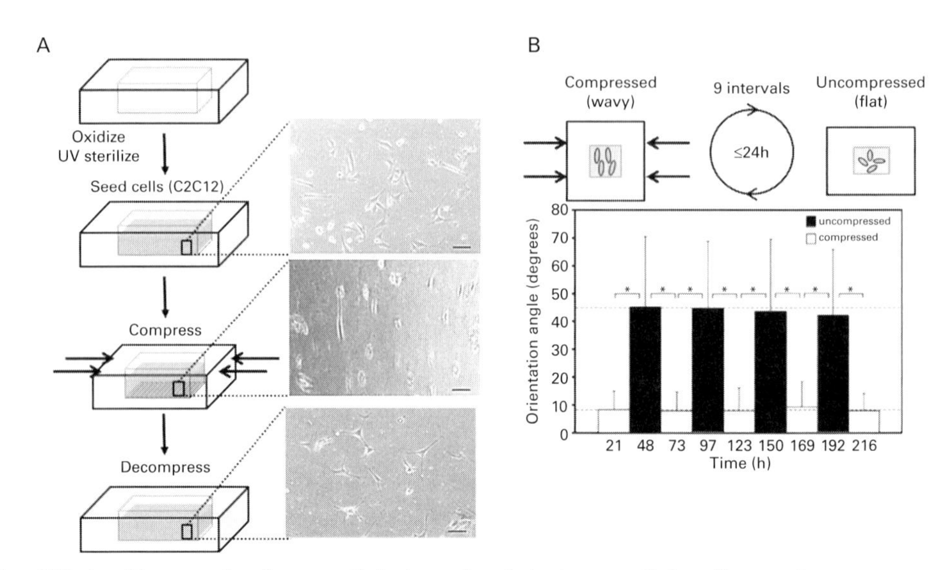

Figure 11.4 **Effects of topography changes of elastomeric substrates on cellular alignment.** (A) Schematic of the process of reversibly introducing microfeatures into the substrate surface. Micrographs on the right show cell behavior on the reconfigurable surface before, during and after compression. Scale bars = 100 μm. (B) Average orientation angles for cells during repeated compression (wavy) and uncompression (flat) (Lam et al. 2008).

Reprinted with permission from Elsevier.

the cells to an isotropic surface. Using this technique, it will be possible to investigate the influence of dynamic changes on cellular anisotropy on hMSC lineage. Kiang et al. (2013) also demonstrated the effect of dynamic and reversible surface topography on vascular smooth muscle cell (VSMCs) morphology. They have synthesized a soft PA hydrogel ($E \sim 1$ kPa) with magnetic nickel microwires with 5 μm in diameter and 20 μm in length. This system can reversibly induce the alignment of microwires by applying magnetic field. Cells responded to acute changes in topography on soft hydrogels, but that prolonged exposure to and dynamic oscillation of a substrate with disordered topographical features did not produce significant changes in morphology. This result implies that the difference roles between acute and chronic response of cell to dynamic topographic change.

As an alternative approach, shape-memory polymers (SMPs) have been extensively developed to create dynamic topographies for directing cell morphologies. SMPs are a class of "smart" materials that have the capability to change from a temporary shape to a memorized permanent shape upon application of an external stimulus. Throughout the last decade, SMPs have represented a cheap and efficient alternative to well-known metallic shape-memory alloys as a consequence of the relative ease of manufacture and programming. Among them, thermally induced SMPs are the most extensively investigated group of SMPs. Davis et al. (2011) first applied SMP substrate to investigate cell behavior. They used commercially available, non-cytotoxic optical adhesive, Norland Optical Adhesive 63 (NOA-63), which is a polyurethane-based thiol-ene crosslinked polymer system, as a dynamic cell culture substrate. A flat substrate was first embossed with microgrooves with amplitude of 25.6 ± 0.8 μm to produce a temporary topography. The temporary topography was then erased by heating around a glass transition temperature (T_g) of the NOA-63. Alignment of mouse embryonic fibroblasts (C3 H/10T1/2) along the grooves of the temporary topography was significantly reduced upon transition to a flat permanent topography. The study demonstrated that SMPs enable a high degree of control over the activation of the surface shape-memory effect. However, the large, irregular dimensions of the surface patterns limited the degree of control over fibroblast cell morphology. From this regard, Le et al. (2011) developed biocompatible shape-memory surfaces that can accommodate diverse, well-defined and biologically relevant surface transformations under physiological conditions. This dynamic cell culture substrate was prepared by crosslinking a methacrylate end-functionalized PCL macromonomer in a mold to produce a permanent shape and then mechanically compressed the desired shape to form a temporary shape. These substrates showed the potential to transition between two predefined surface topography by changing the temperature from 28°C to 40°C. In fact, hMSC morphology switched from highly aligned to stellate shaped in response to a surface transformation between a 3 μm × 5 μm groove array and planar surface.

We have also been developing a SMP system with dynamically tunable nanopatterns to direct cell fate (Fig. 11.5). The shape-memory nanopatterns were prepared by chemically crosslinking semi-crystalline PCL in a mold to show shape-memory effects over its melting temperature (T_m = 33°C). Permanent surface patterns were first generated by crosslinking the PCL macromonomers in a mold, and temporary surface patterns were then embossed onto the permanent patterns. The temporary surface patterns could be easily triggered to transition quickly to the permanent surface patterns by a 37°C heat treatment (Fig. 11.5A). One of the great advantages of PCL over other temperature-responsive polymers is that surface properties such as wettability and charge are independent of temperature. Using this substrate, we have successfully demonstrated time-dependent cell orientation changes by inducing nanotopographical transition from grooves with a height of 300 nm to flat (Ebara et al. 2012). Upon transition from the grooved topography to a flat surface, cell alignment was lost and random cell migration and growth ensued. We have also succeeded in inducing a 90° rotation of the cell orientation by using a shape-memory nanogrooves, the direction of which was transitioned 90° to the temporary grooves (Fig. 11.5B) (Ebara et al. 2014a). Interestingly, 90% of cells did not change their direction 1 hour after the topographic transition. By 36 hours, however, 70% of cells realigned parallel to the permanent grooves that emerged (Fig. 11.5C). To understand the effects of pattern dimension (nm vs μm) on the interlude between the topographic transition of shape-memory nanopatterns and cell response on them, we have also monitored time-dependent changes in surface nanotopographic features associated with shape-memory transition as well as the cell morphology or alignment (Ebara et al. 2014b). Holographic microscope revealed that the application of heat to PCL SMP quickly and completely transitioned temporary surface patterns substrate to permanent patterns within 30 seconds. However, it took more than 2 hours and 8 hours for cells on substrate with 500 nm and 2000 nm grooves to

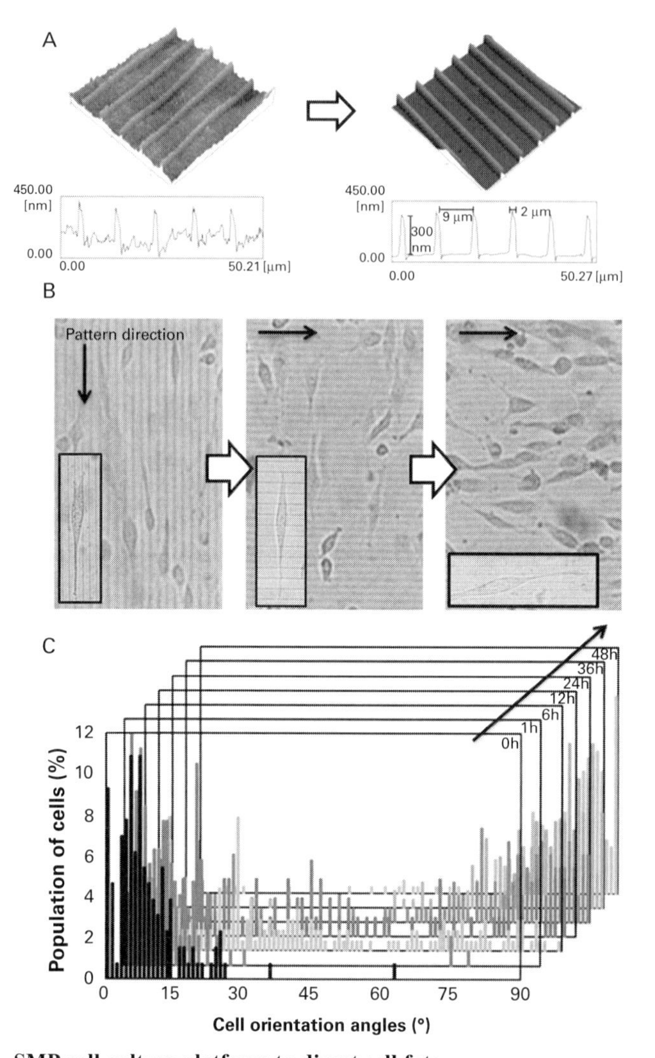

Figure 11.5 **SMP cell culture platform to direct cell fate.**
(A) Shape-memory transition from a memorized temporal pattern (left) to the original permanent pattern (right) observed by AFM. (B) Phase contrast images of NIH3T3 fibroblasts seeded on the fibronectin-coated PCL films with the temporal grooved surface. Cells were cultured at 32°C for 24 h (left). Then, the cells were subjected to a 37°C heat treatment for 1 h (middle). The cells were then allowed to equilibrate at 32°C for 48 h (right). (C) Histograms of cell orientation angle on the PCL film before and after shape-memory activation. Cell orientation angles were quantified by the analysis of phase contrast images and defined as the angle against the temporary groove direction.
Reproduced from Ebara et al. (2014a).

induce 90°-rotation of the cell orientation, respectively. This different alignment behavior can be explained by the different adhesion strength and reorganization of cytoskeletal proteins on nano versus micropatterns. To our best knowledge, we first revealed that dynamic control of geometrical shape exert a dramatic effect on the realignment of adhered cells even using the same material. These dynamic changes in nanotopography created on PCL substrate can also influence the intracellular localization of YAP/TAZ in cardiac progenitor cells (Mosqueira et al. 2014). Following surface change from nanotopography (300 nm of line ridge, 500 nm of groove, 120 nm of height) to flat by switching temperature to 32–37°C, a significant decrease in nuclear expression of YAP/TAZ could be detected after 90 minutes. However, the percentage of YAP nuclear positive cells returned to the original values after 180 min. Given the selective relocalization of YAP/TAZ in response to stiffness change as described in Section 11.2, these proteins could be sensitive to dynamic change in surface nanotopography. In the future, these approaches will also enable not only unprecedented observations of time-dependent cell–substrate interactions without the need for invasive forces against intact adhered cells but also direct manipulation of cell function and fate.

11.4 Switchable surface traps

Cell adhesion process plays an important role in various physiological processes such as migration, proliferation, differentiation, and death. Therefore, the technologies to exercise spatiotemporal control over the adhesion, detachment and migration of cells are desirable for further development of biomedical engineering and biotechnology. The most commonly used dynamic surfaces to date have been designed to respond to changes in temperature. Critical insight into the temperature-responsive cell detachment mechanism was gained by the Okano group's seminal work on poly(*N*-isopropylacrylamide) (PNIPAAm)-grafted surfaces. In 1990, a novel detachment technique for actively adhered cells was demonstrated using PNIPAAm-grafted tissue culture polystyrene (TCPS) dishes (Yamada et al. 1990). Many attachment-dependent cells can adhere on PNIPAAm-grafted surface at 37°C but spontaneously detached from the surface by reducing temperature to room temperature which is below the lower critical solution temperature (LCST; 32°C) of PNIPAAm, as a result of rapid hydration of the grafted polymer. Such dynamic cell adhesion–detachment modulation on the surfaces grafted with thermo-responsive PNIPAAm chains is a novel concept because no enzymes or chelators are required to detach actively adhered cells. The most striking feature is that this system allows the preservation of intact membranes and adhesive proteins, because only interactions between adhesive proteins on the basal side of the cultured cells and the grafted PNIPAAm on the surfaces are disrupted. Kushida et al. (1999) demonstrated that confluent cultured cells were harvested as a contiguous cell sheet from the thermo-responsive surfaces while maintaining cell-cell junctions and basal extracellular matrices (ECMs). Since adhesive proteins derived from cells are also harvested at the same time, the cell sheets can readily adhere to various surfaces including culture dishes, other cell sheets and host tissues. Now this technology has been recognized as cell sheet engineering that is a unique nanotechnology for scaffold-free tissue reconstruction with clinical application in regenerative medicine (Egami et al. 2014).

Although biological systems are dominated by noncovalent interactions, which are generally weaker but more sensitive to the environments, specific interactions such as integrins with ECM components play an important role in many cellular functions, such as regulation of the cell morphology, growth, differentiation, and motility, because anchorage-dependent cells adhere to a substrate via a series of steps beginning with the binding of cell membrane receptors to substrate ligands, followed by cytoskeletal and cytoplasmic reorganization. From these perspectives, we have proposed a novel technique that explores spontaneous and noninvasive cell detachment behavior from substrates. We have developed a novel approach to observe dynamic affinity changes in Arg-Gly-Asp (RGD)-integrin biding using RGD-immobilized temperature-responsive nano layer surfaces (Ebara et al. 2003). RGD was successfully immobilized on NIPAAm copolymer's side chains via reactive groups of comonomer, 2-carboxyisopropylacrylamide (CIPAAm) (Aoyagi et al. 2000). The designed surface can bind cells at 37°C and spontaneously cause the detachment of cells at 20°C. This phenomenon arises from the specific design of the grafted polymer, which protects peptides from integrin access below the LCST. The tight coupling between cells and peptides on the surfaces produces a delay in the detachment of cells from the surfaces (Ebara et al. 2004a; Ebara et al. 2004b). Indeed, the time required for cell detachment is strongly related to the cell adhesion strength according to the trend: RGD < RGDS < GRGD < GRGDS. This approach provides a quick and simple method for examining time-dependent affinity changes between cells and peptides. Also, we have successfully investigated the synergistic effect of Pro–His–Ser–Arg–Asn (PHSRN) on integrin-mediated cell binding using the same technique (Ebara et al. 2008). PHSRN has been found in fibronectin (FN) and is thought to synergistically enhance the cell-adhesive activity of the RGD sequence. The surfaces dramatically retards cell detachment below the LCST only when the peptide sequences were specifically designed with the optimal distance between PHSRN and RGDS, as observed in native FN. The simplicity, ease of use, sensitivity, and noninvasive nature of this technique make it a particularly attractive alternative to more complex methods based on the application of forces to detach cells from substrates. We have also focused on the carbohydrate moiety of glycoproteins, which is one of the most abundant and important biomolecules. The carbohydrates are not only a major source of metabolic energy but are signal biomolecules in a wide range of molecular recognition phenomena. We constructed a temperature-responsive "on-off" surface capture and release system for hepatocytes using temperature-responsive copolymer brushes comprising NIPAAm and 2-lactobionamidoethyl methacrylate (LAMA), which is known as a sugar-based monomer (Fig. 11.6) (Idota et al. 2012). Although NIH3T3 fibroblasts did not adhere on the surface of polymer brushes containing LAMA under serum-free conditions, the surface promoted HepG2 cell adhesion at 37°C through a specific interaction with functional sugar moieties. Moreover, the adhered HepG2 successfully detached from the surface when temperature was reduced to 25°C. Therefore, this system can enable the realization of label-free high-throughput cell separation for stem cell biology, cancer diagnostics, and regenerative medicine.

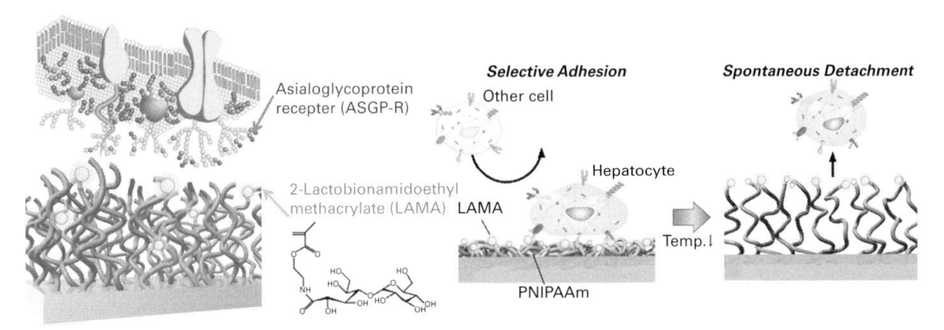

Figure 11.6 **Temperature-responsive cell culture platform.**
Schematic illustration of an "on-off" switchable surface trap for selective adhesion/detachment of HepG2 hepatocyte using a temperature-responsive glycopolymer brush surface (Idota et al. 2012). The brush surfaces facilitate the adhesion of HepG2 at 37°C under non-serum conditions, while no adhesion is observed for NIH3T3 fibroblasts. When temperature is decreased to 25°C, almost all HepG2 cells come off from the brush.

Although the majority of studies to date have focused on temperature-responsive systems, emerging concepts are also focusing on systems that respond to other stimuli. For example, light is able to present precise control over temporal and spatial signals. Photochemical reactions are often used to induce switching the surface physicochemical properties involving either irreversible photocleavage or reversible photoisomerization. Nitrobenzyl ester derivatives are the most commonly used photocleavable protecting group which undergo photolysis with irradiation of near-UV light (~350 nm). Nakanishi et al. (2006) reported a methodology to spatiotemporally control cell adhesion using a photocleavable nitrobenzyl group (Fig. 11.7A).They prepared the surface modified with an alkylsliloxane having a photocleavable 2-nitrobenzyl functional group for patterning of human embryonic kidney 293 (HEK 293) cell. First, the substrate underwent physical adsorption of BSA to make its surface inert to cell adhesion. The substrate was then illuminated with UV light of 365 nm under a fluorescence microscope to release the 2-nitrobenzyl group, leading to desorption of BSA from the surface. The following addition of fibronectin to the substrate produced its selective deposition onto the region from which BSA was desorbed by UV illumination. They also applied this technique to spatiotemporal control of cell migration system (Fig. 11.7B) (Nakanishi et al. 2007). Cell adhesive narrow (5 μm) or wide (25 μm) path was created to induce cell migration after single NIH3T3 fibroblast arrangement. The control of NIH3T3 migration in single cell level can be achieved by simultaneously using patterned illumination technology. They clearly revealed that the size and shape of cell-adhesive path determined the type of formed protrusions. Switchable substrates can be also used for the analysis of collective migration. Rolli et al. (2012) also reported the collective migration of Madin-Darby canine kidney (MDCK) cells from precisely controlled initial adhesion geometry by using photoswitchable adhesive substrates. They revealed the appearance of the leader cell was dependent on the geometry and size of formed cluster and culture time. The photoisomerization process of spiropyrans has been also applied to produce switchable

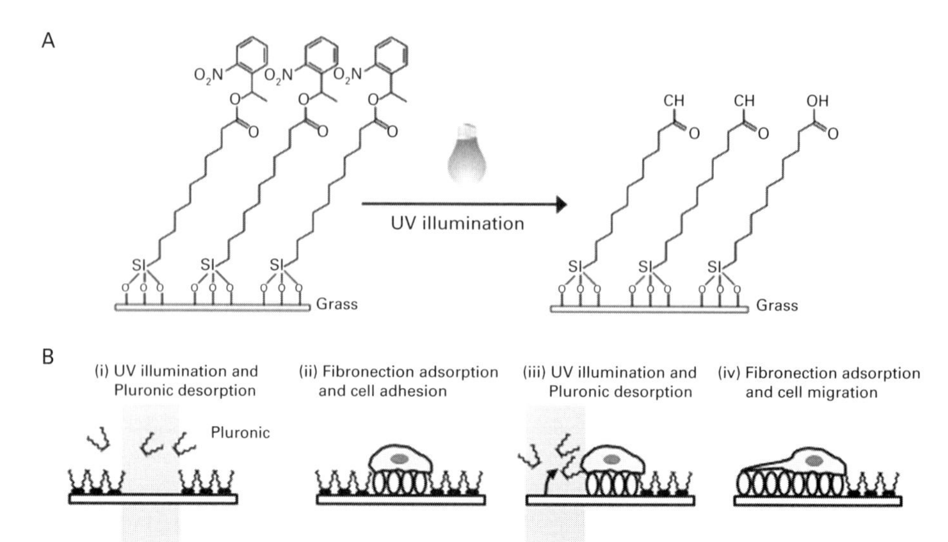

Figure 11.7 **Photo-induced spatiotemporal control of cell adhesion.**
(A) The photochemical reaction on the substrate surface by UV illumination at 365 nm (Nakanishi et al. 2006). Reprinted with permission from Elsevier. (B) Schematic illustrations of the placement of single cells followed by the induction of cell migration on the substrate (Nakanishi et al. 2007).
Reprinted with permission from the American Chemical Society.

surface with photo-tunable cell adhesion. Edashiro et al. (2005) synthesized temperature responsive PNIPAAm having photoresponsive nitrospiropyran chromophores as side chains. Since nitrospiropyran is isomerized into a colored zwitterionic structure (opened-ring form) by UV light irradiation and is isomerized back into a colorless nonionic structure (closed-ring form) by visible light, it provides the cell culture surface with reversible photoresponsive property. Cell adhesion of the surface was drastically enhanced by the irradiation with UV light (wavelength: 365 nm), whereas most cells were removed from the nonirradiated region. Auernheimer et al. (2005), on the other hand, reported the possibility to switch cell adhesion properties using an azobenzene molecule which undergoes isomerization from trans to cis conformations state under UV irradiation (340–380 nm) and back to trans conformation on visible light irradiation (450–490 nm). A set of cyclic RGD peptides containing a photoswitchable 4-[(4-aminophenyl)azo]benzocarbonyl unit were synthesized, and poly(methyl methacrylate) (PMMA) surfaces were modified with these peptides. Control of adhesion stimulation by irradiation with 366 nm or 450 nm light could be achieved. They clearly demonstrated that synthesized peptides lead to enhanced mouse osteoblasts (MC3T3 E1) adhesion on PMMA disks in their trans- form, whereas the adhesion efficiency was decreased by UV irradiation at 366 nm.

Electrochemical reactions have also been used to regulate adhesion and migration of cell. Yeo et al. reported pioneering work that electroactive substrates enable directly switch ligand activities in response to electrical potential (Yeo et al. 2003). They designed a self-assembled monolayer (SAM) that incorporates an *O*-silyl hydroquinone moiety to present cell adhesive RGD peptide. These dynamic substrates with circular pattern (220 μm in diameter) of nonelectroactive SAM allowed for selective release of Swiss 3T3 fibroblasts by applying electrical potential because *O*-silyl hydroquinone ether is electroactive and provides for selective release of the RGD peptide from the substrate by electrochemical oxidation. They successfully demonstrated subsequent reattachment of RGD peptide to the oxidized quinone through Dields-Alder reaction that enables switching back to cell-adhesive substrate. Lamb and Yousaf (2011) developed a novel strategy for switching an immobilized molecular structure on the substrate surface based on sequential two orthogonal click reactions of Huisgen cycloaddition and benzoquinone-oxime chemistry. In this dynamic system, the oxime linkage plays a role of redox active site that is able to switch the activity of RGD peptide presented to cell by in situ changing peptide conformation from cyclic to linear. This switchable surface can dynamically modulate various cell behaviors including cell adhesion morphology and migration rate. They also prepared electroactive gradient surfaces for spatial and temporal control of mammalian cell behavior (Luo and Yousaf 2011). These electrochemically switchable surface systems would be used as a model substrate to monitor a range of cell behaviors including adhesion, migration and polarization and as a tumor invasion model system. As another approach in electrochemical control system, Kakegawa et al. (2013) developed an electrochemically detachable, biocompatible oligopeptide-modified cell culture surface. They focused gold-thiolate bond to form SAM on gold electrode because this bond can be reductively cleaved by applying negative electrical potential. Cell-repulsive oligopeptide, CGGGKEKEKEK and cell-adhesive CGGGKEKEKEKGRGDSP were designed with a terminal cysteine residue to mediate binding to a gold surface, and these are spontaneously formed SAMs via gold-thiolate bond and intermolecular electrostatic interaction. Adhered Swiss 3T3 fibroblasts were completely and rapidly (within two minutes) detached due to desorption of the oloigopeptide layer by applying negative electrical potential. Therefore, this electrochemical control system of cell detachment may be a useful tool for tissue-engineering and regenerative medicine applications.

11.5 Conclusion and future work

Although our understanding of biochemical and physicochemical cues in the cellular microenvironment has been constantly improved, there are only a few examples of systems where those cues can be dynamically controlled. From this regards, stimuli-responsive substrates are becoming one of the most powerful tools available to lead to a new generation of dynamic systems that are capable of responding to external stimuli or cellular signals with spatial precision. In the future, new developments of independent, dynamic regulation of multiple parameters during tissue formation are desired. In addition, materials are being designed with spatial heterogeneity, to replicate properties in native tissue structures. Moreover, development of dynamic materials capable of incorporating a wide range of biologically relevant molecules has the potential to provide useful therapeutic benefits because it has been difficult for natural materials to produce highly organized constructs with tunable properties. Since an ongoing challenge is to control stem cell differentiation, novel strategies for substrate design will facilitate the creation of new class of dynamic substrates in the future.

References

Aoyagi, T., Ebara, M., Sakai, K., Sakurai, Y. and Okano, T. (2000). "Novel bifunctional polymer with reactivity and temperature sensitivity" *Journal of Biomaterials Science, Polymer Edition* **11**: 101–110.

Auernheimer, J., Dahmen, C., Hersel, U., Bausch, A. and Kessler, H. (2005). "Photoswitched cell adhesion on surfaces with RGD peptides." *Journal of the American Chemical Society* **127**: 16107–16110.

Davis, K. A., Burke, K. A., Mather, P. T. and Henderson, J. H. (2011). "Dynamic cell behavior on shape memory polymer substrates." *Biomaterials* **32**: 2285–2293.

Deforest, C. A. and Anseth, K. S. (2011). "Cytocompatible click-based hydrogels with dynamically tunable properties through orthogonal photoconjugation and photocleavage reactions." *Nature Chemistry* **3**: 925–931.

Ebara, M., Akimoto, M., Uto, K., et al. (2014a). "Focus on the interlude between topographic transition and cell response on shape-memory surfaces." *Polymer* **55**: 5961–5968.

Ebara, M., Uto, K., Idota, N., Hoffman, J. M. and Aoyagi, T. (2012). "Shape-memory surface with dynamically tunable nano-geometry activated by body heat." *Advanced Materials* **24**: 273–278.

Ebara, M., Uto, K., Idota, N., Hoffman, J. M. and Aoyagi, T. (2014b). "The taming of the cell: shape-memory nanopatterns direct cell orientation." *International Journal of Nanomedicine* **9**(Supplement 1): 117–126.

Ebara, M., Yamato, M., Aoyagi, T., et al. (2004a). "Immobilization of cell-adhesive peptides to temperature-responsive surfaces facilitates both serum-free cell adhesion and noninvasive cell harvest." *Tissue Engineering* **10**: 1125–1135.

Ebara, M., Yamato, M., Aoyagi, T., et al. (2004b). "Temperature-responsive cell culture surfaces enable 'on–off' affinity control between cell integrins and RGDS ligands." *Biomacromolecules* **5**: 505–510.

Ebara, M., Yamato, M., Aoyagi, T., et al. (2008). "The effect of extensible PEG tethers on shielding between grafted thermo-responsive polymer chains and integrin–RGD binding." *Biomaterials* **29**: 3650–3655.

Ebara, M., Yamato, M., Hirose, M., et al. (2003). "Copolymerization of 2-carboxyisopropylacrylamide with N-isopropylacrylamide accelerates cell detachment from grafted surfaces by reducing temperature." *Biomacromolecules* **4**: 344–349.

Edahiro, J.-I., Sumaru, K., Tada, Y., et al. (2005). "In situ control of cell adhesion using photoresponsive culture surface." *Biomacromolecules* **6**: 970–974.

Egami, M., Haraguchi, Y., Shimizu, T., Yamato, M. and Okano, T. (2014). "Latest status of the clinical and industrial applications of cell sheet engineering and regenerative medicine." *Archives of Pharmacal Research* **37**: 96–106.

Frey, M. T. and Wang, Y.-L. (2009). "A photo-modulatable material for probing cellular responses to substrate rigidity." *Soft Matter* **5**: 1918–1924.

Georges, P. C., Hui, J.-J., Gombos, Z., et al. (2007). "Increased stiffness of the rat liver precedes matrix deposition: implications for fibrosis." *American Journal of Physiology-Gastrointestinal and Liver Physiology* **296**(6): G1147–G1156.

Guvendiren, M. and Burdick, J. A. (2012). "Stiffening hydrogels to probe short- and long-term cellular responses to dynamic mechanics." *Nat Commun* **3**: 792.

Idota, N., Ebara, M., Kotsuchibashi, Y., Narain, R. and Aoyagi, T. (2012). "Novel temperature-responsive polymer brushes with carbohydrate residues facilitate selective adhesion and collection of hepatocytes." *Science and Technology of Advanced Materials* **13**: 064206.

Jiang, F. X., Yurke, B., Schloss, R. S., Firestein, B. L. and Langrana, N. A. (2010a). "Effect of dynamic stiffness of the substrates on neurite outgrowth by using a DNA-crosslinked hydrogel." *Tissue Eng Part A* **16**: 1873–1889.

Jiang, F. X., Yurke, B., Schloss, R. S., Firestein, B. L. and Langrana, N. A. (2010b). "The relationship between fibroblast growth and the dynamic stiffnesses of a DNA crosslinked hydrogel." *Biomaterials* **31**: 1199–1212.

Kakegawa, T., Mochizuki, N., Sadr, N., Suzuki, H. and Fukuda, J. (2014). "Cell-adhesive and cell-repulsive zwitterionic oligopeptides for micropatterning and rapid electrochemical detachment of cells." *Tissue Eng Part A* **19**: 290–298.

Khetan, S., Guvendiren, M., Legant, W. R., et al. (2013). "Degradation-mediated cellular traction directs stem cell fate in covalently crosslinked three-dimensional hydrogels." *Nat Mater* **12**: 458–465.

Kiang, J. D., Wen, J. H., Del Álamo, J. C. and Engler, A. J. (2013). "Dynamic and reversible surface topography influences cell morphology." *Journal of Biomedical Materials Research Part A* **101**A: 2313–2321.

Kirschner, C. M. and Anseth, K. S. (2013). "In situ control of cell substrate microtopographies using photolabile hydrogels." *Small* **9**: 578–584.

Kloxin, A. M., Kasko, A. M., Salinas, C. N. and Anseth, K. S. (2009). "Photodegradable hydrogels for dynamic tuning of physical and chemical properties." *Science* **324**: 59–63.

Kushida, A., Yamato, M., Konno, C., et al. (1999). "Decrease in culture temperature releases monolayer endothelial cell sheets together with deposited fibronectin matrix from temperature-responsive culture surfaces." *Journal of Biomedical Materials Research* **45**: 355–362.

Lam, M. T., Clem, W. C. and Takayama, S. (2008). "Reversible on-demand cell alignment using reconfigurable microtopography." *Biomaterials* **29**: 1705–1712.

Lamb, B. M. and Yousaf, M. N. (2011). "Redox-switchable surface for controlling peptide structure." *Journal of the American Chemical Society* **133** 8870–8873.

Le, D. M., Kulangara, K., Adler, A. F., Leong, K. W. and Ashby, V. S. (2011). "Dynamic topographical control of mesenchymal stem cells by culture on responsive poly(ε-caprolactone) surfaces." *Advanced Materials* **23**: 3278–3283.

Luo, W. and Yousaf, M. N. (2011). "Tissue morphing control on dynamic gradient surfaces." *Journal of the American Chemical Society* **133**: 10780–10783.

Mosqueira, D., Pagliari, S., Uto, K., et al. (2014). "Hippo pathway effectors control cardiac progenitor cell fate by acting as dynamic sensors of substrate mechanics and nanostructure." *Acs Nano* **8**: 2033–2047.

Nakanishi, J., Kikuchi, Y., Inoue, S., et al. (2007). "Spatiotemporal control of migration of single cells on a photoactivatable cell microarray." *Journal of the American Chemical Society* **129**: 6694–6695.

Nakanishi, J., Kikuchi, Y., Takarada, T., et al. (2006). "Spatiotemporal control of cell adhesion on a self-assembled monolayer having a photocleavable protecting group." *Analytica Chimica Acta* **578**: 100–104.

Rolli, C. G., Nakayama, H., Yamaguchi, K., et al. (2012). "Switchable adhesive substrates: Revealing geometry dependence in collective cell behavior." *Biomaterials* **33**: 2409–2418.

Uto, K., Ebara, M. and Aoyagi, T. (2014). "Temperature-responsive poly(ε-caprolactone) cell culture platform with dynamically tunable nano-roughness and elasticity for control of myoblast morphology." *International Journal of Molecular Sciences* **15**: 1511–1524.

Yamada, N., Okano, T., Sakai, H., et al. (1990). "Thermo-responsive polymeric surfaces; control of attachment and detachment of cultured cells." *Makromolekulare Chemie, Rapid Communications* **11**: 571–576.

Yang, C., Tibbitt, M. W., Basta, L. and Anseth, K. S. (2014). "Mechanical memory and dosing influence stem cell fate." *Nat Mater* **13**: 645–652.

Yeo, W.-S., Yousaf, M. N. and Mrksich, M. (2003). "Dynamic interfaces between cells and surfaces: electroactive substrates that sequentially release and attach cells." *Journal of the American Chemical Society* **125**: 14994–14995.

Young, J. L. and Engler, A. J. (2011). "Hydrogels with time-dependent material properties enhance cardiomyocyte differentiation in vitro." *Biomaterials* **32**: 1002–1009.

Yuan, H., Kononov, S., Cavalcante, F. S. A., et al. 2000. "Effects of collagenase and elastase on the mechanical properties of lung tissue strips." *Journal of Applied Physiology* **89**: 3–14.

Part II

Recent progress in cell mechanobiology

12 Forces of nature

Understanding the role of mechanotransduction in stem cell differentiation

Andrew W. Holle, Jennifer L. Young, and Yu Suk Choi

12.1 Introduction

A stem cell's fate is determined by a multitude of signals. Stem cells, both embryonic and adult, are guided throughout the process of differentiation by soluble cues, such as morphogens and signaling molecules, as well as by physical cues, such as the mechanical microenvironment. While chemical cues have been appreciated for some time, mechanical cues have been shown to be equally important in cell fate choices in more recent years. In this chapter, we will highlight the important aspects of mechanobiology of stem cells in the body, both during embryogenesis (during which time the majority of cells are at some stage of stem-ness) and in adult tissues, and discuss approaches for harnessing mechanical properties in stem cell differentiation in culture.

12.2 Mechanobiology in vivo

12.2.1 Mechanical control of embryogenesis

Embryonic development is an elaborately orchestrated event that comprises signaling molecules, genes, morphogens, and mechanical stimuli. Despite great efforts made over the past century, the whole story of this process still remains largely incomplete. Understanding and defining all the factors involved throughout the development of tissues and organs in the embryo could be harnessed for use in embryonic stem cell (ESC) differentiation, disease pathology and treatment, and as well in the development of specific tissues or full organ systems in culture. While the role of chemical cues in development and tissue patterning have been largely investigated, physical forces shaping the embryo have only recently become more widely explored as key players in this process.

As defined by Mammoto and Ingber (2010), mechanical forces involved in the developing embryo include spring forces (Shin et al. 2007; Sanders et al. 1996); osmotic pressure (Horner and Wolfner 2008); surface tension (Foty and Steinberg 2005; Krieg et al. 2008); tensional forces, traction, and prestress (Ingber 2006); and shear stress (le Noble et al. 2004; North et al. 2009; Adamo et al. 2009). These mechanical forces have been shown to be essential to the development of many tissue and organ systems, including muscle (Kahn et al. 2009), brain (Anava et al. 2009; Moore et al. 2009; Wilson et al. 2007), cartilage and bone (Ohashi et al. 2002; Stokes et al. 2002) and the heart (Forouhar et al. 2006; Hove et al. 2003; Voronov et al. 2004).

At the cellular level, the mechanical microenvironment influences cells in a variety of manners, for example the presence of external forces, micro- and macroscale physical and compositional properties of the extracellular matrix (ECM), and cell shape. Physical properties of tissues are transmitted to cells via actomyosin-based contraction of cells produced by pulling against the ECM to which they are attached. These exogenous stresses alter internal cytoskeletal tension and subsequently contribute to changes in cellular behavior such as proliferation, migration, survival, and differentiation (Wozniak and Chen 2009; Discher et al. 2005; Mammoto and Ingber 2010; Chen et al. 1997). These changes in cellular functions occur via the activation of specific mechanotransduction signaling pathways, and many have been identified in the developing embryo and differentiation of embryonic stem cells; however, an in-depth discussion is out of the scope of this chapter. While tissue level forces are equally important in embryogenesis, we choose here to focus on molecular scale mechanical aspects due to their direct role in embryonic stem cell differentiation.

12.2.2 Micromechanical properties of the embryo

As cells sort and differentiate throughout development, they secrete and assemble ECM, giving rise to tissue-specific stiffness, for example: brain 0.1–1 kPa (Flanagan et al. 2002), muscle 8–17 kPa, and bone 25–40 kPa (Engler et al. 2006). Tissue stiffness can arise from alterations in cell-cell and cell-matrix adhesions, secretion and assembly of ECM, the composition of extracellular components, and cytoskeletal tension (i.e., the stiffness of a cell). As mentioned previously, cells can sense the stiffness of their surrounding microenvironment and respond in a force-dependent manner. In the embryo, tissue stiffness has been shown to be an important player driving development.

Early experiments stemming from notochord elongation and straightening in *Xenopus* early tail-bud embryos undergoing gastrulation identified the important role of tissue stiffness in preventing buckling by surrounding tissues (Adams et al. 1990). Additionally, a three-fold stiffening from stage 10+ to 11.5 in the *Xenopus* embryo along the involuting marginal zone during gastrulation was observed and believed to be essential to prevent distortion of the tissue (Moore et al. 1995). On an organ-level, stiffness has also been shown to be both temporally and spatially oriented. In the development of the heart, for example, early microindentation experiments linked the importance of material properties and residual stress present in the myocardium and cardiac jelly in the outer curvature with the morphogenetic process of cardiac looping (Zamir and Taber 2004a; Zamir and Taber 2004b). These researchers postulated that the heart would increase stiffness throughout development as sarcomere formation and tissue organization progress. It was indeed later discovered that the heart does not begin as a contractile ~10 kPa material, but rather originates from soft mesoderm, <500 Pa (Krieg et al. 2008), undergoing ~nine-fold myocardial stiffening throughout development in the chick embryo (Engler et al. 2008; Young and Engler 2011;

Majkut et al. 2013). Consequently, cellular functions have been shown to be affected by matrix stiffness, both in vivo and in vitro, which will be subsequently discussed.

The sculpting of tissues is dependent on highly dynamic changes in cell shape collectively coordinating precise cellular movements. A delicate balance between intracellular contractility and adhesion to the ECM governs cell shape, and thus is an important factor when examining the mechanical microenvironment in development. In the *Drosophila* embryo, intercalation of cells in germ-band elongation requires shrinking of intercellular contacts along the dorsal-ventral axis, mediated by nonmuscle myosin II and cortical tension (Bertet et al. 2004). In turn, originally hexagonal-shaped cells morph into pentagonal- or quadrilateral-shaped, but this process is reversed once new intercellular contacts are formed (Blanchard et al. 2009). Additionally, during heart looping, myocardial cells alter their apical surface area to increase or decrease depending on whether they reside in convex or concave surfaces, respectively (Manasek et al. 1972). Lastly, adhesion to the ECM, which is also dependent on tissue stiffness, modulates integrin recruitment and focal adhesion turnover, guiding cellular migration and differentiation throughout development (Schwartz 2010).

12.2.3 Embryonic stem cell mechanobiology

Cell-generated forces involved in embryogenesis fall under two categories: internal (stemming from cytoskeletal tension) and external (stemming from forces outside of the cell body, such as contraction against the ECM). Perhaps the first experiment displaying cell-generated forces in vitro was visualized when highly contractile cells wrinkled the thin silicone sheet on which they were plated (Harris et al. 1980). In vivo, cellular contractility has been shown to play an important role in the branching morphogenesis of the lung airway epithelium. As cytoskeletal tension is disrupted, development of the lung is greatly hindered due to altered force generation in resident cells (Moore et al. 2005). Indeed, these forces contribute to a wide variety of cellular functions, including proliferation (Wang and Riechmann 2007), migration (Montell 2003), survival (Toyama et al. 2008), and most importantly for this chapter, differentiation (Desprat et al. 2008; Kinney et al. 2014).

Embryonic stem cells are a particularly attractive cell source due to their pluripotent nature, which lends to their potential usefulness in regenerative medicine strategies, as well as their ability to model early embryonic development in culture. Much attention has been focused on the ability of chemical factors (Finley et al. 1999; D'Amour et al. 2005; Nostro et al. 2008) and ECM components (Takito and Al-Awqati 2004; Taylor-Weiner et al. 2013) to guide ESC cell fate into clinically relevant populations. However, with the knowledge that embryogenesis is a largely physical process, researchers are also focusing on the importance of mechanical properties in guiding ESC differentiation.

Examining mechanical contributions to the differentiation of ESCs in vivo has its limitations and thus most experiments are performed on an organ- or whole embryo-level, with tissue-level applied forces, such as fluid sheer stress and compression/

extension (although particle tracking micro-rheology has been utilized to visualize cell mechanics in vivo [Daniels et al. 2006]). One study on gastrulation demonstrated that uniaxial compression alone of the *Drosophila* embryo caused up-regulation of premesodermal transcription factor *Twist*, thereby suggesting that *Twist* is mechanosensitive (Farge 2003; Desprat et al. 2008). On the scale of cellular mechanosensation, however, the influence of mechanical properties is best achieved in culture where their interactions with the microenvironment can be precisely controlled. In one such example, albeit at tissue-level force, cyclic stretch of mouse ESCs embedded in a collagen matrix enhanced mesodermal marker expression, which was diminished when myosin II, responsible for transmitting external forces to the cell, was inhibited (Dado-Rosenfeld et al. 2014).

Cytoskeletal organization is also dependent on nano- and microscale physical features of the microenvironment (e.g., topography and adhesive ligands, which produce alterations of the cytoskeleton, clustering of integrins, and formation of focal adhesions). ECM composition affects integrin binding, driving cell proliferation, migration, and differentiation (Glukhova and Thiery 1993; Holly et al. 2000). While difficult to fully characterize in vivo, nanoscale influences can be easily examined in vitro. In human ESCs, nanotopography has been shown to influence morphology, proliferation and differentiation through cytoskeletal reorganization (Gerecht et al. 2007; Smith et al. 2009). It is clear that, when studying embryogenesis and development, it is important to take into consideration the complex interplay between biochemical processes (not discussed here), micromechanical properties of tissues, and cellular contractility. As one might suspect, these general themes are still present in adult tissues, and thereby influence adult stem cell differentiation in a similar manner.

12.2.4 Mechanical influences in adult tissues

Once embryogenesis ceases, mechanical forces remain an important facet in cell and tissue maintenance and function throughout an organism's life. The microenvironment and physical characteristics of different tissues and organs within the body are, as one would suspect, extremely diverse, and it is these specific properties of individual tissues that guide resident cell function as well as differentiation in recruited adult stem cells. Pressure waves displacing the stereocilia of hair cells in the inner ear or fluid shear stress and contraction of the heart are perhaps the most obvious examples of mechanical forces present in the adult (Chalfie 2009; Lammerding et al. 2004), but at the molecular scale, tissue cells respond to forces present in the microenvironment in a similar manner as those in the developing embryo.

In the body, there exists great variation in the mechanical properties of tissues. The stiffness of brain tissue is between 0.1 kPa and 1 kPa (Flanagan et al. 2002), that of muscle tissue between 8 kPa and 17 kPa, and that of demineralized bone tissue between 25 kPa and 40 kPa (Engler et al. 2006). In turn, it has been shown that adult stem cells do sense these mechanical differences (Engler et al. 2006). With this knowledge, it is not surprising that alterations in the microenvironment, for example in disease, can greatly interfere with normal cellular mechanotransduction, thereby disrupting behavior. The origin of many diseases is mechanical in nature, stemming from the disruption in force transmission between the ECM and the cellular cytoskeleton (Jaalouk and Lammerding 2009).

In myocardial infarction, measurements of the local tissue stiffness within the fibrotic scar show that tissue rigidity increases ~fivefold compared to normal heart muscle (Berry et al. 2006). Glial scarring as a result of acute insults to the central nervous system also alters the mechanical properties of the brain tissue in a similar fashion (Saha et al. 2008). In Duchenne muscular dystrophy, force transmission between the cellular cytoskeleton and the ECM is disrupted by a mutation in the gene dystrophin, causing progressive muscle degeneration (Heydemann and McNally 2007). Cancer is another disease that has mechanics at its roots. Malignant transformation, tumorigenesis and invasive metastasis have all been linked to changes in the ECM (Huang and Ingber 2005; Suresh, 2007). Specifically, tumors display increased ECM stiffness, and this can in turn affect mechanosensitive signaling, cytoskeletal tension and cellular contractility, and integrin clustering (Paszek et al. 2005).

Clearly, mechanical properties of adult tissue must be precisely controlled in order to maintain proper function and homeostasis in the body. This is an important concept in both endogenous stem cell homing, as well as in regenerative therapeutic treatments, in which stem cells are transplanted to the site of injury or disease.

12.2.5 Adult stem cell mechanobiology

Adult stem cells are multipotent cells that reside in tissues and play an integral role in tissue maintenance throughout an organism's life. In the body, hematopoietic stem cells egress from the bone marrow and enter the circulation, homing to the desired tissue microenvironments in order to engraft and differentiate (Katayama et al. 2006). In culture, mesenchymal stem cells (MSCs) are widely popular because of their abundance, clinical applicability, lack of teratoma formation, and ability to differentiate into multiple tissue types, namely adipocytic, chondrocytic, osteocytic (Pittenger et al. 1999), neurocytic (Deng et al. 2006), and myocytic lineages (Gang et al. 2004). However, the role of mechanics in stem cell differentiation was not fully appreciated until nearly a decade ago, when MSCs were shown to commit to lineages by tissue-specific substrate stiffness as the sole cue (Engler et al. 2006).

Cytoskeletal tension and cell shape have also been shown to dominate chemical cues, as a small cell spread area induces adipogenesis when MSCs were cultured in osteogenic medium. Conversely, when allowed to spread onto a larger area, but in adipogenic medium, MSCs became osteogenic (McBeath et al. 2004). In vivo, similar results displaying tissue-level sensitivity in differentiating MSCs have been observed. As previously mentioned, the post-infarct myocardium exhibits fibrotic scarring that renders the tissue several-fold stiffer than healthy myocardium (Berry et al. 2006), and when MSCs were injected into this scarred region in a therapeutic approach, ossification was observed (Breitbach et al. 2007), consistent with in vitro studies (Engler et al. 2006).

As mentioned in development, the nano- and microscale properties of the ECM can influence a wide variety of cellular functions, including differentiation. In the adult, this still holds true, as stem cells mobilize and interact intricately with the surrounding microenvironment as part of the regenerative response. Nanoscale topography, molecular conformations, and protein fiber diameters of tissues are extremely diverse among systems (Brody et al. 2006; Abrams et al. 2000), and these parameters have been shown to influence adult stem cell differentiation in a variety of studies (Yim et al. 2007; Christopherson et al. 2009). Cytoskeletal organization and structure can also be influenced by nanoscale perturbations in adhesive structures, directly influencing integrin clustering and focal adhesion turnover (Arnold et al. 2004).

These forays into in vitro experimentation on stem cells in direct analogy to the niches they inhabit in vivo during both development and adult homeostasis foreshadows Section 12.3, in which the biochemical basis for stem cell mechanotransduction is elaborated on, and a number of nano- and microscale approaches to stem cell engineering are described.

12.3 Nano- and microscale techniques for harnessing stem cell mechanobiology

After surveying the multitude of force modalities present in the body during and after development, it is apparent that efficient and accurate techniques are required to mimic these conditions with the ultimate goal of more completely controlling stem cell self-renewal, proliferation, and differentiation. In order to do this, the synergistic relationship between the mechanical properties of a stem cell's extracellular environment and its own internal mechanosensing mechanisms must be optimized. While controlling the differentiation of stem cells is one engineering challenge, keeping them in a state of pluripotency is another. Often, self-renewal processes rely on chemical additives like leukemia inhibitory factor (LIF) for ESCs. However, it is becoming clear that similar to differentiation, self-renewal may also be stimulated solely through the substrate.

12.3.1 Mechanotransduction mechanisms of stem cells

Four methods of substrate-based stem cell mechanotransduction have been observed and tested (Murphy et al. 2014). Because mechanotransduction is a process that begins with the cell interacting with its extracellular environment and finishes with altered gene expression, it is intuitive that the four mechanosensing methods form a chain from the cell membrane, where initial attachment between the cell and ECM occur, to its nucleus, where master regulation occurs (Fig. 12.1). These four systems work in synergy, as the cellular contractility provided by the Rho/ROCK pathway supplies the force necessary to activate mechanosensitive channels and force-sensitive proteins, ultimately resulting in differential signals sent to nuclear attenuators.

12.3.1.1 Mechanosensitive channels

In the plasma membrane of any cell exist transmembrane proteins, spanning the gap from the cytoplasm to the extracellular environment. Ion channels, which are a subset of

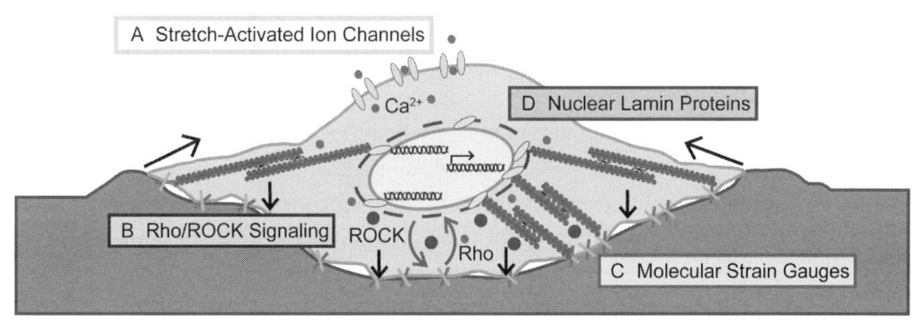

Figure 12.1 **Mechanosensing mechanisms affecting stem cell differentiation.**
(A) As the cell generates contractility, strain on the plasma membrane increases, resulting in the opening of mechanosensitive ion channels and an increase in calcium ion flux. (B) These calcium ions are able to support the positive feedback loop of contractility initiated by RhoA and ROCK proteins, which associate closely with actin and myosin. (C) As a result of increased contractility, the amount of force borne by the molecular strain gauges connecting the cytoskeleton to the deforming ECM also increases, causing differential protein unfolding. (D) Protein cascades initiated by this terminate at the nuclear envelope, where lamin proteins can attenuate multiple signals from the cytosol and cytoskeleton and effect gene expression that stimulates differentiation.

these transmembrane proteins, allow for the diffusion of a range of ions into or out of the cell. These channels can be "opened" in a number of ways, including electric gradient (voltage gated ion channels), protein binding (ligand gated ion channels), temperature, light, and importantly for stem cell mechanotransduction, mechanical force (mechanosensitive channels).

Discovered in 1984 (Guharay and Sachs 1984; Brehm et al. 1984), mechanosensitive channels represent a class of ion channels that can be activated by mechanical stress alone (Sachs 2010). In MSCs, activation of the mechanically sensitive calcium ion channel TRPV4 has been shown to induce SOX9-dependent chondrogenesis (Muramatsu et al. 2007). The mechanosensitive ion channel TRPC1 (Tai et al. 2009), which has been shown to be important during neurogenesis, was found to contribute to calcium influx in proliferation, which allows embryonic rat neural stem cells to self-renew (Fiorio Pla et al. 2005). A family member of TRPC1 is TRPC3, which has also been shown to regulate stem cell mechanotransduction in a subpopulation of ASCs, as its expression is required for differentiation (Poteser et al. 2008). This TRP family of mechanosensitive channels has been implicated in RhoA cytoskeletal activation (Beech 2005), hinting at the gestalt nature of the multiple mechanosensitive mechanisms utilized by stem cells.

12.3.1.2 Rho/ROCK activity

It is well established that stem cells must be able to contract against their environment in order to differentiate. Two of the most important regulators of cellular contractility are RhoA and ROCK, and their role is especially apparent in stem cell mechanotransduction, especially in conjunction with free calcium ions often available because of mechanosensitive ion channels. Both RhoA and ROCK are upregulated in differentiation-primed spread stem cells compared to unspread stem cells. Inhibition of RhoA signaling in MSCs results in a loss of osteogenic potential and an increase in adipogenic potential. This phenomenon is likely due to a disruption of downstream ROCK signaling, which allows for myosin-based cell contractility (McBeath et al. 2004). ROCK itself has been found to be upregulated in MSCs growing on osteogenically favorable 42 kPa substrates versus softer 7 kPa substrates, and ROCK inhibitors used on cells plated on the stiff substrates caused a reduction of osteogenic differentiation markers (Shih et al. 2011). Blocking the activity of myosin II, which results in a loss of cellular contractility initiated by RhoA, renders MSCs unable to differentiate via mechanotransduction (Engler et al. 2006). Other studies have shown that a small degree of tension is required to keep hMSCs multipotent, with a complete loss of actin/myosin interaction resulting in adipogenesis (McMurray et al. 2011). However, this degree of tension may be below the threshold necessary to activate force-sensitive linker proteins, also known as molecular strain gauges.

12.3.1.3 Molecular strain gauges

The positive feedback loop of RhoA activation and mechanosensitive ion channel opening provides a piconewton level of contractile force. This force, which is transmitted through linker proteins connecting integrins to the cytoskeleton, ultimately allows the cell to deform its extracellular matrix. Indeed, integrins have been shown to be necessary for this mechanical cascade, as α2-integrin is upregulated on stiff substrates during osteogenic differentiation, and its loss causes a reduction in osteogenic capacity of MSCs (Shih et al. 2011).

However, it is force-sensitive linker proteins that bear the brunt of the excess force when the ECM stops deforming, resulting in differential protein unfolding and thus environmental sensing. The focal adhesion protein talin has been shown to respond to physiological force by unfolding, discretely exposing up to eleven binding sites for the focal adhesion protein vinculin (del Rio et al. 2009). Upon binding to talin, vinculin undergoes a conformation change (Golji et al. 2011), exposing a binding site for MAPK1 (Holle et al. 2013), which once activated can translocate to the nucleus and activate transcription factors (Roux and Blenis 2004). Perturbing these pathways is one way to influence stem cell differentiation, as vinculin loss has been shown to inhibit the ability of hMSCs to differentiate on myogenically favorable substrates (Holle et al. 2013).

Other force-sensitive linker protein systems may play greater roles in other differentiation regimes. The focal adhesion protein FAK, which plays a role in general cell adhesion and migration, undergoes phosphorylation under force (Pasapera et al. 2010), and is a component of several integrin signaling pathways, has also been shown to play a role in stem cell mechanotransduction. In mice bred to have their FAK genes conditionally inactivated, the abolishment of FAK led to a loss of mechanically induced osteogenesis from bone marrow stem cells (Leucht et al. 2007). In vitro, FAK inhibition and resulting ERK1/2 effects has been shown to reduce differentiation of MSCs on osteogenically favorable substrates (Shih et al. 2011), although FAK phosphorylation has also been shown to stimulate myogenesis in MSCs (Teo et al. 2013). The focal adhesion protein p130Cas, which can also undergo force sensitive unfolding and phosphorylation (Sawada et al. 2006), has been shown to stimulate the differentiation of epithelial progenitor cells (Tornillo et al. 2013).

12.3.1.4 Nuclear lamin proteins

Once cellular contractility has had an opportunity to open mechanosensitive channels, activate the Rho/ROCK pathway, and unfold molecular strain gauges, these signals must be passed to the nucleus in order to effect changes in gene expression. Differentiating MSCs on engineered substrates have been shown to acetylize their histones in response to force, resulting in increased gene transcription and resulting differentiation (Li et al. 2011). In order to do this, a set of lamin proteins must function in and around the nuclear cytoskeleton, anchoring transnuclear membrane proteins capable of transmitting forces or signals from the cytosol into the nucleus. Accordingly, simply reducing the amount of lamin A/C in hMSCs results in a loss of mechanosensitive differentiation (Li et al. 2011). Lamin A/C has also been shown to be upregulated on stiff substrates and downregulated on soft substrates, resulting in changes in mechanosensitive hMSC differentiation (Swift et al. 2013), further confirming its role in the stem cell mechanotransduction process. The transcription factors YAP and TAZ have also been identified as nuclear relays connecting the signals brought about by extracellular mechanics, as they are required for any stiffness-based stem cell differentiation (Dupont et al. 2011).

12.3.2 Methods for controlling mechanosensitive stem cell differentiation

One of the main goals of a significant amount of stem cell engineering is to develop micro- and nano-based techniques that can more efficiently control stem cell behavior than traditional soluble factors. Often this is directly aimed at controlling fate decisions: if stem cells can be quickly and efficiently differentiated into desired lineages, they can be utilized by tissue engineers for a wide range of therapeutic implants, injections, or treatments. While controlling the differentiation of stem cells is one engineering challenge, keeping them in a state of pluripotency is an equally important one. Often, self-renewal processes rely on chemical additives like leukemia inhibitory Ffactor (LIF) for ESCs. However, it is becoming clear that similar to differentiation, self-renewal may also be stimulated solely through the substrate. See Fig. 12.2.

12.3.2.1 Substrate stiffness

For materials intended for use as stem cell culture substrates, the stiffness, or bulk modulus, is perhaps one of the most studied physical characteristics. In order to control the mechanical properties of a cellular substrate, polymeric hydrogels are often employed. Polyacrylamide hydrogels crosslinked with bis-acrylamide are a popular option for stem cell substrates due to their biocompatibility, ease of chemical functionalization, and bulk modulus customizability ranging from less than 1 kPa to over

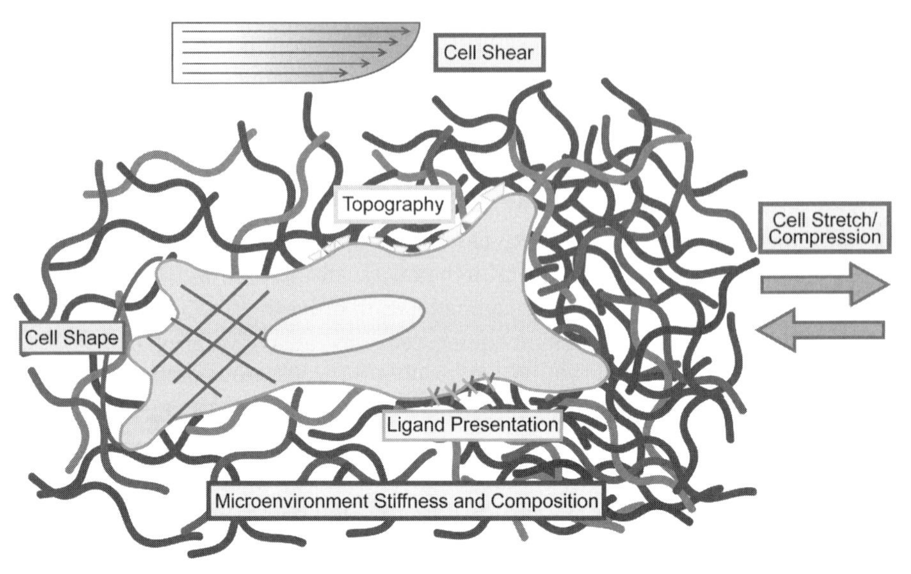

Figure 12.2 **The physical microenvironment and micro- and nano-based techniques to differentiate stem cells.**
Often, micro- and nano-based techniques used to control stem cell behavior aim to more closely mimic the stem cell niche during natural in vivo differentiation. Methods to achieve this include alterations to substrate stiffness, confinement of stem cells to a desired shape or pattern, the use of bio-inspired nanotopography, or the application of active stresses and strains.

50 kPa, spanning a wide swath of physiological stiffnesses (Tse and Engler 2010). Other hydrogels have been used to control the mechanical microenvironment of stem cells, including hyaluronic acid ranging from 4–100 kPa (McMurray et al. 2011), alginate ranging from 1–160 kPa (Huebsch et al. 2010), poly-l-lactide ranging from 35–330 kPa (Forte et al. 2008), and polyethylene glycol (PEG) ranging from 2–42 kPa (Gilbert et al. 2010).

MSCs have been shown to specify their lineage as a direct result of the stiffness of their substrate. Soft matrices of about 1 kPa, mimicking native brain tissue stiffness, will encourage neurogenesis. Firm matrices of about 11 kPa, mimicking native muscle tissue stiffness, will encourage myogenesis. Finally, stiff matrices of about 34 kPa, mimicking native demineralized bone tissue, will encourage osteogenesis (Engler et al. 2006). Adipose-derived stem cells (ASCs) have also been shown to possess stiffness sensitivity in relation to differentiation, as ASCs plated onto myogenically favorable substrates express myogenic differentiation markers, form multi-nucleated myotubes, and remain fused after replating (Choi et al. 2012a). These ASCs also are capable of differentiating into adipocytes when plated onto soft matrices of about 1 kPa in a similar manner as MSCs (Young et al. 2013). A number of other progenitor cells, including muscle stem cells (Gilbert et al. 2010), neural stem cells (Leipzig and Shoichet 2009), and cardiac progenitor cells have been shown to utilize extracellular stiffness to control self-renewal or differentiation.

ESC differentiation in culture can either be focused on early (i.e., germ layer) differentiation events or terminal differentiation. Similar to hMSCs (Engler et al. 2006), mechanical influences on ESC differentiation have largely been examined by plating cells on substrates of varying stiffness. For early differentiation events on mouse ESCs (mESCs), one study showed that the expression of early mesendoderm genes was higher on stiffer substrates made of PDMS, although the range of substrate stiffnesses investigated was relatively high (41 kPa to 2.7 MPa) (Evans et al. 2009). In another study, mESCs were found to display mechanosensitive differentiation within a lower stiffness range of alginate hydrogels (450 Pa to 1.3 kPa) with endoderm being most prevalent (Candiello et al. 2013). Moreover, when osteogenic differentiation was examined on PDMS surfaces, the stiffest substrates were found to enhance osteogenic markers as well as bone nodule formation (Evans et al. 2009). In human ESCs (hESCs), early differentiation events were also shown to be affected by substrate stiffness, with stiffer PDMS substrates (~1 MPa) contributing to mesodermal differentiation (Eroshenko et al. 2013). The mESCs cultured on soft substrates of approximately 1 kPa, which is also the material stiffness of the stem cells, were able to remain undifferentiated for up to fifteen passages in the absence of LIF. This may be due to the lack of traction force exerted by the mESCs on these matrices, resulting in a failure to activate any of the potential mechanosensitive differentiation pathways mentioned earlier (Chowdhury et al. 2010a).

While experiments in two dimensions can provide useful information on cellular interactions and subsequent responses, three-dimensional cultures can better recapitulate the in vivo microenvironment, and this is especially important for embryonic cells. In three-dimensional synthetic scaffolds, hESCs have been shown to differentiate into the three germ lineages in a substrate-stiffness dependent manner. Substrates with stiffnesses higher than 6 MPa did not support hESC differentiation, while cells in 1.5–6 MPa scaffolds became mesodermal, cells in 0.1–1 MPa became endodermal, and cells in scaffolds with stiffnesses less than 100 kPa became ectodermal (Zoldan et al. 2011). Other studies have confirmed the tendency of stem cells to differentiate toward the lineage represented by a given stiffness in three dimensions. Neural stem cells have been encapsulated within hydrogel spheres comprised of alginate with bulk moduli ranging from 0.1 kPa to 10 kPa; in these systems, neural stem cell proliferation decreased as the stiffness increased, while neuronal markers were found to be highest on the 0.1 kPa scaffold, where the stiffness was comparable to that of brain tissue in vivo (Banerjee et al. 2009). In MSCs, three-dimensional substrate stiffness has been shown to elicit differentiation in a similar pattern as two-dimensional substrate stiffness (Huebsch et al. 2010; Khetan et al. 2013), although the signaling mechanism required to do so may be different. Interestingly, it has been shown that the differentiation capability of hMSCs in three dimensions depends on the degree to which the ECM can be degraded – easily degradable substrates support cell spreading, focal adhesion formation, and osteogenesis, while crosslinked, nondegradable substrates downregulate contractility and encourage adipogenesis (Khetan et al. 2013).

Other studies, recognizing that physiologically relevant stiffnesses are not always static, have analyzed the role of stiffness gradients in stem cell biology. Engineered gradients of approximately 1 kPa per millimeter have been shown to cause hMSCs to migrate from 1 kPa regions to stiffer 11 kPa regions, then differentiate myogenically (Tse and Engler 2011). This migration along a mechanical gradient, known as durotaxis, has been investigated further in hASCs. Mechanically patterned polyacrylamide hydrogels containing alternating regions of soft and stiff substrate were found to encourage hASCs to undergo durotaxis towards the stiffer region, align, and undergo myogenesis (Choi et al. 2012b).

While materials with stiffness gradients mimic the gradients found in nature at a given time, other materials take advantage of dynamic stiffness changes found in the body, most often during development, as mentioned earlier. The use of materials that can change their stiffness over time can thus encourage the stem cell behavior that is observed during development. Thiolated hyaluronic acid hydrogels have been used to accomplish this goal by stiffening over time as a result of a slow reaction between acrylate groups and free thiols in a manner designed to mimic the natural stiffening found in the developing heart. Precardiac progenitor cells cultured on these materials were found to more efficiently differentiate into mature cardiomyocytes than on static materials, suggesting time as a further variable for future stiffness-based substrate fabrication (Young and Engler 2011). Additionally, it has been shown that hESC sensitivity to substrate mechanics is temporally regulated. Specifically, hESC substrate stiffness-dependent differentiation to cardiomyocytes occurs at early stages of mesodermal induction, in which polyacrylamide hydrogels of intermediate stiffness (50 kPa) produced the greatest fraction of mature myocytes (Hazeltine et al. 2014).

While recent publications have cast new light onto the role of ECM ligand tethering on mechanosensitive stem cell differentiation (Trappmann et al. 2012), it has been reaffirmed that stiffness alone, and not tethering or porosity, regulates the differentiation of hMSCs on two-dimensional matrices (Wen et al. 2014).

12.3.2.2 Cell shape and area

Cell shape has been implicated in a number of cellular processes historically, and stem cells appear to respond to shape and area in a number of interesting ways (Folkman and Moscona 1978; Manasek et al. 1972; Ingber 1991). While cell shape changes are usually an outcome of a cellular process, this phenomenon can work in reverse by confining stem cells to a predefined shape. Micropatterned substrates have been used to exert control over the amount and shape of cell spreading, with outcomes typically mirroring in vivo observations – lineages whose differentiated cells are fully spread will be induced on shapes allowing for full spreading and vice versa. Plating MSCs on square regions of fibronectin in a nonadhesive background, with the size of the regions varying from 1,024 square microns to 10,000 square microns, resulted in the observation that adipogenic differentiation occurred only on the small regions, osteogenic differentiation only occurred on the large regions, and intermediate sized regions resulted in a mix of differentiated cells (McBeath et al. 2004). MSCs cultured with an actin-disrupting agent, which resulted in the cells becoming more rounded, correlated with a reduction in osteogenic potential and an increase in adipogenic potential (McBeath et al. 2004). The hMSCs confined to narrow, 20-micron strips will become highly elongated and express myogenic markers, which may be a function of forced nuclear shape rearrangement (Tay et al. 2010). In epidermal stem cells, differentiation is induced when cells are plated onto small 20-micron diameter islands, as opposed to larger 50-micron diameter islands (Connelly et al. 2010).

12.3.2.3 Nanotopography

While an entire stem cell is on the scale of microns, the mechanosensitive machinery found in focal adhesions is often on a nanometer scale. Thus, increasing research has focused on the nanotopography of substrates, and how stem cells interpret these cues. Perhaps the simplest metric of nanotopography is surface roughness, or the amount of physical heterogeneity in the substrate. It has been shown that rougher surfaces tend to encourage hESC differentiation, while smooth surfaces are more favorable for self-renewal (Chen et al. 2012).

Aligned nanotubes perpendicular to the surface have been used to control substrate nanotopography. When plated onto substrates comprising titanium oxide tubes with 30-nanometer diameters, hMSCs were able to attach and proliferate without differentiating, while those plated onto nanotubes with 70- to 100-nanometer diameters caused the stem cells to elongate and differentiate toward an osteogenic lineage (Oh et al. 2009; Zhao et al. 2012). When supplemented with osteogenic media, however, hMSCs will preferentially differentiate on much smaller (30- to 50-nanometer) diameter nanotubes (Park et al. 2007). Hexagonal and honeycomb configurations of nanopillars with 30-nanometer diameters were found to support self-renewal in hESCs regardless of their inter-pillar spacing, although the honeycomb configuration was slightly more favorable for self-renewal (Kong et al. 2013).

In some models, even the simple switch from nano-order to nano-disorder is enough to drive differentiation. The use of ordered, 120-nanometer diameter, 100-nanometer-deep nanopits was sufficient to maintain hMSC self renewal, but when disordered pits of the same size were produced, hMSCs upregulated a number of osteogenic differentiation markers (Dalby et al. 2007), perhaps as a result of more heterogeneous distances between cell attachment points.

Nanogratings consist of long, parallel nanogrooves in a substrate, and are of importance for alignment of a number of cell types. Perhaps due to synergistic cooperation between alignment and cell shape, neurogenic markers have been shown to be upregulated in hMSCs plated on 350-nanometer-wide nanogrates (Yim et al. 2007). In contrast, 250-nanometer nanogrates have been used to induce toward muscle, another commonly “aligned” morphology lineage, in hMSCs (Teo et al. 2013). Nanogratings have also been use to align hESCs and reduce their proliferation (Gerecht et al. 2007), which was later found to encourage neurogenic differentiation of hESCs in the absence of growth factors (Lee et al. 2010). Even the reprogramming of somatic cells into iPSCs has been enhanced by culture on parallel nanogrates through mechanomodulation of cellular epigenetics (Downing et al. 2013).

Nanofiber scaffolds have also been examined, as they have the potential to present synthetic replicas of the proteins found in the ECM in vivo. Indeed, the use of poly-L-lactide (PLLA) nanofibers with 150-nanometer diameters in both two- and three-dimension scaffolds encouraged the osteogenic differentiation of hESCs (Smith et al. 2010). However, nanofibrillar surfaces have also been used to enhance hESC self-renewal (Nur et al. 2006), although the fiber diameters and pore diameters of these substrates is significantly different from those that elicit differentiation (Schindler et al. 2005).

12.3.2.4 Active perturbations

The application of active stresses to the exterior of stem cells has been one way to move past passive substrate mechanics. Studies have examined the mechanotransduction of active cyclic strain on hMSCs, finding that 10% cyclic uniaxial strain upregulated myogenic differentiation markers, while equiaxial strain of the same magnitude did not (Park et al. 2004). A combination of alignment-inducing microgrooves with a parallel 5% cyclic uniaxial strain was sufficient to induce upregulation of myogenic differentiation markers, while perpendicular strain of the same magnitude was not (Kurpinski et al. 2006). The hESCs, on the other hand, respond to a 10% biaxial cyclic strain by remaining pluripotent (Saha et al. 2006), although a 4–12% cyclic strain was also shown to induce myogenesis in mouse ESCs (Shimizu et al. 2008). However, 1% and 15% cyclic strains fail to induce myogenesis in MSCs, suggesting complex mechanisms on a cell type–specific basis (Yang et al. 2000). Hydrostatic pressure, often analogous to compressive stress, has also been used to perturb stem cells, often in the context of the chondrogenesis, mimicking the high pressure, fluid filled nature of that niche. Cyclic compressive loading of hMSCs in a model designed to mimic the load found in cartilage in vivo resulted in the upregulation of chondrogenic factors over long term cultures (Mauck et al. 2007; Huang et al. 2004), while hydrostatic pressure alone has been shown to effect chondrogenesis in certain matrices (Steward et al. 2012). Fluid shear stress, analogous to the myriad of fluid flow found in development, has been used extensively to effect differentiation of stem cells to a number of lineages. While it is not considered a nano or micro technique for stem cell mechanotransduction and is out of the purview of this chapter, it has been reviewed thoroughly here (Stolberg and McCloskey 2009).

Active perturbations as a result of the attachment of magnetic beads to MSC membranes, followed by application of force via a magnetic field, have been shown to allow greater ion flow into cells and subsequent expression of osteogenic genes (Kanczler et al. 2010). As material substrates are not required for this, the technique can be scaled up greatly (Dobson et al. 2006). Attachment of magnetic beads to the membrane of ESCs, followed by the application of stresses as a function of the beads moving in a magnetic field, caused the stem cells to spread out, exert higher traction forces, and begin the differentiation process (Chowdhury et al. 2010b). This was shown to be a function of the "soft" nature of the ESCs, as stiffer cells were unable to respond to the moving beads. Whereas one of the earlier mentioned methods of mechanotransduction relied on the unfolding of force-sensitive proteins, it has been speculated that the soft nature of ESCs allows for the propagation of active stresses directly to molecular strain gauges or mechanosensitive ion channels without the need for cellular contractility (Holle and Engler 2010).

Despite the vast nature of the nano- and microscale techniques used to control stem cell mechanotransduction and differentiation, a common thread can be found, in which generally, the most efficient way to follow nature's footsteps is to replicate her trail as much as we can, whether it be through substrate stiffness selection, cell shape modification, or the application of active forces. Once this trail can be optimized, a new generation of stem cell based therapies, utilizing a wide range of engineered stem cells and their progeny, can be realized.

References

Abrams, G. A., Goodman, S. L., Nealey, P. F., Franco, M. and Murphy, C. J. (2000). "Nanoscale topography of the basement membrane underlying the corneal epithelium of the rhesus macaque." *Cell Tissue Res* **299**: 39–46.

Adamo, L., Naveiras, O., Wenzel, P. L., et al. (2009). "Biomechanical forces promote embryonic haematopoiesis." *Nature* **459**: 1131–1135.

Adams, D. S., Keller, R., and Koehl, M. A. (1990). "The mechanics of notochord elongation, straightening and stiffening in the embryo of Xenopus laevis." *Development* **110**: 115–130.

Anava, S., Greenbaum, A., Ben Jacob, E., Hanein, Y., and Ayali, A. (2009). "The regulative role of neurite mechanical tension in network development." *Biophys J* **96**: 1661–1670.

Arnold, M., Cavalcanti-Adam, E. A., Glass, R., et al. (2004). "Activation of integrin function by nanopatterned adhesive interfaces." *Chemphyschem* **5**: 383–388.

Banerjee, A., Arha, M., Choudhary, S., et al. (2009). "The influence of hydrogel modulus on the proliferation and differentiation of encapsulated neural stem cells." *Biomaterials* **30**: 4695–4699.

Beech, D. J. (2005). "TRPC1: store-operated channel and more." *Pflugers Arch* **451**: 53–60.

Berry, M. F., Engler, A. J., Woo, Y. J., et al. (2006). "Mesenchymal stem cell injection after myocardial infarction improves myocardial compliance." *Am J Physiol Heart Circ Physiol* **290**: H2196–H2203.

Bertet, C., Sulak, L. and Lecuit, T. (2004). "Myosin-dependent junction remodelling controls planar cell intercalation and axis elongation." *Nature* **429**: 667–671.

Blanchard, G. B., Kabla, A. J., Schultz, N. L., et al. (2009). "Tissue tectonics: morphogenetic strain rates, cell shape change and intercalation." *Nat Methods* **6**: 458–464.

Brehm, P., Kullberg, R. and Moody-Corbett, F. (1984). "Properties of non-junctional acetylcholine receptor channels on innervated muscle of Xenopus laevis." *J Physiol* **350**: 631–648.

Breitbach, M., Bostani, T., Roell, W., et al. (2007). "Potential risks of bone marrow cell transplantation into infarcted hearts." *Blood* **110**: 1362–1369.

Brody, S., Anilkumar, T., Liliensiek, S., et al. (2006). "Characterizing nanoscale topography of the aortic heart valve basement membrane for tissue engineering heart valve scaffold design." *Tissue Eng* **12**: 413–421.

Candiello, J., Singh, S. S., Task, K., Kumta, P. N. and Banerjee, I. (2013). "Early differentiation patterning of mouse embryonic stem cells in response to variations in alginate substrate stiffness." *J Biol Eng* **7**: 9.

Chalfie, M. (2009). "Neurosensory mechanotransduction." *Nat Rev Mol Cell Biol* **10**: 44–52.

Chen, C. S., Mrksich, M., Huang, S., Whitesides, G. M. and Ingber, D. E. (1997). "Geometric control of cell life and death." *Science* **276**: 1425–1428.

Chen, W., Villa-Diaz, L. G., Sun, Y., et al. (2012). "Nanotopography influences adhesion, spreading, and self-renewal of human embryonic stem cells." *ACS Nano* **6**: 4094–4103.

Choi, Y. S., Vincent, L. G., Lee, A. R., et al. (2012a). "Mechanical derivation of functional myotubes from adipose-derived stem cells." *Biomaterials* **33**: 2482–2491.

Choi, Y. S., Vincent, L. G., Lee, A. R., et al. (2012b). "The alignment and fusion assembly of adipose-derived stem cells on mechanically patterned matrices." *Biomaterials* **33**: 6943–6951.

Chowdhury, F., et al. (2010a). "Soft substrates promote homogeneous self-renewal of embryonic stem cells via downregulating cell-matrix tractions." *PLoS One* **5**: e15655.

Chowdhury, F., et al. (2010b). "Material properties of the cell dictate stress-induced spreading and differentiation in embryonic stem cells." *Nat Mater* **9**: 82–88.

Christopherson, G. T., Song, H. and Mao, H. Q. (2009). "The influence of fiber diameter of electrospun substrates on neural stem cell differentiation and proliferation." *Biomaterials* **30**: 556–564.

Connelly, J. T., Gautrot, J. E., Trappmann, B., et al. (2010). "Actin and serum response factor transduce physical cues from the microenvironment to regulate epidermal stem cell fate decisions." *Nat Cell Biol* **12**: 711–718.

D'Amour, K. A., Agulnick, A. D., Eliazer, S., et al. (2005). "Efficient differentiation of human embryonic stem cells to definitive endoderm." *Nat Biotechnol* **23**: 1534–1541.

Dado-Rosenfeld, D., Tzchori, I., Fine, A., Chen-Konak, L. and Levenberg, S. (2014). "Tensile forces applied on a cell-embedded three-dimensional scaffold can direct early differentiation of embryonic stem cells toward the mesoderm germ layer." *Tissue Eng Part A* **21**(1–2): 124–143.

Dalby, M. J., Gadegaard, N., Tare, R., et al. (2007). "The control of human mesenchymal cell differentiation using nanoscale symmetry and disorder." *Nat Mater* **6**: 997–1003.

Daniels, B. R., Masi, B. C. and Wirtz, D. (2006). "Probing single-cell micromechanics in vivo: the microrheology of C. elegans developing embryos." *Biophys J* **90**: 4712–6719.

del Rio, A., Perez-Jimenez, R., Liu, R., et al. (2009). "Stretching single talin rod molecules activates vinculin binding." *Science* **323**: 638–641.

Deng, J., Petersen, B. E., Steindler, D. A., Jorgensen, M. L. and Laywell, E. D. (2006). "Mesenchymal stem cells spontaneously express neural proteins in culture and are neurogenic after transplantation." *Stem Cells* **24**: 1054–1064.

Desprat, N., Supatto, W., Pouille, P. A., Beaurepaire, E. and Farge, E. (2008). "Tissue deformation modulates twist expression to determine anterior midgut differentiation in Drosophila embryos." *Dev Cell* **15**: 470–477.

Discher, D. E., Janmey, P., and Wang, Y. L. (2005). "Tissue cells feel and respond to the stiffness of their substrate." *Science* **310**: 1139–1143.

Dobson, J., Cartmell, S. H., Keramane, A. and El Haj, A. J. (2006). "Principles and design of a novel magnetic force mechanical conditioning bioreactor for tissue engineering, stem cell conditioning, and dynamic in vitro screening." *IEEE Trans Nanobioscience* **5**: 173–177.

Downing, T. L., Soto, J., Morez, C., et al. (2013). "Biophysical regulation of epigenetic state and cell reprogramming." *Nat Mater* **12**: 1154–1162.

Dupont, S., Morsut, L., Aragona, M., et al. (2011). "Role of YAP/TAZ in mechanotransduction." *Nature* **474**: 179–183.

Engler, A. J., Carag-Krieger, C., Johnson, C. P., et al. (2008). "Embryonic cardiomyocytes beat best on a matrix with heart-like elasticity: scar-like rigidity inhibits beating." *J Cell Sci* **121**: 3794–3802.

Engler, A. J., Sen, S., Sweeney, H. L. and Discher, D. E. (2006). "Matrix elasticity directs stem cell lineage specification." *Cell* **126**: 677–689.

Eroshenko, N., Ramachandran, R., Yadavalli, V. K. and Rao, R. R. (2013). "Effect of substrate stiffness on early human embryonic stem cell differentiation." *J Biol Eng* **7**: 7.

Evans, N. D., Minelli, C., Gentleman, E., et al. (2009). "Substrate stiffness affects early differentiation events in embryonic stem cells." *Eur Cell Mater* **18**: 1–13, discussion 13–14.

Farge, E. (2003). "Mechanical induction of Twist in the Drosophila foregut/stomodeal primordium." *Curr Biol* **13**: 1365–1377.

Finley, M. F., Devata, S. and Huettner, J. E. (1999). "BMP-4 inhibits neural differentiation of murine embryonic stem cells." *J Neurobiol* **40**: 271–287.

Fiorio Pla, A., Maric, D., Brazer, S. C., et al. (2005). "Canonical transient receptor potential 1 plays a role in basic fibroblast growth factor (bFGF)/FGF receptor-1-induced Ca2+ entry and embryonic rat neural stem cell proliferation." *J Neurosci* **25**: 2687–2701.

Flanagan, L. A., Ju, Y. E., Marg, B., Osterfield, M., and Janmey, P. A. (2002). "Neurite branching on deformable substrates." *Neuroreport* **13**: 2411–2415.

Folkman, J., and Moscona, A. (1978). "Role of cell shape in growth control." *Nature* **273**: 345–349.

Forouhar, A. S., Liebling, M., Hickerson, A., et al. (2006). "The embryonic vertebrate heart tube is a dynamic suction pump." *Science* **312**: 751–753.

Forte, G., Carotenuto, F., Pagliari, F., et al. (2008). "Criticality of the biological and physical stimuli array inducing resident cardiac stem cell determination." *Stem Cells* **26**: 2093–2103.

Foty, R. A. and Steinberg, M. S. (2005). "The differential adhesion hypothesis: a direct evaluation." *Dev Biol* **278**: 255–263.

Gang, E. J., Jeong, J. A., Hong, S. H., et al. (2004). "Skeletal myogenic differentiation of mesenchymal stem cells isolated from human umbilical cord blood." *Stem Cells* **22**: 617–624.

Gerecht, S., Bettinger, C. J., Zhang, Z., et al. (2007). " The effect of actin disrupting agents on contact guidance of human embryonic stem cells." *Biomaterials* **28**: 4068–4077.

Gilbert, P. M., Havenstrite, K. L., Magnusson, K. E., et al. (2010). "Substrate elasticity regulates skeletal muscle stem cell self-renewal in culture." *Science* **329**: 1078–1081.

Glukhova, M. A. and Thiery, J. P. (1993). "Fibronectin and integrins in development." *Semin Cancer Biol* **4**: 241–249.

Golji, J., Lam, J. and Mofrad, M. R. (2011). "Vinculin activation is necessary for complete talin binding." *Biophys J* **100**: 332–340.

Guharay, F. and Sachs, F. (1984). "Stretch-activated single ion channel currents in tissue-cultured embryonic chick skeletal muscle." *J Physiol* **352**: 685–701.

Harris, A. K., Wild, P. and Stopak, D. (1980). "Silicone rubber substrata: a new wrinkle in the study of cell locomotion." *Science* **208**: 177–179.

Hazeltine, L. B., Badur, M. G., Lian, X., et al. (2014). "Temporal impact of substrate mechanics on differentiation of human embryonic stem cells to cardiomyocytes." *Acta Biomater* **10**: 604–612.

Heydemann, A. and McNally, E. M. (2007). "Consequences of disrupting the dystrophin-sarcoglycan complex in cardiac and skeletal myopathy." *Trends Cardiovasc Med* **17**: 55–59.

Holle, A. W. and Engler, A. J. (2010). "Cell rheology: Stressed-out stem cells." *Nat Mater* **9**: 4–6.

Holle, A. W., Tang, X., Vijayraghavan, D., et al. (2013). "In situ mechanotransduction via vinculin regulates stem cell differentiation." *Stem Cells* **31**: 2467–2477.

Holly, S. P., Larson, M. K. and Parise, L. V. (2000). "Multiple roles of integrins in cell motility." *Exp Cell Res* **261**: 69–74.

Horner, V. L. and Wolfner, M. F. (2008). "Mechanical stimulation by osmotic and hydrostatic pressure activates Drosophila oocytes in vitro in a calcium-dependent manner." *Dev Biol* **316**: 100–109.

Hove, J. R., Koster, R. W., Forouhar, A. S., et al. (2003). "Intracardiac fluid forces are an essential epigenetic factor for embryonic cardiogenesis." *Nature* **421**: 172–177.

Huang, C. Y., Hagar, K. L., Frost, L. E., Sun, Y. and Cheung, H. S. (2004). "Effects of cyclic compressive loading on chondrogenesis of rabbit bone-marrow derived mesenchymal stem cells." *Stem Cells* **22**: 313–323.

Huang, S. and Ingber, D. E. (2005). "Cell tension, matrix mechanics, and cancer development." *Cancer Cell* **8**: 175–176.

Huebsch, N., Arany, P. R., Mao, A. S., et al. (2010). "Harnessing traction-mediated manipulation of the cell/matrix interface to control stem-cell fate." *Nat Mater* **9**: 518–526.

Ingber, D. (1991). "Extracellular matrix and cell shape: potential control points for inhibition of angiogenesis." *J Cell Biochem* **47**: 236–241.

Ingber, D. E. (2006). "Cellular mechanotransduction: putting all the pieces together again." *FASEB J* **20**: 811–827.

Jaalouk, D. E. and Lammerding, J. (2009). "Mechanotransduction gone awry." *Nat Rev Mol Cell Biol* **10**: 63–73.

Kahn, J., Shwartz, Y., Blitz, E., et al. (2009). "Muscle contraction is necessary to maintain joint progenitor cell fate." *Dev Cell* **16**: 734–743.

Kanczler, J. M., Sura, H. S., Magnay, J., et al. (2010). "Controlled differentiation of human bone marrow stromal cells using magnetic nanoparticle technology." *Tissue Eng Part A* **16**: 3241–3250.

Katayama, Y., Battista, M., Kao, W. M., et al. (2006). "Signals from the sympathetic nervous system regulate hematopoietic stem cell egress from bone marrow." *Cell* **124**: 407–421.

Khetan, S., Guvendiren, M., Legant, W. R., et al. (2013). "Degradation-mediated cellular traction directs stem cell fate in covalently crosslinked three-dimensional hydrogels." *Nat Mater* **12**: 458–465.

Kinney, M. A., Saeed, R. and McDevitt, T. C. (2014). "Mesenchymal morphogenesis of embryonic stem cells dynamically modulates the biophysical microtissue niche." *Sci Rep* **4**: 4290.

Kong, Y. P., Tu, C. H., Donovan, P. J. and Yee, A. F. (2013). "Expression of Oct4 in human embryonic stem cells is dependent on nanotopographical configuration." *Acta Biomater* **9**: 6369–6380.

Krieg, M., Arboleda-Estudillo, Y., Puech, P. H., et al. (2008). "Tensile forces govern germ-layer organization in zebrafish." *Nat Cell Biol* **10**: 429–636.

Kurpinski, K., Chu, J., Hashi, C., and Li, S. (2006). "Anisotropic mechanosensing by mesenchymal stem cells." *Proc Natl Acad Sci USA* **103**: 16095–16100.

Lammerding, J., Kamm, R. D. and Lee, R. T. (2004). "Mechanotransduction in cardiac myocytes." *Ann N Y Acad Sci* **1015**: 53–70.

le Noble, F., Moyon, D., Pardanaud, L., et al. (2004). "Flow regulates arterial-venous differentiation in the chick embryo yolk sac." *Development* **131**: 361–375.

Lee, M. R., Kwon, K. W., Jung, H., et al. (2010). "Direct differentiation of human embryonic stem cells into selective neurons on nanoscale ridge/groove pattern arrays." *Biomaterials* **31**: 4360–4366.

Leipzig, N. D. and Shoichet, M. S. (2009). "The effect of substrate stiffness on adult neural stem cell behavior." *Biomaterials* **30**: 6867–6878.

Leucht, P., Kim, J. B., Currey, J. A., Brunski, J. and Helms, J. A. (2007). "FAK-Mediated mechanotransduction in skeletal regeneration." *PLoS One* **2**: e390.

Li, Y., Chu, J. S., Kurpinski, K., et al. (2011). "Biophysical regulation of histone acetylation in mesenchymal stem cells." *Biophys J* **100**: 1902–1909.

Majkut, S., Idema, T., Swift, J., et al. (2013). "Heart-specific stiffening in early embryos parallels matrix and myosin expression to optimize beating." *Curr Biol* **23**: 2434–2439.

Mammoto, T. and Ingber, D. E. (2010). "Mechanical control of tissue and organ development." *Development* **137**: 1407–1420.

Manasek, F. J., Burnside, M. B. and Waterman, R. E. (1972). "Myocardial cell shape change as a mechanism of embryonic heart looping." *Dev Biol* **29**: 349–371.

Mauck, R. L., Byers, B. A., Yuan, X. and Tuan, R. S. (2007). "Regulation of cartilaginous ECM gene transcription by chondrocytes and MSCs in 3D culture in response to dynamic loading." *Biomech Model Mechanobiol* **6**: 113–125.

McBeath, R., Pirone, D. M., Nelson, C. M., Bhadriraju, K. and Chen, C. S. (2004). "Cell shape, cytoskeletal tension, and RhoA regulate stem cell lineage commitment." *Dev Cell* **6**: 483–495.

McMurray, R. J., Gadegaard, N., Tsimbouri, P. M., et al. (2011). "Nanoscale surfaces for the long-term maintenance of mesenchymal stem cell phenotype and multipotency." *Nat Mater* **10**: 637–644.

Montell, D. J. (2003). "Border-cell migration: the race is on." *Nat Rev Mol Cell Biol* **4**: 13–24.

Moore, K. A., Polte, T., Huang, S., et al. (2005). "Control of basement membrane remodeling and epithelial branching morphogenesis in embryonic lung by Rho and cytoskeletal tension." *Dev Dyn* **232**: 268–281.

Moore, S. W., Biais, N., and Sheetz, M. P. (2009). "Traction on immobilized netrin-1 is sufficient to reorient axons." *Science* **325**: 166.

Moore, S. W., Keller, R. E., and Koehl, M. A. (1995). "The dorsal involuting marginal zone stiffens anisotropically during its convergent extension in the gastrula of Xenopus laevis." *Development* **121**: 3131–3140.

Muramatsu, S., Wakabayashi, M., Ohno, T., et al. (2007). "Functional gene screening system identified TRPV4 as a regulator of chondrogenic differentiation." *J Biol Chem* **282**: 32158–32167.

Murphy, W. L., McDevitt, T. C. and Engler, A. J. (2014). "Materials as stem cell regulators." *Nat Mater* **13**(3): 547–557.

North, T. E., Goessling, W., Peeters, M., et al. (2009). "Hematopoietic stem cell development is dependent on blood flow." *Cell* **137**: 736–748.

Nostro, M. C., Cheng, X., Keller, G. M., and Gadue, P. (2008). "Wnt, activin, and BMP signaling regulate distinct stages in the developmental pathway from embryonic stem cells to blood." *Cell Stem Cell* **2**: 60–71.

Nur, E. K. A., Ahmed, I., Kamal, J., Schindler, M. and Meiners, S. (2006). "Three-dimensional nanofibrillar surfaces promote self-renewal in mouse embryonic stem cells." *Stem Cells* **24**: 426–433.

Oh, S., Brammer, K. S., Li, Y. S., et al. (2009). "Stem cell fate dictated solely by altered nanotube dimension." *Proc Natl Acad Sci USA* **106**: 2130–2135.

Ohashi, N., Robling, A. G., Burr, D. B. and Turner, C. H. (2002). "The effects of dynamic axial loading on the rat growth plate." *J Bone Miner Res* **17**: 284–292.

Park, J., Bauer, S., von der Mark, K. and Schmuki, P. (2007). "Nanosize and vitality: TiO_2 nanotube diameter directs cell fate." *Nano Lett* **7**: 1686–1691.

Park, J. S., Chu, J. S., Cheng, C., et al. (2004). "Differential effects of equiaxial and uniaxial strain on mesenchymal stem cells." *Biotechnol Bioeng* **88**: 359–868.

Pasapera, A. M., Schneider, I. C., Rericha, E., Schlaepfer, D. D. and Waterman, C. M. (2010). "Myosin II activity regulates vinculin recruitment to focal adhesions through FAK-mediated paxillin phosphorylation." *J Cell Biol* **188**: 877–890.

Paszek, M. J., Zahir, N., Johnson, K. R., et al. (2005). "Tensional homeostasis and the malignant phenotype." *Cancer Cell* **8**: 241–254.

Pittenger, M. F., Mackay, A. M., Beck, S. C., et al. (1999). " Multilineage potential of adult human mesenchymal stem cells." *Science* **284**: 143–147.

Poteser, M., Graziani, A., Eder, P., et al. (2008). "Identification of a rare subset of adipose tissue-resident progenitor cells, which express CD133 and TRPC3 as a VEGF-regulated Ca2+ entry channel." *FEBS Lett* **582**: 2696–2702.

Roux, P. P. and Blenis, J. (2004). "ERK and p38 MAPK-activated protein kinases: a family of protein kinases with diverse biological functions." *Microbiol Mol Biol Rev* **68**: 320–344.

Sachs, F. (2010). "Stretch-activated ion channels: what are they?" *Physiology (Bethesda)* **25**: 50–56.

Saha, K., Keung, A. J., Irwin, E. F., et al. (2008). "Substrate modulus directs neural stem cell behavior." *Biophys J* **95**: 4426–4438.

Saha, S., Ji, L., de Pablo, J. J. and Palecek, S. P. (2006). "Inhibition of human embryonic stem cell differentiation by mechanical strain." *J Cell Physiol* **206**: 126–137.

Sanders, M. C., Way, M., Sakai, J. and Matsudaira, P. (1996). "Characterization of the actin cross-linking properties of the scruin-calmodulin complex from the acrosomal process of Limulus sperm." *J Biol Chem* **271**: 2651–2657.

Sawada, Y., Tamada, M., Dubin-Thaler, B. J., et al. (2006). "Force sensing by mechanical extension of the Src family kinase substrate p130Cas." *Cell* **127**: 1015–1026.

Schindler, M., Ahmed, I., Kamal, J., et al. (2005). "A synthetic nanofibrillar matrix promotes in vivo-like organization and morphogenesis for cells in culture." *Biomaterials* **26**: 5624–5631.

Schwartz, M. A. (2010). "Integrins and extracellular matrix in mechanotransduction." *Cold Spring Harb Perspect Biol* **2**: a005066.

Shih, Y. R., Tseng, K. F., Lai, H. Y., Lin, C. H. and Lee, O. K. (2005). "Matrix stiffness regulation of integrin-mediated mechanotransduction during osteogenic differentiation of human mesenchymal stem cells." *J Bone Miner Res* **26**: 730–738.

Shimizu, N., Yamamoto, K., Obi, S., et al. (2008). "Cyclic strain induces mouse embryonic stem cell differentiation into vascular smooth muscle cells by activating PDGF receptor beta." *J Appl Physiol (1985)* **104**: 766–772.

Shin, J. H., Tam, B. K., Brau, R. R., et al. (2007). "Force of an actin spring." *Biophys J* **92**: 3729–8733.

Smith, L. A., Liu, X., Hu, J. and Ma, P. X. (2009). "The influence of three-dimensional nanofibrous scaffolds on the osteogenic differentiation of embryonic stem cells." *Biomaterials* **30**: 2516–2522.

Smith, L. A., Liu, X., Hu, J. and Ma, P. X. (2010). "The enhancement of human embryonic stem cell osteogenic differentiation with nano-fibrous scaffolding." *Biomaterials* **31**: 5526–5535.

Steward, A. J., Thorpe, S. D., Vinardell, T., et al. (2012). "Cell-matrix interactions regulate mesenchymal stem cell response to hydrostatic pressure." *Acta Biomater* **8**: 2153–2159.

Stokes, I. A., Mente, P. L., Iatridis, J. C., Farnum, C. E. and Aronsson, D. D. (2002). "Growth plate chondrocyte enlargement modulated by mechanical loading." *Stud Health Technol Inform* **88**: 378–381.

Stolberg, S. and McCloskey, K. E. (2009). "Can shear stress direct stem cell fate?" *Biotechnol Prog* **25**: 10–19.

Suresh, S. (2007). "Biomechanics and biophysics of cancer cells." *Acta Biomater* **3**: 413–438.

Swift, J., Ivanovska, I. L., Buxboim, A., et al. (2013). "Nuclear lamin-A scales with tissue stiffness and enhances matrix-directed differentiation." *Science* **341**: 1240104.

Tai, Y., Feng, S., Du, W. and Wang, Y. (2009). "Functional roles of TRPC channels in the developing brain." *Pflugers Arch* **458**: 239–283.

Takito, J. and Al-Awqati, Q. (2009). "Conversion of ES cells to columnar epithelia by hensin and to squamous epithelia by laminin." *J Cell Biol* **166**: 1093–1102.

Tay, C. Y., Yu, H., Pal, M., et al. (2010). "Micropatterned matrix directs differentiation of human mesenchymal stem cells towards myocardial lineage." *Exp Cell Res* **316**: 1159–1168.

Taylor-Weiner, H., Schwarzbauer, J. E. and Engler, A. J. (2013)."Defined extracellular matrix components are necessary for definitive endoderm induction." *Stem Cells* **31**: 2084–2094.

Teo, B. K., Wong, S. T., Lim, C. K., et al. (2013). "Nanotopography modulates mechanotransduction of stem cells and induces differentiation through focal adhesion kinase." *ACS Nano* **7**: 4785–4798.

Tornillo, G., Elia, A. R., Castellano, I., et al. (2013). "p130Cas alters the differentiation potential of mammary luminal progenitors by deregulating c-Kit activity." *Stem Cells* **31**: 1422–1433.

Toyama, Y., Peralta, X. G., Wells, A. R., Kiehart, D. P. and Edwards, G. S. (2008). "Apoptotic force and tissue dynamics during Drosophila embryogenesis." *Science* **321**: 1683–1686.

Trappmann, B., Gautrot, J. E., Connelly, J. T., et al. (2012). "Extracellular-matrix tethering regulates stem-cell fate." *Nat Mater* **11**: 642–649.

Tse, J. R. and Engler, A. J. (2010). "Preparation of hydrogel substrates with tunable mechanical properties." *Curr Protoc Cell Biol* Chapter 10: Unit 10. 16.

Tse, J. R. and Engler, A. J. (2011). "Stiffness gradients mimicking in vivo tissue variation regulate mesenchymal stem cell fate." *PLoS One* **6**: e15978.

Voronov, D. A., Alford, P. W., Xu, G. and Taber, L. A. (2004). "The role of mechanical forces in dextral rotation during cardiac looping in the chick embryo." *Dev Biol* **272**: 339–350.

Wang, Y. and Riechmann, V. (2007). "The role of the actomyosin cytoskeleton in coordination of tissue growth during Drosophila oogenesis." *Curr Biol* **17**: 1349–1355.

Wen, J. H., Vincent, L. G., Fuhrmann, A., et al. (2014). "Interplay of matrix stiffness and protein tethering in stem cell differentiation." *Nat Mater* **13**(10): 979–987.

Wilson, N. R., Ty, M. T., Ingber, D. E., Sur, M. and Liu, G. (2007). "Synaptic reorganization in scaled networks of controlled size." *J Neurosci* **27**: 13581–13589.

Wozniak, M. A. and Chen, C. S. (2014). "Mechanotransduction in development: a growing role for contractility." *Nat Rev Mol Cell Biol* **10**: 34–43.

Yang, Y., Beqaj, S., Kemp, P., Ariel, I. and Schuger, L. (2000). "Stretch-induced alternative splicing of serum response factor promotes bronchial myogenesis and is defective in lung hypoplasia." *J Clin Invest* **106**: 1321–1330.

Yim, E. K., Pang, S. W. and Leong, K. W. (2007). "Synthetic nanostructures inducing differentiation of human mesenchymal stem cells into neuronal lineage." *Exp Cell Res* **313**: 1820–1829.

Young, D. A., Choi, Y. S., Engler, A. J. and Christman, K. L. (2013). "Stimulation of adipogenesis of adult adipose-derived stem cells using substrates that mimic the stiffness of adipose tissue." *Biomaterials* **34**: 8581–8588.

Young, J. L. and Engler, A. J. (2011). "Hydrogels with time-dependent material properties enhance cardiomyocyte differentiation in vitro." *Biomaterials* **32**: 1002–1009.

Zamir, E. A. and Taber, L. A. (2004a). "Material properties and residual stress in the stage 12 chick heart during cardiac looping." *J Biomech Eng* **126**: 823–830.

Zamir, E. A. and Taber, L. A. (2004b). "On the effects of residual stress in microindentation tests of soft tissue structures." *J Biomech Eng* **126**: 276–283.

Zhao, L., Liu, L., Wu, Z., Zhang, Y. and Chu, P. K. (2012). "Effects of micropitted/nanotubular titania topographies on bone mesenchymal stem cell osteogenic differentiation." *Biomaterials* **33**: 2629–2641.

Zoldan, J., Karagiannis, E. D., Lee, C. Y., et al. (2011). "The influence of scaffold elasticity on germ layer specification of human embryonic stem cells." *Biomaterials* **32**: 9612–9621.

13 Mechanobiological stimulation of tissue engineered blood vessels

Kyle G. Battiston, J. Paul Santerre, and Craig A. Simmons

13.1 Introduction

Cardiovascular disease, and coronary artery disease in particular, is one of the leading causes of death worldwide (Seifu et al. 2013). While less advanced arterial disease can be treated by medication, angioplasty, or stents, significant vessel occlusion due to plaque deposition or intimal hyperplasia requires surgical intervention in the form of a bypass graft. The preferred source of grafts for these procedures is from autologous sources, such as the long saphenous vein, internal mammary artery, or radial artery. However, due to progressive vascular disease in these vessels or use in previous procedures, autologous vessels are unavailable in approximately one third of cases. Synthetic nondegradable materials like Dacron® and Goretex® perform well in large-diameter vascular graft applications; however, they perform poorly relative to autologous vessels when used to bypass small-diameter blood vessels such as the coronary artery, due to complications such as thrombosis and intimal hyperplasia at the anastomoses. Furthermore, while monocytes and macrophages are thought to be critical to supporting healing in the initial stages after implantation, the use of nondegradable materials is also thought to promote a chronic inflammatory response by macrophages and foreign-body giant cells that can compromise graft performance and inhibit long-term tissue remodeling and healing (Zilla et al. 2007; Hagerty et al. 2000).

A promising alternative approach to meet the need for an abundant supply of effective small-diameter vascular grafts is to engineer responsive, living blood vessels. Current vascular tissue engineering approaches use combinations of cells, biomaterial scaffolds, and bioreactor systems to achieve engineered vessel properties similar to those of native tissues (Fig. 13.1). Independent of the types of cells or scaffold materials used, mechanical stimulation in bioreactors has emerged as an effective method to mature engineered vascular tissues prior to implantation. In this chapter, we review this aspect of vascular tissue engineering – mechanobiological stimulation – and the role of microtechnologies in informing mechanobiological stimulation strategies. We further limit our focus to cells within the vascular wall that are candidates for tissue engineering – primarily vascular smooth muscle cells (VSMCs), but also monocytes/macrophages – and not the endothelial cells (ECs) that line the vascular lumen. The reader is referred to other excellent articles for comprehensive reviews of the effects of mechanical forces on endothelial cells in engineered tissues (Vara et al. 2011) and of other cell sources and biomaterial aspects of vascular tissue engineering (Seifu et al. 2013; Huang and Niklason 2014)).

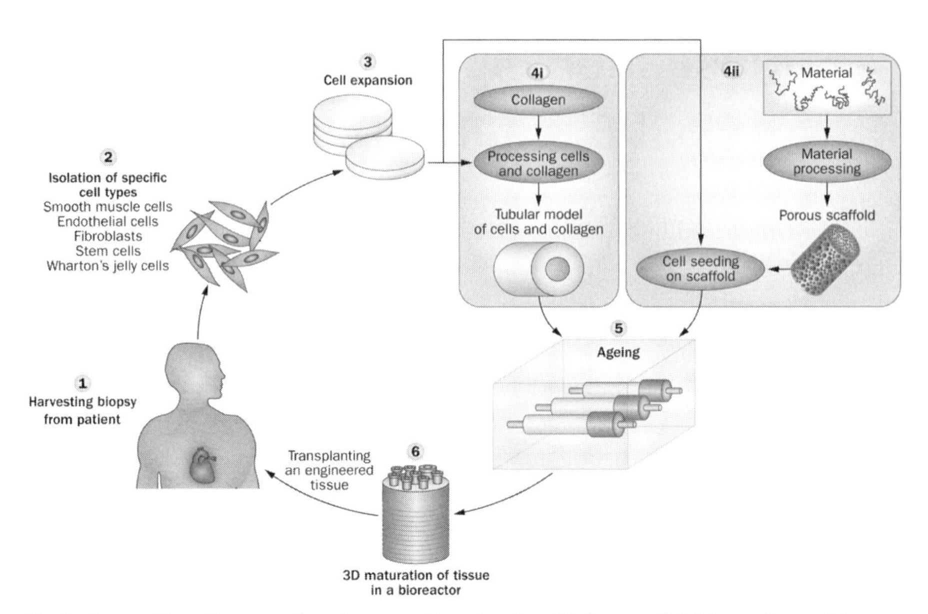

Figure 13.1 **Typical paradigm for vascular tissue engineering in which a scaffold is used to guide tissue regeneration.**
(1–3) Autologous cells are harvested from the patient, selected, and expanded in vitro. Alternatively, non-autologous sources (e.g., mesenchymal stem cells and induced pluripotent stem cells) may be prove useful. (4) Cells are seeded onto (4i) native extracellular matrix protein or (4ii) synthetic scaffolds are (5) grown in culture and (6) matured in a bioreactor, which may apply mechanical, chemical, or other stimuli to aid in tissue development prior to implantation back into the patient. Reprinted by permission from Macmillan Publishers Ltd: *Nature Reviews Cardiology*, 2013, Seifu D.G. et al. *Nature Reviews Cardiology* 10: 410–421, copyright 2013.

13.2 Vascular tissue engineering

Tissue engineered vascular tissue grafts aim to achieve similar mechanical, cellular, and physiological properties as native vasculature. Native arteries and veins have a hierarchical structure consisting of three distinct layers: the *tunica intima*, *tunica media*, and *tunica adventitia*. The *tunica intima* consists of a monolayer of endothelial cells that reside on a basal lamina consisting primarily of collagen IV and laminin, which is supported by an internal elastic lamina containing microfibrils and collagen fibers (Wagenseil and Mecham 2009). The *tunica media* is rich in VSMCs found within an extracellular matrix consisting primarily of elastin and collagens I and III. The outermost layer of the arterial wall, the *tunica adventitia*, contains myofibroblasts in a collagen-rich extracellular matrix (ECM) that is separated from the *tunica media* by an external elastic lamina (Wagenseil and Mecham 2009). The *tunica intima* provides the vessel with its nonthrombogenic properties, while the *tunica media* is associated with vessel contractility and mechanical strength.

While some efforts to tissue engineer a blood vessel using purely cell-based approaches have had good success (Laflamme et al. 2006; McAllister et al. 2009), the challenge of achieving adequate mechanical properties for implantation has driven most approaches to use biomaterial scaffolds in combination with cells to engineer a highly cellularized, ECM-rich tissue that is equivalent to the media layer, the primary load-bearing layer of the vasculature. Several different biomaterial scaffolds have been explored for developing a tissue engineered vascular graft, including modifications to Goretex® and Dacron® to improve their performance, decellularized xenografts, natural biopolymers, and synthetic polymers (primarily polyurethanes and polyesters) (Seifu et al. 2013). Regardless of the biomaterial system, tissue composition, structure, and mechanical properties can be improved by applying biomechanical forces to the developing tissue. Of course, this mimics what occurs in vivo when pulsatile blood flow causes the media to be strained circumferentially by 5–15%, thereby imparting cyclic stretch on the VSMCs embedded in this layer. Strain exposure affects VSMC phenotype, including regulation of cell signaling pathways, proliferation, ECM production, and migration (Lehoux et al. 2006). It is this mechanoregulation of cell function that is exploited to engineer functional tissues, either through the inherent biomechanical properties of the scaffold or via externally applied strains to engineered constructs.

13.2.1 Biomaterial stiffness and vascular smooth muscle cell response

The elastic properties of the matrix to which a cell is adhered play an important role in governing a wide variety of cell responses, including stem cell differentiation, cell phenotypic expression, and migration, among others. VSMCs have been shown to be sensitive to substrates of stiffness varying from 0.3 to 500 kPa, which covers the range of healthy to pathological tissue.

Of particular relevance to tissue engineering, VSMC phenotype (synthetic vs proliferative) is regulated by substrate stiffness. For example, on two-dimensional (2-D) polyethylene glycol (PEG) substrates with tensile moduli ranging from 13.7 to 423.9 kPa, VSMC proliferation was increased on stiffer substrates, suggesting a shift toward a synthetic phenotype (Peyton et al. 2006). This was further supported by a decrease in VSMC differentiation markers with increasing stiffness, including reduced association of calponin and caldesmon with α-actin fibrils. This response mirrors the phenotypic shift of VSMCs in atherosclerosis, which is also associated with a stiffer ECM. This response has also been demonstrated with polyacrylamide gels having modulus values varying from 19 to 84 kPa, suggesting that this response is not an artifact of the material chosen, but is reflective of the response to substrate modulus (Brown et al. 2010; Robinson et al. 2012). Notably, responses to substrate modulus appear to be cell-type dependent. For example, on PEG hydrogels of 0.3 to 13.7 kPa, fibroblasts showed increases in proliferation on stiffer substrates but ECs showed the opposite trend (Robinson et al. 2012).

Additionally, cellular response to matrix stiffness is sensitive to other aspects of the microenvironment. For example, while VSMCs on 135 kPa polyacrylamide gels spread more and had greater focal adhesion protein expression than VSMCs on 25 kPa gels, this effect was attenuated when VSMCs were cultured at higher density and allowed to form cell-cell contacts (Sazonova et al. 2011). Furthermore, when comparing 1.79 MPa versus 0.05 MPa PDMS substrates, cells were sensitive to differences in modulus with respect to attachment and spreading only in the absence of serum (Brown et al. 2005). ECM coating can also regulate VSMC response to substrates of different stiffness. When cultured on acrylamide/bisacrylamide hydrogels of 1 to 308 kPa, the migration speed of cells on surfaces coated with 0.8 μg/cm^2 fibronectin was greatest for 51.9 kPa, while when the surface was coated with 8.0 μg/cm^2 fibronectin, the maximum migration speed was observed at 21.6 kPa (Peyton and Putnam 2005).

Collectively, these fundamental studies suggest that biomaterial elasticity is a design variable that could be used to regulate cell function and tissue maturation. However, these studies also clearly demonstrate the complex interplay between mechanical and non-mechanical microenvironmental cues in regulating cell responses, and that these integrated effects must be considered if one wants to engineer biomaterials that predictively guide tissue regeneration.

13.2.2 Externally applied mechanical strain and engineered vascular tissue responses

Bioreactor systems are often used in tissue engineering to impart mechanical stimuli to cells, with the goal of supporting growth, specific phenotypes, oxygenation, and other critical factors (Riehl et al. 2012). For vascular tissue engineering strategies targeting the medial (VSMC-rich) layer of a graft, bioreactors are typically designed to apply mechanical strain to cell-seeded constructs either using pulsatile flow, such that cells experience both mechanical strain and shear stress, or cyclic stretch systems.

13.2.2.1 Pulsatile perfusion bioreactors

Pulsatile perfusion bioreactors have been successfully used to accelerate the development of vascular tissue in vitro (Fig. 13.2). Parameters that have been varied using this approach to gauge the effect on VSMCs include flow rate, shear stress, pressure, distension, and pulse frequency. In a seminal study by Niklason and Langer, an implantable tissue engineered vascular graft was generated using a pulsatile perfusion bioreactor for a period of 8 weeks, exposing VSMCs seeded on a polyglycolic acid mesh to 5% radial strain at 2.75 Hz with exposure to shear stress of 0.3 dynes/cm^2 (6 mL/min) (Niklason et al. 1999). This dynamic culture increased collagen production and the mechanical strength of the tissue engineered construct, but not VSMC number. Other systems have shown that pulsatile loading with pressures ranging from 60/40 mmHg to 120/70 mmHg, distension from 3% to 10%, and flow rates from 5 to 3000 mL/min in various combinations can increase cell number, promote contractile marker expression, and increase ECM deposition (Isenberg et al. 2006; Zhang et al. 2009; Opitz et al. 2004; Engbers-Buijtenhuijs et al. 2006; Xu et al. 2005; Jeong et al. 2005; Hahn et al. 2007; Song et al. 2011). Fluid flow in these systems, when reported, is described as laminar with Reynolds numbers ranging from 96 to 249 (Engbers-Buijtenhuijs et al. 2006; Zhang et al. 2009). In studies with rat aortic SMCs, turbulent flow has been shown promote cell growth (Rosati and Garay 1991; Shigematsu et al. 2000), although no studies have involved the study of turbulent versus laminar flow in the context of vascular tissue engineering.

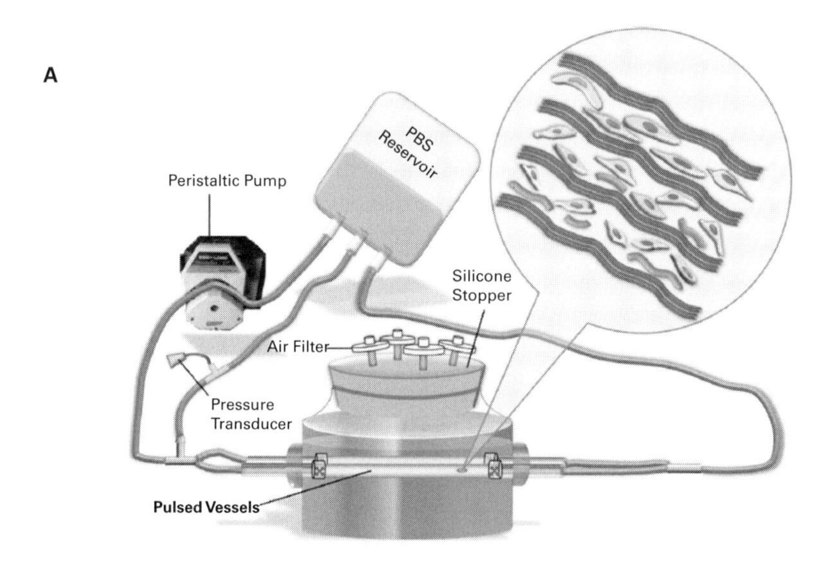

Figure 13.2 **Examples of pulsatile perfusion bioreactors to mechanically condition tissue engineered blood vessels.**
(A) Schematic of the simple pulsatile bioreactor used in the seminal work by Niklason et al. (2005). A peristaltic pump creates cyclic circumferential stretching by flowing saline through the tubing. Pressure upstream of the vascular grafts in monitored with a pressure transducer. The inset shows vascular smooth muscle cells in a synthesized collagen matrix with remnants of a biodegradable

In typical pulsatile perfusion bioreactor studies, it is difficult to identify how individual mechanical factors regulate VSMC response because a wide range of factors (e.g., shear stress, flow rate, pressure, distension) are varied simultaneously. Further, perfusion not only imposes mechanical forces on cells, but also dictates the transport of nutrients, gases, and waste products to and from cells. Indeed, improved mass transport and increased aerobic respiration, as demonstrated by glucose consumption, medium pH, and pCO_2 values (Engbers-Buijtenhuijs et al. 2006), has been suggested to be responsible for increased proliferation under pulsatile perfusion.

Controlled studies in which individual mechanical factors have been varied provide some insight into how pulsatile perfusion bioreactors regulate VSMC response. In a study by Crapo and Wang (2011), poly(glycerol sebacate) scaffolds were seeded with baboon arterial VSMCs in a modified bioreactor such that pressure could be controlled independently of flow rate. Constructs were exposed to 10 ± 5 mmHg hydrostatic pressure for the entire 21-day culture period, or to 60 ± 10 mmHg that was gradually increased to 120 ± 20 mmHg. Constructs exposed to increased hydrostatic pressure saw increased production of soluble and insoluble elastin as well as higher burst pressure, suggesting that exposure to hydrostatic pressure is one mechanism through which pulsatile perfusion supports a synthetic VSMC phenotype. In another study by Wayman et al. (2008), shear stress was controlled independently of mechanical strain for porcine carotid arteries cultured in a custom-designed bioreactor. Circumferential stress was varied from 50 to 150 kPa while keeping shear stress constant at physiological levels of 15 dynes/cm^2, or alternatively shear stress was varied from 7.5 to 22.5 dynes/cm^2 while circumferential stress was maintained at 100 kPa. High circumferential strain, but not shear stress, resulted in medial and adventitial layer proliferation as well as enhanced ^{3}H-proline incorporation (increased matrix synthesis). While this study suggests mechanical strain can increase VSMC growth and matrix production, it should also be noted that the VSMCs were not exposed directly to shear stress due to the presence of a complete endothelium. Furthermore, the presence of ECs can modify VSMC response to shear stress (Redmond et al. 2001). Specifically, the effects observed in the presence of ECs are not transferrable to pulsatile perfusion bioreactor systems where VSMCs in monoculture are exposed directly to flow. Studies on the effects of VSMCs in monoculture to shear stress suggest that this stimulus can increase proliferation-inducing cytokines

Caption for Figure 13.2 (cont.)

synthetic scaffold. (B) Schematic of a pulse duplicator bioreactor used to apply physiological flow to engineered vascular grafts ("engineered media equivalents" (eME)). Physiological flow is created by superimposing a steady component generated by the peristaltic pump (2) and pulse dampeners (3) with a pulsatile component generated by the syringe pump (4). On the right is a schematic of the flow chambers in which the vascular grafts were mounted for flow conditioning.

(A) reproduced with permission from Huang et al. (2014) with kind permission from Springer Science and Business Media; (B) reproduced with permission from Isenberg et al. (2006) with kind permission from Springer Science and Business Media.

and growth factors such as fibroblast growth factor-2 and platelet derived growth factor (PDGF) (3–25 dynes/cm^2); induce VSMC alignment; either inhibit (laminar, steady flow) (Fitzgerald et al. 2008) or promote (pulsatile, oscillatory flow) VSMC proliferation; and reduce contractile marker expression, all of which suggest that shear stress promotes a synthetic VSMC phenotype (Shi and Tarbell 2011).

13.2.2.2 Mechanical stretch bioreactors

Unlike pulsatile perfusion systems, cyclic stretch bioreactors are largely able to isolate the effects of mechanical strain from flow, and allow for a more fundamental understanding of the mechanical factors regulating VSMC response. Furthermore, these systems represent a more physiologically relevant biomechanical stimulation for VSMCs, since VSMCs in vivo do not directly experience the effects of shear stress due to the presence of the endothelium. Isolating effects of strain from flow is achieved with most bioreactor systems by using a distensible mandrel (e.g., silicone, latex) as a support for the tissue engineered construct. In these systems, distension of the mandrel transfers the strain to the cell-seeded material without exposure to the pressurizing medium (typically air or water). In addition to strain magnitude, other factors that can be varied with these systems include frequency and duty cycle. In a study by Wang et al. (2010), a pulsatile perfusion bioreactor was used to apply strain (5% radial strain, 1.25 Hz, 100% duty cycle) for eight weeks to adipose stem cell-derived VSMCs seeded on a polyglycolic acid mesh that was mounted on a silicone tubing, resulting in increased mechanical properties (ultimate tensile strength, elastic modulus, suture retention strength, burst pressure), cell number, and collagen content. Similar results have been achieved with human aortic SMCs embedded in a collagen gel and exposed to 10% radial strain (1 Hz) for four to eight days (Seliktar et al. 2000). In a detailed study by Isenberg and Tranquillo (2003) using rat aortic SMCs embedded in a collagen gel, the various factors involved in the application of cyclic strain (percent strain, duty cycle, frequency) were varied in order to determine the optimal conditions for enhancing construct mechanical properties and promoting tissue growth. Circumferential strain had significant effects on elastic modulus (10% > 5% = 2.5% > static) and ultimate tensile strength (UTS) (2.5% = 5% > 10% > static), as did duty cycle with strain held constant at 5% and relaxation time at 1.75 sec while varying the stretch time (UTS: 6.7% = 12.5% > 18% = 22% = static; EM: 6.7% > 12.5% > 18% > 22% > static). The effect of frequency was also investigated by maintaining stretch time constant at 0.25 sec and strain at 5%, while varying relaxation time, demonstrating effects on UTS (0.25 Hz > 0.5 Hz = 1 Hz > static) and EM (0.25 Hz = 1 Hz > 0.5 Hz > static). In an important follow-up study by Syedain et al. (2008) using human dermal fibroblasts embedded in a tubular fibrin gel, constructs were either subjected to a single strain rate over three weeks (2.5%, 5%, 10%, or 15%), or had their strain rate increased in either two (ICD2: 5% → 10% → 15%) or four (ICD4: 5% → 7.5% → 12.5% → 15%) equal time steps from 5% to 15%. In all cases, a 12.5% duty cycle and 0.5 Hz frequency were used. They found that incrementally changing the strain level maximized UTS and EM, with four equal time intervals achieving the best results. Within the constant cyclic strain conditions, trends were also observed for UTS and EM (15% = 10% > 5% = 2.5% = static). While increasing cell number and collagen content were observed with increasing cyclic strain level, ICD2 and ICD4 conditions were the only ones to show increases in collagen produced when normalized to cell number. Such studies provide support for the use of mechanical strain as a means to enhance vascular tissue formation in vitro.

Combining the effects of biomechanical stimulation with biochemical stimulation has also been used to further optimize the production of vascular tissue. Rat aortic SMCs exposed to mechanical strain (10% strain, 1 Hz) showed increases in cell number. Exposure to PDGF resulted in an increase in cell number in general, but when exposed to PDGF, mechanical stimulation had a negative effect on cell number, while exposure to transforming growth factor (TGF)-β1 had a drastic negative effect on cell number and also eliminated any benefits of the observed mechanical effects (Stegemann and Nerem 2003). TGF-β1 is a promising growth factor for use in vascular tissue engineering as it can promote elastin synthesis, the lack of which is one of the main limitations of many tissue engineered vascular constructs. However, Syedain and Tranquillo (2011) demonstrated that stimulation of fibrin constructs embedded with human dermal fibroblasts with TGF-β1 (1 ng/mL) in conjunction with mechanical strain (0.5 Hz, 12.5% duty cycle, strain increased from 5% → 7.5% → 12.5% → 15% over 3 weeks) *prevented* the upregulation of collagen synthesis and improved mechanical properties observed without biochemical stimulation over seven weeks. However, TGF-β1-stimulated cultures did demonstrate the desirable property of increasing elastin production. By limiting TGF-β1 exposure to the final two weeks of the seven-week culture period, Syedain et al. (2008) were able to utilize the initial 5 weeks of mechanical conditioning to increase the mechanical properties (via increased collagen deposition) while the final two weeks involving strain in combination with TGF-β1 stimulation were sufficient to enhance elastin production without a negative impact on the construct's mechanical properties. Combining mechanical strain with growth factor supplementation, or other means of introducing stimulatory factors, thus has the potential to further optimize the production of tissue-engineered vascular grafts.

Independent of mechanical strain variables, the underlying substrate on which VSMCs are cultured can play a role in the ability of VSMCs to sense mechanical stimuli. Neonatal rat VSMCs cultured on 2-D silicone elastomers coated with ECM proteins and exposed to a strain rate of 1 Hz demonstrated ECM protein-dependent effects (Wilson et al. 1995). VSMC proliferation was enhanced with the application of strain on substrates coated with fibronectin and collagen (fibronectin > collagen), but not on substrates coated with elastin, laminin, or polylysine. Furthermore, when substrates were coated with a mixture of laminin and vitronectin, proliferation increased with a corresponding increase in the concentration of adsorbed vitronectin on the surface. Promotion of proliferation by fibronectin-coated surfaces under strain could be blocked through soluble RGD peptides as well as blocking specific integrins (β_3 and $\alpha_3\beta_5$), suggesting the critical role of these integrins in mechanotransduction. Such studies have further applicability to the application of strain with different biomaterials, as differences in the adsorbed protein layer in addition to the surface chemistry itself can potentially result in changes to the ability of adherent VSMCs to respond to strain.

While not as widely studied as VSMCs, monocytes/macrophages and other immune cells are also affected by mechanical strain. Because macrophages contribute to neovessel formation following the implantation of tissue engineered vascular grafts (Hibino et al. 2011), these cells will be exposed to mechanical strain in such situations, and how their phenotype is affected may play a role in their subsequent polarization and ability to contribute to vascular remodeling. U937 cells, a macrophage-like cell line, have been shown to be responsive to both uniaxial and biaxial strain. Uniaxial strain was shown to decrease DNA levels on polyRGD, but not collagen I coated surfaces, with cells aligning in the direction of the applied strain. Biaxial and uniaxial strains were both shown to upregulate intracellular esterase activity, IL-6 but not IL-8 production, and total protein production by U937 cells (Matheson et al. 2007; Matheson et al. 2006). Exposure of peritoneal macrophages to 5% strain (1 Hz) suppressed their ability to phagocytose latex particles after 24 and 48 hours (Miyazaki and Hayashi 2001), while other studies have shown that biaxial strain at 1, 2, and 3% (1 Hz) promotes macrophage scavenger receptor (CD36) expression, which may suggest increased phagocytic capacity (Sakamoto et al. 2001). These studies highlight the importance of considering the magnitude of the applied strain, rather than solely the presence of mechanical stimulation itself. Using human peripheral blood mononuclear cells seeded on polycaprolactone bis-urea nanofibrous strips, Ballotta et al. (2014) further demonstrated the importance of strain magnitude on macrophage phenotype. Following one day of exposure to strain, 7% strain upregulated several pro- and antiinflammatory cytokines, including MCP-1, IL-6, and IL-10 in addition to MMP9, while 12% strain upregulated SDF1α expression. These effects appear to be evident only initially following the application of strain, as no effects were observed after 2 days of culture. With regards to phenotype polarization, exposure to no strain as well as 7% strain resulted in an increase in immunoregulatory macrophages over time (increase CD206/CCR7 ratio), while 12% strain promoted a decrease in this ratio. Increasing strain magnitude was also associated with decreased cell numbers (0% > 7% > 12%). Studies with dendritic cells, which like macrophages can be derived from monocytes, further demonstrate the strain-responsiveness of immune cells, as well as the dependence of these effects on adsorbed proteins (Lewis et al. 2013). Exposure of cells to 3% or 10% strain resulted in increased apoptotic and necrotic cells. CD86 expression was also shown to increase on cells exposed to 3% strain on collagen and fibrinogen, but not laminin, coated surfaces after one hour, while IL-12 release was only increased in the presence of fibrinogen coating.

In consideration of the important interacting roles of VSMCs, monocyte/macrophages, and mechanical stimulation in vascular tissue regeneration, we have investigated the effects of cyclic stretch on co-cultures of human coronary artery VSMCs and monocytes on tubular (3 mm ID) degradable polyurethane scaffolds with a unique combination of nonionic polar, hydrophobic, and ionic chemistry (Battiston 2015). Scaffolds were mounted on distensible silicone tubing and subjected to dynamic biomechanical stimulation (10% circumferential strain, 1 Hz) for up to four weeks (Fig. 13.3). We found that mechanical strain and monocyte co-culture

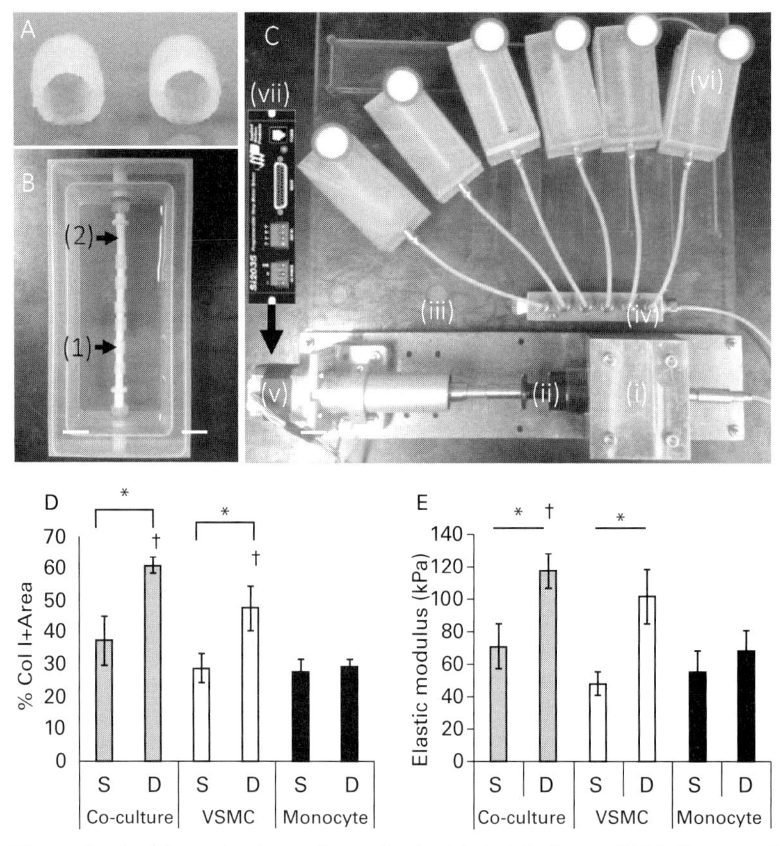

Figure 13.3 Example of a bioreactor to apply cyclic stretch to tubular scaffolds for vascular tissue engineering.

(A) Tubular porous polyurethane scaffolds. (B) Close-up of chamber with (1) scaffolds mounted on (2) distensible silicone tubing. (C) Picture of bioreactor set-up, showing (i) syringe holder, (ii) 50 cc syringe, (iii) bioreactor tray, (iv) manifold, (v) actuator, (vi) bioreactor chamber with syringe filter for gas exchange, and (vii) controller. Dynamic mechanical stimulation increased (D) the amount of type I collagen and (E) the elastic modulus of the constructs for conditions with vascular smooth muscle cells (VSMC) alone or co-cultured with monocytes. While the inclusion of monocytes did not significantly influence collagen amount or elastic modulus, co-culture did have a positive effect on the synthesis of other extracellular matrix proteins (see the text for details). $^*p < 0.05$; † $p < 0.05$ versus dynamic monocyte.

Reproduced with permission from Battiston (2015).

had complementary and non-mitigating effects on VSMC growth. Co-culture increased deposition of sulphated glycosaminoglycans and elastin, while dynamic culture increased collagen I and III synthesis and increased the tissue elastic modulus and tensile strength versus static controls. Unexpectedly, the macrophages produced significant amounts of vascular extracellular matrix components, including collagen I, collagen III, elastin, and GAGs. The combination of the immunomodulatory effects of this polyurethane formulation and the demonstrated effect of biomechanical strain to augment human vascular tissue production make this system promising for small diameter graft applications.

13.3 Insights on mechanoregulation of vascular smooth muscle cells from microtechnologies

While reasonable progress toward functional tissue-engineered vascular grafts has been made using bioreactor-based mechanical stimulation, there clearly remains room for improvement and optimization, particularly for more demanding clinical situations. Recent advances in micro and nanotechnologies and their application to vascular cells have provided insights into the effects of various biophysical stimuli on VSMC function. They also enable higher-throughput investigations, which opens the possibility of interrogating the effects of combinations of microenvironmental stimuli on cell function to inform biomaterial and bioreactor design. Here we review a sampling of studies that have used microtechnologies to probe cell mechanobiological responses of relevance to vascular tissue engineering.

13.3.1 Mechanical stimulation of VSMCs

Traditionally, cellular responses to mechanical strain are probed using flexible membrane systems, such as the commercially available Flexcell systems in which cells seeded on a flexible membrane are exposed to mechanical strain by the deformation of the membrane using vacuum pressure. By combining microfabrication with this conventional system, Tan et al. (2008) were able to systematically investigate the effect of anisotropic biaxial strain on VSMC response. Microgrooves (10–30 μm) were produced on flexible membranes oriented either circumferentially or radially. Because VSMCs aligned in the microgrooves, cell orientation could be controlled such that strain was applied along the cells minor axis (circumferential grooves) or major axis (radial grooves), with cell anisotropy strain index (CASI) defined as the ratio of cells major/minor axis (CASI < 1 for circumferential grooves, CASI > 1 for radial grooves). Using this microdevice, Tan et al. (2008) demonstrated that application of strain increased VSMC proliferation, but in a manner dependent on cell orientation with respect to the direction of applied strain. In general, application of strain increased VSMC proliferation. Compared to equibiaxial strain, cell proliferation was reduced on membranes with circumferential microgrooves, with a corresponding decrease in CASI, whereas proliferation and CASI were increased on membranes with radial microgrooves. Mechanical stimulation is commonly used to promote VSMC proliferation in tissue engineering strategies. This study thus suggests that in order to maximize the effect of mechanical strain on cell proliferation, tissue-engineering scaffolds should be designed in a manner that promotes VSMC elongation in the direction of the applied strain.

Micropatterning has also been used to control cell orientation to investigate VSMC response to shear stress (Li et al. 2013). Micropatterns of high molecular weight hyaluroanan were prepared on titanium surfaces using a PDMS stamp to orient human umbilical artery SMCs perpendicular or parallel to the direction of applied shear stress (3, 9, or 15 dyn/cm^2). Orientation of VSMCs perpendicular versus parallel to the direction of shear stress resulted in increased VSMC apoptosis, increased cell number (only for 3 dyn/cm^2), increased cell spreading (only for 3 dyn/cm^2), increased contractile marker expression, and decreased TGF-β1 release. Regardless of cell orientation relative to the applied shear stress, an increase in shear stress was associated with increased contractile marker expression and TGF-β1 release, while increasing shear stress had opposite effects on apoptosis when cells were oriented perpendicular (increased apoptosis) or parallel (decreased apoptosis) relative to the direction of fluid flow.

Microfabrication techniques can be used to control shear stress, τ, by varying channel height in a microfluidic device according to the equation $\tau = 6\mu Q/h^2w$, where μ is the viscosity of the fluid medium, Q is the volumetric flow rate, h is the height of the channel and w is the width of the channel. High-throughput devices can be prepared using this approach by fabricating channels with varying height to assess multiple shear stresses simultaneously using a single flow rate. This approach has been used by Abaci et al. (2012) to investigate the effect of shear stress (0.01 vs 10 dyn/cm^2) on smooth muscle-like cells derived from embryonic stem cells (ESCs), which indicated greater cell detachment in the presence of shear but no effect on contractile marker expression or cell alignment.

Microfabrication techniques can also be used to control the spacing of cells either in monoculture or co-culture to investigate how they respond to shear stress as a result of proximity to other cell populations (Yeh et al. 2011). Microchips were micropatterned to introduce gaps of 50, 100, 200, or 500 μm between adjacent populations of SMCs or SMCs and ECs. This study indicated that shear stress delays the onset of migration in co-cultured cells and that increasing gap size decreases migration velocity and distance. However, this study did not systematically vary gap size and shear rates, which could provide further insight into the ability of two adjacent cell populations to communicate under the application of shear.

As discussed in Chapter 3, microfabrication techniques have also provided insight into single-cell responses to mechanical stimuli. For example, Fu and co-workers used a micropost array membrane made of PDMS (1.83 μm post diameter, 0.7–14.97 μm height, 4 μm center-to-center post distance) to assess VSMC contractility through micropost deformation under mechanical stimulation (Mann et al. 2012). Coronary artery SMCs were seeded on microposts and subjected to static strain of 6% or 15% for 60 min. By monitoring micropost deformation, 6% static strain was shown to induce higher maximum contractility and require more time to reach maximum contractility compared to 15% static strain. Due to the high spatial specificity of this approach, peripheral focal adhesions were also shown to contribute more significantly to total contractility. Individual cell response to mechanical stress has also been achieved using micropost arrays (1.8 μm diameter post, 6.4 μm in a hexagonal array with lattice constant of 4 μm) in combination with magnetic Ni nanowires (Lin et al. 2012). Magnetic Ni nanowires allow for the application of strong forces and torques to cells in the presence of external magnetic fields and field gradients. Bovine pulmonary artery SMCs were seeded on microposts coated with fibronectin and nanowires were allowed to either externally bind to SMCs or be internalized by SMCs. Externally bound nanowires interact with the cortical actin network, while internalized nanowires interact with the intracellular cytoskeleton. Increases in cell contractility in the presence of an applied magnetic field could be blocked completely for externally bound nanowires by inhibiting calcium channels, but only diminished and did not eliminate response in the presence of internalized nanowires. Lin et al. (2012) thus concluded that cortical stimulation via adhesion receptors is fully dependent on calcium channels, but internal cytoskeletal stimulation may occur partially through a separate mechanism.

13.3.3 Array-based systems to probe the effects of combinations of mechanobiological stimuli on cell function

A clear advantage of microtechnologies is that their miniaturization can enable high-throughput experimentation via array-based systems. In the context of tissue engineering, array of microdevices can be used to probe the effects of combinations of microenvironmental factors, including biomaterial mechanical and chemical properties, soluble signals, and applied mechanical stimuli, on cell fate and function; this can inform the design of instructive biomaterials and bioreactors systems. The technical challenges and features of microdevice arrays that incorporate mechanical stimulation are discussed in Chapter 4.

While these systems have yet to be applied to investigate VMSC responses, we have preliminarily demonstrated their utility using fibroblastic cells from heart valves (valvular interstitial cells (VICs)), which have similarities to VSMCs. In a study by Moraes et al. (2013), VICs were isolated from two distinct layers of porcine aortic valves (fibrosa vs ventricularis) and subjected to combinations of matrix adhesion protein (type I collagen vs fibronectin), mechanical strain (0, 3%, or 12% strain), and biochemical stimulus (with or without 5 ng/mL TGF-β1) to assess their effects on myofibroblast differentiation. We found that mechanical stimulation, matrix proteins and soluble cues produced integrated and distinct responses in layer-specific VIC populations. Strikingly, myofibroblast differentiation was most significantly influenced by the layer the cells came from, despite the presence of potent mechanobiological cues such as applied strain and TGF-β1 (Fig. 13.4). These results demonstrate that spatially distinct VIC subpopulations respond differentially to microenvironmental cues, with implications for valve tissue engineering and pathobiology. This study also demonstrates the utility of microdevice arrays to enable rapid identification of biological phenomena arising from systematically manipulating the cellular microenvironment; this microtechnology therefore may be useful in screening mechanosensitive cell cultures with applications in drug screening, tissue engineering and fundamental cell biology. In emerging work, we have

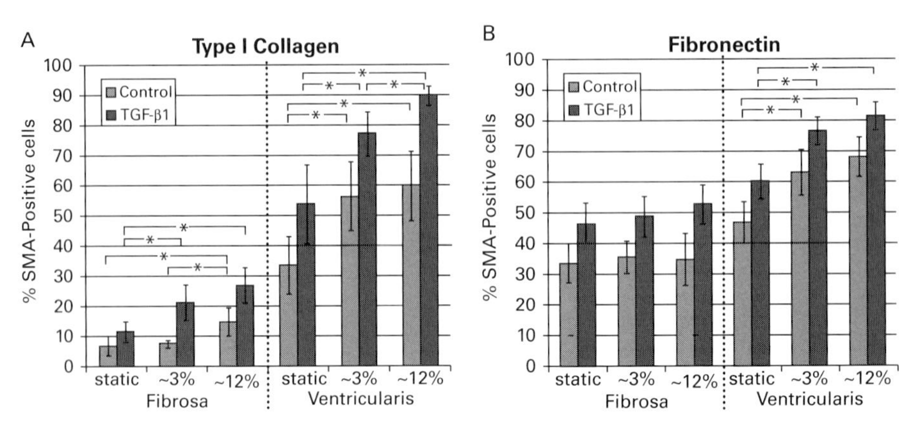

Figure 13.4 **Influence in myofibroblast differention.**
Microdevice arrays (see Fig. 4.1) were used to probe myofibroblast differentiation of side-specific valvular interstitial cells in response to combinations of extracellular matrix proteins (Panel (a) – type I collagen; Panel (b) – fibronectin), dynamic mechanical strain (static, 3%, and 12% strain), and TGF-β1 (0 or 5 ng/mL). Myofibroblasts were identified as cells expressing α-SMA in stress fibers. These experiments were facilitated by the high-throughput nature of the microdevice arrays and demonstrated cell-specific responses and complex interactions between multiple microenvironmental cues. Selected statistical comparisons are highlighted (*$p < 0.01$).
Adapted from Moraes et al. (2013) by permission of The Royal Society of Chemistry.

developed microtechnologies that enable combinatorial mechanobiological stimulation of cells in 3-D microtissue constructs, making these 3-D microdevice array systems more directly relevant to tissue engineering applications (Liu et al. in press).

13.4 Conclusions

Mechanical stimulation has emerged as an important component of strategies aimed at engineering load-bearing tissues, particularly those of the cardiovascular and musculoskeletal systems. In vascular tissue engineering, there has been good progress in mechanically stimulating engineered vascular graft tissue using bioreactors that mimic complex physiological loads or apply individual mechanical stimuli. Moving forward, major advances to achieve functional engineered tissues for a wide variety of clinical applications will likely come from more comprehensive understanding of how the mechanobiological environment regulates cell function and tissue formation. To that end, there are important roles for micro and nanotechnologies in more precisely defining potent mechanobiological cues and the mechanisms by which they act and in enabling systematic interrogation of combinations of stimuli. Moving forward, these insights are expected to define design criteria for biomaterials, bioreactors, and tissue engineering systems that *predictively* guide functional tissue regeneration.

References

Abaci, H. E., Devendra, R., Soman, R., Drazer, G. and Gerecht, S. (2012). "Microbioreactors to manipulate oxygen tension and shear stress in the microenvironment of vascular stem and progenitor cells." *Biotechnol Appl Biochem* **59**: 97–105.

Ballotta, V., Driessen-Mol, A., Bouten, C. V. and Baaijens, F. P. (2014). "Strain-dependent modulation of macrophage polarization within scaffolds." *Biomaterials* **35**: 4919–4928.

Battiston, K. G. (2015). *Evaluating the Use of Monocytes with a Degradable Polyurethane for Vascular Tissue Regeneration*. PhD dissertation, University of Toronto.

Brown, X. Q., Bartolak-Suki, E., Williams, C., Walker, M. L., Weaver, V. M. and Wong, J. Y. (2010). "Effect of substrate stiffness and PDGF on the behavior of vascular smooth muscle cells: implications for atherosclerosis." *J Cell Physiol* **225**: 115–122.

Brown, X. Q., Ookawa, K. and Wong, J. Y. (2005). "Evaluation of polydimethylsiloxane scaffolds with physiologically-relevant elastic moduli: interplay of substrate mechanics and surface chemistry effects on vascular smooth muscle cell response." *Biomaterials* **26**: 3123–3129.

Crapo, P. M. and Wang, Y. (2011). "Hydrostatic pressure independently increases elastin and collagen co-expression in small-diameter engineered arterial constructs." *J Biomed Mater Res A* **96**: 673–681.

Engbers-Buijtenhuijs, P., Buttafoco, L., Poot, A. A., Dijkstra, P. J., de Vos, R. A., Sterk, L. M., Geelkerken, R. H., et al. (2006). "Biological characterisation of vascular grafts cultured in a bioreactor." *Biomaterials* **27**: 2390–2397.

Fitzgerald, T. N., Shepherd, B. R., Asada, H., Teso, D., Muto, A., Fancher, T., Pimiento, J. M., et al. (2008). "Laminar shear stress stimulates vascular smooth muscle cell apoptosis via the Akt pathway." *J Cell Physiol* **216**: 389–395.

Hagerty, R. D., Salzmann, D. L., Kleinert, L. B. and Williams, S. K. (2010). "Cellular proliferation and macrophage populations associated with implanted expanded polytetrafluoroethylene and polyethyleneterephthalate." *J Biomed Mater Res* **49**: 489–497.

Hahn, M. S., Mchale, M. K., Wang, E., Schmedlen, R. H. and West, J. L. (2007). "Physiologic pulsatile flow bioreactor conditioning of poly(ethylene glycol)-based tissue engineered vascular grafts." *Ann Biomed Eng* **35**: 190–200.

Hibino, N., Yi, T., Duncan, D. R., Rathore, A., Dean, E., Naito, Y., Dardik, A., et al. (2011). "A critical role for macrophages in neovessel formation and the development of stenosis in tissue-engineered vascular grafts." *FASEB J* **25**: 4253–4263.

Huang, A. H. and Niklason, L. E. (2014). "Engineering of arteries in vitro." *Cell Mol Life Sci* **71**: 2103–2118.

Isenberg, B. C. and Tranquillo, R. T. (2003). "Long-term cyclic distention enhances the mechanical properties of collagen-based media-equivalents." *Ann Biomed Eng* **31**: 937–949.

Isenberg, B. C. Williams, C. and Tranquillo, R. T. (2006). "Endothelialization and flow conditioning of fibrin-based media-equivalents." *Ann Biomed Eng* **34**: 971–985.

Jeong, S. I., Kwon, J. H., Lim, J. I., Cho, S. W., Jung, Y., Sung, W. J., Kim, S. H., et al. (2005). "Mechano-active tissue engineering of vascular smooth muscle using pulsatile perfusion bioreactors and elastic PLCL scaffolds." *Biomaterials* **26**: 1405–1411.

Laflamme, K., Roberge, C. J., Pouliot, S., D'orleans-Juste, P., Auger, F. A. and Germain, L. (2006). "Tissue-engineered human vascular media produced in vitro by the self-assembly approach present functional properties similar to those of their native blood vessels." *Tissue Eng* **12**: 2275–2281.

Lehoux, S., Castier, Y. and Tedgui, A. (2006). "Molecular mechanisms of the vascular responses to haemodynamic forces." *J Intern Med* **259**: 381–392.

Lewis, J. S., Dolgova, N. V., Chancellor, T. J., Acharya, A. P., Karpiak, J. V., Lele, T. P. and Keselowsky, B. G. (2013). "The effect of cyclic mechanical strain on activation of dendritic cells cultured on adhesive substrates." *Biomaterials* **34**: 9063–9070.

Li, J., Zhang, K., Yang, P., Liao, Y., Wu, L., Chen, J., Zhao, A., et al. (2013). "Research of smooth muscle cells response to fluid flow shear stress by hyaluronic acid micro-pattern on a titanium surface." *Exp Cell Res* **319**: 2663–2672.

Lin, Y. C., Kramer, C. M., Chen, C. S. and Reich, D. H. (2012). "Probing cellular traction forces with magnetic nanowires and microfabricated force sensor arrays." *Nanotechnology* **23**: 075101.

Liu, H., Usprech, J., Sun, Y. and Simmons, C. A. (in press). "A microfabricated platform with hydrogel arrays for 3D mechanical stimulation of cells." *Acta Biomaterialia*.

Mann, J. M., Lam, R. H., Weng, S., Sun, Y. and Fu, J. (2012). "A silicone-based stretchable micropost array membrane for monitoring live-cell subcellular cytoskeletal response." *Lab Chip* **12**: 731–740.

Matheson, L. A., Maksym, G. N., Santerre, J. P. and Labow, R. S. (2006). "The functional response of U937 macrophage-like cells is modulated by extracellular matrix proteins and mechanical strain." *Biochem Cell Biol* **84**: 763–773.

Matheson, L. A., Maksym, G. N., Santerre, J. P. and Labow, R. S. (2007). "Differential effects of uniaxial and biaxial strain on U937 macrophage-like cell morphology: influence of extracellular matrix type proteins." *J Biomed Mater Res A* **81**: 971–981.

Mcallister, T. N., Maruszewski, M., Garrido, S. A., Wystrychowski, W., Dusserre, N., Marini, A., Zagalski, K., et al. (2009). "Effectiveness of haemodialysis access with an autologous tissue-engineered vascular graft: a multicentre cohort study." *Lancet* **373**: 1440–1446.

Miyazaki, H. and Hayashi, K. (2001). "Effects of cyclic strain on the morphology and phagocytosis of macrophages." *Biomed Mater Eng* **11**: 301–309.

Moraes, C., Likhitpanichkul, M., Lam, C. J., Beca, B. M., Sun, Y. and Simmons, C. A. (2013). "Microdevice array-based identification of distinct mechanobiological response profiles in layer-specific valve interstitial cells." *Integr Biol (Camb)* **5**(4): 673–680.

Niklason, L. E., Gao, J., Abbott, W. M., Hirschi, K. K., Houser, S., Marini, R. and Langer, R. (1999). "Functional arteries grown in vitro." *Science* **284**: 489–493.

Opitz, F., Schenke-Layland, K., Richter, W., Martin, D. P., Degenkolbe, I., Wahlers, T. and Stock, U. A. (2004). "Tissue engineering of ovine aortic blood vessel substitutes using applied shear stress and enzymatically derived vascular smooth muscle cells." *Ann Biomed Eng* **32**: 212–222.

Peyton, S. R. and Putnam, A. J. (2005). "Extracellular matrix rigidity governs smooth muscle cell motility in a biphasic fashion." *J Cell Physiol* **204**: 198–209.

Peyton, S. R., Raub, C. B., Keschrumrus, V. P. and Putnam, A. J. (2006). "The use of poly(ethylene glycol) hydrogels to investigate the impact of ECM chemistry and mechanics on smooth muscle cells." *Biomaterials* **27**: 4881–4893.

Redmond, E. M., Cullen, J. P., Cahill, P. A., Sitzmann, J. V., Stefansson, S., Lawrence, D. A. and Okada, S. S. (2001). "Endothelial cells inhibit flow-induced smooth muscle cell migration: role of plasminogen activator inhibitor-1." *Circulation* **103**: 597–603.

Riehl, B. D., Park, J. H., Kwon, I. K. and Lim, J. Y. (2012). "Mechanical stretching for tissue engineering: two-dimensional and three-dimensional constructs." *Tissue Eng Part B Rev* **18**: 288–300.

Robinson, K. G., Nie, T., Baldwin, A. D., Yang, E. C., Kiick, K. L. and Akins, R. E., Jr. (2012). "Differential effects of substrate modulus on human vascular endothelial, smooth muscle, and fibroblastic cells." *J Biomed Mater Res A* **100**: 1356–1367.

Rosati, C. and Garay, R. (1991). "Flow-dependent stimulation of sodium and cholesterol uptake and cell growth in cultured vascular smooth muscle." *J Hypertens* **9**: 1029–1033.

Sakamoto, H., Aikawa, M., Hill, C. C., Weiss, D., Taylor, W. R., Libby, P. and Lee, R. T. (2011). "Biomechanical strain induces class a scavenger receptor expression in human monocyte/macrophages and THP-1 cells: a potential mechanism of increased atherosclerosis in hypertension." *Circulation* **104**: 109–114.

Sazonova, O. V., Lee, K. L., Isenberg, B. C., Rich, C. B., Nugent, M. A. and Wong, J. Y. (2011). "Cell-cell interactions mediate the response of vascular smooth muscle cells to substrate stiffness." *Biophys J* **101**: 622–630.

Seifu, D. G., Purnama, A., Mequanint, K. and Mantovani, D. (2013). "Small-diameter vascular tissue engineering." *Nat Rev Cardiol* **10**: 410–421.

Seliktar, D., Black, R. A., Vito, R. P. and Nerem, R. M. (2000). "Dynamic mechanical conditioning of collagen-gel blood vessel constructs induces remodeling in vitro." *Ann Biomed Eng* **28**: 351–362.

Shi, Z. D. and Tarbell, J. M. (2011). "Fluid flow mechanotransduction in vascular smooth muscle cells and fibroblasts." *Ann Biomed Eng* **39**: 1608–1619.

Shigematsu, K., Yasuhara, H., Shigematsu, H. and Muto, T. (2000). "Direct and indirect effects of pulsatile shear stress on the smooth muscle cell." *Int Angiol* **19**: 39–46.

Song, Y., Wennink, J. W., Kamphuis, M. M., Sterk, L. M., Vermes, I., Poot, A. A., Feijen, J. et al. (2011). "Dynamic culturing of smooth muscle cells in tubular poly(trimethylene carbonate) scaffolds for vascular tissue engineering." *Tissue Eng Part A* **17**: 381–387.

Stegemann, J. P. and Nerem, R. M. (2003). "Phenotype modulation in vascular tissue engineering using biochemical and mechanical stimulation." *Ann Biomed Eng* **31**: 391–402.

Syedain, Z. H. and Tranquillo, R. T. (2011). "TGF-beta1 diminishes collagen production during long-term cyclic stretching of engineered connective tissue: implication of decreased ERK signaling." *J Biomech* **44**: 848–855.

Syedain, Z. H., Weinberg, J. S. and Tranquillo, R. T. (2008). "Cyclic distension of fibrin-based tissue constructs: evidence of adaptation during growth of engineered connective tissue." *Proc Natl Acad Sci USA* **105**: 6537–6542.

Tan, W., Scott, D., Belchenko, D., Qi, H. J. and Xiao, L. (2008). "Development and evaluation of microdevices for studying anisotropic biaxial cyclic stretch on cells." *Biomedical Microdevices* **10**: 869–882.

Vara, D. S., Punshon, G., Sales, K. M., Hamilton, G. and Seifalian, A. M. (2011). "Haemodynamic regulation of gene expression in vascular tissue engineering." *Curr Vasc Pharmacol* **9**: 167–187.

Wagenseil, J. E. and Mecham, R. P. (2009). "Vascular extracellular matrix and arterial mechanics." *Physiol Rev* **89**: 957–989.

Wang, C., Cen, L., Yin, S., Liu, Q., Liu, W., Cao, Y. and Cui, L. (2010). "A small diameter elastic blood vessel wall prepared under pulsatile conditions from polyglycolic acid mesh and smooth muscle cells differentiated from adipose-derived stem cells." *Biomaterials* **31**: 621–630.

Wayman, B. H., Taylor, W. R., Rachev, A. and Vito, R. P. (2008). "Arteries respond to independent control of circumferential and shear stress in organ culture." *Ann Biomed Eng* **36**: 673–684.

Wilson, E., Sudhir, K. and Ives, H. E. (1995). "Mechanical strain of rat vascular smooth muscle cells is sensed by specific extracellular matrix/integrin interactions." *Journal of Clinical Investigation* **96**: 2364–2372.

Xu, J., Ge, H., Zhou, X., Yang, D., Guo, T., He, J., Li, Q., et al. (2005). "Tissue-engineered vessel strengthens quickly under physiological deformation: application of a new perfusion bioreactor with machine vision." *J Vasc Res* **42**: 503–508.

Yeh, C. H., Tsai, S. H., Wu, L. W. and Lin, Y. C. (2011). "Using a co-culture microsystem for cell migration under fluid shear stress." *Lab Chip* **11**: 2583–2590.

Zhang, X., Wang, X., Keshav, V., Johanas, J. T., Leisk, G. G. and Kaplan, D. L. (2009). "Dynamic culture conditions to generate silk-based tissue-engineered vascular grafts." *Biomaterials* **30**: 3213–3223.

Zilla, P., Bezuidenhout, D. and Human, P. (2007). "Prosthetic vascular grafts: wrong models, wrong questions and no healing." *Biomaterials* **28**: 5009–5027.

14 Bone cell mechanobiology using micro- and nano-techniques

Chao Liu, Kevin Middleton, and Lidan You

14.1 Bone structure at the micro- and nanoscale

14.1.1 Bone matrix

The mineralized matrix of bone gives it unique mechanical properties compared to other tissues. The Young's modulus of bone is in the order of tens of gigapascals (Rho et al. 1997). This hard matrix is formed by calcium crystal woven between collagen fibers (Fratzl et al. 2004). Bone matrix is a composite material with an inorganic component, the hydroxyapatite, embedded in the organic components, which are collagen I fibrils with noncollagenous proteins (Robinson 1952). The noncollagenous proteins include osteopontin, bone sialoprotein, osteonectin, and osteocalcin (Roach 1994). Although the roles of these noncollagenous proteins are still not fully understood, it has been shown that they alter the crystallization dynamics of calcium ions and, thus, are used to control bone mineralization (Neuman et al. 1982; Roach 1994). More recently, it has been suggested that osteocytes could utilize these proteins to alter the mineralized matrix in the lacuna-canalicular system (LCS) (Teti and Zallone 2009).

14.1.2 Lacuna-canalicular system and osteocytes

Bone tissue contains a 3-D network of interconnected pores called the lacunae, which are oval in shape and have a cross-sectional area of 25–60 μm^2 depending on their age and location (Cane et al. 1982). The structure connecting the lacunae are the canaliculi. They are tunnels with a cross-sectional diameter of approximately 200 nm (You et al. 2004). Figure 14.1 shows the LCS of a cross section of human cortical bone.

Osteocytes are the cells that populate the LCS. Osteocytes are in contact with each other, as well as cells on the periosteal and endosteal surface of the bone through their extensive cellular processes, which are 14 to 100 nm in diameter (You et al. 2004; Weinger and Holtrop 1974; King and Holtrop 1975). The membranes of cells have been shown to be connected by gap junctions (Doty 1981). This interconnected network of osteocytes is ideally suited to sense any mechanical loads and damage within the bone.

14.1.3 The pericellular matrix

Osteocytes, residing in the lacuna-canalicular space, are not in direct contact with the mineralized bone matrix at the majority of their membrane (Fig. 14.1). There is a layer

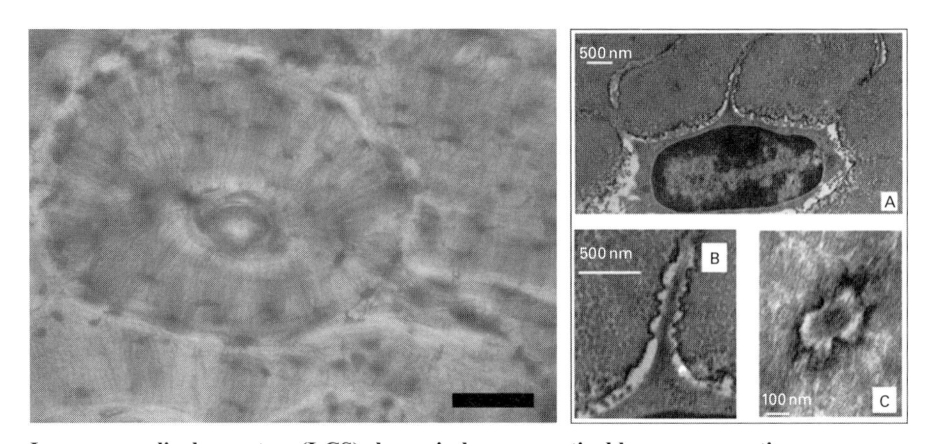

Figure 14.1 **Lacuna-canalicular system (LCS) shown in human cortical bone cross section.** (1) Cross section of human cortical bone showing the lacuna-canalicular network, which is occupied by osteocytes. The osteocytes form an interconnected network through their processes. Scale bar = 50 μm. (2) TEM micro-photograph of an osteocyte in lacuna (A), longitudinal (B) and cross-section (C) of cell process showing protrusion of mineralized bone matrix in contact with osteocyte process. (2) From *The Anatomical Record*.

Reproduced with permission from John Wiley & Sons (McNamara et al. 2009).

of pericellular matrix separating the osteocyte and the bone matrix (Sauren et al. 1992), which has been measured to be 30–50 nm (You et al. 2001). The pericellular matrix consists primarily of albumin (Owen and Triffitt 1976) and proteoglycans (Sauren et al. 1992). The structure of these proteoglycans tethers the osteocyte membrane to the mineralized bone matrix. This matrix is also supported by transverse fibrils (Shapiro et al. 1995) that provide additional drag force to the cells in the presence of fluid flow, amplifying cellular strain up to 100 fold compared to tissue level strain (You et al. 2001).

There are also areas of the osteocyte cell membrane that are in direct contact with the mineralized bone matrix in the canaliculi. Conical mineralized protrusions have been shown to be in direct contact with the cell membrane (Fig. 14.1), and these attachment sites are integrin $\alpha v\beta 3$ positive (McNamara et al. 2009).

14.1.4 Mimicking cell distribution and pericellular environment in vitro

Recently, many groups have been trying to better mimic the spatial and geometric organization of the bone cell network for in vitro experiment through the use of micro-contact printing (Fig. 14.2A) (Singhvi et al. 1994; You et al. 2008a; Guo et al. 2006; Lu et al. 2012). This seeding method also promotes the formation of unified cellular connections and can better control the amount of cell spreading. Additionally, it has been observed that bone cells are more responsive to fluid shear stress when in these networks as compared to the more standard random distribution of cells. This is believed to be caused by the increased cell spreading that results in an increased cellular tension, affecting the degree of biochemical signaling (You et al. 2008a).

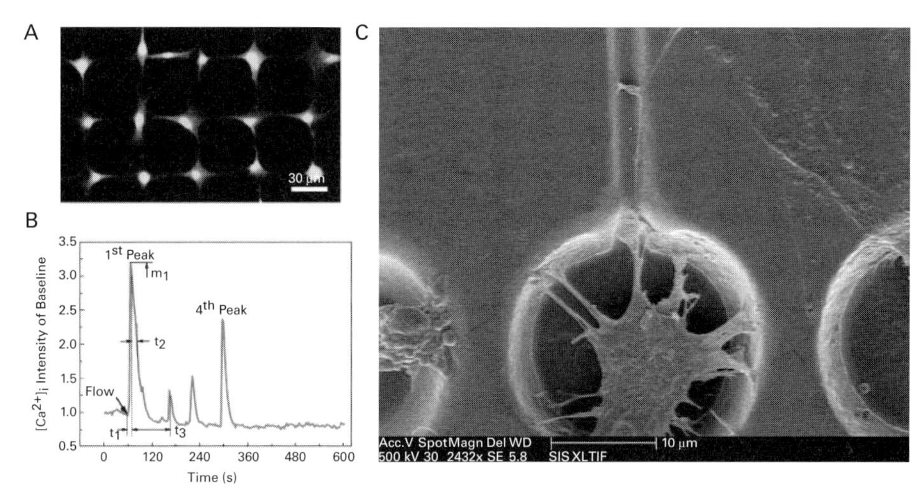

Figure 14.2 **Micro-scale experiments using bone cells to improve physiological relevance.** (A) Image of MLO-Y4 (osteocyte-like) cells grown as a microcontact printed network. (B) Typical $[Ca^{2+}]_i$ response of cells to fluid shear stress. Flow is applied after 1 minute of baseline measurement, then the time to first peak, t_1, height of first peak, m_1, time to midline, t_2, and time to second peak, t_3, are all quantified. Reproduced with permission from John Wiley & Sons (Lu et al. 2012). (C) SEM image of MLO-Y4 cells seeded within in vitro LCS. Cell protrusions can clearly be seen interacting with the PDMS wall. Reproduced with permission from Springer (You et al. 2008c).

To further model the LCS in vitro, You et al. (2008c) have developed a very promising microfluidic device. In this system, osteocyte-like MLO-Y4 cells were individually micropipetted within a lacunar well (D = 20 µm, H = 10 µm) that has two adjacent "canaliculi" channels (w = 1 µm). Due to the limitations of the fabrication method the "canalicular" size is limited to 1 µm, an order of magnitude larger than the in vivo canalicular dimensions. However, with the emergence of nanomilling technologies it could be possible to prepare a device with even more physiological dimensions. Despite these large dimensions, cellular protrusions were still able to interact with the PDMS wall (Fig. 14.2C), reminiscent to what is observed in vivo. This device provides a unique opportunity to study osteocyte mechanotransduction in a device that is morphologically similar to the LCS; however, due to the small cell numbers, this system is mainly viable for single cell PCR or intracellular calcium experiments. This system is also limited by the intensive micropipetting technique that is used to seed the device. This system has also yet to be used experimentally to apply mechanical stimulation to cells and still requires validation for mechanotransduction studies.

14.2 Physiological level of loading and its effect on bone cells

It is the unique structure of the bone matrix which gives rise to the unique forces experienced by bone cells. Macroscopically, the forces on bone could be characterized as compressive, tensile, or torsional. At the cellular level, the forces translate into matrix stretch, fluid pressure, and fluid flow shear (Fig. 14.3). Additionally the frequency of these forces could have an impact on cellular response. Since osteocytes are the cell type that resides within the bone, they will be the focus of this section.

Currently, fluid flow shear stress seems to be the dominant force that osteocytes sense and respond to, compared to pressure (Klein-Nulend et al. 1995) and substrate stretch (Mullender et al. 2004). Numerous studies have investigated the mechanism of detection and response to fluid flow shear in osteocytes (Weinbaum et al. 1994; You et al. 2001; Price et al. 2011; Wang et al. 2007; Malone et al. 2007). However, studies on the mechanism of detection is still lacking for other modes of mechanical loading.

14.2.1 Deformation of the mineralized matrix

The level of strain in high impact loading is uniform across various species such as dogs, horses, buffalos, and elephants (Rubin and Lanyon 1982). Due to the stiffness of the matrix, the levels of deformation produced by physiological level of loading are in the order of 0.04–0.3% (Rubin and Lanyon 1982), and rarely above 0.1% (Fritton et al. 2000). In isolated osteoblasts, prostaglandin E2 (PGE2) expression is not changed below 0.7% strain (Murray and Rushton 1990). At 0.5% strain, intracellular calcium and osteopontin mRNA were unchanged in osteoblasts (hFOB 1.19) in vitro (You et al. 2000). Based on these results, it is unlikely that the physiological level of strain in bone is the dominant type of loading to bone cells. However, it has been proposed that, mineralized protrusions within the lacuna could amplify the strain in an area of 5–10 μm^2 (Nicolella et al. 2004).

14.2.2 Vibration

Though strains at the same frequency of physiological loading (0–5 Hz) do not seem to induce responses from osteocytes, higher frequencies have shown to induce response from osteocytes. Vibration at 60 Hz with <1g of acceleration has reduced receptor activator of nuclear κ-β ligand (RANKL) expression and PGE2 release in an osteocyte cell line in vitro (Lau et al. 2010).

14.2.3 Pressure force

Since fluid is essentially incompressible, a significant pressure force is present around osteocytes in loaded bone. Analytical modeling has estimated the pressure force is 40 times that of the blood pressure difference (Zhang et al. 1998). Oscillatory loading of bone with 0–14 MPa at 1 Hz was calculated to induce 0.27 MPa fluid pressure at the lacunar-canalicular porosity (Zhang et al. 1998). More recent studies found that the hydraulic permeability of the bone tissue was smaller than the previous model assumed, leading to an even higher estimation of the hydraulic pressure build-up (~5 MPa) around osteocytes (Cowin et al. 2009; Gailani et al. 2009; Gardinier et al. 2010). Also, significant amplification of this pressure occurs at a higher frequency (order of tens of hertz) (Wang et al. 1999).

The mechanism through which osteocytes sense pressure loading is unclear. However, the cytoskeleton has been implicated as a possible structure for this function as it has been shown that osteocyte microtubule structure is affected by cyclic pressure with peaks of 68 kPa (Liu et al. 2010). Constant hydraulic pressure of 0.8 MPa with rest insertion (1 min loading, 14 min rest, for 4–12 hrs) has induced increase in mRNA levels for matrix metalloproteinase-1 and -3 (MMP-1 and MMP-3) in an osteoblastic cell line (Tasevski et al. 2005), which would hint at extracellular matrix turnover.

14.2.4 Fluid flow shear and drag force

Mechanical loading at the organ level creates tissue level pressure gradients in the bone matrix. This pressure gradient will drive the flow of interstitial fluid, creating shear stress on the surface of the osteocyte (Weinbaum et al. 1994) and drag force in the pericellular matrix (You et al. 2001). The movement of fluid in the lacuna-canalicular space have been characterized by tracer studies (Knothe Tate and Knothe 2000; Knothe Tate et al. 2000; Mak et al. 2000; Tami et al. 2003; Wang et al. 2005), and a fluorescent recovery after photo bleaching study (Price et al. 2011). From avian studies, the shear stress experienced by osteocytes could be up to 2 Pa (Mi et al. 2005a; Mi et al. 2005b). Recent studies performed on mouse bone estimates the shear stress to be up to 5 Pa (Price et al. 2011).

How osteocytes translate shear stress to biochemical signals is still not clear. One theory is that the fluid drag force could cause conformational changes in proteins on the cell membrane; also the cytoskeleton could transduce this force further to affect proteins in the cell cytoplasm (You et al. 2001). The force experienced by osteocyte would be amplified at regions with adhesion points that connect the cell membrane directly to the mineralized matrix (Wang et al. 2007). Osteocyte processes would be attached to these protrusions by focal adhesion proteins, specifically $\alpha_v\beta_3$ integrins (McNamara et al. 2009). Stretch activated ion channels could be activated through the force amplification mechanism mentioned above to release PGE2 and nitric oxide (NO) (Rawlinson et al. 1996).

The cytoskeleton has been shown to play a role in prostaglandin release in chicken osteocytes (Ajubi et al. 1996) and NO release (McGarry et al. 2005) in osteoblast and osteocyte cell lines subjected to pulsatile fluid flow in vitro. Actin filaments have been shown to form stress fibers in osteoblast and osteocyte cell lines subjected to unidirectional flow in vitro (Ponik et al. 2007).

Another theory of how osteocytes sense fluid flow is through primary cilia, as they are deformed by physiological level of fluid flow in osteoblast (MC3T3-E1 cell line) and induce intracellular calcium fluctuations (Malone et al. 2007).

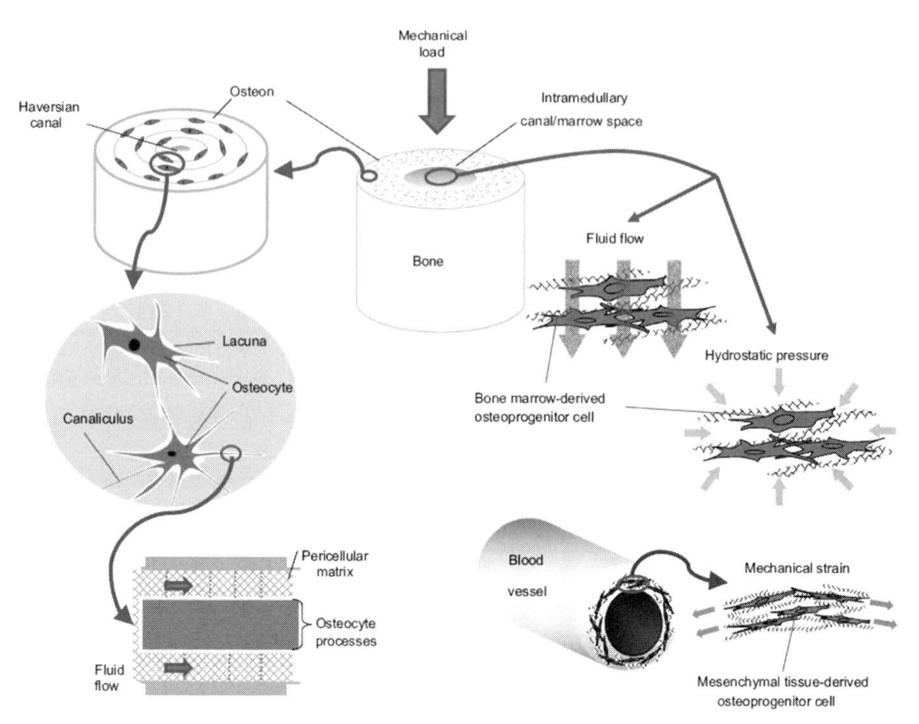

Figure 14.3 **Mechanical loading on the bone results in mechanical stimuli to cells that reside in the bone.**
Reproduced with permission from Elsevier BV (Chen et al. 2010).

14.2.5 In vitro cell stimulation methods

Current studies to investigate how fluid shear stress affects the signaling of bone cells typically occur within parallel plate flow chambers (Klein-Nulend et al. 1995). These studies apply stimulation to the cells of interest in the form of fluid shear stress, and the flow that is typically applied is steady, pulsatile, or oscillatory in nature (Jacobs et al. 1998). Although oscillatory fluid flow is the most physiologically relevant, the sinusoidal method through which it is typically applied is not. Additionally, these systems are limited in terms of the information obtainable from them. The large volume-to-cell ratio heavily dilutes the strength of the signal that the cell produces, which can severely affect cellular response when investigating paracrine and autocrine signaling. Additionally, low half-life signals, such as NO, are likely to degrade to an insignificant amount before they can interact with nearby cells or their release can be analysed by most cell signal protocols. These systems also require large volumes of reagents and cell numbers, preventing the use of primary osteocytes. Many of these issues can be solved through the use microfluidic flow systems. Kou et al. (2011) have developed and used a multi-shear microfluidic device to study the calcium response of osteoblasts to fluid shear stress. The device was fabricated using soft-lithography out of PDMS. To induce a different shear stress in each cell channel, different channel widths of 400 μm, 800 μm,

1600 μm, and 3200 μm were used. Based on computational models of their system, they could simultaneously apply fluid shear stress to cells ranging from 0.03 Pa to 0.3 Pa, providing a method to investigate how cells respond to different shear stresses in tandem.

Recently, interest in understanding how bone cells respond to hydrostatic pressure has increased (Gardinier et al. 2009b; Liu et al. 2010). Liu et al. (2010) applied cyclic hydraulic pressure (CHP) within a pressure chamber with a peak pressure of 68 kPa at 0.5 Hz and with a triangular waveform. The osteocytes within this experiment showed a substantial increase in the RANKL to osteoprotegerin (OPG) ratio after 2 hours, an increase in cyclooxygenase-2 (COX-2) mRNA expression after 1 hour, and showed the release of intracellular calcium. These results suggest that CHP may play a critical role in the bone remodeling process, and warrants further investigation.

Another unique stimulation technique used with bone cells is through nanoindentation, which was used to better understand how bone cells signal amongst each other when undergoing mechanical stimulation, specifically the role that both the adenosine triphosphate (ATP) pathway and gap junctions play in propagating signals (Huo et al. 2010). As has been previously discussed, MC3T3-E1 pre-osteoblastic cells were patterned through microcontact printing on a glass slide. The cells were loaded with Fluo-4AM as an intercellular calcium indicator and imaged on a fluorescent microscope. One cell in the center of the network was indented with an atomic force microscopy (AFM) probe at a force of 60 nN, which was determined to be the critical force to induce mechanical stimulation of the cells without causing them to burst. Three groups were studied, an untreated control, gap junction blocked, and ATP blocked. There was no significant difference in the calcium propagation between the untreated group and gap junction group, but a significant decrease was observed in the ATP blocked group, suggesting that the ATP pathway plays the dominant role in mediating intercellular calcium propagation. Additionally, the use of AFM for single cell indentation provides an improvement over previous methods, which applied the force using a micropipette as it provided a controllable force to the cells. However, cellular response is highly dependent of the force applied, and it has yet to be determined what physiologically relevant force is necessary.

Song et al. (2010) have also developed an interesting system for mechanically stimulating MC3T3-E1 cells. This system used alternating pulsed magnetic fields to repetitively impact the cells with beads. The researchers used magnetic particles of 4.5 μm, 7.6 μm, and 8.4 μm, and these beads were pulsed to impact with the cells at frequencies of 60 Hz, 1 kHz, and 1 MHz. Loading with this system was applied at different phases of the cell life cycle, specifically G1, S, and G2, and the cells were investigated for their growth rate and viability. This loading system, however, does not provide a physiologically relevant loading scheme. However, it is meant as a culturing system to promote increased cell viability and population growth rates in hopes of applying it to increase the number of primitive stem cells for graft development.

14.3 Mechanobiology of bone at the micro- and nano-scale

Bone is an organ that is traditionally thought to have two functions: structural support of the body and provide a store of calcium and phosphates. More recently, evidence has shown that bone could also act as an endocrine organ by secreting systemic signals that regulate mineral homeostasis (Fukumoto and Martin 2009).

14.3.1 Random remodeling and bone homeostasis

Random remodeling has the effect of modulating calcium and phosphate homeostasis by liberating or storing of dissolved ions into or from the bone matrix. Random remodeling is thought to be hormonally regulated, as reviewed by Lawrence Raisz (Raisz 1999).

14.3.2 Targeted remodeling from microdamage

In contrast to random remodeling, targeted remodeling also occurs in bone to repair damaged tissue which in part leads to the adaptation to changes in mechanical loads. This area is reviewed by David Burr (Burr 2002).

Bone contains many micrometer cracks or microdamage under normal loading conditions (Burr et al. 1997). The repair of these microdamage sites is done by targeted remodeling in which bone is being resorbed and new bone is laid down around the damage site.

Computational modeling has shown that microdamage could change the mechanical property around the lacuna; and the resulting change in strain level could be sensed by osteocytes (Prendergast and Huiskes 1996). Also, it has been shown that osteocytes undergo apoptosis near the damage sites (Verborgt et al. 2000). It is hypothesized that, at the same time, they would release signals to initiate and attract the basic multicellular unit (BMU) to remodel the damaged area (Martin 2007). In vivo, the expressions of remodeling associated genes in tissue with microdamage have a distinct temporal profile (Kidd et al. 2010).

This repair process has micrometer accuracy (Kidd et al. 2010), where the damage area is remodeled by a BMU. But the role of each cell type and/or local populations still needs to be investigated. In a rat model, osteocytes have been observed to release RANKL (pro-osteoclastogenic) and VEGF (pro-angiogenic) factors in response to microdamage induced by fatigue loading (Kennedy et al. 2012). From the same study, the population of osteocytes that releases these signals has shown to be localized, within 300 μm of the damage site.

14.3.3 Basic multicellular unit

During the remodeling process a collection of cells maintain a structure that destroys old bone matrix then lay down new matrix (Fig. 14.4). Together with the blood vessels that supply nutrients and progenitor cells, it is called the basic multicellular unit (BMU)

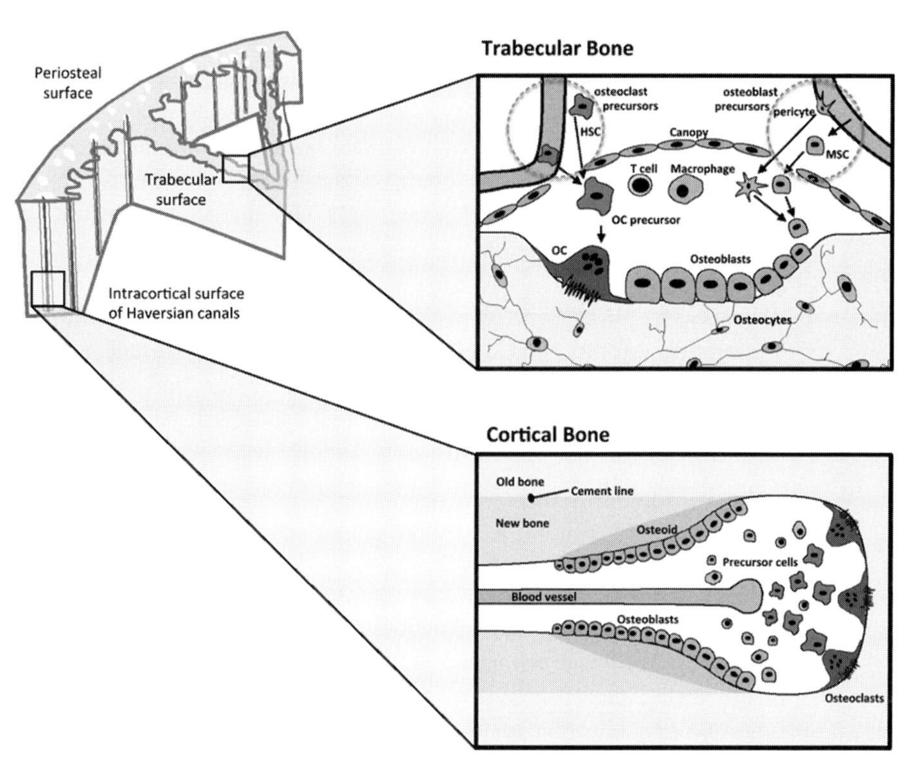

Figure 14.4 **Basic multicellular unit (BMU) in trabecular and cortical bone.**
Multiple types of cells are involved in bone remodeling in a structure that is micrometer in length scale. Osteoclasts (OC) are differentiated from circulating hematopoietic precursor cells (HSC). Osteoblast precursors originates from mesenchymal stem cells (MSCs).
Reproduced with permission from Macmillan Publishers Ltd. (Sims and Martin L 2014).

(Frost 1990). The BMU typically has a cross-sectional diameter of 200 μm in cortical bone (Taylor et al. 2007). It is within this length scale that multiple cell types: bone resorbing, forming, endothelial, and progenitor cells maintain this structure for up to weeks through various signaling events.

14.3.3.1 Resorption

Bone remodeling starts with the breakdown of the mineralized bone matrix by osteoclasts (Boyle et al. 2003). They are multinucleated cells differentiated from circulating precursor cells of a hematopoietic lineage. Mature osteoclasts form a sealed space that has a low pH and various enzymes to dissolve the minerals and digest the organic components of bone respectively. The differentiation of osteoclasts from precursor cells is supported by signaling molecules secreted by osteoblasts (Lacey et al. 1998) and osteocytes (Zhao et al. 2002), both are cells in direct contact with the mineralized matrix.

Before the osteoclast precursors fuse and differentiate they direct the osteoblasts to migrate away, exposing the matrix beneath (Perez-Amodio et al. 2004). The signaling events involved in this process are still unknown.

14.3.3.2 Formation

As the osteoclasts penetrate further into the bone, they are followed by osteoblasts, which deposit an organic matrix. Some osteoblasts are embedded into this matrix and further differentiate into osteocytes as the organic matrix mineralizes. In the case of cancellous bone, a layer of osteoblastic cells has shown to form a canopy over the BMU during remodeling (Hauge et al. 2001).

14.3.3.3 Angiogenesis

Blood vessel formation, or angiogenesis, is also critical in the formation and propagation of the BMU. New blood vessels bring osteoclast and osteoblast precursor cells to the remodeling site. Vascular endothelial growth factor (VEGF) is secreted by osteocytes and would induce angiogenesis. The expression of VEGF in osteocytes could be altered with mechanical loading as well (Cheung et al. 2011).

14.3.4 Intracellular signaling in mechanobiology

Immediately after mechanical loading, the mechanical force would trigger changes in structures within the cell, which in turn would be transduced into biochemical signals. In osteocytes, the purinergic pathway, calcium ions and related channels, and the cytoskeleton are the main players in this process.

14.3.4.1 Purinergic pathway

ATP has been shown to be released through hemichannels, which is upregulated under fluid shear stress in osteocytes (Genetos et al. 2007). ATP has been shown to be required in inducing PGE2 release in the osteoblast cell line through purinergic (P2) receptors (Genetos et al. 2005); the release of ATP peaks within one minute of fluid shear application and returns to baseline after five minutes.

In the MC3T3-E1 osteoblast cell line, ATP release increased under unidirectional fluid shear and cyclic hydraulic pressure loading by sixfold and fourfold, respectively (Gardinier et al. 2009a).

14.3.4.2 Calcium ion

Intracellular calcium ion concentration, $[Ca^{2+}]$, is an early messenger of signal transduction in bone cells, especially stimulated by mechanical loading (Hung et al. 1995).

Isolated rat calvaria osteoblasts showed up to a fourfold increase in $[Ca^{2+}]$ when exposed to fluid shear of up to 7 Pa (Hung et al. 1995). Under steady fluid flow (1.2 Pa for 2.5 min), chicken calvaria primary osteocytes and osteoblast both showed elevated $[Ca^{2+}]$; and osteoblast response is focal adhesion dependent, while osteocyte response is not (Kamioka et al. 2006).

From in vitro studies on the osteoblast cell line, $[Ca^{2+}]$ fluctuations after nanoindentation have also been shown to propagate form one cell to the surrounding cells connected by gap junctions (Guo et al. 2006) as has previously been discussed.

Physiological levels of substrate strain did not induce changes in intracellular calcium concentration (You et al. 2000) from in vitro studies of human osteoblast cells, rat osteoblasts, and the mouse osteocyte cell line (MLO-Y4).

14.3.4.3 Nitric oxide

NO is also a fast acting signal that is important in inflammatory and mechanical loading induced bone turnover (Van'T Hof and Ralston 2001). NO is released by osteocytes subjected to fluid shear stress (Kleinnulend et al. 1995; Johnson et al. 1996). NO has been shown to be involved in mechanical loading induced bone formation in a rat model (Fox et al. 1996).

14.3.4.4 Cytoskeleton

In osteocyte-like (MLO-Y4) cells, 24 hours of steady flow was needed to elicit stress fiber formation; after 24 hours of oscillatory fluid flow the number of dendritic processes increased (Ponik et al. 2007). Under unidirectional fluid shear stress osteocyte-like (MLO-Y4) cells showed increased dendricity and elongation of dendrites that depends on the E11 protein (Zhang et al. 2006).

The effect of pressure on osteoblast-like cells (MG-63) has been studied by Tasevski et al. (Tasevski et al. 2005). The hydrostatic pressure was varied between 0 and 0.8 MPa. The pressure was applied for 1 minute on, then 14 minutes off, for durations from 4 to 12 hours. The mRNA levels for MMP-1 and MMP-3 were significantly increased ($p < 0.001$) in cells exposed to cyclic pressure under serum-free conditions for 4–12 hours, and the mRNA levels for MMP-3, but not MMP-1, were significantly enhanced in cells subjected to static pressure. The changes in MMP-1 and MMP-3 may indicate extra cellular matrix turnover, and remodeling of bone.

The different responses to cyclic and static loading suggest that cells have different mechanosensing pathways for frequency of loading, some of which are responsible for detecting oscillatory patterns in mechanical loading. Also, the different responses may lead to bone remodeling that results in different bone structure and composition.

Osteocyte-like (MLO-Y4) cells showed increased force traction on beads attached to the cell body, up to 30 pN. The force was enough for the activation of integrins, suggesting a mechanical feedback loop. AFM experiments have shown increased elastic modulus in osteocytes after mechanical loading (Zhang et al. 2008). This stiffening response was related to changes in material properties of the cell, suggesting that the cells actively change their cytoskeleton in response to a mechanical load.

14.3.5 Cellular signaling in bone cell mechanobiology

To coordinate the bone remodeling process, bone cells release a variety of signals to each other and circulating cells. Released small molecules, or cytokines, play key roles in BMU formation and maintanence.

Signaling through secondary messengers, such as calcium ions, ATP, and nitric oxide, is also likely as gap junctions have been shown to exist between osteoblasts and osteocytes (Doty 1981). These gap junctions could allow transport of these messengers, and signaling through electrical gradient and receptor activation.

Mechanical coupling could also be an avenue of signaling between cells. It has been shown that osteoblasts and osteocytes are connected by tight junctions (Weinger and Holtrop 1974).

14.3.5.1 RANKL and OPG

At the early stage of BMU formation, osteoclast differentiation is initiated and maintained by RANKL and macrophage-colony stimulating factor (M-CSF). RANKL is secreted by cells of the osteoblast lineage (Udagawa et al. 1999). In cancellous bone, osteocytes have been suggested to be the main source of RANKL (Xiong and O'Brien 2012). Also produced in osteoblast lineage cells, OPG is a decoy receptor that binds to RANKL, thus inhibiting its effect (Hofbauer et al. 2000). Changes in the RANKL/OPG ratio are indicative of changes in osteoclastogenesis. It is also proposed that osteocytes generate a gradient of RANKL and OPG, which direct the spatial organization of the BMU (Ryser et al. 2009). This model still needs to be validated with experiments. Micro- and nanotechniques, which are introduced in this chapter, will enable the experimental validation of this theory.

RANKL/OPG ratio is decreased with physiological level of cyclic hydraulic pressure (Liu et al. 2010) and oscillatory fluid shear stress (You et al. 2008b) in MLO-Y4 osteocyte cell line.

14.3.5.2 Prostaglandin E2

PGE2 is one of the small molecules that have been implicated to have auto- and/or paracrine effect during bone remodeling. It has a stimulatory effect on both bone resorption and formation (Raisz et al. 1993; Bergmann and Schoutens 1995).

From in vitro studies using cell lines, PGE2 release has been shown to be regulated by oscillatory fluid shear loading in MLO-Y4 osteocytes (Genetos et al. 2007) and MC3T3-E1 osteoblasts (Malone et al. 2007). Extended (24 hr) fluid shear loading showed similar results (Ponik et al. 2007)

Under pulsatile fluid shear of up to 2.4 Pa, MLO-Y4 osteocytes and 2T3 osteoblasts increased PGE2 concentration in media by up to 1000 pg/ml and 200 pg/ml, respectively, 120 min after loading (Kamel et al. 2010).

COX-2, an enzyme for the production of PGE2, is upregulated under unidirectional fluid shear and cyclic hydraulic pressure loading by six- and threefolds, respectively, in MC3T3-E1 osteoblast cell line (Gardinier et al. 2009a).

14.3.5.3 Sclerostin and Wnt

Sclerostin, produced in osteocytes, is a negative regulator of bone formation (Van Bezooijen et al. 2002). Expression of sclerostin occurs mostly in osteocytes; and its level is dependent on changes of mechanical loading in vivo (Robling et al. 2006).

Activation of the Wnt signaling pathway is one of the known effects of auto- or paracrine signaling of PGE2. It leads to translocation of β-catenin to the nucleus, down-regulation of Sclerostin and the release of Wnt, which acts in an auto- and/or paracrine manner to further amplify the signal. Bonewald and Johnson (2008) have reviewed this pathway in osteocytes,.

The Wnt pathway is regulated by mechanical loading. Pulsatile fluid flow shear stress induced β-catenin nuclear translocation in MLO-Y4 osteocytes and 2T3 osteoblasts by similar amounts (Kamel et al. 2010).

14.3.6 Future directions

Macroscopically, increased mechanical loading leads to increases in bone strength, and vice versa for reduced mechanical loading. The signaling between osteocytes, osteoblasts, and osteoclasts is complex and multidirectional. A recent review by Matsuo and Irie (2008) has summarized the effect of signaling between osteoclasts and osteoblasts. Osteocyte signals to other cell types have not been as extensively studied. Much work still needs to be done in this area.

Also new signals are being discovered. One of the molecules that warrant further study is the sphingosine 1-phosphate (S1P). Mature osteoclasts express elevated sphingosine kinase 1 (SPHK1), which is involved in the production of S1P. S1P links osteoclast and osteoblast activity by recruiting osteoblast precursors and promoting the survival of mature osteoblasts (Pederson et al. 2008). However, the role of S1P under mechanical loading is unclear.

To better model how the different bone cells communicate with one another as it pertains to bone remodeling, Middleton and You have developed a microfluidic system (Fig. 14.5) that consists of three cell culture channels for cellular co-culture (in a paper titled "A microfluidic system to study the effects of mechanically loaded osteocytes on osteoclast precursor recruitment and formation," presented at the *American Society of Bone and Mineral Research* annual meeting, Houston, September 12–15, 2014). The cell channels are separated by high resistance side channels to prevent convective fluid flow while still allowing for solute transfer. This system is currently being used to study how osteocytes communicate with osteoclast precursors to promote differentiation and recruitment. The two outer channels are seeded with MLO-Y4 cells while the central channel is seeded with RAW264.7 osteoclast precursor cells. One channel of osteocytes is mechanically stimulated once per day for one hour with a steady shear stress of 1 Pa. For the remainder of the time, all cells receive a nonstimulatory perfusion flow of 1 μl/min to replace media. This produces a dynamic gradient of stimulated and unstimulated signals across the central channel. Preliminary results from this work suggest that osteoclast precursors migrate and differentiate preferentially towards the unstimulated osteocytes.

14.4 Summary of the advantages of micro- and nanotechniques for bone mechanotransduction studies

The advancement of microfluidics provides a multitude of advantages over typical platforms currently used for bone mechanotransduction studies. These systems allow

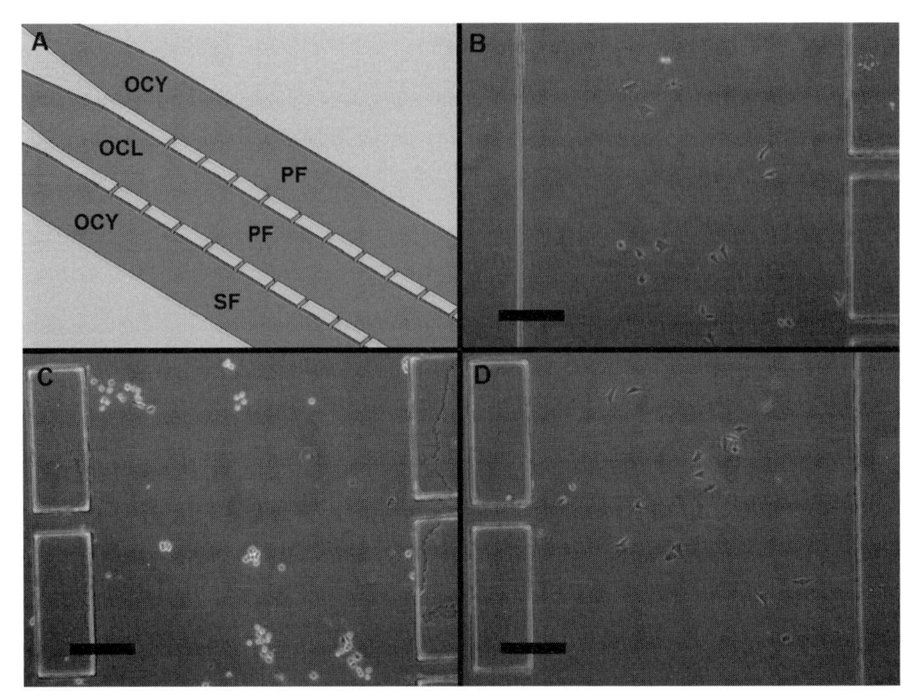

Figure 14.5 **Microfluidic system to monitor bone cells.**
(A) Schematic of microfluidic system to study (C) osteoclast precursor (OCL) response to a dynamically developed signal gradient from (B) osteocytes (OCY) undergoing a stimulatory steady flow (SF) and (D) unstimulated osteocytes receiving a weak perfusion flow (PF) to replace media. Scale bars are 200 μm.

for a significantly increased throughput, which will prove invaluable not only for drug testing, but also for more rapidly investigating the multitude of complex signaling pathways and interactions involved (Feng et al. 2009). As well, because of the small dimensions involved, smaller volumes of expensive reagents can be used, substantially reducing experimental costs.

More specifically for bone research, osteocytes reside within a microenvironment in the order of tens of microns. More refined is the canaliculus of bone, which has a diameter in the order of a few hundred nanometres. Through the use of micro- and nanotechnologies, it is feasible that novel in vitro platforms can create a better replication of the physical in which the cells reside, and could allow us to better understand how the specific physical characteristics of this environment plays a role in bone cell mechanotransduction (You et al. 2008c).

This spatial arrangement also promotes the co-culture of these different cells, as different cell populations are able to be seeded within close range of each other. This allows for an improvement on current methods for cell-cell signaling, especially for bone cells. Bone cell signaling occurs over relatively short length scales, typically in the order of a few hundred microns. By compartmentalizing different cell types at these short distances from each other, it is possible to observe cellular response to dynamic mechanically transduced signals. Additionally, because of the reduced distances, it is possible to investigate the effects and interactions of low half-life signals, such as NO.

Another critical advantage of microfluidics is the ability to study primary osteocytes; since they reside deep within the bone matrix, it makes these cells very difficult to investigate in vivo, as well as being difficult and time consuming to isolate for in vitro studies (Rath Stern et al. 2012; Rath Stern, Stern et al. 2012). Additionally, as osteocytes are terminally differentiated, they are unable to proliferate, limiting the effective number of cells for study. The MLO-Y4 cell line was developed to mitigate some of these issues (Kato et al. 1997); however, they lack the ability to produce significant levels of important signals characteristic of osteocytes, such as Sclerostin (Rath Stern et al. 2012). Microfluidics, however, requires very low cell numbers due to the small dimensions used. Because of this, significant mechanotransduction studies can be performed within microfluidic systems using primary osteocytes.

References

Ajubi, N. E., Klein-Nulend, J., Nijweide, P. J., Vrijheid-Lammers, T., Alblas, M. J. and Burger, E. H. 1996. "Pulsating fluid flow increases prostaglandin production by cultured chicken osteocytes–a cytoskeleton-dependent process." *Biochemical and Biophysical Research Communications* **225**: 62–68.

Bergmann, P. and Schoutens, A. 1995. "Prostaglandins and bone." *Bone* **16**: 485–488.

Bonewald, L. F. and Johnson, M. L. 2008. "Osteocytes, mechanosensing and Wnt signaling." *Bone* **42**: 606–15.

Boyle, W. J., Simonet, W. S. and Lacey, D. L. 2003. "Osteoclast differentiation and activation." *Nature* **423**: 337–342.

Burr, D. 2002. "Targeted and nontargeted remodeling." *Bone* **30**: 2–4.

Burr, D. B., Forwood, M. R., Fyhrie, D. P., Martin, R. B., Schaffler, M. B. and Turner, C. H. 1997. "Bone microdamage and skeletal fragility in osteoporotic and stress fractures." *Journal of Bone and Mineral Research* **12**: 6–15.

Cane, V., Marotti, G., Volpi, G., Zaffe, D., Palazzini, S., Remaggi, F. and Muglia, M. A. 1982. "Size and density of osteocyte lacunae in different regions of long bones." *Calcified Tissue International* **34**: 558–563.

Chen, J.-H., Liu, C., You, L. and Simmons, C. A. 2010. "Boning up on Wolff's Law: mechanical regulation of the cells that make and maintain bone." *Journal of Biomechanics* **43**: 108–118.

Cheung, W. Y., Liu, C., Tonelli-Zasarsky, R. M. L., Simmons, C. A. and You, L. 2011. "Osteocyte apoptosis is mechanically regulated and induces angiogenesis in vitro." *Journal of Orthopaedic Research* **29**: 523–530.

Cowin, S. C., Gailani, G. and Benalla, M. 2009. "Hierarchical poroelasticity: movement of interstitial fluid between porosity levels in bones." *Philosophical Transactions of the Royal Society A: Mathematical, Physical and Engineering Sciences* **367**: 3401–3444.

Doty, S. B. 1981. "Morphological evidence of gap junctions between bone cells." *Calcified Tissue International* **33**: 509–512.

Feng, X., Du, W., Luo, Q. and Liu, B. F. 2009. "Microfluidic chip: next-generation platform for systems biology." *Anal Chim Acta* **650**: 83–97.

Fox, S., Chambers, T. and Chow, J. 1996. "Nitric oxide is an early mediator of the increase in bone formation by mechanical stimulation." *American Journal of Physiology-Endocrinology and Metabolism* **33**: E955.

Fratzl, P., Gupta, H. S., Paschalis, E. P. and Roschger, P. 2004. "Structure and mechanical quality of the collagen-mineral nano-composite in bone." *Journal of Materials Chemistry* **14**: 2115–2123.

Fritton, S. P., McLeod, K. J. and Rubin, C. T. 2000. "Quantifying the strain history of bone: spatial uniformity and self-similarity of low-magnitude strains." *Journal of Biomechanics* **33**: 317–325.

Frost, H. M. 1990. "Skeletal structural adaptations to mechanical usage (SATMU): 1. Redefining Wolff's law: the bone modeling problem." *Anatomical Record* **226**: 403–413.

Fukumoto, S. and Martin, T. J. 2009. "Bone as an endocrine organ." *Trends in Endocrinology and Metabolism* **20**: 230–236.

Gailani, G. B., Benalla, M., Mahamud, R., Cowin, S. C. and Cardoso, L. L. 2009. "Experimental protocol for the measurement of the permeability of a single osteon." *J Biomech Eng* **131**: 101007.

Gardinier, J. D., Majumdar, S., Duncan, R. L. and Wang, L. 2009a. "Cyclic hydraulic pressure and fluid flow differentially modulate cytoskeleton re-organization in MC3T3 osteoblasts." *Cellular and Molecular Bioengineering* **2**: 133–143.

Gardinier, J. D., Majumdar, S., Duncan, R. L. and Wang, L. 2009b. "Cyclic hydraulic pressure and fluid flow differentially modulate cytoskeleton re-organization in MC3T3 osteoblasts." *Cell Mol Bioeng* **2**: 133–143.

Gardinier, J. D., Townend, C. W., Jen, K.-P., Wu, Q., Duncan, R. L. and Wang, L. 2010. "In situ permeability measurement of the mammalian lacunar-canalicular system." *Bone* **46**: 1075–1081.

Genetos, D. C., Geist, D. J., Liu, D., Donahue, H. J. and Duncan, R. L. 2005. "Fluid shear-Induced ATP secretion mediates prostaglandin release in MC3T3-E1 osteoblasts." *Journal of Bone and Mineral Research* **20**.

Genetos, D. C., Kephart, C. J., Zhang, Y., Yellowley, C. E. and Donahue, H. J. 2007. "Oscillating fluid flow activation of gap junction hemichannels induces ATP release from MLO-Y4 osteocytes." *Journal of Cellular Physiology* **212**: 207–214.

Guo, X. E., Takai, E., Jiang, X., Xu, Q., Whitesides, G. M., Yardley, J. T., Hung, C. T., et al. 2006. "Intracellular calcium waves in bone cell networks under single cell nanoindentation." *MCB Molecular and Cellular Biomechanics* **3**: 95–107.

Hauge, E. M., Qvesel, D., Eriksen, E. F., Mosekilde, L. and Melsen, F. 2001. "Cancellous bone remodeling occurs in specialized compartments lined by cells expressing osteoblastic markers." *Journal of Bone and Mineral Research* **16**: 1575–1582.

Hofbauer, L. C., Khosla, S., Dunstan, C. R., Lacey, D. L., Boyle, W. J. and Riggs, B. L. 2000. "The roles of osteoprotegerin and osteoprotegerin ligand in the paracrine regulation of bone resorption." *Journal of Bone and Mineral Research* **15**: 2–12.

Hung, C. T., Pollack, S. R., Reilly, T. M. and Brighton, C. T. 1995. "Real-time calcium response of cultured bone cells to fluid flow." *Clinical Orthopaedics and Related Research*: 256–269.

Huo, B., Lu, X. L., Costa, K. D., Xu, Q. and Guo, X. E. 2010. "An ATP-dependent mechanism mediates intercellular calcium signaling in bone cell network under single cell nanoindentation." *Cell Calcium* **47**: 234–41.

Jacobs, C. R., Yellowley, C. E., Davis, B. R., Zhou, Z., Cimbala, J. M. and Donahue, H. J. 1998. "Differential effect of steady versus oscillating flow on bone cells." *Journal of Biomechanics* **31**: 969–976.

Johnson, D. L., Mcallister, T. N. and Frangos, J. A. 1996. "Fluid flow stimulates rapid and continuous release of nitric oxide in osteoblasts." *American Journal of Physiology-Endocrinology And Metabolism* **34**: E205.

Kamel, M. A., Picconi, J. L., Lara-Castillo, N. and Johnson, M. L. 2010. "Activation of β-catenin signaling in MLO-Y4 osteocytic cells versus 2T3 osteoblastic cells by fluid flow shear stress and PGE2: Implications for the study of mechanosensation in bone." *Bone* **47**: 872–881.

Kamioka, H., Sugawara, Y., Murshid, S. A., Ishihara, Y., Honjo, T. and Takano-Yamamoto, T. 2006. "Fluid shear stress induces less calcium response in a single primary osteocyte than in a single osteoblast: implication of different focal adhesion formation." *Journal of Bone and Mineral Research* **21**: 1012–1021.

Kato, Y., Windle, J. J., Koop, B. A., Mundy, G. R. and Bonewald, L. F. 1997. "Establishment of an osteocyte-like cell line, MLO-Y4." *Journal of Bone and Mineral Research* **12**: 2014–2023.

Kennedy, O. D., Herman, B. C., Laudier, D. M., Majeska, R. J., Sun, H. B. and Schaffler, M. B. 2012. "Activation of resorption in fatigue-loaded bone involves both apoptosis and active pro-osteoclastogenic signaling by distinct osteocyte populations." *Bone* **50**: 1115–1122.

Kidd, L. J., Stephens, A. S., Kuliwaba, J. S., Fazzalari, N. L., Wu, A. C. K. and Forwood, M. R. 2010. "Temporal pattern of gene expression and histology of stress fracture healing." *Bone* **46**: 369–378.

King, G. J. and Holtrop, M. E. 1975. "Actin-like filaments in bone cells of cultured mouse calvaria as demonstrated by binding to heavy meromyosin." *Journal of Cell Biology* **66**: 445–451.

Klein-Nulend, J., van der Plas, A., Semeins, C. M., Ajubi, N. E., Frangos, J. A., Nijweide, P. J. and Burger, E. H. 1995. "Sensitivity of osteocytes to biomechanical stress in vitro." *FASEB Journal* **9**: 441–445.

Kleinnulend, J., Semeins, C., Ajubi, N., Nijweide, P. and Burger, E. 1995. "Pulsating fluid flow increases nitric oxide (NO) synthesis by osteocytes but not periosteal fibroblasts-correlation with prostaglandin upregulation." *Biochemical and Biophysical Research Communications* **217**: 640–648.

Knothe Tate, M. L. and Knothe, U. 2000. "An ex vivo model to study transport processes and fluid flow in loaded bone." *Journal of Biomechanics* **33**: 247–254.

Knothe Tate, M. L., Steck, R., Forwood, M. R. and Niederer, P. 2000. "In vivo demonstration of load-induced fluid flow in the rat tibia and its potential implications for processes associated with functional adaptation." *Journal of Experimental Biology* **203**: 2737–2745.

Kou, S., Pan, L., van Noort, D., Meng, G., Wu, X., Sun, H., Xu, J. and Lee, I. 2011. "A multishear microfluidic device for quantitative analysis of calcium dynamics in osteoblasts." *Biochem Biophys Res Commun* **408**: 350–355.

Lacey, D., Timms, E., Tan, H.-L., Kelley, M., Dunstan, C., Burgess, T., Elliott, R., et al. 1998. "Osteoprotegerin ligand is a cytokine that regulates osteoclast differentiation and activation." *Cell* **93**: 165–176.

Lau, E., Al-Dujaili, S., Guenther, A., Liu, D., Wang, L. and You, L. 2010. "Effect of low-magnitude, high-frequency vibration on osteocytes in the regulation of osteoclasts." *Bone* **46**: 1508–1515.

Liu, C., Zhao, Y., Cheung, W. Y., Gandhi, R., Wang, L. and You, L. 2010. "Effects of cyclic hydraulic pressure on osteocytes." *Bone* **46**: 1449–1456.

Lu, X. L., Huo, B., Chiang, V. and Guo, X. E. 2012. "Osteocytic network is more responsive in calcium signaling than osteoblastic network under fluid flow." *J Bone Miner Res* **27**: 563–574.

Mak, A. F. T., Qin, L., Hung, L. K., Cheng, C. W. and Tin, C. F. 2000. "A histomorphometric observation of flows in cortical bone under dynamic loading." *Microvascular Research* **59**: 290–300.

Malone, A. M. D., Anderson, C. T., Tummala, P., Kwon, R. Y., Johnston, T. R., Stearns, T. and Jacobs, C. R. 2007. "Primary cilia mediate mechanosensing in bone cells by a calcium-independent mechanism." *Proceedings of the National Academy of Sciences of the United States of America* **104**: 13325–13330.

Martin, R. 2007. "Targeted bone remodeling involves BMU steering as well as activation." *Bone* **40**: 1574–1580.

Matsuo, K. and Irie, N. 2008. "Osteoclast-osteoblast communication." *Archives of Biochemistry and Biophysics* **473**: 201–209.

McGarry, J. G., Klein-Nulend, J. and Prendergast, P. J. 2005. "The effect of cytoskeletal disruption on pulsatile fluid flow-induced nitric oxide and prostaglandin E2 release in osteocytes and osteoblasts." *Biochemical and Biophysical Research Communications* **330**: 341–348.

McNamara, L., Majeska, R., Weinbaum, S., Friedrich, V. and Schaffler, M. 2009. "Attachment of osteocyte cell processes to the bone matrix." *The Anatomical Record* **292**: 355–363.

Mi, L. Y., Basu, M., Fritton, S. P. and Cowin, S. C. 2005a. "Analysis of avian bone response to mechanical loading. Part two: Development of a computational connected cellular network to study bone intercellular communication." *Biomechanics and Modeling in Mechanobiology* **4**: 132–146.

Mi, L. Y., Fritton, S. P., Basu, M. and Cowin, S. C. 2005b. "Analysis of avian bone response to mechanical loading-Part one: Distribution of bone fluid shear stress induced by bending and axial loading." *Biomechanics and Modeling in Mechanobiology* **4**: 118–131.

Mullender, M., El Haj, A. J., Yang, Y., Van Duin, M. A., Burger, E. H. and Klein-Nulend, J. 2004. "Mechanotransduction of bone cells in vitro: mechanobiology of bone tissue." *Medical and Biological Engineering and Computing* **42**: 14–21.

Murray, D. and Rushton, N. 1990. "The effect of strain on bone cell prostaglandin E2 release: a new experimental method." *Calcified Tissue International* **47**: 35–39.

Neuman, W. F., Neuman, M. W., Diamond, A. G., Menanteau, J. and Gibbons, W. S. 1982. "Blood: bone disequilibrium. VI. Studies of the solubility characteristics of brushite: apatite mixtures and their stabilization by noncollagenous proteins of bone." *Calcified Tissue International* **34**: 149–157.

Nicolella, P., Moravits, D. M., Lankford, J. and Bonewald, L. F. 2004. "Bone matrix strain is amplified at osteocyte lacunae in cortical bone." *Journal of Bone and Mineral Research* **19**: S72–S72.

Owen, M. and Triffitt, J. 1976. "Extravascular albumin in bone tissue." *The Journal of Physiology* **257**: 293–307.

Pederson, L., Ruan, M., Westendorf, J. J., Khosla, S. and Oursler, M. J. 2008. "Regulation of bone formation by osteoclasts involves Wnt/BMP signaling and the chemokine sphingosine-1-phosphate." *Proceedings of the National Academy of Sciences* **105**: 20764–20769.

Perez-Amodio, S., Beertsen, W. and Everts, V. 2004. "(Pre-)osteoclasts induce retraction of osteoblasts before their fusion to osteoclasts." *Journal of Bone and Mineral Research* **19**: 1722–1731.

Ponik, S. M., Triplett, J. W. and Pavalko, F. M. 2007. "Osteoblasts and osteocytes respond differently to oscillatory and unidirectional fluid flow profiles." *Journal of Cellular Biochemistry* **100**: 794–807.

Prendergast, P. J. and Huiskes, R. 1996. "Microdamage and osteocyte-lacuna strain in bone: a microstructural finite element analysis." *Journal of Biomechanical Engineering-Transactions of the Asme* **118**: 240–246.

Price, C., Zhou, X., Li, W. and Wang, L. 2011. "Real-time measurement of solute transport within the lacunar-canalicular system of mechanically loaded bone: direct evidence for load-induced fluid flow." *Journal of Bone and Mineral Research* **26**: 277–285.

Raisz, L. G. 1999. "Physiology and pathophysiology of bone remodeling." *Clinical Chemistry* **45**: 1353–1358.

Raisz, L. G., Pilbeam, C. C. and Fall, P. M. 1993. "Prostaglandins: mechanisms of action and regulation of production in bone." *Osteoporosis International* **3**: 136–140.

Rath Stern, A., Stern, M., Van Dyke, M., Jähn, K., Prideaux, M. and Bonewald, L. 2012. "Isolation and culture of primary osteocytes from the long bones of skeletally mature and aged mice." *BioTechniques*: 52.

Rawlinson, S. C. F., Pitsillides, A. A. and Lanyon, L. E. 1996. "Involvement of different ion channels in osteoblasts' and osteocytes' early responses to mechanical strain." *Bone* **19**: 609–614.

Rho, J.-Y., Tsui, T. Y. and Pharr, G. M. 1997. "Elastic properties of human cortical and trabecular lamellar bone measured by nanoindentation." *Biomaterials* **18**: 1325–1330.

Roach, H. 1994. "Why does bone matrix contain non-collagenous proteins? The possible roles of osteocalcin, osteonectin, osteopontin and bone sialoprotein in bone mineralisation and resorption." *Cell Biology International* **18**: 617–628.

Robinson, R. A. 1952. "An electron-microscopic study of the crystalline inorganic component of bone and its relationship to the organic matrix." *The Journal of Bone and Joint Surgery* **34**: 389–476.

Robling, A. G., Bellido, T. M. and Turner, C. H. 2006. "Mechanical loading reduces osteocyte expression of sclerostin protein." *Journal of Bone and Mineral Research* **21**: S72–S72.

Rubin, C. T. and Lanyon, L. E. 1982. "Limb mechanics as a function of speed and gait: a study of functional strains in the radius and tibia of horse and dog." *Journal of Experimental Biology* **101**: 187–211.

Ryser, M. D., Nigam, N. and Komarova Sr., S. V. 2009. "Mathematical modeling of spatio-temporal dynamics of a single bone multicellular unit." *Journal of Bone and Mineral Research* **24**: 860–870.
Sauren, Y. M., Mieremet, R. H., Groot, C. G. and Scherft, J. P. 1992. "An electron microscopic study on the presence of proteoglycans in the mineralized matrix of rat and human compact lamellar bone." *The Anatomical Record* **232**: 36–44.
Shapiro, F., Cahill, C., Malatantis, G. and Nayak, R. C. 1995. "Transmission electron-microscopic demonstration of vimentin in rat osteoblast and osteocyte cell-bodies and processes using the immunogold technique." *Anatomical Record* **241**: 39–48.
Sims, N. A. and Martin, T. J. 2014. "Coupling the activities of bone formation and resorption: a multitude of signals within the basic multicellular unit." *BoneKey Reports*: 3.
Singhvi, R., Kumar, A., Lopez, G. P., Stephanopoulos, G. N., Wang, D. I. C., Whitesides, G. M. and Ingber, D. E. 1994. "Engineering cell shape and function." *Science* **264**: 696–698.
Song, S. H., Choi, J. and Jung, H. I. 2010. "A microfluidic magnetic bead impact generator for physical stimulation of osteoblast cell." *Electrophoresis* **31**: 2762–2770.
Tami, A. E., Schaffler, M. B. and Knothe Tate, M. L. 2003. "Probing the tissue to subcellular level structure underlying bone's molecular sieving function." *Biorheology* **40**: 577–590.
Tasevski, V., Sorbetti, J. M., Chiu, S. S., Shrive, N. G. and Hart, D. A. 2005. "Influence of mechanical and biological signals on gene expression in human MG-63 cells: evidence for a complex interplay between hydrostatic compression and vitamin D3 or TGF-beta 1 on MMP-1 and MMP-3 mRNA level." *Biochemistry and Cell Biology* **83**: 96.
Taylor, D., Hazenberg, J. G. and Lee, T. C. 2007. "Living with cracks: damage and repair in human bone." *Nature Materials* **6**: 263–268.
Teti, A. and Zallone, A. 2009. "Do osteocytes contribute to bone mineral homeostasis? Osteocytic osteolysis revisited." *Bone* **44**: 11–16.
Udagawa, N., Takahashi, N., Jimi, E., Matsuzaki, K., Tsurukai, T., Itoh, K., Nakagawa, N., et al. 1999. "Osteoblasts/stromal cells stimulate osteoclast activation through expression of osteoclast differentiation factor/RANKL but not macrophage colony-stimulating factor." *Bone* **25**: 517–523.
Van't Hof, R. J. and Ralston, S. H. 2001. "Nitric oxide and bone." *Immunology* **103**: 255–261.
Van Bezooijen, R. L., Winkler, D., Hayes, T., Karperien, M., Visser, A., van der Wee-Pals, L., Hamersma, H., et al. 2002. "Sclerostin: an osteocyte-expressed BMP antagonist the inhibits bone formation by mature osteoblasts." *Journal of Bone and Mineral Research* **17**: S144–S144.
Verborgt, O., Gibson, G. J. and Schaffler, M. B. 2000. "Loss of osteocyte integrity in association with microdamage and bone remodeling after fatigue in vivo." *Journal of Bone and Mineral Research* **15**: 60–67.
Wang, L., Fritton, S. P., Cowin, S. C. and Weinbaum, S. 1999. "Fluid pressure relaxation depends upon osteonal microstructure: modeling an oscillatory bending experiment." *Journal of Biomechanics* **32**: 663–672.
Wang, L., Wang, Y., Han, Y., Henderson, S. C., Majeska, R. J., Weinbaum, S. and Schaffler, M. B. 2005. "In situ measurement of solute transport in the bone lacunar-canalicular

system." *Proceedings of the National Academy of Sciences of the United States of America* **102**: 11911–11916.

Wang, Y., McNamara, L. M., Schaffler, M. B. and Weinbaum, S. 2007. "A model for the role of integrins in flow induced mechanotransduction in osteocytes." *Proceedings of the National Academy of Sciences of the United States of America* **104**: 15941–15946.

Weinbaum, S., Cowin, S. C. and Zeng, Y. 1994. "A model for the excitation of osteocytes by mechanical loading-induced bone fluid shear stresses." *J Biomech* **27**: 339–360.

Weinger, J. M. and Holtrop, M. E. 1974. "An ultrastructural study of bone cells: the occurrence of microtubules, microfilaments and tight junctions." *Calcified Tissue Research* **14**: 15–29.

Xiong, J. and O'Brien, C. A. 2012. "Osteocyte RANKL: new insights into the control of bone remodeling." *Journal of Bone and Mineral Research* **27**: 499–505.

You, J., Yellowley, C. E., Donahue, H. J., Zhang, Y., Chen, Q. and Jacobs, C. R. 2000. "Substrate deformation levels associated with routine physical activity are less stimulatory to bone cells relative to loading-induced oscillatory fluid flow." *Journal of Biomechanical Engineering* **122**: 387–393.

You, L., Cowin, S. C., Schaffler, M. B. and Weinbaum, S. 2001. "A model for strain amplification in the actin cytoskeleton of osteocytes due to fluid drag on pericellular matrix." *Journal of Biomechanics* **34**: 1375–1386.

You, L., Temiyasathit, S., Coyer, S. R., García, A. J. and Jacobs, C. R. 2008a. "Bone cells grown on micropatterned surfaces are more sensitive to fluid shear stress." *Cellular and Molecular Bioengineering* **1**: 182–188.

You, L., Temiyasathit, S., Lee, P., Kim, C. H., Tummala, P., Yao, W., Kingery, W., et al. 2008b. "Osteocytes as mechanosensors in the inhibition of bone resorption due to mechanical loading." *Bone* **42**: 172–179.

You, L., Temiyasathit, S., Tao, E., Prinz, F. and Jacobs, C. R. 2008c. "3D microfluidic approach to mechanical stimulation of osteocyte processes." *Cellular and Molecular Bioengineering* **1**: 103–107.

You, L. D., Weinbaum, S., Cowin, S. C. and Schaffler, M. B. 2004. "Ultrastructure of the osteocyte process and its pericellular matrix." *Anatomical Record Part A-Discoveries in Molecular Cellular and Evolutionary Biology* **278**: 505–513.

Zhang, D., Weinbaum, S. and Cowin, S. 1998. "On the calculation of bone pore water pressure due to mechanical loading." *International Journal of Solids and Structures* **35**: 4981.

Zhang, K., Barragan-Adjemian, C., Ye, L., Kotha, S., Dallas, M., Lu, Y., Zhao, S., et al. 2006. "E11/gp38 selective expression in osteocytes: regulation by mechanical strain and role in dendrite elongation." *Molecular and Cellular Biology* **26**: 4539–4552.

Zhang, X., Liu, X., Sun, J., He, S., Lee, I. and Pak, H. K. 2008. "Real-time observations of mechanical stimulus-induced enhancements of mechanical properties in osteoblast cells." *Ultramicroscopy* **108**: 1338–1341.

Zhao, S., Kato, Y., Zhang, Y., Harris, S., Ahuja, S. S. and Bonewald, L. F. 2002. "MLO-Y4 osteocyte-like cells support osteoclast formation and activation." *Journal of Bone and Mineral Research* **17**: 2068–2079.

15 Molecular mechanisms of cellular mechanotransduction in wound healing

Vincent F. Fiore, Dwight M. Chambers, and Thomas H. Barker

15.1 Introduction

Tissue mechanics emerge from the ensemble physical behaviors of the biopolymer networks that constitute multicellular tissues and the cells resident therein. The physical properties of these networks (viscoelasticity, strength, porosity, dimensionality) are in large part attributable to the natural hydrogel surrounding cells, the extracellular matrix (ECM). The ECM is a mixed polymeric network of structural glycoproteins, water-swollen proteoglycans, and mineralized inorganics with a tissue- and function-specific composition. This scaffold is dynamic – cells respond to and alter their environment over time, giving rise to regenerative or pathologic outcomes that dictate tissue function. Cells achieve their environmental sensitivity to the physical traits of the ECM using their cytoskeleton and associated protein complexes, called focal adhesions. Fibroblasts, in particular, are exquisitely sensitive to the rigidity of their underlying matrix, and the physio/pathologic outcomes of wound repair and fibrosis have been linked to their rigidity-dependent cellular processes. Intriguingly, as fibroblasts are also the main ECM synthesizing, assembling and remodeling cell type, there is a complex and dynamic bidirectional communication system between the cell and its environment. How does the cell respond to physical properties of the ECM, how are these properties sensed at the molecular scale, and how do these responses influence the outcome of wound repair processes? These will be the topics of this chapter.

15.2 The extracellular matrix

Cells reside in a structured, complex, and dynamic matrix composed of extracellular proteins (e.g., collagen, elastin, fibronectin, fibrin, and laminin), glycosaminoglycans/proteoglycans (e.g., heparan, hyaluronic acid, chondroitin, and keratan sulfates), and inorganic precipitates (e.g., pyro-phosphates and calcium), collectively referred to as the extracellular matrix (ECM). The composition of this matrix varies in a tissue-specific fashion while providing important stimuli to cells and tissues as a whole in both physiology and pathophysiology; The flavors of this ECM-cellular communication are diverse, stretching across biochemistry (e.g., type and density of receptor ligands presented, sequestration, and release of growth factors) and biophysics (e.g., stiffness, viscosity, porosity, topology), and this communication is bilateral as cells dynamically shape, degrade, and elaborate their surrounding matrix.

While we will focus on mechanical properties of the ECM in this section, it is important to briefly note important properties of the ECM as they relate to regulating cellular phenotype. One well-studied property of the ECM is how it presents various ligands that cell surface receptors recognize. Studies have shown that ECM composition and compositional density can affect the presentation of various ECM ligands, which is capable of directing cell phenotype in physiology and disease. (Brown et al. 2011; Brown et al. 2013; Chaudhuri et al. 2014). Furthermore, the ability of the ECM to bind, sequester, and release trophic factors has long been recognized (Ingber et al. 1987; Baird et al. 1988) and is currently an active area of research in the field of regenerative medicine (Martino et al. 2011; Martino et al. 2013). Additionally, the ECM creates a three-dimensional space around cells, whose porosity and nanoscale topology can influence cellular phenotype and differentiation (Sohier et al. 2014; Zhang et al. 2010), though special effort has been required in three dimensions to try and decouple these network properties from concurrent changes in other ECM network parameters, such as stiffness and ligand density (Miroshnikova et al. 2011; Ananthanarayanan et al. 2011).

The stiffness of the extracellular matrix is also an important characteristic for understanding cell-matrix interactions in both physiology and pathophysiology (Engler et al. 2004; Engler et al. 2006; Markowski et al. 2012; Brown et al. 2011; Brown et al. 2013; Tan et al. 2014; Parker et al. 2014; Rubashkin et al. 2014). The stiffness of matrix can be studied on several distinct spatial scales – from the macro-rheological properties of an ECM hydrogel, to the fibular bundles of ECM components that cells interact with directly, to the single molecule limit. In the larger physical domains, the mechanical properties of the ECM emerge from the network of interactions between individual ECM components and resident cells. From the viewpoint of polymer science, ECM polymer concentration and crosslinking density, as well as the rigidity of individual polymer units, are commonly associated with changes in the overall network's mechanical properties (Tse and Engler 2010; Trappmann and Chen 2013; Stabenfeldt et al. 2012 and Brown et al. 2015). For simplicity, the macroscale elastic modulus of the network can be estimated as k_bT/ζ^3, where k_bT is thermal energy and ζ represents the mesh size; thus, the modulus is inversely correlated with the mesh size. Studies that have examined the stiffness of cellular and acellular ECMs also show that cellular contractility may also play an important role in determining the overall stiffness of a tissue. For example, in the healthy lung, the total cellular contribution to stiffness is approximately 15%, while in the fibrotic lung of idiopathic pulmonary fibrosis, the cellular contribution is significantly enhanced (Booth et al. 2012). Furthermore, cellular contractility within strain-stiffening polymer networks locally stiffens the network and can transmit such stresses to neighboring elements (Winer et al. 2009).

In the single molecule limit, sensitive techniques like atomic force microscopy reveal the viscoelastic properties of ECM components. While many ECM molecules have been studied, the mechanics of fibronectin have been extensively studied and mechanistic models developed to explain its behavior. Fibronectin is composed of tandem repeats of three domain types: type I, type II, or type III immunoglobulin-like domains. While type I and type II domains are stabilized by intra-domain disulfide linkages, type III domains are simple beta-sandwich structures that have been shown to unfold under physiologically relevant forces, ranging from 80 pN to 200 pN of force (Krammer et al. 1999; Smith et al. 2007; Li et al. 2005; Leahy et al. 1996). Additionally there is evidence that post-translational modification of ECM molecules, like fibronectin in a reduced environment, can alter the mechanics of the individual fibers considerably (Oberhauser et al. 2002; Vadillo-Rodriguez et al. 2013).

ECM stiffness is also strongly coupled to other biochemical features such as ligand presentation and growth factor signaling. While the domain unfolding hierarchy of type III repeats in fibronectin is intrinsically stochastic, one of the weakest domains is FNIII-10, the 10th type III domain, which contains the RGD cell binding sequence and is adjacent to FNIII-9, which contains the PHSRN cell binding site (the so-called synergy site) (Oberhauser et al. 2002; Ng et al. 2008). Physical separation of these sites due to insertion of a physical spacer within the linker region has been shown to affect integrin specificity between $\alpha V\beta 3$ and $\alpha 5\beta 1$ and correspondingly drive cellular phenotypes such as epithelial-to-mesenchymal transition (Martino et al. 2009; Oberhauser et al. 2002; Markowski et al. 2012; Brown et al. 2013; Altroff et al. 2003). ECM stiffness has also been shown to affect soluble trophic factor signaling directly. Stiff extracellular matrices are able to provide the physical resistance for cells to mechanically activate TGF-β from its latent complex via contractile stresses through integrin, while softer matrices may shield the biochemical complex from cellular stresses and inhibit activation (Wipff et al. 2007). Additionally, there is evidence that mechanical distension of fibronectin can expose hydrophobic binding sites that can alter the molecules affinity for carrier molecules like albumin (Little et al. 2009).

15.3 Fibroblast biology and function in normal physiology

15.3.1 Fibroblast populations and diversity

As is evidenced by the name, the main effector cell in tissue wound healing and fibrosis is the fibroblast. Fibroblasts are stroma resident cells responsible for much of the assembly and remodeling of ECM during development and adult homeostasis. Fibroblasts are defined as "cells that (1) synthesize and secrete a complex array of structural (e.g., collagens and fibronectin) and nonstructural (e.g., matricellular family of molecules such as thrombospondins, SPARC, and osteopontin) ECM molecules, (2) actively organize and remodel the ECM through production of proteinases, and (3) converse with nearby cells through paracrine, autocrine, and other forms of communication" (Sorrell and Caplan 2009, pg.163). Fibroblasts are also identified by their ability to migrate in response to directional and mechanical cues and respond to the local mechanical environment (i.e., mechanotransduction). It is therefore apparent that fibroblasts are central to tissue and organ physiology through their response and organization of extracellular cues.

As the defining characteristics of fibroblasts are typically their spindle-shaped morphology, ability to adhere to plastic culture surfaces, and the absence of other lineage-specific markers, the diversity of such cells is typically slighted. Fibroblasts from distinct stromal tissues can exhibit significant phenotypic heterogeneity. For example skin fibroblasts differ from those from the lung in terms of morphology, proliferation rates, and synthesis of cytokines and ECM constituents (Sorrell and Caplan 2009). Comparison of genome-wide mRNA expression profiles of fibroblasts from distinct anatomical locations has revealed divergence among a wide array of genes, from lipid metabolism to cell migration (Chang et al. 2002). This is perhaps unsurprising, as different anatomical locations have evolved with unique functions, and so fibroblasts within these compartments may be specialized to perform distinct roles. However, fibroblasts within the same stromal compartment can exhibit phenotypic differences as well. Clonal derivatives from skin, intestinal tissue, and lung have been demonstrated to differ in ECM component synthesis and assembly, epithelial-mesenchymal crosstalk, and secretion and responsiveness to cytokines (Sorrell and Caplan 2009; Fritsch et al. 1999). This is further complicated by the fact that these cells can undergo phenotypic plasticity within their lifespan, dependent on microenvironmental cues. For example, carcinoma-associated fibroblasts (CAFs) can exhibit significant functional differences between subpopulations promoting metastatic processes versus nonmalignant growth: caveolin-1 expressing CAFs are proficient at Rho-mediated matrix alignment and stiffening to promote stromal invasion of tumor cells; whereas caveolin-1 lacking CAFs are diminished in this capacity (Goetz et al. 2011). In the skin, distinct fibroblast lineages play unique roles in defining dermal architecture during skin development and repair. Fibroblasts arising from a single progenitor will go on to create two distinct fibroblast lineages in the mouse: one forms the upper dermis, including the dermal papilla and supports hair follicle growth and regeneration, whereas the reticular/hypodermal subpopulation will go on to synthesize the majority of fibrillar ECM and are the progenitors for pre-adipocytes within the hypodermis (Driskell et al. 2013). During wound healing, the initial wave of regeneration depends on cells of the reticular/hypodermal lineage elaborating a collagenous ECM, while upper cells become active later in the process to facilitate re-epithelialization and follicular morphogenesis.

Efforts have also been made to identify markers of fibroblast subpopulations during the wound repair process. Fibroblasts that are actively involved in matrix secretion and generating contractile forces have long been identified through their expression of alpha-smooth muscle actin (α-SMA) (Klingberg et al. 2013). Additionally, fibroblast subpopulations characterized by surface expression of glycoprotein Thy-1 (CD90) are an intriguing example of phenotypic diversity with implications in wound repair. In vitro, fibroblasts lacking expression of Thy-1 undergo enhanced proliferation and activation of pro-fibrotic cytokines in response to fibrogenic growth factors, such as platelet-derived growth factor-A (PDGF-A) and connective tissue growth factor (CTGF), and exhibit differential signaling responses to pro-inflammatory cytokines tumor necrosis factor-α (TNF-α) and interleukin-1β (IL-1β) (Hagood et al. 2001; Hagood et al. 2002; Hagood et al. 1999; Zhou et al. 2004). Thy-1 also regulates the activity of Rho, via negative regulation of Src family kinase activity and activation of p190GAP, leading to alterations in stress fiber, focal adhesion assembly, and migration on rigid substrates (Barker et al. 2004a) and is essential for thrombospondin-1-mediated focal adhesion disassembly (Barker et al. 2004b). Fibroblasts lacking Thy-1 undergo enhanced myofibroblast differentiation in response to fibrogenic cytokines, as evidenced by expression of myogenic regulatory factors MyoD and myocardin and collagen gel contraction (Sanders et al. 2007). Of foremost clinical significance, Thy-1 expression is absent in fibroblasts within the fibroblastic foci IPF patients, whereas the majority of lung fibroblasts from healthy lung interstitium are Thy-1^{pos} (Hagood et al. 2005; Sanders et al. 2008). Furthermore, Thy-1–/– mice display more severe fibrosis in response to bleomycin challenge, a commonly used model for the induction of lung fibrosis, and regions of the interstitium undergoing fibrotic remodeling in wild-type mice are predominated by Thy-1 negative fibroblasts (Hagood et al. 2005). Therefore, subpopulations of fibroblasts appear to play central roles in the pathophysiology of lung fibrosis and perhaps in progressive fibroproliferative disorders more generally.

15.3.2 Fibroblasts in wound healing

As previously discussed, fibroblast populations are central to the physiological wound-healing process. Wound repair in the skin is a well-studied and clinically relevant model system and can be used conceptually to understand the role of the fibroblast in wound healing. The fibroblastic response to injury can be divided schematically into three distinct phenotypes: recruitment, differentiation, and resolution. During the recruitment phase, tissue injury activates both the construction of a provisional fibrin/fibronectin matrix and the release of platelet-derived cytokines that recruit fibroblasts, immunological and endothelial cells (Nurden et al. 2008). Recruited immunological and fibroblastic populations engage in mutual paracrine and autocrine stimulation through a variety of inflammatory cytokines (e.g., IL-1/6/13/33, TNF-α, TGF-β, PDGF) leading to a period of fibroblastic proliferation (Kendall and Feghali-Bostwick 2014). There is also evidence that the stiffness of the provisional matrix may directly stimulate fibroblast proliferation (Liu et al. 2010). After the provisional tissue is deposited, these fibroblasts undergo differentiation into an activated subtype, commonly called the "myofibroblast." Myofibroblasts are highly synthetic and exert considerable contractile forces, which progressively increase as the myofibroblasts organize newly expressed α smooth muscle actin (α-SMA) into contractile filaments (Klingberg et al. 2013). This differentiation process is central to wound healing and discussed in greater detail below. Myofibroblasts both (1) secrete a new, permanent matrix, initially rich in a mixture of collagens that is progressively remodeled into a collagen-I rich matrix, along with some elastin; and (2) remodel and remove the provisional matrix, through expression of matrix metalloproteinases (MMPs), like MMP-9 and MMP-13 (Hattori et al. 2009). In the resolution phase of injury, the dermal wound is closed through cellular contractile forces in the newly elaborated matrix, resulting in the final scar. The end of this phase is punctuated by the apoptosis of the resident myofibroblastic population (Kapanci et al. 1995).

15.4 Tissue stiffening as an outcome and driver of fibrosis

Fibrosis is a common alteration of tissue structure-function in human diseases, with chronic fibroproliferative processes contributing in some aspect to nearly 45% of deaths in the developed world (Wynn 2007). During fibrosis, connective tissue is assembled and remodeled by activated fibroblasts in an effort to restore tissue integrity. However, unlike in regulated wound healing where such wounds in the adult typically heal as resolved scars (Gurtner et al. 2008), significant accumulation of such scar tissue within parenchymal regions disrupts tissue architecture and results in altered tissue/organ function. Fibrosis is a principal cause of tissue dysfunction in major vital organs, such as the lungs, liver, and heart, and represents a major cause of human mortality worldwide, however the molecular and biophysical mechanisms underlying fibrogenesis, especially in the context of chronic fibroproliferative disorders, are incompletely understood.

During fibrogenesis, major changes to the biochemical and biophysical properties of the ECM occur. The abundance of connective tissue deposited by fibroblasts results in an increase in the amount of fibrous ECM protein, predominantly type I–III collagens (Tomasek et al. 2002), thereby altering both the amount, type, and topology of ligands presented in the cellular microenvironment. The increase in fibrous ECM protein is concomitant with an increase in tissue rigidity, as the amount of collagen I scales linearly with tissue elastic modulus over three orders of magnitude and is largely responsible for the tensile strength of tissues (Swift et al. 2013). Additionally, dense fibrotic pockets are often characterized by their increased concentration of collagen VI (Betz et al. 1993; Specks et al. 1995). Enzymatic or nonenzymatic (i.e., glycation) crosslinking of ECM fibers can act further enhance tissue rigidity and increase network strength.

Studies have demonstrated that tissue compliance decreases as a consequence of disease progression in chronic fibroproliferative disorders, such as idiopathic pulmonary fibrosis (IPF). This was demonstrated at the whole organ level through lung pressure-volume measurements from IPF patients, where lung compliance was seen to decrease by more than 30% and was the most strongly correlated pulmonary function parameter with the extent of fibrosis (Sansores et al. 1996). Similarly, bulk tensile stress testing of lung tissue strips from rats after bleomycin-induced fibrosis showed a 2-fold increase, from approximately 5 to 10 kPa, as a result of fibrosis (Ebihara et al. 2000; Dolhnikoff et al. 1999). Recently, work to measure the rigidity of tissue at the cell scale (10^{-5}–10^{-7} m) has been performed on a variety of tissues. These studies are qualitatively similar with results gathered at bulk scales, however important insights from the microscale have been gained. For instance although average rigidity values have been obtained for lung, heart, breast, skin, and liver, it is clear that spatial heterogeneities exist even within normal tissue architectures. This seems to be exaggerated in fibrosis, where even larger spatial variations in tissue rigidity are observed. In a recent atomic force microscopy study of lung cores from healthy individual or patients with IPF, mean tissue stiffness increased from 1.96 kPa (range: 0.4–8 kPa) in health cores to 16.52 kPa (range: 0.3–110 kPa) in diseased cores (Booth et al. 2012). What is the significance and origin of such heterogeneity? Presumably distinct tissue structures (e.g., varied collagen content, degree of crosslinking enzyme expression) give rise to microenvironments with distinct mechanics.

How do the mechanics of the ECM drive profibrotic cell phenotype? Recent studies demonstrate that the rigidity of the ECM regulate a host of cellular processes involved in fibrosis (Engler et al. 2004; Liu et al. 2010; Levental et al. 2009; Huang et al. 2012; Brown et al. 2013; Booth et al. 2012). As discussed above, fibroblasts increase their expression of α-SMA, which promotes further activation of cytoskeletal machinery (Hinz et al. 2001; Goffin et al. 2006). Enhanced polymerization of F-actin leads to translocation of the cytoplasmic G-actin-bound transcription factor MRTF-A into the nucleus, where it complexes with serum response factor and drives α-SMA gene expression and myofibroblast differentiation (Huang et al. 2012; Velasquez et al. 2013). Both fibroblasts and type II alveolar epithelial cells activate the pro-fibrotic cytokine TGF-β to a greater extent (Wipff et al. 2007). Also, ECM stiffness promotes alveolar epithelial-to-mesenchymal transition through enhanced αvβ6 integrin-mediated activation of latent TGF-β from ECM (Brown et al. 2013). Active TGF-β is central not only to myofibroblast differentiation, but is also a potent anti-apoptotic signal, a significant phenotype found in pathological myofibroblasts (Popova et al. 2010; Kulasekaran et al. 2009; Horowitz et al. 2008). Importantly, cytoskeletal contractility and Rho-mediated signaling are necessary for stiffness-induced EMT and latent TGF-β activation.

Matrix stiffening promotes fibroblast proliferation and matrix synthesis through suppression of COX-2 and prostaglandin E_2 (PGE_2) expression, linking matrix stiffness and fibroblast proliferation (Liu et al. 2010). In contrast, physiologic stiffness enhances COX-2 and PGE_2 expression, which promotes fibroblast quiescence and inhibits assembly of mature cytoskeletal structures. One of the common features of these studies is that ECM stiffening results in phenotypic changes that further promote cytoskeletal activity, profibrotic ECM remodeling, and fibroblast proliferation (Parker et al. 2014; Liu et al. 2010; Arora et al. 1999). This demonstrates a positive feedback loop between the biophysical state of the ECM and the activation state of the cytoskeleton that, in the case of fibrosis, may lead to system instability (i.e., chronic fibrotic remodeling). Interestingly, how the microenvironment is initially remodeled during the wound healing/fibrotic processes that result in changes from an initially physiologic rigidity regime to a disease-associated and profibrotic state is unknown. However, it must be stated the majority of studies investigating the physiology and disease-associated cell behavior in vitro have been performed on rigid glass or plastic materials. In fact, simply removing mesenchymal cells from their in vivo microenvironment results in dramatic activation of the cytoskeleton, hypertrophy, and changes in gene expression reminiscent of myofibroblast differentiation associated with fibrosis (Balestrini et al. 2012).

While stiffness of the cellular microenvironment clearly plays a significant role in defining the resident cell phenotype (Engler et al. 2006; del Rio et al. 2009; Paszek et al. 2005; Wipff et al. 2007), the origins of the increased tissue stiffness are still poorly understood. It is likely that stiffness emanates from the residential cell population and/or the over-production or activation of crosslinking enzymes. Fibroblasts that have differentiated down a contractile, myofibroblastic pathways are known to exhibit significant contractile force on the ECM (Wipff et al. 2007). Such cell-derived forces are capable of stressing nearby strain-stiffening ECM polymers, leading to increased local stiffness. Enhanced levels of crosslinker within the microenvironment, such as lysyl oxidase enzymes, are sufficient to crosslink and stiffen the tumor stroma (Levental et al. 2009). Tissue transglutaminases, long known to catalyze ECM crosslinking, have also recently been shown to result in tissue stiffening (Santhanam et al. 2010). In fact, successful reduction of bleomycin-induced pulmonary fibrosis in a murine model, along with a marked reduction of activated fibroblasts and decreased TGFβ signaling, was achieved by inhibiting the matrix crosslinking enzyme lysyl oxidase-like-2 (Barry-Hamilton et al. 2010; Levental et al. 2009). However, the cellular mechanisms underlying how and under what external cues cells indeed stiffen their surrounding microenvironment, from relatively compliant to the rigidity associated with mature scar tissue, should be the focus of future investigation.

15.5 Cellular responses to ECM mechanics

Of the many parameters characterizing the material properties of biological tissue, stiffness or rigidity is perhaps the most widely investigated. As a measure of the extent to which an object resists elastic deformation under an applied force, the elastic modulus has emerged as a critical regulator of cell behavior (Discher et al. 2005). Investigation into the cellular responses to ECM rigidity has yielded multiple phenotypic outputs that are modulated by this parameter, including migration, differentiation, proliferation, and apoptosis. As may be expected, many of these phenotypes are intimately connected with the cytoskeleton, which is largely responsible for a cell's own mechanical properties and its ability to perform mechanical work (Yeung et al. 2005; Ingber 2003). Although the molecular and biophysical details of how a cell senses and responds to its mechanical environment are ongoing, a generalizable description is that a cell probes the mechanics of the ECM by generating forces within the cytoskeleton that are transmitted to the ECM through protein complexes (focal adhesions) that are also biochemically sensitive to these forces (Moore et al. 2010).

Tissue cells in vitro exhibit multiple modes of cytoskeletal contractility leading to force generation on the ECM (Ponti et al. 2004; Hu et al. 2007). These modes are highly regulated both in space and time and are dependent on both intrinsic (e.g., retrograde actin flow, myosin filament assembly and activity) and extrinsic

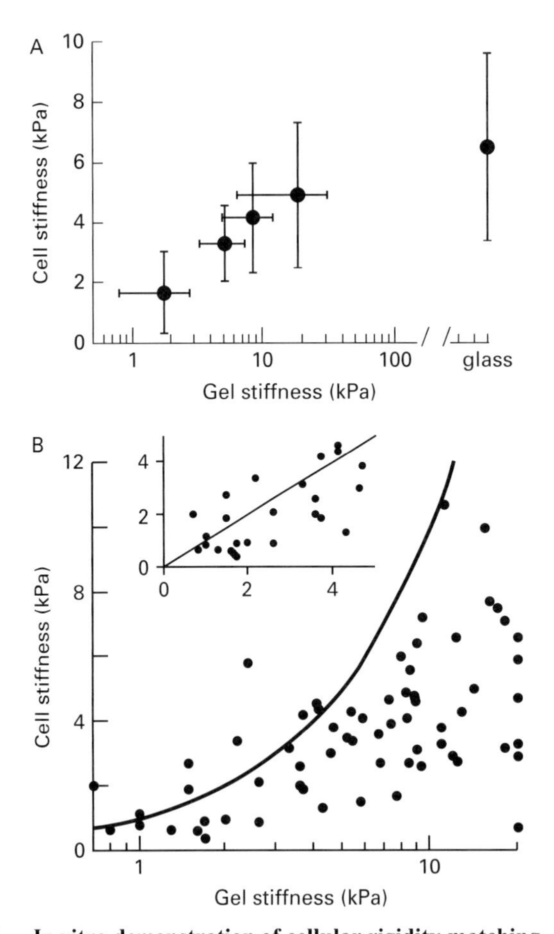

Figure 15.1 **In vitro demonstration of cellular rigidity matching.**
Fibroblasts plated on surfaces with defined elastic moduli stiffen themselves in response to increasing substrate stiffness. This effect saturates around 20 kPa.

Reprinted from *Biophysical Journal* 2007, 93: 4453–4461, Solon et al. Fibroblasts adaptation and stiffness matching to soft elastic substrates. Copyright 2007 with permission from Elsevier.

(ECM stiffness, biochemical composition) factors. For example, under low tension, actin filaments are not assembled into stress fibers and instead exhibit contractile network-arrays that are highly efficient in cell migration (Aratyn-Schaus et al. 2011). This tension is controlled both by the stiffness of the underlying ECM and the ability of myosin to reorganize and tense actin filament networks. In response to a stiff ECM, contractile-network arrays can give rise to highly contractile, parallel actin stress fibers, and this in turn can generate maximal tension on the ECM. Thus, cells exhibit precisely tuned intracellular machinery, namely the actinomyosin cytoskeleton, to respond to extracellular stimuli such as changes in the mechanical microenvironment, which ultimately give rise to distinct motile or contractile phenotypes.

Cells on stiff substrates not only assemble bundles of F-actin filaments and generate larger traction forces, but also increase the size of cell-matrix adhesions, structures that can be directly correlated, during the initial stages of growth and maturation, to the cell's application of force on the matrix (Pelham and Wang 1997; Stricker et al. 2011). Cells increase their applied traction forces and spread area in response to increasing substrate stiffness (Saez et al. 2005; Yeung et al. 2005). These responses occur up to a point (~20 kPa), at which effective saturating levels are reached. A consequence of the assembly of cytoskeletal structures, such as stress fibers and focal adhesions, and enhanced contractility on stiff substrates, is that cells also increase their cortical stiffness. Thus, cells change their internal structure and contractility in attempt to "match" that of their surrounding environment (Fig. 15.1) (Solon et al. 2007; Tee et al. 2011).

15.6 Molecular signal transduction through focal adhesions

15.6.1 Focal adhesions: local sites of mechanical signal transduction

Integrin-based adhesions are the physical attachment sites between cells and the ECM (Hynes 2002). Forces originating from the cytoskeleton or external environment are linked to the ECM through the transmembrane integrins and their associated molecules that cluster to form adhesions, more specifically termed focal adhesions, focal complexes, or nascent adhesions depending on their molecular composition, size, and/or historical commonplace (Geiger et al. 2009). In this chapter, we refer to these structures more generally as focal adhesions (FAs). The cytoplasmic tails of integrin serve as scaffolds for the binding of integrin-associated proteins, including cytoskeletal binding and adapter proteins, enzymes such as kinases and phosphatases, and small GTPases and their regulators (Gardel et al. 2010). Thereby, FAs serve both as a macromolecular complex that physically associates the cytoskeleton and ECM and as a biochemical signaling hub. FAs are mechanosensitive organelles, in that they grow and change composition in response to mechanical force (Riveline et al. 2001; Balaban et al. 2001). Specific proteins are recruited to adhesions in a force-dependent manner, and these proteins may elicit specific downstream responses (Pasapera et al. 2010; Kuo et al. 2011; Gardel et al. 2010). For example, nascent complexes, including scaffolding molecules such as paxillin, form at the leading edge, while myosin II-mediated tension and structural templating lead to recruitment of accessory proteins such as zyxin and α-actinin, promoting growth and "maturation" of the complex (Oakes et al. 2012; Choi et al. 2008; Pasapera et al. 2010). Mechanical tension can activate downstream signaling modules such as focal adhesion kinase (FAK), Src and Rho family GEFs, as discussed below, however the details of these processes at the molecular scale are still under investigation (Friedland et al. 2009; Wang et al. 2005).

Downstream of ECM ligation, a plethora of integrin-specific and tension-dependent signaling responses have been discovered (Kuo et al. 2011; Schiller et al. 2013). Intriguingly, exogenous forces applied to integrins initiate similar signaling pathways to those stimulated by internally generated forces. Initial studies on adhesion reinforcement demonstrated that cells are able to sense the restraining force applied to fibronectin-coated beads and respond by a localized and proportional strengthening of cytoskeletal linkages

(Choquet et al. 1997). Force-bearing FA proteins, such as vinculin, are also recruited and/or activated at the site of exogenous force application, further demonstrating the structural importance of adhesion strengthening and mechanotransduction (Galbraith et al. 2002). These cytoskeletal rearrangements result in localized stiffening of the cell at the cell-ECM interface (Wang et al. 1993).

One family of signaling molecules heavily involved in integrin mechanosignaling is the Src family of tyrosine kinases (SFKs), a family of nine related non-receptor tyrosine kinases with homology to the proto-oncogene c-Src (Thomas and Brugge 1997). SFKs are activated downstream of integrins where they play a critical role in FA assembly and signaling (Moore et al. 2010). For example, one well-characterized interaction involves the complex formation of SFKs with active FAK, resulting in phosphorylation of additional tyrosine residues in FAK, which serve as docking sites for subsequent proteins (Shattil 2005). SFKs modulate multiple other downstream pathways including those involving phosphoinositide (PtdIns) 3-kinase, mitogen-activated protein (MAP) kinases, and Rho GTPases (Shattil 2005). In particular, activation of the SFK member Fyn is required for the force-dependent formation of focal complexes and reinforcement of $\alpha v\beta 3$ integrin-cytoskeleton connections (von Wichert et al. 2003). Intriguingly, the stretchactivated substrate of Fyn, p130Cas (further discussed below), and its phosphorylation is dependent on substrate rigidity, and so a mechanism of force-dependent Fyn phosphorylation of p130Cas with rigidity-dependent displacement has been proposed (Jiang et al. 2006; Kostic and Sheetz 2006). SFK activation is one of the initial responses in cell-ECM mechanical connectivity, with the activation of SFKs occurring within 300 milliseconds of force application to fibronectin-coated beads (Wang et al. 2005; Na et al. 2008).

Another major pathway involved in integrin mechanosignaling is the GTPase RhoA, a critical signaling hub in the assembly of stress fibers and FAs. RhoA is a Ras-related member of the Rho family of small GTPases that promotes the assembly of FAs and actin stress fibers, whereas inhibition with dominant-negative mutants or pharmacological inhibitors results in disassembly of stress fibers and FAs (Ridley and Hall 1992; Nobes and Hall 1995). RhoA induces stress fibers and focal adhesions by stimulating contractility, as active Rho elevates myosin light chain (MLC) phosphorylation in a Rho-associated kinase (ROCK)-dependent manner (Chrzanowska-Wodnicka and Burridge 1996). This is in contrast to other Rho family members, such as Rac and Cdc42, which have distinct, but interdependent functions with respect to cytoskeletal organization (Nobes and Hall 1995; Machacek et al. 2009). In addition, RhoA mediates actin assembly though an mDia-dependent pathway, both of which are necessary for the proper assembly of stress fibers (Watanabe et al. 1997). The requirement for ROCK-mediated myosin II contraction, activated downstream of RhoA, in tension-induced adhesion assembly was demonstrated; ROCK activity could be bypassed by external force application to induce FA assembly (Riveline et al. 2001). Correspondingly, tension activates RhoA, however the molecular mechanisms upstream of this were unknown (Bhadriraju et al. 2007; Wozniak et al. 2003). Recently two guanine nucleotide exchange factors (GEFs), LARG and GEF-H1, were demonstrated to regulate RhoA activation in response to force via their

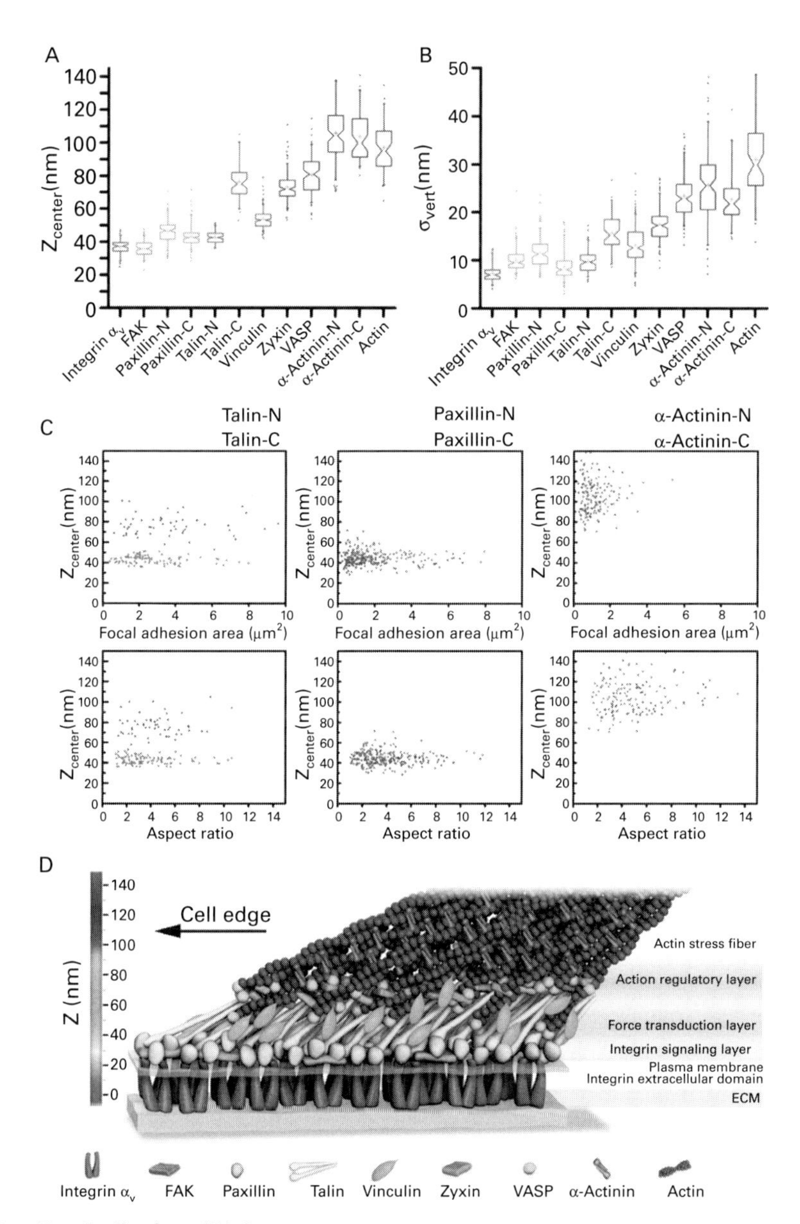

Figure 15.2 Focal adhesions (FAs).

Focal adhesions (FAs) are a multiprotein complex with defined substructure that transduce mechanical forces between the cell and the extracellular space. Using fusion proteins, the position and distribution of various FA proteins has been mapped allowing visualization of this complex 3-D structure.

activation and recruitment to the adhesion complex (Guilluy et al. 2011). These studies further illuminate the mechanochemical responses to mechanical tension that mediate cytoskeletal remodeling and cellular adaptation.

Perhaps obfuscating the molecular mechanisms that regulate mechanosensing and mechanotransduction within FAs is the large number of FA-associated proteins and interactions. Recent proteomics analysis points to upwards of 900 proteins that may associate with integrin-ECM complexes at some point during their lifecycle (Kuo et al. 2011; Humphries et al. 2009). How these proteins associate in space and time is critical for overall FA function and, thus, biochemical signal generation, force transmission, and mechanotransduction. For example, the scaffolding molecule paxillin is prominent during initial complex formation (i.e., nascent adhesions), as is the integrin-activating and actin-bridging protein talin, whereas other structural molecules such as α-actinin, zyxin, and tensin associate later in the FA assembly cycle (Geiger and Yamada 2011; Gardel et al. 2010). Whereas the structural molecule vinculin is present in nascent adhesions, in more mature adhesions it can switch to an active conformation, linking talin and actin to stabilize force transmission (Dumbauld et al. 2013; Thievessen et al. 2013). Single-molecule localization microscopy has identified the nanoscale localization of FA components, which shows signaling, force transmitting, and actin network regulating molecules in distinct laminae with respect to the substratum (Fig. 15.2)(Kanchanawong, 2010). In space, these dynamic but coordinated interactions build a hierarchical network of proteins that link the cytoskeleton to the ECM, resembling a stratified structure of protein-protein interactions wherein innumerous opportunities for mechanotransductory events exist (Kanchanawong et al. 2010).

15.6.2 Protein conformational changes as a mechanotransducer

For the reasons stated above, the FA has been the site of intense investigation into molecular mechanotransduction mechanisms. In general, two dominant mechanistic archetypes – physical destabilization of protein structure revealing cryptic binding sites / destroying binding sites and forced alterations to the receptor-ligand reaction landscape – have emerged as possible transduction mechanisms. Conformation changes which reveal or destroy protein binding sites have been demonstrated with various structural FA- and cytoskeleton-associated proteins such as talin (del Rio et al. 2009), filamin A (Ehrlicher et al. 2011), spectrins (Johnson et al. 2007), integrin (Kong et al. 2009; Kong et al. 2013), and others in single molecule studies. These studies conceptually parallel studies conducted on the conformation of ECM proteins discussed in Section 15.2 (e.g., stretch-mediated destabilization of fibronectin's 9th and 10th type III repeats). For example, force causes cryptic exposure of up to 11 vinculin-binding motifs within a normally shielded hydrophobic stretch of talin (del Rio et al. 2009). Force also exposes SFK-binding motifs within p130Cas, resulting in substrate phosphorylation and a potential mechanism for signal transduction through Rap1 to other downstream targets (Sawada et al. 2006). Additionally, conformational changes in the 21st and 23rd domains of filamin A have been shown to switch its affinity for the cytoplasmic tail of the β7 integrin and FilGAP, a GTPase activating protein whose downstream target, Rac, modulates cell morphology and bleb formation (Fig. 15.3) (Ehrlicher et al. 2011). However within the FAs, a multitude of proteins, binding interactions, and complex architecture make resolving mechano-signaling mechanisms at the molecular level a challenge. To that end, innovative strategies have had to be developed to validate these single molecule studies in vivo. Fluorescent reporters/FRET strategies have been used successfully to demonstrate the in vivo relevance of vinculin, talin, and filamin A conformational changes (Nakamura et al. 2014; Margadant et al. 2011; Grashoff et al. 2010). Moreover, the health relevance of these mechanotransductory mechanisms is demonstrated by exciting recent findings that have implicated the glutathionylation of cryptic cysteines in titin in human cardiomyocytes and connected mechanical changes seen in that molecule after glutathionylation to remodeling in dilated cardiomyopathy (Alegre-Cebollada et al. 2014).

Conformational changes are also able to change the ligand-receptor binding profiles of various molecules. Integrins are dynamic molecular machines, existing in multiple conformations that control ligand recognition and affinity, cytoplasmic binding partner accessibility, and forces transmitted through the receptor-ligand complex. In a low-affinity state, integrin is bent, resulting in close juxtaposition of

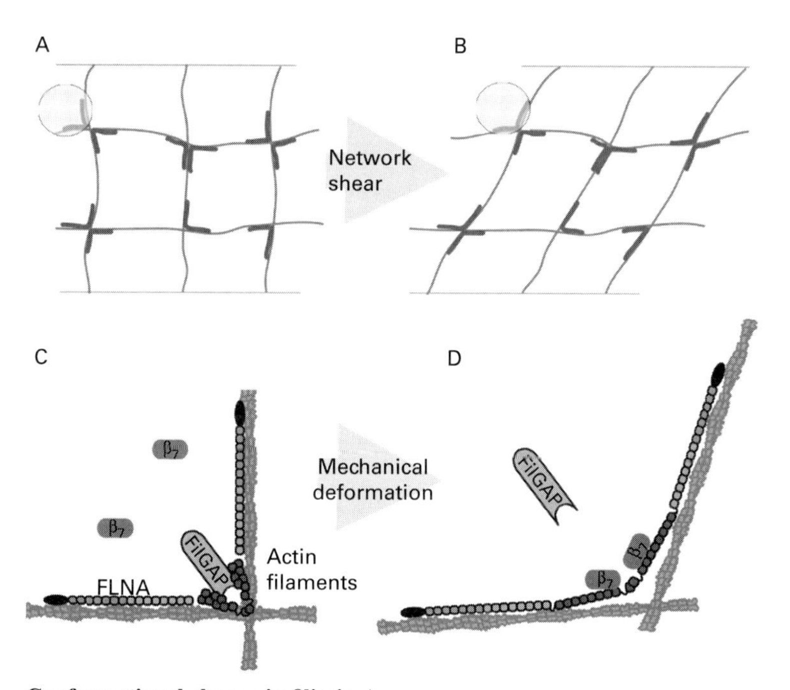

Figure 15.3 **Conformational change in filimin A.**
Filamin A (FLNa) undergoes conformational changes in response to strain, destabilizing the interaction between FilGAP and FLNa's 21st domain and stabilizing interactions between FLNa's 23rd domain and the tail of the beta-7 integrin. This conformational switch enables strain dependent activation of Rac kinase in a FilGAP-dependent fashion.

Reprinted by permission from Macmillan Publishers Ltd, from *Nature* 2011, 478: 260–263, Ehrlicher et al. Mechanical strain in actin networks regulates FilGAP and integrin binding to filamin A. Copyright 2011.

the ligand-recognizing headpiece to the plasma membrane (Takagi et al. 2002; Luo et al. 2007). A cytoplasmic salt bridge between Asp-723 in the β integrin subunit and Arg-995 in the α integrin subunit (i.e., membrane-proximal clasp) forms to stabilize this conformation and competes for the binding of cytoplasmic integrin-activating proteins (Vinogradova et al. 2002). In a simple description, when integrin-activating proteins, such as talin, bind to the cytoplasmic face and disrupt inter-subunit bonds, they facilitate global conformation changes (e.g., extension) of the protein (Garcia-Alvarez et al. 2003; Calderwood et al. 1999). Interestingly, this global conformation change is linked to enhanced ligand-binding affinity, coupling global conformational changes to local conformation shifts within the ligand recognition site (Takagi et al. 2002; Kim et al. 2011). The significance of these conformational changes has been recently verified in single molecules using ICAM-1 and $\alpha_L\beta_2$ as a representative system. Using a biomembrane force probe, the conformation of integrin and bond kinetics of the $\alpha_L\beta_2$-ICAM-1 bond were measured in real time, and conformational changes were found to predict functional changes in bond kinetics and force stabilization of the $\alpha_L\beta_2$-ICAM-1 bond (Chen et al. 2012).

15.6.3 Protein-protein bond dynamics as a mechanotransducer

Noncovalent protein-protein bonds are the physical basis for force transmission between the cell and ECM. As such, the kinetics and kinematics of these interactions play an important role in the dynamics of macromolecular protein complex formation, timing and duration of signal propagation, and FA complex disassembly. The transmission of forces through FA molecules depends on their engagement of constituent binding partners to the immobile integrin-ECM interface. This can be observed by measuring the motion of such molecules within the FA structure, with the relative position and motion of molecules largely corresponding to motor force transmission through a hierarchical clutch (Hu et al. 2007; Kanchanawong et al. 2010). For example, vinculin and α-actinin move faster within the complex and are highly correlated with actin motion, while integrins and adapters like FAK and paxillin are more stationary and immediately proximal to the plasma membrane, indicating a spatial hierarchy of transient connections (Hu et al. 2007). Engagement of this clutch through tripartite interactions within the FA transmits actomyosin forces to the extracellular space. Higher actin flow rates are associated with weak or nascent connections to the ECM, and thus a weaker connection between actin and the ECM. As some of the bonds experience lower rupture forces than others, the recruitment of reinforcing interactions may help to strengthen clutch engagement, with the resultant actomyosin forces applied to the ECM (Gardel et al. 2008). For instance, the interaction between fibronectin-bound integrins and actin was measured in vivo to rupture at forces of ~2 pN, and this was critically dependent on talin (Jiang et al. 2003). However vinculin, by virtue of its direct binding to both talin and the actin cytoskeleton can strengthen such interactions, leading to reinforcement of the mechanical connection (Galbraith et al. 2002; Humphries et al. 2007; Dumbauld et al. 2013). Importantly both vinculin and talin have been described to sustain tension in vivo

(Margadant et al. 2011; Grashoff et al. 2010). Therefore, both the binding kinetics and conformation of force-bearing proteins appear to have integral roles in the stability of FAs under tension, the stability, macromolecular composition, and morphology of which is linked to their signaling output.

Most biochemical bonds are weakened by force application, so called slip bonds. In this case, the probability of bond survival decreases exponentially with the level of applied force. However, a subclass of protein-protein interactions are strengthened by force application; specifically, the off-rate of the bound complex is decreased under force, typically within a discrete force regime, and these bonds are called "catch bonds" (Evans and Calderwood 2007). Interestingly, most catch bonds have been found between adhesion receptors and their ligands, suggesting functional selectivity among protein-protein interfaces that routinely and necessarily bears forces. The bond between α5β1 integrin and the cell-binding domain of fibronectin shows a catch bond behavior – bond lifetime is increased within the force regime of 10–30 pN before yielding to slip bond behavior (Kong et al. 2009). Furthermore, a lower bound of molecular tensions on integrin ligands (1–20 pN) have been demonstrated within the focal adhesions of living cells (Liu et al. 2014; Liu et al. 2013). Interestingly, integrin-generated forces have been recently shown to cause unbinding of biotin-streptavidin bonds (Jurchenko et al. 2014), which is the strongest noncovalent bond in nature with a $\sim 10^6$ larger disassociation constant than αvβ3-RGD in solution. How is such a bond broken by integrin-generated forces? Interactions between components within the functional macromolecular adhesion complex, as well as strengthening at the single-bond level, are likely to play a critical role in stabilizing such integrin-ligand bonds. Further support of this comes from evidence that FA formation requires integrin ligands stabilize a force of ~50–60 pN of steady-state tension (Wang and Ha 2013). All together, these data clearly show that receptor affinity change under force is a relevant process by which molecular bonds are stabilized and altered during cellular adhesion.

More complex force-dependent interactions have also been described at the single-molecule level. For example, the interaction between integrin α5β1 and the cell-binding domain of fibronectin can also undergo "cyclic mechanical reinforcement," where prior force application to a receptor-ligand complex is sufficient to induce high affinity binding between successively unloaded interactions (Kong et al. 2013). This mechanical reinforcement is structurally linked to the ability of the hybrid domain of the β1 subunit to change conformations, creating a higher affinity receptor to fibronectin. Intriguingly, α5β1 is able to use a distinct biophysical mechanism to enhance bond stability under force to a different ligand, Thy-1. In the presence of a co-receptor, the heparin sulfate proteoglycan syndecan-4, Thy-1 co-engages both receptors via its distinct binding motifs. After force application to a threshold level (approximately 15 pN), the complex becomes mechanically stiffer as the co-receptor is engaged and the lifetime of the complex is also enhanced, presumably due to conformational changes within the complex (Fiore et al. 2014). These are called "dynamic catch bonds." How might a single receptor exhibit so many complex interaction features during adhesion and mechanosignal transduction? Presumably, distinct conformations are linked with differential mechanoresponsiveness under unique physiological settings, the details of which remain for future study.

15.6.4 Membrane platforms and protein complex spatial localization

What are less well known are processes that mediate FA assembly and mechanotransduction upstream of integrin-ECM binding. However, as these complexes assemble in a rapid yet well-organized manner, additional regulatory mechanisms that contribute to the timing and configuration of these complexes are not surprising. Once integrins bind an immobilized ligand, FA assembly proceeds from a relatively stable position, microscopically, although FA translocation or frictional slippage does occur (Aratyn-Schaus and Gardel 2010; Smilenov et al. 1999). However preceding immobilization, integrin complexes are ambulatory within the plasma membrane, where they can undergo directional mobility at the leading edge to "probe" for ligand-binding sites or undergo Brownian motion within membrane compartments (Galbraith et al. 2007; Wiseman et al. 2004). For examples, α5 integrins group in clusters of 3–5 molecules in non-FA regions, where they are highly mobile (Wiseman et al. 2004). At the nanoscopic level, single integrins receptors are highly mobile even within FA structures, exhibiting intermittent modes of confinement, diffusion and immobilization depending on their tripartite interaction with the ECM and the cytoskeleton (Rossier et al. 2012). Furthermore, it is known that simply clustering integrins in the absence of ligation or force recruits downstream signaling molecules requisite for early ligand-induced signal generation, such as Src family kinases (SFKs) (Shattil 2005; Boettiger 2012). Intriguingly, not all early signaling intermediates directly bind to integrin receptors, necessitating alternate scaffolding molecules to bridge these interactions.

Lipid rafts are functional subdomains of proteins and lipids that serve as signaling platforms within the plasma membrane (Lingwood and Simons 2010). Typically, they are enriched in cholesterol, sphingolipids, and GPI-anchored proteins and have distinct mobility characteristics within the plasma membrane (Sharma et al. 2004; Goswami et al. 2008). These domains can facilitate the clustering and association of cooperative signaling molecules, thus organizing functional modules within the lipid bilayer. Such mechanisms have been hypothesized to facilitate protein sorting, endocytosis, transmembrane signal transduction, cell-matrix adhesion, and other critical cellular processes (Simons and Ikonen 1997). In artificial membranes, differences in lipid and/or protein composition can give rise to the coexistence of compositionally distinct equilibrium structures, such as liquid-ordered and liquid-disordered phases (van den Bogaart et al. 2011); it has been hypothesized that similar segregation principles might operate in living cells, however cell membranes are composed of extremely diverse molecular constituents that are under non-equilibrium conditions (Simons and Gerl 2010). Much of the current compositional and functional understanding of these membrane subdomains stems from molecular association with detergent-resistant membrane (DRM) fractions, as defined by extraction with cold, non-ionic detergents or disruption of these subdomains by perturbation of membrane cholesterol (Lingwood and Simons 2007). However, recent advances in superresolution microscopy and molecular imaging have demonstrated the existence of such structures within live cells and their exceptionally dynamic characteristics (Sharma et al. 2004; Eggeling et al. 2009; Goswami et al. 2008; van den Bogaart et al. 2011).

It has been long hypothesized that lipid rafts play a critical role in integrin signaling and FAs, as many adhesion-associated signaling molecules, such as SFKs, FAK, and Rac associate with DRM fractions (Palazzo et al. 2004; del Pozo et al. 2004; Shima et al. 2003). FAs also have a higher level of liquid-ordered membrane than do surrounding regions of the FA, suggesting a significant presence of cholesterol-rich membrane domains (Gaus et al. 2006). Intriguingly, the majority of integrins in the nonstimulated or ligated state are in nonraft fractions (Leitinger and Hogg 2002). It was recently demonstrated using superresolution microscopy that inactive integrins exist in nanodomains that are spatially separate, but adjacent to nanodomains of GPI-anchored proteins (van Zanten et al. 2009). These domains coalesce upon integrin binding and are disrupted by cholesterol depletion, which also functionally impairs integrin signaling. Therefore, it appears that under basal conditions, integrins have some affinity toward nanodomains of GPI-anchored proteins, but are not continuous. What is the role of such proximal coupling of these distinct plasma membrane domains? One importance for such a complex might be to spatially couple associated signaling molecules to inactive receptors, such that in response to ligand-induced activation, receptors can rapidly generate downstream signals; potentially critical for receptors that lack intrinsic enzymatic activity, such as integrins. Intriguingly, multiple lipid raft-associated outer membrane proteins, including uPAR, GPI-80, and CD47, have been shown to laterally associate with integrin and alter its signaling function (Yoshitake et al. 2003; Wei et al. 2005; Brown and Frazier 2001; Wei et al. 1996). Although the molecular details of these complexes are not known, it invites questions as to whether these complexes have a similar structure-function, and why lipid raft-associated (more specifically GPI-anchored) proteins are utilized in this function?

References

Alegre-Cebollada, J., Kosuri, P., Giganti, D., Eckels, E., Rivas-Pardo, J. A., Hamdani, N., Warren, C. M., et al. (2014). "S-glutathionylation of cryptic cysteines enhances titin elasticity by blocking protein folding." *Cell* **156**: 1235–1246.

Altroff, H., Choulier, L. and Mardon, H. J. (2003). "Synergistic activity of the ninth and tenth FIII domains of human fibronectin depends upon structural stability." *J Biol Chem* **278**: 491–497.

Ananthanarayanan, B., Kim, Y. and Kumar, S. (2011). "Elucidating the mechanobiology of malignant brain tumors using a brain matrix-mimetic hyaluronic acid hydrogel platform." *Biomaterials* **32**: 7913–7923.

Aratyn-Schaus, Y. and Gardel, M. L. (2010). "Transient frictional slip between integrin and the ECM in focal adhesions under myosin II tension." *Curr Biol* **20**: 1145–1153.

Aratyn-Schaus, Y., Oakes, P. W. and Gardel, M. L. (2011). "Dynamic and structural signatures of lamellar actomyosin force generation." *Mol Biol Cell* **22**: 1330–1339.

Arora, P. D., Narani, N. and Mcculloch, C. A. (1999). "The compliance of collagen gels regulates transforming growth factor-beta induction of alpha-smooth muscle actin in fibroblasts." *Am J Pathol* **154**: 871–882.

Baird, A., Schubert, D., Ling, N. and Guillemin, R. (1988). "Receptor- and heparin-binding domains of basic fibroblast growth factor." *Proc Natl Acad Sci USA* **85**: 2324–2328.

Balaban, N. Q., Schwarz, U. S., Riveline, D., Goichberg, P., Tzur, G., Sabanay, I., Mahalu, D., et al. (2001). "Force and focal adhesion assembly: a close relationship studied using elastic micropatterned substrates." *Nat Cell Biol* **3**: 466–472.

Balestrini, J. L., Chaudhry, S., Sarrazy, V., Koehler, A. and Hinz, B. (2012). "The mechanical memory of lung myofibroblasts." *Integr Biol (Camb)* **4**: 410–421.

Barker, T. H., et al. (2004a). "Thy-1 regulates fibroblast focal adhesions, cytoskeletal organization and migration through modulation of p190 RhoGAP and Rho GTPase activity." *Exp Cell Res* **295**: 488–496.

Barker, T. H., et al. (2004b). "Thrombospondin-1-induced focal adhesion disassembly in fibroblasts requires Thy-1 surface expression, lipid raft integrity, and Src activation." *J Biol Chem* **279**: 23510–23516.

Barry-Hamilton, V., Spangler, R., Marshall, D., McCauley, S., Rodriguez, H. M., Oyasu, M., Mikels, A., et al. (2010). "Allosteric inhibition of lysyl oxidase-like-2 impedes the development of a pathologic microenvironment." *Nat Med* **16**: 1009–1017.

Betz, P., Nerlich, A., Wilske, J., Tubel, J., Penning, R. and Eisenmenger, W. (1993). "Immunohistochemical localization of collagen types I and VI in human skin wounds." *Int J Legal Med* **106**: 31–34.

Bhadriraju, K., Yang, M., Alom Ruiz, S., Pirone, D., Tan, J. and Chen, C. S. (2007). "Activation of ROCK by RhoA is regulated by cell adhesion, shape, and cytoskeletal tension." *Exp Cell Res* **313**: 3616–3623.

Boettiger, D. (2012). "Mechanical control of integrin-mediated adhesion and signaling." *Curr Opin Cell Biol* **24**: 592–599.

Booth, A. J., Hadley, R., Cornett, A. M., Dreffs, A. A., Matthes, S. A., Tsui, J. L., Weiss, K., et al. (2012). "Acellular normal and fibrotic human lung matrices as a culture system for in vitro investigation." *Am J Respir Crit Care Med* **186**: 866–876.

Brown, A. C., Baker, S. R., Douglas, A. M., Keating, M., Alvarez-Elizondo, M. B., Botvinick, E. L., Guthold, M., Barker, T. H. (2015). "Molecular interference of fibron's divalent polymerization mechanism enables modulation of multiscale material properties." *Biomaterials* **49**: 27–36.

Brown, A. C., Fiore, V. F., Sulchek, T. A. and Barker, T. H. (2013). "Physical and chemical microenvironmental cues orthogonally control the degree and duration of fibrosis-associated epithelial-to-mesenchymal transitions." *J Pathol* **229**: 25–35.

Brown, A. C., Rowe, J. A. and Barker, T. H. (2011). "Guiding epithelial cell phenotypes with engineered integrin-specific recombinant fibronectin fragments." *Tissue Eng Part A* **17**: 139–150.

Brown, E. J. and Frazier, W. A. (2001). "Integrin-associated protein (CD47) and its ligands." *Trends Cell Biol* **11**: 130–135.

Calderwood, D. A., Zent, R., Grant, R., Rees, D. J., Hynes, R. O. and Ginsberg, M. H. (1999). "The Talin head domain binds to integrin beta subunit cytoplasmic tails and regulates integrin activation." *J Biol Chem* **274**: 28071–28074.

Chang, H. Y., Chi, J. T., Dudoit, S., Bondre, C., Van De Rijn, M., Botstein, D. and Brown, P. O. (2002). "Diversity, topographic differentiation, and positional memory in human fibroblasts." *Proc Natl Acad Sci USA* **99**: 12877–12882.

Chaudhuri, O., Koshy, S. T., Branco Da Cunha, C., Shin, J. W., Verbeke, C. S., Allison, K. H. and Mooney, D. J. (2014). "Extracellular matrix stiffness and composition jointly regulate the induction of malignant phenotypes in mammary epithelium." *Nat Mater* **13**: 970–978.

Chen, W., Lou, J., Evans, E. A. and Zhu, C. (2012). "Observing force-regulated conformational changes and ligand dissociation from a single integrin on cells." *J Cell Biol* **199**: 497–512.

Choi, C. K., Vicente-Manzanares, M., Zareno, J., Whitmore, L. A., Mogilner, A. and Horwitz, A. R. (2008). "Actin and alpha-actinin orchestrate the assembly and maturation of nascent adhesions in a myosin II motor-independent manner." *Nat Cell Biol* **10**: 1039–1050.

Choquet, D., Felsenfeld, D. P. and Sheetz, M. P. (1997). "Extracellular matrix rigidity causes strengthening of integrin-cytoskeleton linkages." *Cell* **88**: 39–48.

Chrzanowska-Wodnicka, M. and Burridge, K. (1996). "Rho-stimulated contractility drives the formation of stress fibers and focal adhesions." *J Cell Biol* **133**: 1403–1415.

Del Pozo, M. A., Alderson, N. B., Kiosses, W. B., Chiang, H. H., Anderson, R. G. and Schwartz, M. A. (2004). "Integrins regulate Rac targeting by internalization of membrane domains." *Science* **303**: 839–842.

Del Rio, A., Perez-Jimenez, R., Liu, R., Roca-Cusachs, P., Fernandez, J. M. and Sheetz, M. P. (2009). "Stretching single talin rod molecules activates vinculin binding." *Science* **323**: 638–641.

Discher, D. E., Janmey, P. and Wang, Y. L. (2005). "Tissue cells feel and respond to the stiffness of their substrate." *Science* **310**: 1139–1143.

Dolhnikoff, M., Mauad, T. and Ludwig, M. S. (1999). "Extracellular matrix and oscillatory mechanics of rat lung parenchyma in bleomycin-induced fibrosis." *Am J Respir Crit Care Med* **160**: 1750–1757.

Driskell, R. R., Lichtenberger, B. M., Hoste, E., Kretzschmar, K., Simons, B. D., Charalambous, M., Ferron, S. R., et al. (2013). "Distinct fibroblast lineages determine dermal architecture in skin development and repair." *Nature* **504**: 277–281.

Dumbauld, D. W., Lee, T. T., Singh, A., Scrimgeour, J., Gersbach, C. A., Zamir, E. A., Fu, J., et al. (2013). "How vinculin regulates force transmission." *Proc Natl Acad Sci USA* **110**: 9788–9793.

Ebihara, T., Venkatesan, N., Tanaka, R. and Ludwig, M. S. (2000). "Changes in extracellular matrix and tissue viscoelasticity in bleomycin-induced lung fibrosis. Temporal aspects." *Am J Respir Crit Care Med* **162**: 1569–1576.

Eggeling, C., Ringemann, C., Medda, R., Schwarzmann, G., Sandhoff, K., Polyakova, S., Belov, V. N., et al. (2009). "Direct observation of the nanoscale dynamics of membrane lipids in a living cell." *Nature* **457**: 1159–1162.

Ehrlicher, A. J., Nakamura, F., Hartwig, J. H., Weitz, D. A. and Stossel, T. P. (2011). "Mechanical strain in actin networks regulates FilGAP and integrin binding to filamin A." *Nature* **478**: 260–263.

Engler, A. J., Griffin, M. A., Sen, S., Bonnemann, C. G., Sweeney, H. L. and Discher, D. E. (2004). "Myotubes differentiate optimally on substrates with tissue-like stiffness: pathological implications for soft or stiff microenvironments." *J Cell Biol* **166**: 877–887.

Engler, A. J., Sen, S., Sweeney, H. L. and Discher, D. E. (2006). "Matrix elasticity directs stem cell lineage specification." *Cell* **126**: 677–689.
Evans, E. A. and Calderwood, D. A. (2007). "Forces and bond dynamics in cell adhesion." *Science* **316**: 1148–1153.
Fiore, V. F., Ju, L., Chen, Y., Zhu, C. and Barker, T. H. (2014). "Dynamic catch of a Thy-1-alpha5beta1+syndecan-4 trimolecular complex." *Nat Commun* **5**: 4886.
Friedland, J. C., Lee, M. H. and Boettiger, D. (2009). "Mechanically activated integrin switch controls alpha5beta1 function." *Science* **323**: 642–644.
Fritsch, C., Orian-Rousseaul, V., Lefebvre, O., Simon-Assmann, P., Reimund, J. M., Duclos, B. and Kedinger, M. (1999). "Characterization of human intestinal stromal cell lines: response to cytokines and interactions with epithelial cells." *Exp Cell Res* **248**: 391–406.
Galbraith, C. G., Yamada, K. M. and Galbraith, J. A. (2007). "Polymerizing actin fibers position integrins primed to probe for adhesion sites." *Science* **315**: 992–995.
Galbraith, C. G., Yamada, K. M. and Sheetz, M. P. (2002). "The relationship between force and focal complex development." *J Cell Biol* **159**: 695–705.
Garcia-Alvarez, B., De Pereda, J. M., Calderwood, D. A., Ulmer, T. S., Critchley, D., Campbell, I. D., Ginsberg, M. H. et al. (2003). "Structural determinants of integrin recognition by talin." *Mol Cell* **11**: 49–58.
Gardel, M. L., Sabass, B., Ji, L., Danuser, G., Schwarz, U. S. and Waterman, C. M. (2008). "Traction stress in focal adhesions correlates biphasically with actin retrograde flow speed." *J Cell Biol* **183**: 999–1005.
Gardel, M. L., Schneider, I. C., Aratyn-Schaus, Y. and Waterman, C. M. (2010). "Mechanical integration of actin and adhesion dynamics in cell migration." *Annu Rev Cell Dev Biol* **26**: 315–333.
Gaus, K., Le Lay, S., Balasubramanian, N. and Schwartz, M. A. (2006). "Integrin-mediated adhesion regulates membrane order." *J Cell Biol* **174**: 725–734.
Geiger, B., Spatz, J. P. and Bershadsky, A. D. (2009). "Environmental sensing through focal adhesions." *Nat Rev Mol Cell Biol* **10**: 21–33.
Geiger, B. and Yamada, K. M. (2011). "Molecular architecture and function of matrix adhesions." *Cold Spring Harb Perspect Biol* **3**.
Goetz, J. G., Minguet, S., Navarro-Lerida, I., Lazcano, J. J., Samaniego, R., Calvo, E., Tello, M., et al. (2011). "Biomechanical remodeling of the microenvironment by stromal caveolin-1 favors tumor invasion and metastasis." *Cell* **146**: 148–163.
Goffin, J. M., Pittet, P., Csucs, G., Lussi, J. W., Meister, J. J. and Hinz, B. (2006). "Focal adhesion size controls tension-dependent recruitment of alpha-smooth muscle actin to stress fibers." *J Cell Biol* **172**: 259–268.
Goswami, D., Gowrishankar, K., Bilgrami, S., Ghosh, S., Raghupathy, R., Chadda, R., Vishwakarma, R., et al. (2008). "Nanoclusters of GPI-anchored proteins are formed by cortical actin-driven activity." *Cell* **135**: 1085–1097.
Grashoff, C., Hoffman, B. D., Brenner, M. D., Zhou, R., Parsons, M., Yang, M. T., Mclean, M. A. (2010). "Measuring mechanical tension across vinculin reveals regulation of focal adhesion dynamics." *Nature* **466**: 263–266.
Guilluy, C., Swaminathan, V., Garcia-Mata, R., O'Brien, E. T., Superfine, R. and Burridge, K. (2011). "The Rho GEFs LARG and GEF-H1 regulate the mechanical response to force on integrins." *Nat Cell Biol* **13**: 722–727.
Gurtner, G. C., Werner, S., Barrandon, Y. and Longaker, M. T. (2008). "Wound repair and regeneration." *Nature* **453**: 314–321.

Hagood, J. S., Lasky, J. A., Nesbitt, J. E. and Segarini, P. (2001). "Differential expression, surface binding, and response to connective tissue growth factor in lung fibroblast subpopulations." *Chest* **120**: 64S–66S.

Hagood, J. S., Mangalwadi, A., Guo, B., Macewen, M. W., Salazar, L. and Fuller, G. M. (2002). "Concordant and discordant interleukin-1-mediated signaling in lung fibroblast thy-1 subpopulations." *Am J Respir Cell Mol Biol* **26**: 702–708.

Hagood, J. S., Miller, P. J., Lasky, J. A., Tousson, A., Guo, B., Fuller, G. M. and Mcintosh, J. C. (1999). "Differential expression of platelet-derived growth factor-alpha receptor by Thy-1(-) and Thy-1(+) lung fibroblasts." *Am J Physiol* **277**: L218–L224.

Hagood, J. S., Prabhakaran, P., Kumbla, P., Salazar, L., Macewen, M. W., Barker, T. H., Ortiz, L. A., et al. (2005). "Loss of fibroblast Thy-1 expression correlates with lung fibrogenesis." *Am J Pathol* **167**: 365–379.

Hattori, N., Mochizuki, S., Kishi, K., Nakajima, T., Takaishi, H., D'armiento, J. and Okada, Y. (2009). "MMP-13 plays a role in keratinocyte migration, angiogenesis, and contraction in mouse skin wound healing." *Am J Pathol* **175**: 533–546.

Hinz, B., Celetta, G., Tomasek, J. J., Gabbiani, G. and Chaponnier, C. (2001). "Alpha-smooth muscle actin expression upregulates fibroblast contractile activity." *Mol Biol Cell* **12**: 2730–2741.

Horowitz, J. C., Rogers, D. S., Simon, R. H., Sisson, T. H. and Thannickal, V. J. (2008). "Plasminogen activation induced pericellular fibronectin proteolysis promotes fibroblast apoptosis." *Am J Respir Cell Mol Biol* **38**: 78–87.

Hu, K., Ji, L., Applegate, K. T., Danuser, G. and Waterman-Storer, C. M. (2007). "Differential transmission of actin motion within focal adhesions." *Science* **315**: 111–115.

Huang, X., Yang, N., Fiore, V. F., Barker, T. H., Sun, Y., Morris, S. W., Ding, Q., et al. (2012). "Matrix stiffness-induced myofibroblast differentiation is mediated by intrinsic mechanotransduction." *Am J Respir Cell Mol Biol* **47**: 340–348.

Humphries, J. D., Byron, A., Bass, M. D., Craig, S. E., Pinney, J. W., Knight, D. and Humphries, M. J. (2009). "Proteomic analysis of integrin-associated complexes identifies RCC2 as a dual regulator of Rac1 and Arf6." *Sci Signal* **2**: ra51.

Humphries, J. D., Wang, P., Streuli, C., Geiger, B., Humphries, M. J. and Ballestrem, C. (2007). "Vinculin controls focal adhesion formation by direct interactions with talin and actin." *J Cell Biol* **179**: 1043–1057.

Hynes, R. O. (2002). "Integrins: bidirectional, allosteric signaling machines." *Cell* **110**: 673–687.

Ingber, D. E. (2003). "Tensegrity I. Cell structure and hierarchical systems biology." *J Cell Sci* **116**: 1157–1173.

Ingber, D. E., Madri, J. A. and Folkman, J. (1987). "Endothelial growth factors and extracellular matrix regulate DNA synthesis through modulation of cell and nuclear expansion." *In Vitro Cell Dev Biol* **23**: 387–394.

Jiang, G., Giannone, G., Critchley, D. R., Fukumoto, E. and Sheetz, M. P. (2003). "Two-piconewton slip bond between fibronectin and the cytoskeleton depends on talin." *Nature* **424**: 334–337.

Jiang, G., Huang, A. H., Cai, Y., Tanase, M. and Sheetz, M. P. (2006). "Rigidity sensing at the leading edge through alphavbeta3 integrins and RPTPalpha." *Biophys J* **90**: 1804–1809.

Johnson, C. P., Tang, H. Y., Carag, C., Speicher, D. W. and Discher, D. E. (2007). "Forced unfolding of proteins within cells." *Science* **317**: 663–666.
Jurchenko, C., Chang, Y., Narui, Y., Zhang, Y. and Salaita, K. S. (2014). "Integrin-generated forces lead to streptavidin-biotin unbinding in cellular adhesions." *Biophys J* **106**: 1436–1446.
Kanchanawong, P., Shtengel, G., Pasapera, A. M., Ramko, E. B., Davidson, M. W., Hess, H. F. and Waterman, C. M. (2010). "Nanoscale architecture of integrin-based cell adhesions." *Nature* **468**: 580–584.
Kapanci, Y., Desmouliere, A., Pache, J. C., Redard, M. and Gabbiani, G. (1995). "Cytoskeletal protein modulation in pulmonary alveolar myofibroblasts during idiopathic pulmonary fibrosis. Possible role of transforming growth factor beta and tumor necrosis factor alpha." *Am J Respir Crit Care Med* **152**: 2163–2169.
Kendall, R. T. and Feghali-Bostwick, C. A. (2011). "Fibroblasts in fibrosis: novel roles and mediators." *Front Pharmacol* **5**: 123.
Kim, C., Ye, F. and Ginsberg, M. H. (2011). "Regulation of integrin activation." *Annu Rev Cell Dev Biol* **27**: 321–345.
Klingberg, F., Hinz, B. and White, E. S. (2013). "The myofibroblast matrix: implications for tissue repair and fibrosis." *J Pathol* **229**: 298–309.
Kong, F., Garcia, A. J., Mould, A. P., Humphries, M. J. and Zhu, C. (2009). "Demonstration of catch bonds between an integrin and its ligand." *J Cell Biol* **185**: 1275–1284.
Kong, F., Li, Z., Parks, W. M., Dumbauld, D. W., Garcia, A. J., Mould, A. P., Humphries, M. J. and Zhu, C. (2013). "Cyclic mechanical reinforcement of integrin-ligand interactions." *Mol Cell* **49**: 1060–1068.
Kostic, A. and Sheetz, M. P. (2006). "Fibronectin rigidity response through Fyn and p130Cas recruitment to the leading edge." *Mol Biol Cell* **17**: 2684–2695.
Krammer, A., Lu, H., Isralewitz, B., Schulten, K. and Vogel, V. (1999). "Forced unfolding of the fibronectin type III module reveals a tensile molecular recognition switch." *Proc Natl Acad Sci USA* **96**: 1351–1356.
Kulasekaran, P., Scavone, C. A., Rogers, D. S., Arenberg, D. A., Thannickal, V. J. and Horowitz, J. C. (2009). "Endothelin-1 and transforming growth factor-beta1 independently induce fibroblast resistance to apoptosis via AKT activation." *Am J Respir Cell Mol Biol* **4**: 484–493.
Kuo, J. C., Han, X., Hsiao, C. T., Yates, J. R., 3rd, and Waterman, C. M. (2011). "Analysis of the myosin-II-responsive focal adhesion proteome reveals a role for β-Pix in negative regulation of focal adhesion maturation." *Nat Cell Biol* **13**: 383–393.
Leahy, D. J., Aukhil, I. and Erickson, H. P. (1996). "2.0 A crystal structure of a four-domain segment of human fibronectin encompassing the RGD loop and synergy region." *Cell* **84**: 155–164.
Leitinger, B. and Hogg, N. (2002). "The involvement of lipid rafts in the regulation of integrin function." *J Cell Sci* **115**: 963–972.
Levental, K. R., Yu, H., Kass, L., Lakins, J. N., Egeblad, M., Erler, J. T., Fong, S. F., et al. (2009). "Matrix crosslinking forces tumor progression by enhancing integrin signaling." *Cell* **139**: 891–906.
Li, L., Huang, H. H., Badilla, C. L. and Fernandez, J. M. (2005). "Mechanical unfolding intermediates observed by single-molecule force spectroscopy in a fibronectin type III module." *J Mol Biol* **345**: 817–826.

Lingwood, D. and Simons, K. (2007). "Detergent resistance as a tool in membrane research." *Nat Protoc* **2**: 2159–2165.

Lingwood, D. and Simons, K. (2010). "Lipid rafts as a membrane-organizing principle." *Science* **327**: 46–50.

Little, W. C., Schwartlander, R., Smith, M. L., Gourdon, D. and Vogel, V. (2009). "Stretched extracellular matrix proteins turn fouling and are functionally rescued by the chaperones albumin and casein." *Nano Lett* **9**: 4158–4167.

Liu, F., Mih, J. D., Shea, B. S., Kho, A. T., Sharif, A. S., Tager, A. M. and Tschumperlin, D. J. (2010). "Feedback amplification of fibrosis through matrix stiffening and COX-2 suppression." *J Cell Biol* **190**: 693–706.

Liu, Y., Medda, R., Liu, Z., Galior, K., Yehl, K., Spatz, J. P., Cavalcanti-Adam, E. A. et al. (2014). "Nanoparticle tension probes patterned at the nanoscale: impact of integrin clustering on force transmission." *Nano Lett* **14**: 5539–5546.

Liu, Y., Yehl, K., Narui, Y. and Salaita, K. (2013). "Tension sensing nanoparticles for mechano-imaging at the living/nonliving interface." *J Am Chem Soc* **135**: 5320–5323.

Luo, B. H., Carman, C. V. and Springer, T. A. (2007). "Structural basis of integrin regulation and signaling." *Annu Rev Immunol* **25**: 619–647.

Machacek, M., Hodgson, L., Welch, C., Elliott, H., Pertz, O., Nalbant, P., Abell, A., et al. (2009). "Coordination of Rho GTPase activities during cell protrusion." *Nature* **461**: 99–103.

Margadant, F., Chew, L. L., Hu, X., Yu, H., Bate, N., Zhang, X. and Sheetz, M. (2011). "Mechanotransduction in vivo by repeated talin stretch-relaxation events depends upon vinculin." *PLoS Biol* **9**: e1001223.

Markowski, M. C., Brown, A. C. and Barker, T. H. (2012). "Directing epithelial to mesenchymal transition through engineered microenvironments displaying orthogonal adhesive and mechanical cues." *J Biomed Mater Res A* **100**: 2119–2127.

Martino, M. M., Briquez, P. S., Guc, E., Tortelli, F., Kilarski, W. W., Metzger, S., Rice, J. J., (2011). "Growth factors engineered for super-affinity to the extracellular matrix enhance tissue healing." *Science* **343**: 885–888.

Martino, M. M., Briquez, P. S., Ranga, A., Lutolf, M. P. and Hubbell, J. A. (2013). "Heparin-binding domain of fibrin(ogen) binds growth factors and promotes tissue repair when incorporated within a synthetic matrix." *Proc Natl Acad Sci USA* **110**: 4563–4568.

Martino, M. M., Mochizuki, M., Rothenfluh, D. A., Rempel, S. A., Hubbell, J. A. and Barker, T. H. (2009). "Controlling integrin specificity and stem cell differentiation in 2D and 3D environments through regulation of fibronectin domain stability." *Biomaterials* **30**: 1089–1097.

Miroshnikova, Y. A., Jorgens, D. M., Spirio, L., Auer, M., Sarang-Sieminski, A. L. and Weaver, V. M. (2011). "Engineering strategies to recapitulate epithelial morphogenesis within synthetic three-dimensional extracellular matrix with tunable mechanical properties." *Phys Biol* **8**: 026013.

Moore, S. W., Roca-Cusachs, P. and Sheetz, M. P. (2010). "Stretchy proteins on stretchy substrates: the important elements of integrin-mediated rigidity sensing." *Dev Cell* **19**: 194–206.

Na, S., Collin, O., Chowdhury, F., Tay, B., Ouyang, M., Wang, Y. and Wang, N. (2008). "Rapid signal transduction in living cells is a unique feature of mechanotransduction." *Proc Natl Acad Sci USA*, **105**: 6626–6631.

Nakamura, F., Song, M., Hartwig, J. H. and Stossel, T. P. (2014). "Documentation and localization of force-mediated filamin A domain perturbations in moving cells." *Nat Commun* **5**: 4656.
Ng, S. P., Billings, K. S., Randles, L. G. and Clarke, J. (2008). "Manipulating the stability of fibronectin type III domains by protein engineering." *Nanotechnology* **19**: 384023.
Nobes, C. D. and Hall, A. (1995). "Rho, rac, and cdc42 GTPases regulate the assembly of multimolecular focal complexes associated with actin stress fibers, lamellipodia, and filopodia." *Cell* **81**: 53–62.
Nurden, A. T., Nurden, P., Sanchez, M., Andia, I. and Anitua, E. (2008). "Platelets and wound healing." *Front Biosci* **13**: 3532–3548.
Oakes, P. W., Beckham, Y., Stricker, J. and Gardel, M. L. (2012). "Tension is required but not sufficient for focal adhesion maturation without a stress fiber template." *J Cell Biol* **196**: 363–374.
Oberhauser, A. F., Badilla-Fernandez, C., Carrion-Vazquez, M. and Fernandez, J. M. (2002). "The mechanical hierarchies of fibronectin observed with single-molecule AFM." *J Mol Biol* **319**: 433–447.
Palazzo, A. F., Eng, C. H., Schlaepfer, D. D., Marcantonio, E. E. and Gundersen, G. G. (2004). "Localized stabilization of microtubules by integrin- and FAK-facilitated Rho signaling." *Science* **303**: 836–839.
Parker, M. W., Rossi, D., Peterson, M., Smith, K., Sikstrom, K., White, E. S., Connett, J. E. (2014). "Fibrotic extracellular matrix activates a profibrotic positive feedback loop." *J Clin Invest* **124**: 1622–1635.
Pasapera, A. M., Schneider, I. C., Rericha, E., Schlaepfer, D. D. and Waterman, C. M. (2010). "Myosin II activity regulates vinculin recruitment to focal adhesions through FAK-mediated paxillin phosphorylation." *J Cell Biol* **188**: 877–890.
Paszek, M. J., Zahir, N., Johnson, K. R., Lakins, J. N., Rozenberg, G. I., Gefen, A., Reinhart-King, C. A., et al. (2005). "Tensional homeostasis and the malignant phenotype." *Cancer Cell* **8**: 241–254.
Pelham, R. J., Jr. and Wang, Y. (1997). "Cell locomotion and focal adhesions are regulated by substrate flexibility." *Proc Natl Acad Sci USA* **94**: 13661–13665.
Ponti, A., Machacek, M., Gupton, S. L., Waterman-Storer, C. M. and Danuser, G. (2004). "Two distinct actin networks drive the protrusion of migrating cells." *Science* **305**: 1782–1786.
Popova, A. P., Bozyk, P. D., Goldsmith, A. M., Linn, M. J., Lei, J., Bentley, J. K. and Hershenson, M. B. (2010). "Autocrine production of TGF-beta1 promotes myofibroblastic differentiation of neonatal lung mesenchymal stem cells." *Am J Physiol Lung Cell Mol Physiol* **298**: L735–L743.
Ridley, A. J. and Hall, A. (1992). "The small GTP-binding protein rho regulates the assembly of focal adhesions and actin stress fibers in response to growth factors." *Cell* **70**: 389–399.
Riveline, D., Zamir, E., Balaban, N. Q., Schwarz, U. S., Ishizaki, T., Narumiya, S., Kam, Z., et al. (2001). "Focal contacts as mechanosensors: externally applied local mechanical force induces growth of focal contacts by an mDia1-dependent and ROCK-independent mechanism." *J Cell Biol* **153**: 1175–1186.
Rossier, O., Octeau, V., Sibarita, J. B., Leduc, C., Tessier, B., Nair, D., Gatterdam, V., (2012). "Integrins beta1 and beta3 exhibit distinct dynamic nanoscale organizations inside focal adhesions." *Nat Cell Biol* **14**: 1057–1067.

Rubashkin, M. G., Cassereau, L., Bainer, R., Dufort, C. C., Yui, Y., Ou, G., Paszek, M. J. (2014). "Force engages vinculin and promotes tumor progression by enhancing PI3K activation of phosphatidylinositol (3,4,5)-triphosphate." *Cancer Res* **74**: 4597–4611.

Saez, A., Buguin, A., Silberzan, P. and Ladoux, B. (2005). "Is the mechanical activity of epithelial cells controlled by deformations or forces?" *Biophys J* **89**: L52–54.

Sanders, Y. Y., Kumbla, P. and Hagood, J. S. (2007). "Enhanced myofibroblastic differentiation and survival in Thy-1(-) lung fibroblasts." *Am J Respir Cell Mol Biol* **36**: 226–235.

Sanders, Y. Y., Pardo, A., Selman, M., Nuovo, G. J., Tollefsbol, T. O., Siegal, G. P. and Hagood, J. S. (2008). "Thy-1 promoter hypermethylation: a novel epigenetic pathogenic mechanism in pulmonary fibrosis." *Am J Respir Cell Mol Biol* **39**: 610–618.

Sansores, R. H., Ramirez-Venegas, A., Perez-Padilla, R., Montano, M., Ramos, C., Becerril, C., Gaxiola, M., et al. (1996). "Correlation between pulmonary fibrosis and the lung pressure-volume curve." *Lung* **174**: 315–323.

Santhanam, L., Tuday, E. C., Webb, A. K., Dowzicky, P., Kim, J. H., Oh, Y. J., Sikka, G., et al. (2010). "Decreased S-nitrosylation of tissue transglutaminase contributes to age-related increases in vascular stiffness." *Circ Res* **107**: 117–125.

Schiller, H. B., Hermann, M. R., Polleux, J., Vignaud, T., Zanivan, S., Friedel, C. C., Sun, Z., et al. 2013. "Beta1– and alphav-class integrins cooperate to regulate myosin II during rigidity sensing of fibronectin-based microenvironments." *Nat Cell Biol* **15**: 625–636.

Sharma, P., Varma, R., Sarasij, R. C., Ira, Gousset, K., Krishnamoorthy, G., Rao, M. et al. (2004). "Nanoscale organization of multiple GPI-anchored proteins in living cell membranes." *Cell* **116**: 577–589.

Shattil, S. J. (2005). "Integrins and Src: dynamic duo of adhesion signaling." *Trends Cell Biol* **15**: 399–403.

Shima, T., Nada, S. and Okada, M. (2003). "Transmembrane phosphoprotein Cbp senses cell adhesion signaling mediated by Src family kinase in lipid rafts." *Proc Natl Acad Sci USA* **100**: 14897–14902.

Simons, K. and Gerl, M. J. (2010). "Revitalizing membrane rafts: new tools and insights." *Nat Rev Mol Cell Biol* **11**: 688–699.

Simons, K. and Ikonen, E. (1997). "Functional rafts in cell membranes." *Nature* **387**: 569–572.

Smilenov, L. B., Mikhailov, A., Pelham, R. J., Marcantonio, E. E. and Gundersen, G. G. (1999). "Focal adhesion motility revealed in stationary fibroblasts." *Science* **286**: 1172–1174.

Smith, M. L., Gourdon, D., Little, W. C., Kubow, K. E., Eguiluz, R. A., Luna-Morris, S. and Vogel, V. (2007). "Force-induced unfolding of fibronectin in the extracellular matrix of living cells." *PLoS Biol*: e268.

Sohier, J., Carubelli, I., Sarathchandra, P., Latif, N., Chester, A. H. and Yacoub, M. H. (2014). "The potential of anisotropic matrices as substrate for heart valve engineering." *Biomaterials* **35**: 1833–1844.

Solon, J., Levental, I., Sengupta, K., Georges, P. C. and Janmey, P. A. (2007). "Fibroblast adaptation and stiffness matching to soft elastic substrates." *Biophys J* **93**: 4453–4461.

Sorrell, J. M. and Caplan, A. I. (2009). "Fibroblasts-a diverse population at the center of it all." *Int Rev Cell Mol Biol* **276**: 161–214.

Specks, U., Nerlich, A., Colby, T. V., Wiest, I. and Timpl, R. (1995). "Increased expression of type VI collagen in lung fibrosis." *Am J Respir Crit Care Med* **151**: 1956–1964.

Stabenfeldt, S. E., Gourley, M., Krishnan, L., Hoying, J. B., Barker, T. H. (2012). "Engineering fobrin polymers through engagement of alternative polymerization mechanisms." *Biomaterials* **33**: 535–544.

Stricker, J., Aratyn-Schaus, Y., Oakes, P. W. and Gardel, M. L. (2011). "Spatiotemporal constraints on the force-dependent growth of focal adhesions." *Biophys J* **100**: 2883–2893.

Swift, J., Ivanovska, I. L., Buxboim, A., Harada, T., Dingal, P. C., Pinter, J., Pajerowski, J. D., et al. (2013). "Nuclear lamin-A scales with tissue stiffness and enhances matrix-directed differentiation." *Science* **341**: 1240104.

Takagi, J., Petre, B. M., Walz, T. and Springer, T. A. (2002). "Global conformational rearrangements in integrin extracellular domains in outside-in and inside-out signaling." *Cell* **110**: 599–611.

Tan, Y., Tajik, A., Chen, J., Jia, Q., Chowdhury, F., Wang, L., Chen, J., et al. (2014). "Matrix softness regulates plasticity of tumour-repopulating cells via H3K9 demethylation and Sox2 expression." *Nat Commun* **5**: 4619.

Tee, S. Y., Fu, J., Chen, C. S. and Janmey, P. A. (2011). "Cell shape and substrate rigidity both regulate cell stiffness." *Biophys J* **100**: L25–L27.

Thievessen, I., Thompson, P. M., Berlemont, S., Plevock, K. M., Plotnikov, S. V., Zemljic-Harpf, A., Ross, R. S., et al. (2013). "Vinculin-actin interaction couples actin retrograde flow to focal adhesions, but is dispensable for focal adhesion growth." *J Cell Biol* **202**: 163–177.

Thomas, S. M. and Brugge, J. S. (1997). "Cellular functions regulated by Src family kinases." *Annu Rev Cell Dev Biol* **13**: 513–609.

Tomasek, J. J., Gabbiani, G., Hinz, B., Chaponnier, C. and Brown, R. A. (2002). "Myofibroblasts and mechano-regulation of connective tissue remodelling." *Nat Rev Mol Cell Biol* **3**: 349–363.

Trappmann, B. and Chen, C. S. (2013). "How cells sense extracellular matrix stiffness: a material's perspective." *Curr Opin Biotechnol* **24**: 948–953.

Tse, J. R. and Engler, A. J. (2010). "Preparation of hydrogel substrates with tunable mechanical properties." *Curr Protoc Cell Biol* Chapter 10: Unit 10.16.

Vadillo-Rodriguez, V., Bruque, J. M., Gallardo-Moreno, A. M. and Gonzalez-Martin, M. L. (2013). "Surface-dependent mechanical stability of adsorbed human plasma fibronectin on Ti6Al4V: domain unfolding and stepwise unraveling of single compact molecules." *Langmuir* **29**: 8554–8560.

van den Bogaart, G., Meyenberg, K., Risselada, H. J., Amin, H., Willig, K. I., Hubrich, B. E., Dier, M., et al. (2011). "Membrane protein sequestering by ionic protein-lipid interactions." *Nature* **479**: 552–555.

van Zanten, T. S., Cambi, A., Koopman, M., Joosten, B., Figdor, C. G. and Garcia-Parajo, M. F. (2009). "Hotspots of GPI-anchored proteins and integrin nanoclusters function as nucleation sites for cell adhesion." *Proc Natl Acad Sci USA* **106**: 18557–18562.

Velasquez, L. S., Sutherland, L. B., Liu, Z., Grinnell, F., Kamm, K. E., Schneider, J. W., Olson, E. N. et al. (2013). "Activation of MRTF-A-dependent gene expression with a small molecule promotes myofibroblast differentiation and wound healing." *Proc Natl Acad Sci USA* **110**: 16850–16855.

Vinogradova, O., Velyvis, A., Velyviene, A., Hu, B., Haas, T., Plow, E. and Qin, J. (2002). "A structural mechanism of integrin alpha(IIb)beta(3) 'inside-out' activation as regulated by its cytoplasmic face." *Cell* **110**: 587–597.
von Wichert, G., Jiang, G., Kostic, A., De Vos, K., Sap, J. and Sheetz, M. P. (2003). "RPTP-alpha acts as a transducer of mechanical force on alphav/beta3-integrin-cytoskeleton linkages." *J Cell Biol* **161**: 143–153.
Wang, N., Butler, J. P. and Ingber, D. E. (1993). "Mechanotransduction across the cell surface and through the cytoskeleton." *Science* **260**: 1124–1127.
Wang, X. and Ha, T. (2013). "Defining single molecular forces required to activate integrin and notch signaling." *Science* **340**: 991–994.
Wang, Y., Botvinick, E. L., Zhao, Y., Berns, M. W., Usami, S., Tsien, R. Y. and Chien, S. (2005). "Visualizing the mechanical activation of Src." *Nature* **434**: 1040–1045.
Watanabe, N., Madaule, P., Reid, T., Ishizaki, T., Watanabe, G., Kakizuka, A., Saito, Y., et al. (1997). "p140mDia, a mammalian homolog of Drosophila diaphanous, is a target protein for Rho small GTPase and is a ligand for profilin." *EMBO J* **16**: 3044–3056.
Wei, Y., Czekay, R. P., Robillard, L., Kugler, M. C., Zhang, F., Kim, K. K., Xiong, J. P., (2005). "Regulation of alpha5beta1 integrin conformation and function by urokinase receptor binding." *J Cell Biol* **168**: 501–511.
Wei, Y., Lukashev, M., Simon, D. I., Bodary, S. C., Rosenberg, S., Doyle, M. V. and Chapman, H. A. (1996). "Regulation of integrin function by the urokinase receptor." *Science* **273**: 1551–1555.
Winer, J. P., Oake, S. and Janmey, P. A. (2009). "Non-linear elasticity of extracellular matrices enables contractile cells to communicate local position and orientation." *PLoS One* **4**: e6382.
Wipff, P. J., Rifkin, D. B., Meister, J. J. and Hinz, B. (2007). "Myofibroblast contraction activates latent TGF-beta1 from the extracellular matrix." *J Cell Biol* **179**: 1311–1323.
Wiseman, P. W., Brown, C. M., Webb, D. J., Hebert, B., Johnson, N. L., Squier, J. A., Ellisman, M. H. et al. (2004). "Spatial mapping of integrin interactions and dynamics during cell migration by image correlation microscopy." *J Cell Sci* **117**: 5521–5534.
Wozniak, M. A., Desai, R., Solski, P. A., Der, C. J. and Keely, P. J. (2003). "ROCK-generated contractility regulates breast epithelial cell differentiation in response to the physical properties of a three-dimensional collagen matrix." *J Cell Biol* **163**: 583–595.
Wynn, T. A. (2007). "Common and unique mechanisms regulate fibrosis in various fibroproliferative diseases." *J Clin Invest* **117**: 524–529.
Yeung, T., Georges, P. C., Flanagan, L. A., Marg, B., Ortiz, M., Funaki, M., Zahir, N., et al. (2005). "Effects of substrate stiffness on cell morphology, cytoskeletal structure, and adhesion." *Cell Motil Cytoskeleton* **60**: 24–34.
Yoshitake, H., Takeda, Y., Nitto, T., Sendo, F. and Araki, Y. (2003). "GPI-80, a beta2 integrin associated glycosylphosphatidylinositol-anchored protein, concentrates on pseudopodia without association with beta2 integrin during neutrophil migration." *Immunobiology* **208**: 391–399.

Zhang, Y., Fan, W., Ma, Z., Wu, C., Fang, W., Liu, G. and Xiao, Y. (2010). "The effects of pore architecture in silk fibroin scaffolds on the growth and differentiation of mesenchymal stem cells expressing BMP7." *Acta Biomater* **6**: 3021–3028.
Zhou, Y., Hagood, J. S. and Murphy-Ullrich, J. E. (2004). "Thy-1 expression regulates the ability of rat lung fibroblasts to activate transforming growth factor-beta in response to fibrogenic stimuli." *Am J Pathol* **165**: 659–669.

16 Micropost arrays as a means to assess cardiac muscle cells

Andrea Leonard, Marita L. Rodriguez, and Nathan J. Sniadecki

16.1 Introduction

Cardiomyocytes are specialized muscle cells that reside within the heart. Their primary function is to provide the mechanical power necessary to pump blood throughout the circulatory system, in order to supply the body with essential oxygen and nutrients. Characterizing the mechanical properties of these cells is important both for assessing their contractile function, as well as determining their adaptive response to applied stimuli. This chapter focuses on the use of arrays of microposts to quantify the contractile properties of cardiomyocytes – for example twitch force, velocity, and power – as well as important kinetic properties such as beating frequency, the beat duration, and the time to peak contraction. In the background sections of this chapter, we discuss cardiomyocyte structure-function relationships and review other approaches for measuring cardiomyocyte contractility. We then describe the micropost technique, give an overview of cardiomyocyte studies with microposts, and provide concluding remarks regarding future uses for this assay.

In vitro force assays allow for the study of cardiomyocyte function in a highly controlled environment. However, the microscopic size of individual cardiomyocytes and the low magnitude of contractile forces that they produce limits the number of assays that can be used for these studies. Adult human ventricular cardiomyocytes, the largest human cardiomyocytes, are 17–25 μm in diameter and 60–140 μm in length (Braunwald 2012), and produce contractile forces on the order of micronewtons. In contrast, immature cardiomyocytes can be 30–40 times smaller, and generate contractile forces on the order of nanonewtons (Yang 2014a). Additionally, the cardiac extracellular environment is a dynamic scaffold that provides cardiomyocytes with biochemical and mechanical cues that regulate their developmental state. However, typical force transducers are composed of stagnant engineered materials, such as metal or plastic, which lack these biochemical and biomechanical cues. Micropost arrays have been used to overcome many of these technical and physiological challenges.

In addition to these challenges, there is a lack of available sources of mature human cardiomyocytes for in vitro studies. However, recent advances in stem cell (SC) research have allowed for the production of unlimited sources of human cardiomyocytes. These cells offer new opportunities for cardiac research, including developmental studies, disease modeling, and drug screening (Burridge 2012). Of particular interest is their potential to be used as a source of replacement cardiomyocytes for damaged or diseased

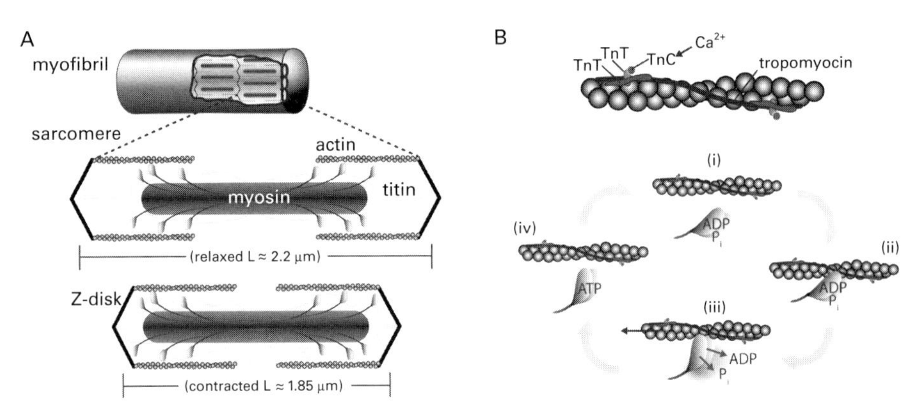

Figure 16.1 **Cardiomyocyte contractility.**
Schematic diagrams of the (A) sarcomere and myofibril structure, and (B) cross-bridge cycle.

cardiac tissue (Braunwald 2012). However, in order for these cells to serve as either a model or a replacement for *in vivo* cardiomyocytes, their physiological properties should first be quantified, to determine how closely these cells are able to replicate the structural and functional characteristics of mature human cardiomyocytes. In an effort to address some of these questions, micropost arrays have been utilized to measure and manipulate the contractile response of human stem cell–derived cardiomyocytes (hSC-CMs).

16.2 Cardiomyocyte structure and contractile mechanism

16.2.1 Cardiomyocyte contractile structure

Myofibrils make up 50–60% of the volume of a cardiomyocyte and are arranged in parallel, giving the cardiomyocyte its rod-like shape. Myofibrils are composed of aligned sarcomeres, which are the fundamental units of contraction within a muscle cell. A sarcomere is composed of actin thin filaments, myosin thick filaments, and a Z-disc at either end (also referred to as Z-line or Z-band) (Fig. 16.1A). During contraction, the overall length of a sarcomere shortens, as a result of myosin heads binding to, and pulling on, neighboring actin thin filaments. This action causes the thin filaments to slide over the thick filaments and generate tension in the muscle. Z-discs separate adjoining sarcomeres and are cross-linked by α-actinin proteins. Actin thin filaments are directly connected to the Z-disc, and myosin thick filaments are tethered to the Z-disc via the spring-like protein titin. The distance between Z-discs is a measure of sarcomere length, which varies during contraction, while Z-disc width is proportional to the number of parallel filaments within the sarcomere.

16.2.2 The cross-bridge cycle and calcium handling

The sliding of the myosin and actin filaments within a sarcomere during contraction occurs as a result of the cross-bridge cycle (Fig. 16.1B). This cycle can be broken up into a series of steps, which rely on the hydrolysis of adenosine triphosphate (ATP) and the presence of Ca^{2+} ions. The cycle is as follows: (i) ATP attached to the myosin head is hydrolyzed, which causes the head to cock; (ii) the cocked head binds to actin and the inorganic phosphate (P_i) is released; (iii) the power stroke ensues, during which the myosin head returns to its bent (low energy) position, and thereby displaces the attached actin filament. During this process, ADP is released. Lastly, (iv) a new ATP binds to myosin, allowing it to release from actin and restart the cross-bridge cycle.

Fluctuations in cardiomyocyte calcium concentrations govern this process. The cycle is initiated by a depolarization wave, which originates at pacemaker cells in the sinoatrial node and reaches the ventricular cardiomyocytes by way of the atrioventricular node and Purkinje fibers. This depolarization causes a small influx of Ca^{2+} into the cytosol through the L-type calcium channels located in transverse tubules (T-tubules). These Ca^{2+} ions trigger a larger release of Ca^{2+} from the sarcoplasmic reticulum (SR) through the ryanodine receptors. At low levels of cytosolic Ca^{2+}, the tropomyosin-troponin complex blocks the myosin-binding site on actin, which restricts cross-bridges from forming. High levels of cytosolic Ca^{2+} promote the binding of Ca^{2+} ions to the TnC subunit of troponin, which induces a conformal change in TnI and exposes the myosin-binding site. During muscle relaxation, Ca^{2+} ions either exit the cell or return to the sarcoplasmic reticulum via the sarco/endoplasmic reticulum Ca^{2+}-ATPase (SERCA) ion pump. As a consequence, Ca^{2+} concentration drops and troponin returns to its inactive state until the next depolarization wave occurs.

16.2.3 Contraction properties as a function of loading conditions

A vital function of the heart is the ability to respond to increased demands for blood flow. The Frank-Starling relationship describes the heart's ability to pump a greater volume of blood with larger force as demand increases, and then rapidly return to its baseline state. Greater volumes of blood generate higher ventricular pressure, and higher ventricular pressure results in increased sarcomere lengths at the cellular level. Early experiments with single cardiomyocytes (see Section 16.3.2) demonstrated that there is a direct relationship between sarcomere length and peak force production, as well as an increased time to peak contraction (Brady 1991).

The exact mechanism for the Frank-Starling relation is not well defined, but could be a combination of several proposed factors. The long-standing hypothesis is that increased myocardial pressure brings the actin and myosin filaments closer together, allowing for better coupling and increased force production. More recently, it has been suggested that titin plays a role in this mechanism (Braunwald 2012). Titin is important in establishing passive tension, and undergoes stretching and relaxation during the contractile cycle. Increased myocardial pressure, which increases sarcomere length, translates to an increased resting tension in titin. In this case, titin acts as a loaded spring, which promotes actin-myosin sliding, and results in quicker production of higher peak forces.

In addition to the response to pressure load, cardiomyocytes also respond to higher stimulation rates with increased force production, that is they exhibit a positive force-frequency relationship (Braunwald 2012). However, at too high stimulation frequencies, contractile forces start to decrease. The decreased force production is due to a breakdown in calcium handling; in other words, the calcium transients aren't able to keep up with the demand.

16.2.4 The structure and calcium handling properties of immature hSC-CMs versus adult cardiomyocytes

While hSC-CMs have the ability to serve as an unlimited source of cells for cardiac research, newly differentiated cardiomyocytes lack both structural and functional maturity. Although hSC-CMs have been observed to spontaneously beat and possess key cardiac markers, such as TnI, the stiffer titin isoform N2B, and α-actinin, they more closely resemble cardiac cells in the fetal heart than those of the fully matured adult heart (Burridge 2012; Yang 2014a). For example, newly differentiated hSC-CMs are small, round, and have a disorganized myofibril structure. Conversely, adult CMs are large, rectangular in shape, and have highly organized and oriented sarcomeres. Like neonatal cardiomyocytes, early stage hSC-CMs have few or no T-tubules (Yang 2014a). An immature cardiomyocyte largely relies upon fluxes of calcium into the cell from an extracellular fluid, rather than the intercellular sarcoplasmic reticulum stores that adult cardiomyocytes rely upon. In general, these properties result in hSC-CMs having unsynchronized and uneven Ca^{2+} dynamics. For a more comprehensive description of the differences between immature hSC-CMs and mature cardiomyocytes, the reader is referred to Yang (2014a).

16.3 Measuring cardiomyocyte contractile force: an overview

Cardiomyocyte contractile forces can be measured on the single cell level, multicellular level, or tissue level. This section of the chapter summarizes the existing methodologies used to determine the contractile properties of these samples, as well as the relative advantages and disadvantages of each technique. A summary of the forces measured using various different techniques is provided in Table 16.1.

16.3.1 Tissue level measurements

Three-dimensional cardiomyocyte tissues can either be harvested from a donor heart or artificially engineered. Primary tissue samples are often millimeters in size, and prepared for mechanical testing by cutting them into thin slices, which are then attached to commercial force transducers (Hasenfuss 1991; Pillekamp 2007a; Pillekamp 2007b; Xi 2010). Engineered tissues samples are generally constructed by seeding cardiomyocytes into an extracellular matrix (ECM) that is cast into a desired shape, using a mold (Eschenhagen 1997; Tulloch 2011; Boudou 2012; Schaaf 2011). These samples can

Table 16.1 Contractile measurements of cardiomyocytes.

Method	Cell type	Contractile measurement		REFS
		Force (μN)	Stress (kPa)	
		Multicellular		
Two-point	Primary tissue	5–148	44	(Hasenfuss 1991; Pillekamp 2007a; Pillekamp 2007b; Xi 2010)
	Engineered	2–200	0.01–0.4	(Eschenhagen 1997; Tulloch 2011; Boudou 2012; Schaaf 2011)
		Single cell: whole-cell or two-point		
Glass pipette cantilever	Adult (rat/guinea pig)	1.2–5.6	53–178	(Fabiato 1975; Shepherd 1990; Brady 1979; Puceat 1990)
	Frog	0.15–0.25	–	(Tarr 1981)
Carbon fiber cantilever	Adult (rat/guinea pig)	0.15–5.7	0.35–41.6	(Le Guennec 1990; Yasuda 2001; Nishimura 2004; Iribe 2007)
Optical fiber	Frog	~0.1	–	(Luo 1991)
Force transducer + piezoelectric translator	Adult (rat/rabbit/human)	2.7–24.5	36.6–51	(Strang 1994; Araujo 1994; Bluh, 1995; van der Velden 1998; Vannier 1996)
Steel foils/cantilever	Adult (rat)	4.3–12	–	(Tasche 1999; Garcia-Webb 2007)
Magnetic bead	Adult (rat)	5–10	–	(Yin 2005)
MEMS	Adult (rat)	5.8–12.6	14.7–23.7	(Lin 2000; Lin 2001)
Micropillars	Neonatal (rat)	0.146–3.5	–	(Tanaka 2006; Kajzar 2008; Taylor 2013)
	hSC-CM	~0.075	–	(Taylor 2013)
Microcantilever	Neonatal (rat)	–	2 – 7	(Park 2005; Park 2006; Kim 2006)
AFM	Adult (rat)	3–12 nN	–	(Li, 2012b)
	hSC-CM	0.1–0.5 nN	–	(Liu 2012a; Wang 2012)
		Single cell: subcellular resolution		
TFM – thin film	Neonatal (rat)	–	0.5–0.8	(Qin 2007)
– PDMS		0.07–1.0	2–5	(Balaban 2001; Hersch 2013)
– PA gel		0.72–1.9	0.05–7.25	(Jacot 2008; Xing; 2014; Hazeltine 2012)
Micropost Array	Adult (rat)	0.067	–	(Zhao 2005; Zhao 2006)
	Neonatal (rat)	0.02–0.243	–	(Rodriguez 2011; Kim 2011; Rodriguez, 2013)
	hSC-CM	0.008–0.018		(Yang 2014b; Rodriguez 2014)

either use commercial force transducers (Eschenhagen 1997; Tulloch 2011) or custom-built force sensing cantilevers (Boudou 2012; Schaaf 2011) to measure their contraction. Although these tissue level approaches arguably replicate the in vivo myocardial environment more closely than single cell studies, the measured contraction can only determine the average or bulk properties of primary or engineered tissues. Additionally, tissue-level force measurements suffer from inhomogeneities between tissues, from the side of the tissue to the middle, and from the outside of the tissue to the core. Therefore measurements are highly dependent on the tissue size.

16.3.2 Single cell measurements

Single cardiomyocyte studies are important because they allow for mechanical studies in the absence of connective tissue and other cell types, which can significantly contribute to the viscoelastic and contractile properties of a sample. Additionally, single cell analysis can be used to study cellular properties governed by subcellular mechanisms.

16.3.2.1 Single whole-cell force measurements

The first experiments to measure contractile forces of single cardiac myocytes employed two-point measuring systems which operated much like a tensile test apparatus (Fig. 16.2A). This approach requires single cardiac myocytes to be isolated, and often skinned in solution, before being physically attached to cantilever force transducers (Brady 1991). This attachment can be achieved by a number of different ways, including: piercing the cell (Fabiato 1975), adherence with poly-L-lysine coated glass (Shepherd 1990; Tarr 1981), using suction pipettes

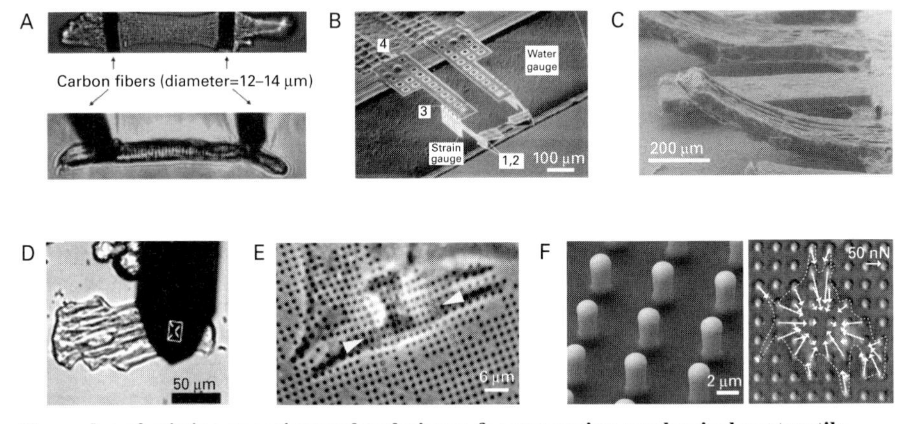

Figure 16.2 Examples of existing experimental techniques for measuring mechanical contractile properties of cardiomyocytes.

(A) Carbon-fiber tensioning force-transducer system (Iribe 2007). (B) MEMS device (Lin 2001). (C) Microcantilever (Kim 2008). (D) Atomic force microscopy (AFM) (Liu 2012b). (E) Traction force microscopy (TFM) (Balaban 2001). (F) Micropost array and cardiomyocyte contracting on micropost array (Rodriguez 2011).

Images reproduced with permission from cited references.

(Brady 1979; Puceat 1990), or via natural adhesion, as is the case with carbon fiber cantilevers (Le Guennec 1990). Cantilever deflections have been measured by laser beam reflection onto a photodiode array (Brady 1979; Fabiato 1975), phototransistors (Shepherd 1990), video image analysis (Tarr 1981; Le Guennec 1990), or with strain gauges (Puceat 1990). In another method, single cardiomyocytes were wrapped over an optical fiber, and deflections in this fiber were used to determine contractile forces (Tung 1986; Luo 1991). More recent incarnations of these two-point force assays attach an individual cell between a piezoelectric translator and a commercial force transducer via fibrin (Copelas 1987), silicone (Strang 1994; Araujo 1994; Bluhm 1995; Metzger 1995; McDonald 1993; van der Velden 1998), or optical glue (Vannier 1996). Experiments aimed at improving the frequency range of these two-point setups attached single cardiomyocytes between tungsten needles adhered to thin steel foils and monitored bending via laser beam reflections (Tasche 1999). Other studies focused on simplifying this cell attachment by using carbon graphite fibers (Yasuda 2001; Nishimura 2004; Iribe 2007). A modular system, consisting of two independently controlled Lorentz force actuators-force transducers, was demonstrated to increase the throughput of single cell measurements (Garcia-Webb 2007).

With the emergence of silicon microfabrication, several other two-point force platforms were developed (Lin 1995; Lin 2000; Lin 2001). In these devices, an individual cell is glued or clamped between two silicon beams (Fig. 16.2B). In the first versions of these systems, beam deflections were measured optically (Lin 2000). Subsequent iterations of this system employed integrated-circuit fabrication to mount strain gauges on the silicon beams, and measure deflections electrically (Lin 2001).

Arguably the largest drawback of the two-point assays discussed so far is the need to manually attach the cardiomyocytes to the force transducing mechanism. Performing this process requires training and takes a substantial amount of time, which limits the overall number of cells that can be tested during a single experiment. One of the first systems to be developed that did not rely upon manual cell attachment, utilized large micropillars made from the silicone rubber polydimethylsiloxane (PDMS) (Tanaka 2006; Kajzar 2008; Taylor 2013). In these experiments, cardiomyocytes were passively attached to these micropillars via an ECM protein coating, which was adsorbed onto the surface of the pillars. A similar approach, using PDMS microcantilevers (Fig. 16.2C), has also been proposed to measure cellular contraction. In this approach, the cardiomyocytes attach along the length of the microcantilevers, as opposed to the tips. However, these studies have primarily focused on the development of cell-driven actuators, where the contraction of the cardiomyocytes drives the deflection of the microcantilevers (Yin 2005; Park 2006; Park 2007).

Several studies have used atomic force microscopy (AFM) to map the mechanical properties of beating cardiomyocytes (Shroff 1995; Domke 1999; Azeloglu 2010; Liu 2012b; Liu 2012a; Wang 2012; Chang 2013; Soufivand 2014) (Fig. 16.2D). Typically, these studies report the mechanical properties in terms of a change in cell height or cell stiffness over the surface of a cardiomyocyte (Shroff 1995; Domke 1999; Azeloglu

2010; Chang 2013; Soufivand 2014). Force measurements, obtained from AFM deflections, have also been reported in several cases (Liu 2012b; Liu 2012a; Wang 2012). However, it is unclear how these force measurements, which report the maximum force at an apical point on the surface of a cardiomyocytes, relate to the total contractile force produced by the cells.

16.3.2.2 Subcellular resolution force measurements

The wrinkling membrane technique (Harris 1980) was, to the best of our knowledge, the first subcellular force assay used to measure the contractile forces produced by cardiomyocytes (Danowski 1992). This study confirmed that cardiomyocytes transmit forces to the substratum through their costameres, an assembly of integrin-associated proteins that are akin to focal adhesions in non-muscle cells, which provide a direct connection between the Z-disc and the ECM (Danowski 1992). Later, traction force microscopy (TFM) was introduced, to more accurately calculate the contractile forces of cells based on the elastic properties of the substrata (Lee 1994; Oliver 1994; Dembo 1996). In TFM, beads are embedded in the elastic substrate and their movements, which are caused by cellular contractions, are tracked. The traction forces are then inferred from the bead displacements using elasticity theory (Fig. 16.2E).

Another method for determining subcellular forces utilizes arrays of microposts (Tan 2003). Unlike micropillar or microcantilever techniques mentioned above, where cardiomyocytes attach to a single beam or span two beams, micropost arrays are designed to have smaller dimension posts and close packing between adjacent microposts. This configuration enables a cardiomyocyte to spread across the tips of multiple microposts, allowing for force measurements at multiple points underneath a single contracting cardiomyocyte (Fig. 16.2F).

16.4 Measuring force, velocity, and power with micropost arrays

16.4.1 Micropost preparation

Arrays of microposts are typically fabricated out of PDMS using a soft lithography process (Tan 2003; Sniadecki 2007b). This process involves: (i) fabrication of a master structure containing an array of microposts; (ii) casting a negative mold from the master; and subsequently (iii) casting the final array of microposts from the negative model (replicating the master) (Fig. 16.3). The dimensions of the microposts are typically 3–4 μm in diameter, 6 μm in center-to-center spacing, and 6–10 μm in height. This technique allows for fabrication of microposts with consistent dimensions, and therefore consistent mechanical properties, from repeated castings of the same master. A detailed description of the fabrication process can be found in Tan (2003).

In order to enable cell attachment on top of the microposts, ECM protein is stamped onto the tips of the posts by microcontact printing (Tan 2003; Sniadecki 2007b). Afterward, the microposts can be coated with a fluorescent marker (such as Alexa Fluor 594 conjugated to bovine serum albumin), which is useful for tracking the location

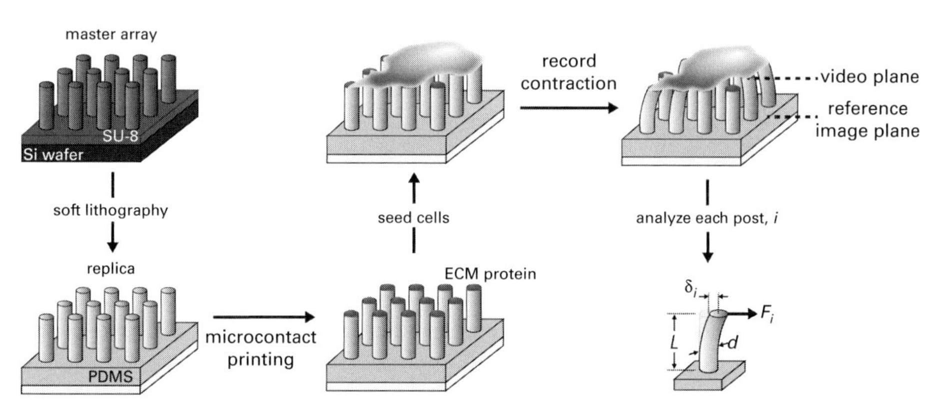

Figure 16.3 **Schematic depicting the fabrication and use of micropost arrays.**

of each micropost. The micropost arrays are then soaked in a surfactant, such as Pluronic F-127, which coats the sidewalls of the posts and base surfaces between the microposts. This coating restricts cell attachment to the tips of the microposts. Finally, the arrays are placed into tissue culture dishes, and the cardiomyocytes are seeded directly onto the microposts at the desired cell density.

16.4.2 Measurements of force, velocity, and power

When an attached cardiomyocyte pulls against the microposts, the top of each post moves relative to its stationary base. By modeling each micropost as a cantilever beam using Euler-Bernoulli beam theory, the following relationship can be used to estimate the force, F_i, applied at the tip of each post, i, beneath the cell:

$$F_i = \left(\frac{3E\pi d^4}{64L^3}\right)\delta_i, \tag{16.1}$$

where E is the Young's Modulus of PDMS, d is the diameter of the micropost, L is its height, and δ_i is its deflection. However, this theory assumes a rigid cantilever base, whereas the base of PDMS microposts have some compliance. Therefore to more accurately calculate the force, a correction factor, which is a function of the micropost dimensions, should be used to account for small rotations at the base of the micropost (Schoen 2010). To measure post displacements, a reference image at the plane near the base of the microposts is obtained. Subsequently, a video at the plane of the tips of the microposts is recorded to obtain the deflection of each tip relative to its base, at each time point (see Fig. 16.3). Videos can be obtained using either phase contrast or fluorescent imaging. There are advantages and disadvantages of each: phase contrast allows for faster image data acquisition; however fluorescent imaging allows for simpler image analysis routines. The velocity, V_i, of a micropost due to the cardiomyocyte contraction can be calculated as the rate of change in its deflection over j time points:

$$V_i = \frac{\delta_{i,j+1} - \delta_{i,j-1}}{t_{j+1} - t_{j-1}} \tag{16.2}$$

Subsequently, the power at post i, P_i, at any given time j is the product of the force and velocity:

$$P_i = F_i V_i \tag{16.3}$$

Although force is most often used as an assessment of cardiomyocyte contractility, power is a more complete metric because it is directly related to the heart's ability to pump blood through the body during the ejection phase of systole (Sonnenblick 1962).

16.5 Micropost arrays in cardiac research

Micropost arrays are particularly attractive for studies on immature cardiomyocytes, such as neonatal cardiomyocytes or hSC-CMs, since they allow for higher spatial resolution compared to conventional two-point methods. Here we highlight several recent studies, which make use of the micropost array technology, revealing the versatility of this methodology.

16.5.1 Substrate stiffness

Previous studies have shown that substrate stiffness can affect the twitch force and velocity produced by individual immature cardiomyocytes, as well as their average spread area, sarcomere length, and expression of cardiac-specific markers (Hersch 2013; Jacot 2008; Jacot 2010; Hazeltine 2012; Engler 2008; Forte 2012). We examined the effect of substrate stiffness on twitch force, velocity, and power, in relationship to measurements of sarcomere lengths, Z-band widths, and intercellular calcium levels in neonatal rat cardiomyocytes (Rodriguez 2011). For these studies, we defined the substrate stiffness in terms of the effective shear modulus of the micropost array, assuming that the array is a homogeneous continuum:

$$G = \frac{\pi E d^4}{32 s^2 L^2}, \tag{16.4}$$

where s is the center-to-center spacing between posts. Measurements of force, velocity, and power were achieved by seeding the cells onto arrays of microposts. In parallel, transients in a cardiomyocyte's calcium levels were assessed via the activity of a calcium-indicating dye. After the live cell experiments, sarcomere length and Z-band width were measured by immunofluorescence.

The main findings of this work were that: (i) twitch force and power increase with increasing substrate stiffness; (ii) these changes were concurrent with increases in average sarcomere length and Z-disc width; as well as (iii) increased intracellular calcium levels (Fig. 16.4). Overall, these results indicate that, within the range of stiffnesses we examined, neonate cardiomyocytes on stiffer substrates develop a higher degree of cytoskeletal maturation and calcium activation, which correlate with enhanced contraction.

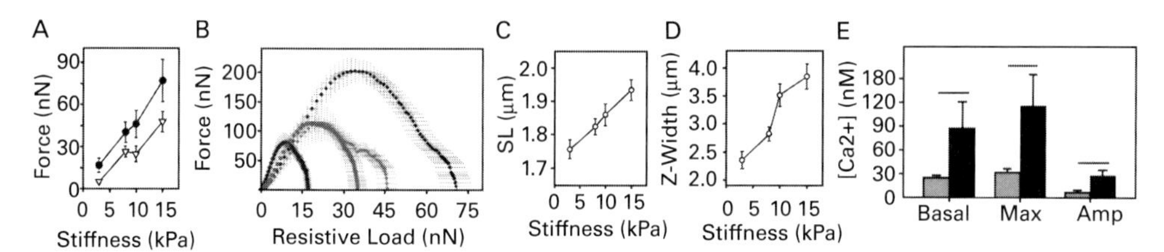

Figure 16.4 **Influence of micropost stiffness on neonate cardiomyocyte.**
(A) Force production, (B) power production, (C) average sarcomere length, (D) average Z-band width, and (E) calcium transients.
Images reproduced with permission from Rodriguez (2011).

16.5.2 Extracellular matrix

The basement membrane of the heart is primarily composed of fibronectin, collagen, and laminin (Parker 2007; Kresh 2011). Previous in vitro work has found that attachment of cardiomyocytes to these proteins is age dependent (Borg 1984; Lundgren 1985b; Lundgren 1985a; Terracio 1991). However, the influence that these matrix proteins have on the attachment and maturation of cardiomyocytes derived from human induced pluripotent stem cells (hiPSC-CM) was unclear.

We examined the effect of ECM protein coating composition on the attachment and maturation of hiPSC-CMs on microposts stamped with laminin, fibronectin, or collagen IV (Rodriguez 2014). When the cells resumed beating five days after seeding them onto the microposts, contractile measurements of force, velocity, and power were taken using arrays of microposts. Following live experiments, the cells were fixed and stained to image their cytoskeletal structure.

These analyses revealed that the hiPSC-CMs seeded onto laminin-stamped microposts attached at a significantly higher rate and spread to a significantly higher degree than cells on fibronectin or collagen IV. Cells on laminin also demonstrated higher contraction and relaxation velocities. Conversely, there were no obvious differences in multinucleation percentage, circularity, sarcomere length, Z-band width, twitch force, twitch power, or beat frequency. These findings indicate that, although the stamped ECM protein had no effect on the majority of the examined cellular properties, the laminin coating resulted in better adherence between the hiPSC-CMs and the microposts and may have led to enhanced contractile maturation of these cells (Fig. 16.5).

16.5.3 Hormonal development

Triiodothyronine (T3), or thyroid hormone, is involved in various different aspects of heart development. In early stages of development, it has been found to regulate the fetal to adult isoform switching of myosin heavy chain (MHC) and titin (Klein 2001; Kruger 2008). During the neonatal state, T3 represses the expression of fetal genes (Klein 2001; Dillmann 2002). And in adult cardiomyocytes, it has been found to regulate cardiac physiology and hypertrophy (Lee 2010; Chattergoon 2012; Ivashchenko 2013). Additionally, in a recent study, T3 treatment was found to modulate the gene expression

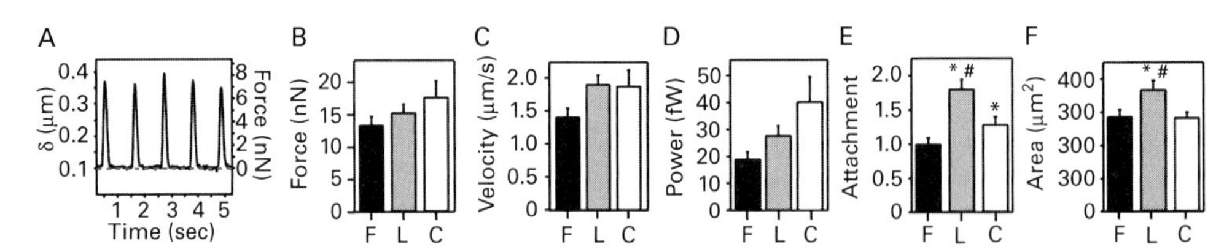

Figure 16.5 **Effect of ECM protein coating on hiPSC-CM maturation.**
(A) Example deflection and force trace for an individual micropost, (B) total twitch force, (C) maximum twitch velocity, (D) total twitch power, (E) normalized attachment, and (F) spread area. For all panels, F denotes fibronectin, L denotes laminin, and C denotes collagen IV.
Images reproduced with permission from Rodriguez (2014).

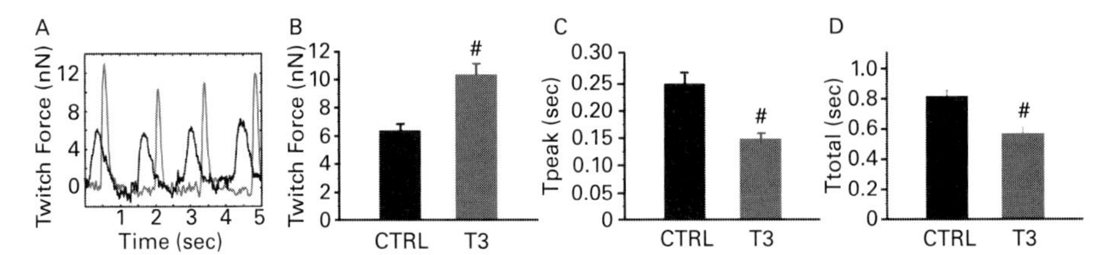

Figure 16.6 **Effect of T3 treatment on hiPSC-CM contractility.**
(A) Representative force traces for control and T3 conditions, (B) total twitch force, (C) time to peak contraction, (D) total twitch time. For all panels, black denotes control and gray denotes T3 treatment.
Images reproduced with permission from Yang (2014b).

profile of hiPSC-CMs to a more adult-like phenotype (Ivashchenko 2013). To determine the effect of T3 on hiPSC-CM maturation, we cultured these cells on microposts in the presence of T3 (Yang 2014b).

For these studies, we determined the contractile forces produced by individual T3-treated cells using the microposts. To more fully characterize the contractile properties of these cells, differences in time to peak contraction, time to 90% relaxation, and twitch duration were also investigated. Functionally, we found that T3-treated hiPSC-CMs exhibited significantly higher twitch forces, a shorter peak to contraction time, and significant decreases in the time to 90% relaxation and total twitch time, when compared to control cells. Structurally, hiPSC-CMs treated with T3 had significantly increased spread areas, anisotropy, and sarcomere lengths. T3 treatment was also found to be associated with reduced cell cycle activity, increases in the rates of calcium release and reuptake, a significant increase in sarcoendoplasmic reticulum ATPase expression, and a significant increase in the maximal mitochondrial respiratory capacity and respiratory reserve capability of hiPSC-CMs. However, there was a decrease in β-MHC expression, which is an indication of cardiomyocyte immaturity. Based on the results summarized in this study (Fig. 16.6), we concluded that T3 enhances hiPSC-CM maturation, and could enhance the utility of hiPSC-CMs for therapeutic, disease modeling, or drug/toxicity screens.

16.5.4 Viral and pharmacological

It has previously been found that externally-applied forces and substrate stiffness have been found to promote cardiomyocyte hypertrophic growth (Simpson 1996; Simpson 1999; Jacot 2008; Engler 2008; Rodriguez 2011); Majkut 2012). It has also been observed that preventing cardiomyocyte contraction suppresses hypertrophic growth and reduces myofibril development (Klein 1985; McDermott 1989). These observations suggest that cardiomyocytes adapt to their mechanical environment by remodeling their myofibril structure, and that this remodeling is dependent upon cellular contraction. To determine whether or not hypertrophic remodeling occurs in response to internal forces, i.e., myofibril forces, we investigated the effect of viral and pharmacological alterations to the force-producing capability of neonate rat cardiomyocytes. The cells were cultured on arrays of microposts to measure cellular contractility, while immunofluorescent staining was used to image their cytoskeletal structure. Enhanced internal forces were achieved virally via adenoviral-mediated overexpression of ribonucleotide reductase (RR) (Regnier 1998a; Regnier 1998b; Regnier 1998c; Regnier 2000; Korte 2011; Nowakowski 2013) and pharmacologically via treatment with EMD 57033 (EMD) (Solaro 1993; Strauss 1994; Kraft 1997; Lipscomb-Allhouse 2001), while decreased internal forces were achieved by treatment with blebbistatin.

We observed that RR-overexpressing and EMD-treated cardiomyocytes produced twitch forces that were similar to those produced by cardiomyocytes cultured under control conditions. However, twitch velocity and twitch power in the treated cells were significantly lower than control cells. Immunofluorescent analysis of α-actinin revealed that RR-overexpression and EMD treatment resulted in cardiomyocytes with reduced spreading and poor myofibril structure. These results suggest that twitch forces act as an internal cue for a tensional homeostasis, which is maintained by altering myofibril structure and, subsequently, cardiac power. This correlation was confirmed by exposing the cells to blebbistatin, which resulted in significant reduction to the contractile and structural properties of these cells. These results are summarized in Fig. 16.7.

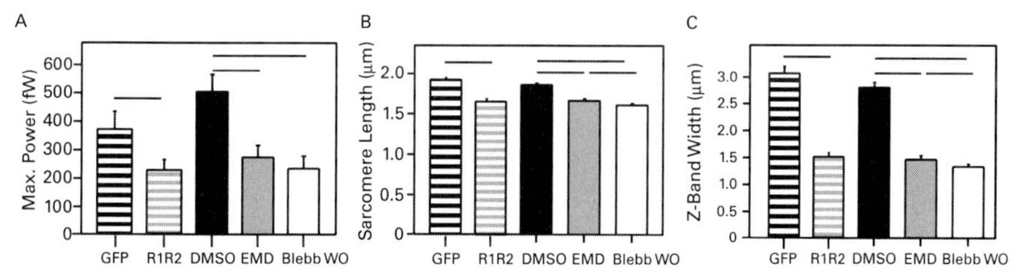

Figure 16.7 **Effect of force-enhancing viral and pharmacological treatments on neonate cardiomyocyte maturation.**
(A) power, (B) sarcomere length, (C) Z-band width. For all panels, black stripes denote GFP control, gray stripes denote GFP and R1R2, solid black denote DMSO control, solid gray denote EMD57033 in DMSO, and solid white denotes conditions following washout of blebbistatin in DMSO.

Data taken from Rodriguez (2013).

16.6 Summary and future outlook

Assessing cardiomyocyte contractility is fundamental to understanding the physiological properties of both individual cardiomyocytes and the heart as a whole. The ability to simply and accurately measure cardiomyocyte contractile properties, including force, velocity, and power, with subcellular resolution can be achieved with the use of micropost arrays. Due to the small size of immature cardiomyocytes and the disorganized structure of their myofibrils, these cells are difficult to study with conventional approaches. Additionally, with microposts, the effective substrate stiffness and ECM composition can be altered, allowing for insightful studies on cardiomyocyte mechanotransduction. While micropost arrays prove to be a versatile experimental tool, improvements could also be made to the existing technology.

Micropost technology would benefit from improvements in spatial resolution, more accurate micropost calibration, control of cell placement, and the ability to impart tension to individual cardiomyocytes. Denser packing of microposts or nanoscale posts would not only improve the spatial resolution of force measurements, but should also improve cell spreading and attachment, since adhesion points are less restrictive. However, the size and spacing of microposts are limited both by the master array micro/nanofabrication process and the PDMS soft lithography process.

Such small-scale posts could also further complicate experimental calibration methods. Currently, the stiffness of microposts is determined either theoretically from dimensions obtain via scanning electron microscopy or experimentally using a calibrated glass micropipette attached to a micromanipulator stage (Sniadecki 2007b), a piezoresistive force cantilever (Kim 2011), or with AFM (Cheng 2010). However, microscale effects, such as an oxide surface layer introduced to PDMS microposts during the cell preparation process, could contribute to additional deviations from the standard theoretical beam theory and corresponding warping correction factors (Schoen 2010). Therefore, micropost technology would benefit from improved or standardized calibration methods. In addition to size and calibration improvements, control over cardiomyocyte placement on the micropost arrays would be beneficial for studying cell-cell interactions as well as effects of cardiomyocyte size and shape restrictions. This could be achieved, in part, by patterned microcontact printing, as already shown for cardiomyocytes on flat surfaces (Bray 2008).

The ability to control cardiomyocyte length and tension during a contractile cycle is essential in determining relationships between contractile function and the cardiovascular demand. External forces could be applied to cardiomyocytes adhered to micropost arrays in a number of ways, such as incorporation of magnetic microposts (Sniadecki 2007a) or by casting microposts directly onto a stretchable substrate (Mann 2012). While the above-mentioned studies would provide key insights into cardiomyocyte contractility and function, we should also simultaneously be working to develop 3-D in vitro assays. It will be interesting to see how these studies compare with 3-D assays in the future.

References

Araujo, A., and Walker, J. W. (1994). Kinetics of tension development in skinned cardiac myocytes measured by photorelease of Ca2+. *Am J Physiol* **267**: H1643–H1653.

Azeloglu, E. U., and Costa, K. D. (2010). Cross-bridge cycling gives rise to spatiotemporal heterogeneity of dynamic subcellular mechanics in cardiac myocytes probed with atomic force microscopy. *Am J Physiol Heart Circ Physiol* **298**: H853–H860.

Balaban, N. Q., Schwarz, U. S., Riveline, D., et al. (2001). Force and focal adhesion assembly: a close relationship studied using elastic micropatterned substrates. *Nature Cell Biology* **3**: 466–472.

Bluhm, W. F., Mcculloch, A. D., and Lew, W. Y. (1995). Active force in rabbit ventricular myocytes. *J Biomech* **28**: 1119–1122.

Borg, T. K., Rubin, K., Lundgren, E., Borg, K., and Obrink, B. (1984). Recognition of extracellular matrix components by neonatal and adult cardiac myocytes. *Dev Biol* **104**: 86–96.

Boudou, T., Legant, W. R., Mu, A. B., et al. (2012). A microfabricated platform to measure and manipulate the mechanics of engineered cardiac microtissues. *Tissue Engineering Part A* **18**: 910–919.

Brady, A. J. (1991). Mechanical properties of isolated cardiac myocytes. *Physiol Rev* **71**: 413–428.

Brady, A. J., Tan, S. T., and Ricchiuti, N. V. (1979). Contractile force measured in unskinned isolated adult rat heart fibres. *Nature* **282**: 728–729.

Braunwald, E., and Bonow, R. O. (2012). *Braunwald's heart disease: a textbook of cardiovascular medicine.* Philadelphia: Saunders.

Bray, M. A., Sheehy, S. P., and Parker, K. K. (2008). Sarcomere alignment is regulated by myocyte shape. *Cell Motility and the Cytoskeleton* **65**: 641–651.

Burridge, P. W., Keller, G., Gold, J. D., and Wu, J. C. (2012). Production of de novo cardiomyocytes: human pluripotent stem cell differentiation and direct reprogramming. *Cell Stem Cell* **10**: 16–28.

Chang, W. T., Yu, D., Lai, Y. C., Lin, K. Y., and Liau, I. (2013). Characterization of the mechanodynamic response of cardiomyocytes with atomic force microscopy. *Anal Chem* **85**: 1395–1400.

Chattergoon, N. N., Giraud, G. D., Louey, S., et al. (2012). Thyroid hormone drives fetal cardiomyocyte maturation. *FASEB J* **26**: 397–408.

Cheng, Q., Sun, Z., Meininger, G. A., and Almasri, M. (2010). Note: Mechanical study of micromachined polydimethylsiloxane elastic microposts. *Rev Sci Instrum* **81**: 106104.

Copelas, L., Briggs, M., Grossman, W., and Morgan, J. P. (1987). A method for recording isometric tension development by isolated cardiac myocytes: transducer attachment with fibrin glue. *Pflugers Arch* **408**: 315–317.

Danowski, B. A., Imanaka-Yoshida, K., Sanger, J. M., and Sanger, J. W. (1992). Costameres are sites of force transmission to the substratum in adult rat cardiomyocytes. *J Cell Biol* **118**: 1411–1420.

Dembo, M., Oliver, T., Ishihara, A., and Jacobson, K. (1996). Imaging the traction stresses exerted by locomoting cells with the elastic substratum method. *Biophys J* **70**: 2008–2022.

Dillmann, W. H. (2002). Cellular action of thyroid hormone on the heart. *Thyroid* **12**: 447–452.

Domke, J., Parak, W. J., George, M., Gaub, H. E., and Radmacher, M. (1999). Mapping the mechanical pulse of single cardiomyocytes with the atomic force microscope. *Eur Biophys J* **28**: 179–186.

Engler, A. J., Carag-Krieger, C., Johnson, C. P., et al. (2008). Embryonic cardiomyocytes beat best on a matrix with heart-like elasticity: scar-like rigidity inhibits beating. *J Cell Sci* **121**: 3794–3802.

Eschenhagen, T., Fink, C., Remmers, U., et al. (1997). Three-dimensional reconstitution of embryonic cardiomyocytes in a collagen matrix: a new heart muscle model system. *FASEB J* **11**: 683–694.

Fabiato, A., and Fabiato, F. (1975). Contractions induced by a calcium-triggered release of calcium from the sarcoplasmic reticulum of single skinned cardiac cells. *J Physiol* **249**: 469–495.

Forte, G., Pagliari, S., Ebara, M., et al. (2012). Substrate stiffness modulates gene expression and phenotype in neonatal cardiomyocytes in vitro. *Tissue Eng Part A* **18**: 1837–1848.

Garcia-Webb, M. G., Taberner, A. J., Hogan, N. C., and Hunter, I. W. (2007). A modular instrument for exploring the mechanics of cardiac myocytes. *Am J Physiol Heart Circ Physiol* **293**: H866–H874.

Harris, A. K., Wild, P., and Stopak, D. (1980). Silicone rubber substrata: a new wrinkle in the study of cell locomotion. *Science* **208**: 177–179.

Hasenfuss, G., Mulieri, L. A., Blanchard, E. M., et al. (1991). Energetics of isometric force development in control and volume-overload human myocardium. Comparison with animal species. *Circulation Research* **68**: 836–846.

Hazeltine, L. B., Simmons, C. S., Salick, M. R., et al. (2012). Effects of substrate mechanics on contractility of cardiomyocytes generated from human pluripotent stem cells. *Int J Cell Biol* **2012**: 508294.

Hersch, N., Wolters, B., Dreissen, G., et al. (2013). The constant beat: cardiomyocytes adapt their forces by equal contraction upon environmental stiffening. *Biol Open* **2**: 351–361.

Iribe, G., Helmes, M., and Kohl, P. (2007). Force-length relations in isolated intact cardiomyocytes subjected to dynamic changes in mechanical load. *Am J Physiol Heart Circ Physiol* **292**: H1487–H1497.

Ivashchenko, C. Y., Pipes, G. C., Lozinskaya, I. M., et al. (2013). Human-induced pluripotent stem cell-derived cardiomyocytes exhibit temporal changes in phenotype. *Am J Physiol Heart Circ Physiol* **305**: H913–H922.

Jacot, J. G., Martin, J. C., and Hunt, D. L. (2010). Mechanobiology of cardiomyocyte development. *J Biomech* **43**: 93–98.

Jacot, J. G., Mcculloch, A. D., and Omens, J. H. (2008). Substrate stiffness affects the functional maturation of neonatal rat ventricular myocytes. *Biophys J* **95**: 3479–3487.

Kajzar, A., Cesa, C. M., Kirchgessner, N., Hoffmann, B., and Merkel, R. (2008). Toward physiological conditions for cell analyses: forces of heart muscle cells suspended between elastic micropillars. *Biophys J* **94**: 1854–1866.

Kim, J., Park, J., Na, K., et al. (2008). Quantitative evaluation of cardiomyocyte contractility in a 3D microenvironment. *J Biomech* **41**: 2396–2401.

Kim, J., Park, J., Ryu, S. K., et al. (2006). Realistic computational modeling for hybrid biopolymer microcantilevers. *Conf Proc IEEE Eng Med Biol Soc* **1**: 2102–2105.
Kim, K., Taylor, R., Sim, J. Y., et al. (2011). Calibrated micropost arrays for biomechanical characterisation of cardiomyocytes. *Micro and Nano Letters* **6**: 317.
Klein, I., Daood, M., and Whiteside, T. (1985). Development of heart cells in culture: studies using an affinity purified antibody to a myosin light chain. *J Cell Physiol* **124**: 49–53.
Klein, I., and Ojamaa, K. (2001). Thyroid hormone and the cardiovascular system. *N Engl J Med* **344**: 501–509.
Korte, F. S., Dai, J., Buckley, K., et al. (2011). Upregulation of cardiomyocyte ribonucleotide reductase increases intracellular 2 deoxy-ATP, contractility, and relaxation. *J Mol Cell Cardiol* **51**: 894–901.
Kraft, T., and Brenner, B. (1997). Force enhancement without changes in cross-bridge turnover kinetics: the effect of EMD 57033. *Biophys J* **72**: 272–281.
Kresh, J. Y., and Chopra, A. (2011). Intercellular and extracellular mechanotransduction in cardiac myocytes. *Pflugers Arch* **462**: 75–87.
Kruger, M., Sachse, C., Zimmermann, W. H., et al. (2008). Thyroid hormone regulates developmental titin isoform transitions via the phosphatidylinositol-3-kinase/ AKT pathway. *Circ Res* **102**: 439–447.
Le Guennec, J. Y., Peineau, N., Argibay, J. A., Mongo, K. G., and Garnier, D. (1990). A new method of attachment of isolated mammalian ventricular myocytes for tension recording: length dependence of passive and active tension. *J Mol Cell Cardiol* **22**: 1083–1093.
Lee, J., Leonard, M., Oliver, T., Ishihara, A., and Jacobson, K. (1994). Traction forces generated by locomoting keratocytes. *J Cell Biol* **127**: 1957–1964.
Lee, Y. K., Ng, K. M., Chan, Y. C., et al. (2010). Triiodothyronine promotes cardiac differentiation and maturation of embryonic stem cells via the classical genomic pathway. *Mol Endocrinol* **24**: 1728–1736.
Lin, G., Palmer, R. E., Pister, K. S., and Roos, K. P. (2001). Miniature heart cell force transducer system implemented in MEMS technology. *IEEE Trans Biomed Eng* **48**: 996–1006.
Lin, G., Pister, K. S. J., and Roos, K. P. (1995). Microscale force-transducer system to quantify isolated heart cell contractile characteristics. *Sensors and Actuators A: Physical* **46**: 233–236.
Lin, G., Pister, K. S. J., and Roos, K. P. (2000). Surface micromachined polysilicon heart cell force transducer. *Journal of Microelectromechanical Systems* **9**: 9–17.
Lipscomb-Allhouse, S., Mulligan, I. P., and Ashley, C. C. (2001). The effects of the inotropic agent EMD 57033 on activation and relaxation kinetics in frog skinned skeletal muscle. *Pflugers Arch* **442**: 171–177.
Liu, J., Sun, N., Bruce, M. A., Wu, J. C., and Butte, M. J. (2012a). Atomic force mechanobiology of pluripotent stem cell-derived cardiomyocytes. *PLoS One* **7**: e37559.
Liu, Y., Feng, J., Shi, L., et al. (2012b). In situ mechanical analysis of cardiomyocytes at nano scales. *Nanoscale* **4**: 99–102.
Lundgren, E., Terracio, L., and Borg, T. K. (1985a). Adhesion of cardiac myocytes to extracellular matrix components. *Basic Res Cardiol* **80**(Suppl 1): 69–74.

Lundgren, E., Terracio, L., Mardh, S. and Borg, T. K. (1985b). Extracellular matrix components influence the survival of adult cardiac myocytes in vitro. *Exp Cell Res* **158**: 371–381.

Luo, C. H., and Tung, L. (1991). Null-balance transducer for isometric force measurements and length control of single heart cells. *IEEE Trans Biomed Eng* **38**: 1165–1174.

Majkut, S. F., and Discher, D. E. (2012). Cardiomyocytes from late embryos and neonates do optimal work and striate best on substrates with tissue-level elasticity: metrics and mathematics. *Biomech Model Mechanobiol* **11**: 1219–1225.

Mann, J. M., Lam, R. H., Weng, S., Sun, Y., and Fu, J. (2012). A silicone-based stretchable micropost array membrane for monitoring live-cell subcellular cytoskeletal response. *Lab Chip* **12**: 731–740.

McDermott, P. J., and Morgan, H. E. (1989). Contraction modulates the capacity for protein synthesis during growth of neonatal heart cells in culture. *Circ Res* **64**: 542–553.

McDonald, K. S., Leiden, J. M., Field, L. J., et al. (1993). Length-dependence of Ca-2+ sensitivity of tension in transgenic mouse myocytes expressing skeletal troponin-C. *Circulation* **88**: 86–86.

Metzger, J. M. (1995). Myosin binding-induced cooperative activation of the thin filament in cardiac myocytes and skeletal muscle fibers. *Biophys J* **68**: 1430–1442.

Nishimura, S., Yasuda, S., Katoh, M., et al. (2004). Single cell mechanics of rat cardiomyocytes under isometric, unloaded, and physiologically loaded conditions. *Am J Physiol Heart Circ Physiol* **287**: H196–H202.

Nowakowski, S. G., Kolwicz, S. C., Korte, F. S., et al. (2013). Transgenic overexpression of ribonucleotide reductase improves cardiac performance. *Proc Natl Acad Sci USA* **110**: 6187–6192.

Oliver, T., Lee, J., and Jacobson, K. 1994. Forces exerted by locomoting cells. *Semin Cell Biol* **5**: 139–147.

Park, J., Kim, I. C., Cha, J., et al. (2007). Mechanotransduction of cardiomyocytes interacting with a thin membrane transducer. *Journal of Micromechanics and Microengineering* **17**: 1162–1167.

Park, J., Kim, J., Roh, D., et al. (2006). Fabrication of complex 3D polymer structures for cell-polymer hybrid systems. *Journal of Micromechanics and Microengineering* **16**: 1614–1619.

Park, J., Ryu, J., Choi, S. K., et al. (2005). Real-time measurement of the contractile forces of self-organized cardiomyocytes on hybrid biopolymer microcantilevers. *Anal Chem* **77**: 6571–6580.

Parker, K. K., and Ingber, D. E. (2007). Extracellular matrix, mechanotransduction and structural hierarchies in heart tissue engineering. *Philos Trans R Soc Lond B Biol Sci* **362**: 1267–1279.

Pillekamp, F., Halbach, M., Reppel, M., et al. (2007a). Neonatal murine heart slices. A robust model to study ventricular isometric contractions. *Cell Physiol Biochem* **20**: 837–846.

Pillekamp, F., Reppel, M., Rubenchyk, O., et al. (2007b). Force measurements of human embryonic stem cell-derived cardiomyocytes in an in vitro transplantation model. *Stem Cells* **25**: 174–180.

Puceat, M., Clement, O., Lechene, P., et al. (1990). Neurohormonal control of calcium sensitivity of myofilaments in rat single heart cells. *Circ Res* **67**: 517–524.

Qin, L., Huang, J. Y., Xiong, C. Y., Zhang, Y. Y., and Fang, J. (2007). Dynamical stress characterization and energy evaluation of single cardiac myocyte actuating on flexible substrate. *Biochem Biophys Res Commun* **360**: 352–356.

Regnier, M., and Homsher, E. (1998a). The effect of ATP analogs on posthydrolytic and force development steps in skinned skeletal muscle fibers. *Biophys J* **74**: 3059–3071.

Regnier, M., Lee, D. M. and Homsher, E. (1998b). ATP analogs and muscle contraction: mechanics and kinetics of nucleoside triphosphate binding and hydrolysis. *Biophys J* **74**: 3044–3058.

Regnier, M., Martyn, D. A. and Chase, P. B. (1998c). Calcium regulation of tension redevelopment kinetics with 2-deoxy-ATP or low [ATP] in rabbit skeletal muscle. *Biophys J* **74**: 2005–2015.

Regnier, M., Rivera, A. J., Chen, Y., and Chase, P. B. (2000). 2-deoxy-ATP enhances contractility of rat cardiac muscle. *Circ Res* **86**: 1211–1217.

Rodriguez, A. G., Han, S. J., Regnier, M., and Sniadecki, N. J. (2011). Substrate stiffness increases twitch power of neonatal cardiomyocytes in correlation with changes in myofibril structure and intracellular calcium. *Biophys J* **101**: 2455–2464.

Rodriguez, A. G., Rodriguez, M. L., Han, S. J., Sniadecki, N. J., and Regnier, M. (2013). Enhanced contractility with 2-deoxy-ATP and EMD 57033 is correlated with reduced myofibril structure and twitch power in neonatal cardiomyocytes. *Integr Biol (Camb)* **5**: 1366–1373.

Rodriguez, M., Graham, B. T., Pabon, L. M., et al. (2014). Measuring the contractile forces of human induced pluripotent stem cell-derived cardiomyocytes with arrays of microposts. *J Biomech Eng* **136**: 051005.

Schaaf, S., Shibamiya, A., Mewe, M., et al. (2011). Human engineered heart tissue as a versatile tool in basic research and preclinical toxicology. *Plos One* **6**.

Schoen, I., Hu, W., Klotzsch, E., and Vogel, V. (2010). Probing cellular traction forces by micropillar arrays: contribution of substrate warping to pillar deflection. *Nano Lett* **10** 182318–182330.

Shepherd, N., Vornanen, M., and Isenberg, G. (1990). Force measurements from voltage-clamped guinea pig ventricular myocytes. *Am J Physiol* **258**: H452–H459.

Shroff, S. G., Saner, D. R., and Lal, R. (1995). Dynamic micromechanical properties of cultured rat atrial myocytes measured by atomic force microscopy. *Am J Physiol* **269**: C286–C292.

Simpson, D. G., Majeski, M., Borg, T. K., and Terracio, L. (1999). Regulation of cardiac myocyte protein turnover and myofibrillar structure in vitro by specific directions of stretch. *Circ Res* **85**: e59–e69.

Simpson, D. G., Sharp, W. W., Borg, T. K., et al. (1996). Mechanical regulation of cardiac myocyte protein turnover and myofibrillar structure. *Am J Physiol* **270**: C1075–C1087.

Sniadecki, N. J., et al. (2007a). Magnetic microposts as an approach to apply forces to living cells. *Proc Natl Acad Sci USA*, **104**: 14553–14558.

Sniadecki, N. J. et al. (2007b). Microfabricated silicone elastomeric post arrays for measuring traction forces of adherent cells. *Methods in Cell Biology* **83**: 313–328.

Solaro, R. J., Gambassi, G., Warshaw, D. M., et al. (1993). Stereoselective actions of thiadiazinones on canine cardiac myocytes and myofilaments. *Circ Res* **73**: 981–990.

Sonnenblick, E. H. (1962). Implications of muscle mechanics in the heart. *Fed Proc* **21**: 975–990.
Soufivand, A. A., Soleimani, M., and Navidbakhsh, M. (2014). Is it appropriate to apply Hertz model to describe cardiac myocytes' mechanical properties by atomic force microscopy nanoindentation? *Micro & Nano Letters* **9**: 153–156.
Strang, K. T., Sweitzer, N. K., Greaser, M. L., and Moss, R. L. (1994). Beta-adrenergic receptor stimulation increases unloaded shortening velocity of skinned single ventricular myocytes from rats. *Circ Res* **74**: 542–549.
Strauss, J. D., Bletz, C., and Ruegg, J. C. (1994). The calcium sensitizer EMD 53998 antagonizes phosphate-induced increases in energy cost of isometric tension in cardiac skinned fibres. *Eur J Pharmacol* **252**: 219–224.
Tan, J. L., Tien, J., Pirone, D. M., et al. (2003). Cells lying on a bed of microneedles: an approach to isolate mechanical force. *Proc Natl Acad Sci USA* **100**: 1484–1489.
Tanaka, Y., Morishima, K., Shimizu, T., et al. (2006). Demonstration of a PDMS-based bio-microactuator using cultured cardiomyocytes to drive polymer micropillars. *Lab Chip* **6**: 230–235.
Tarr, M., Trank, J. W., Leiffer, P., and Shepherd, N. (1981). Evidence that the velocity of sarcomere shortening in single frog atrial cardiac cells is load dependent. *Circ Res* **48**: 200–206.
Tasche, C., Meyhofer, E., and Brenner, B. (1999). A force transducer for measuring mechanical properties of single cardiac myocytes. *Am J Physiol* **277**: H2400–H4008.
Taylor, R. E., Kim, K., Sun, N., et al. (2013). Sacrificial layer technique for axial force post assay of immature cardiomyocytes. *Biomed Microdevices* **15**: 171–181.
Terracio, L., Rubin, K., Gullberg, D., et al. (1991). Expression of collagen binding integrins during cardiac development and hypertrophy. *Circ Res* **68**: 734–744.
Tulloch, N. L., Muskheli, V., Razumova, M. V., et al. (2011). Growth of engineered human myocardium with mechanical loading and vascular coculture. *Circ Res* **109**: 47–59.
Tung, L. (1986). An ultrasensitive transducer for measurement of isometric contractile force from single heart cells. *Pflugers Arch* **407**: 109–115.
van der Velden, J., Klein, L. J., Van Der Bijl, M., et al. (1998). Force production in mechanically isolated cardiac myocytes from human ventricular muscle tissue. *Cardiovasc Res* **38**: 414–423.
Vannier, C., Chevassus, H., and Vassort, G. (1996). Ca-dependence of isometric force kinetics in single skinned ventricular cardiomyocytes from rats. *Cardiovasc Res* **32**: 580–586.
Wang, I. N., Wang, X., Ge, X., et al. (2012). Apelin enhances directed cardiac differentiation of mouse and human embryonic stem cells. *PLoS One* **7**: e38328.
Xi, J., Khalil, M., Shishechian, N., et al. (2010). Comparison of contractile behavior of native murine ventricular tissue and cardiomyocytes derived from embryonic or induced pluripotent stem cells. *FASEB J* **24**: 2739–2751.
Xing, R., Li, S., Liu, K., et al. (2014). HIP-55 negatively regulates myocardial contractility at the single-cell level. *J Biomech* **47**: 2715–2720.
Yang, X., Pabon, L. and Murry, C. E. (2014a). Engineering adolescence: maturation of human pluripotent stem cell-derived cardiomyocytes. *Circ Res* **114**: 511–523.

Yang, X., Rodriguez, M., Pabon, L., et al. (2014b). Tri-iodo-l-thyronine promotes the maturation of human cardiomyocytes-derived from induced pluripotent stem cells. *J Mol Cell Cardiol* **72**: 296–304.

Yasuda, S. I., Sugiura, S., Kobayakawa, N., et al. (2001). A novel method to study contraction characteristics of a single cardiac myocyte using carbon fibers. *Am J Physiol Heart Circ Physiol* **281**: H1442–H1446.

Yin, S., Zhang, X., Zhan, C., et al. (2005). Measuring single cardiac myocyte contractile force via moving a magnetic bead. *Biophys J* **88**: 1489–1495.

Zhao, Y., and Zhang, X. (2005). *Contraction force measurements in cardiac myocytes using PDMS pillar arrays*. 18th IEEE International Conference on Micro Electro Mechanical Systems: 834–837.

Zhao, Y., and Zhang, X. (2006). Cellular mechanics study in cardiac myocytes using PDMS pillars array. *Sensors and Actuators A: Physical* **125**: 398–404.

17 Micro- nanofabrication for the study of biochemical and biomechanical regulation of T cell activation

Hye Mi Kim and Junsang Doh

17.1 Introduction

T cells are a type of white blood cells that play critical roles in antigen-specific immune responses. Naïve T cells, which have never been activated, do not participate in immune responses until they become activated by interacting with antigen presenting cells (APCs). Upon activation, T cells undergo massive proliferation and differentiate into various subsets of effector or memory T cells (Fig. 17.1A). The majority of activated T cells become effector T cells that perform various effector functions such as secreting cytokines, activating other immune cells via direct cell-cell contact, or directly killing infected or transformed host cells (Fig. 17.1B). On the other hand, small fractions of activated T cells differentiate into memory T cells that can mobilize rapid and robust immune responses to secondary infections.

Fates of T cells are influenced by many microenvironmental factors surrounding T cells (Luther and Cyster 2001; Williams and Bevan 2007; Zhu et al. 2010; Murphy and Stockinger 2010). Among them, APCs triggering the activation of T cells convey critical information of local inflamed tissues to T cells, and thus play a critical role in programming T cells (Banchereau and Steinman 1998; Guermonprez et al. 2002). During T-APC interactions, receptors, signaling molecules, and cytoskeletons are polarized toward the interfaces between T-APC and form complex supramolecular clusters known as the immunological synapse (IS) (Cemerski and Shaw 2006; Huppa and Davis 2003; Grakoui et al. 1999). As shown in Fig. 17.2A, a mature IS is composed of three distinct supramolecular adhesion complexes (SMACs) (Monks et al. 1998; Freiberg et al. 2002): in central SMAC (c-SMAC), key signaling molecules such as T cell receptors (TCRs) and CD28 are accumulated; in peripheral SMAC (p-SMAC), lymphocyte function antigen 1 (LFA-1), an adhesion molecule mediating stable T-APC interactions, is enriched; and in distal SMAC (d-SMAC), large glycoproteins such as CD43 and CD45 are localized.

The mature IS is widely considered to be a representative structure of the IS, but indeed it is only one of the many diverse forms of ISs. As shown in Fig. 17.2B, the structures of the IS vary depending on types of T cells and APCs

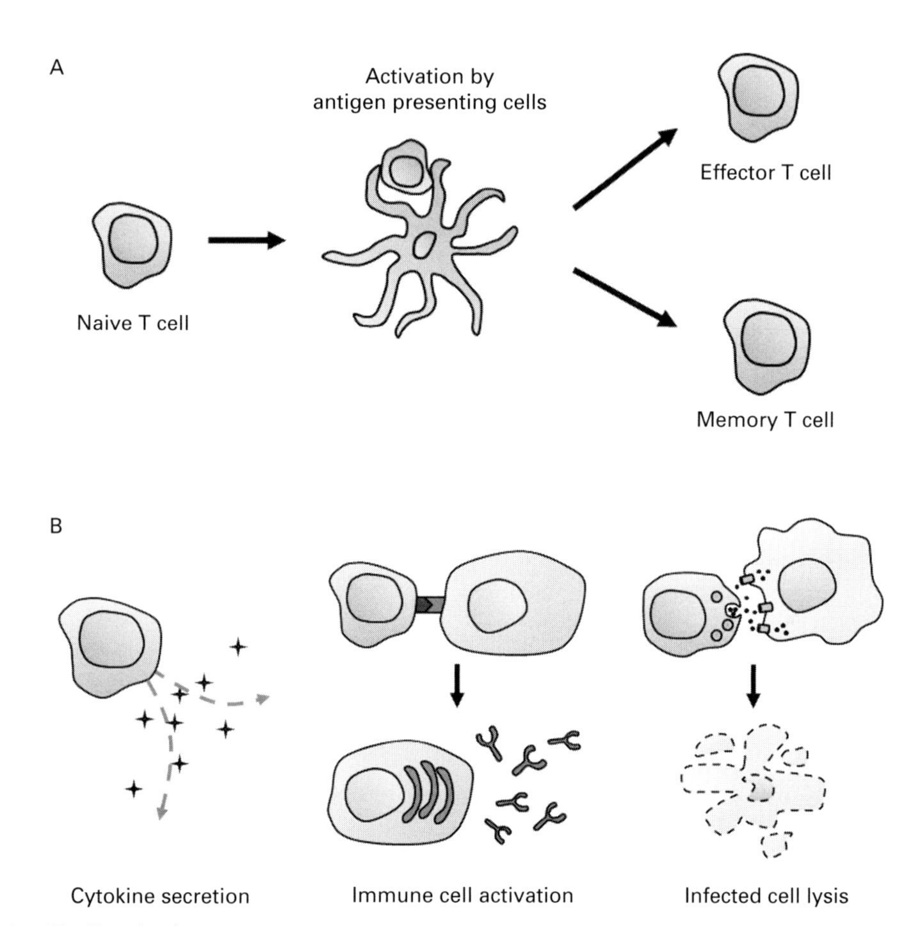

Figure 17.1 **T cell activation.**
(A) Activation and differentiation of T cells. (B) Various effector functions exerted by effector T cells.

(Brossard et al. 2005; Hallman et al. 2002; Thauland et al. 2008; O'Keefe et al. 2004; Purtic et al. 2005). Moreover, the IS is a highly dynamic structure, as revealed by video microscopy. TCR microclusters initially generated at the periphery translocate to the center to form a cSMAC (Varma et al. 2006; Campi et al. 2005; Yokosuka et al. 2005; Krummel et al. 2000). Actin polymerization plays an important role in the generation and translocation of TCR microclusters (Varma et al. 2006; Campi et al. 2005).

Micro/nanofabrication has served as a powerful tool to investigate the IS formed in T cells by allowing dynamic events occurring in various time and length scales to be manipulated and monitored (Irvine and Doh 2007; Jung et al. 2013). In this chapter, we describe four different micro- nanofabrication–based approaches developed to study biochemical and biomechanical regulations of T cell activation.

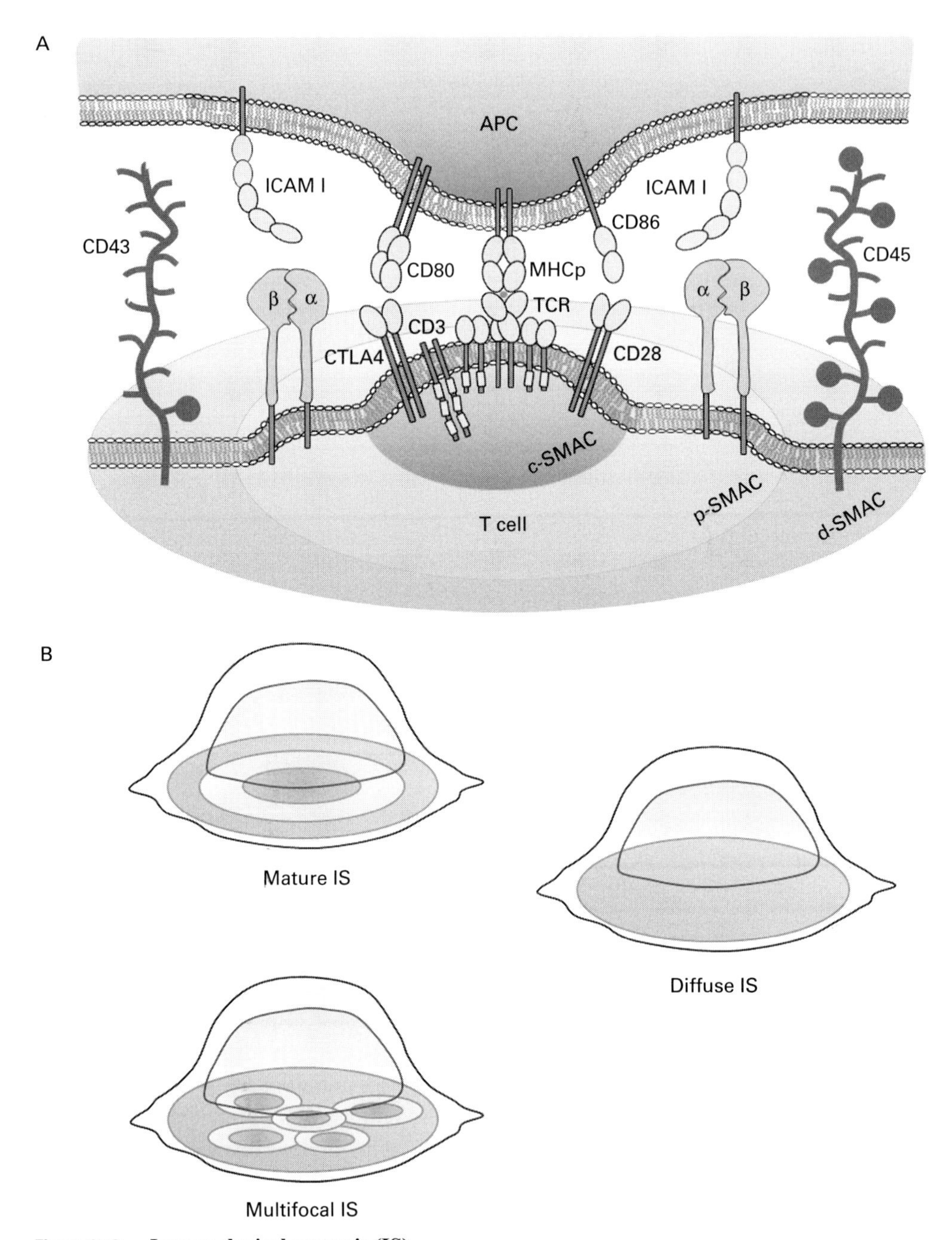

Figure 17.2 **Immunological synapsis (IS).**
(A) Bull's eye structure of a mature IS. (B) Diversity in the structure of the IS.

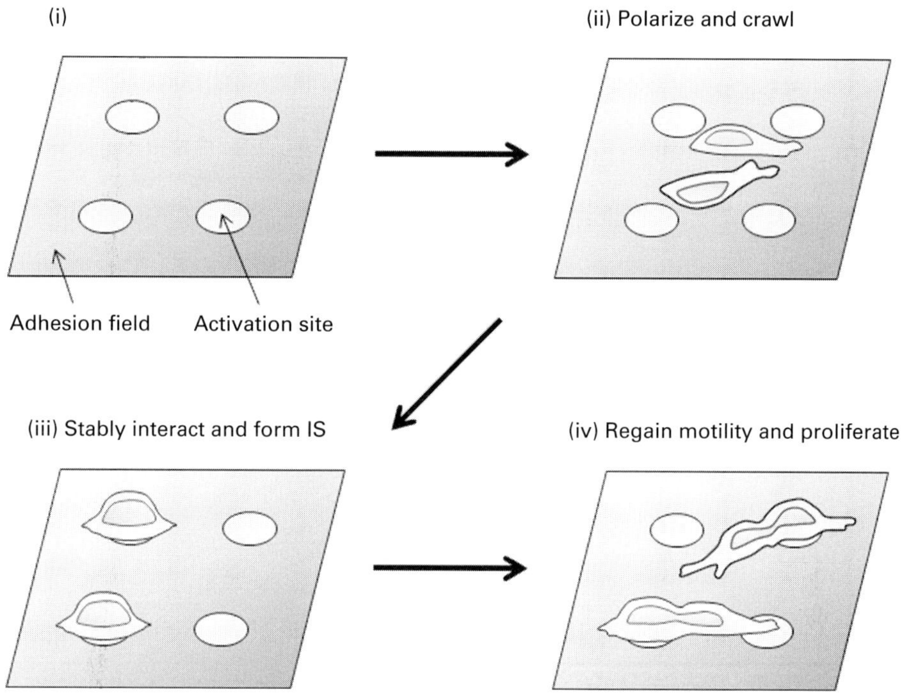

Figure 17.3 **Immunological synapse array (ISA).**
Schematic illustration of the structures of the ISA. (i) Sequential behaviors of T cells on ISAs (ii–iv).

17.2 Protein micropatterned surfaces

As described in Fig. 17.2B, various structures of ISs can be formed during T-APC interactions. To study the effect of the IS structure on T cell activation, we developed an "immunological synapse array (ISA)" by micropatterning key ligands for T cell activation on flat substrates (Doh and Irvine 2006) (Fig. 17.3i). First, anti-CD3 that triggers TCR signaling by crosslinking TCRs was micropatterned on an "activation site." The "adhesion field," the surrounding area of the activation sites, was then backfilled with intercellular adhesion molecule 1 (ICAM-1), which supports adhesion and migration of T cells. Mouse primary CD4+ T cell blasts, which are in vitro activated T cells and thus do not require co-stimulation, were used. T cells landed on the adhesion field and crawled (Fig. 17.3ii) until they encountered the activation sites. When crawling T cells contacted the activation sites, they stopped and formed an IS with the activation sites (Fig. 17.3iii). After several hours of interactions, T cells that stopped on the activation sites regained motility and exhibited 1~2 hours of transient interactions with the activation sites and underwent cell division (Fig. 17.3iv). Of note, these multi-phase behaviors of T cells on the ISAs are somewhat similar to the behaviors of T cells in inflamed lymph nodes (Miller et al. 2004; Mempel et al. 2004).

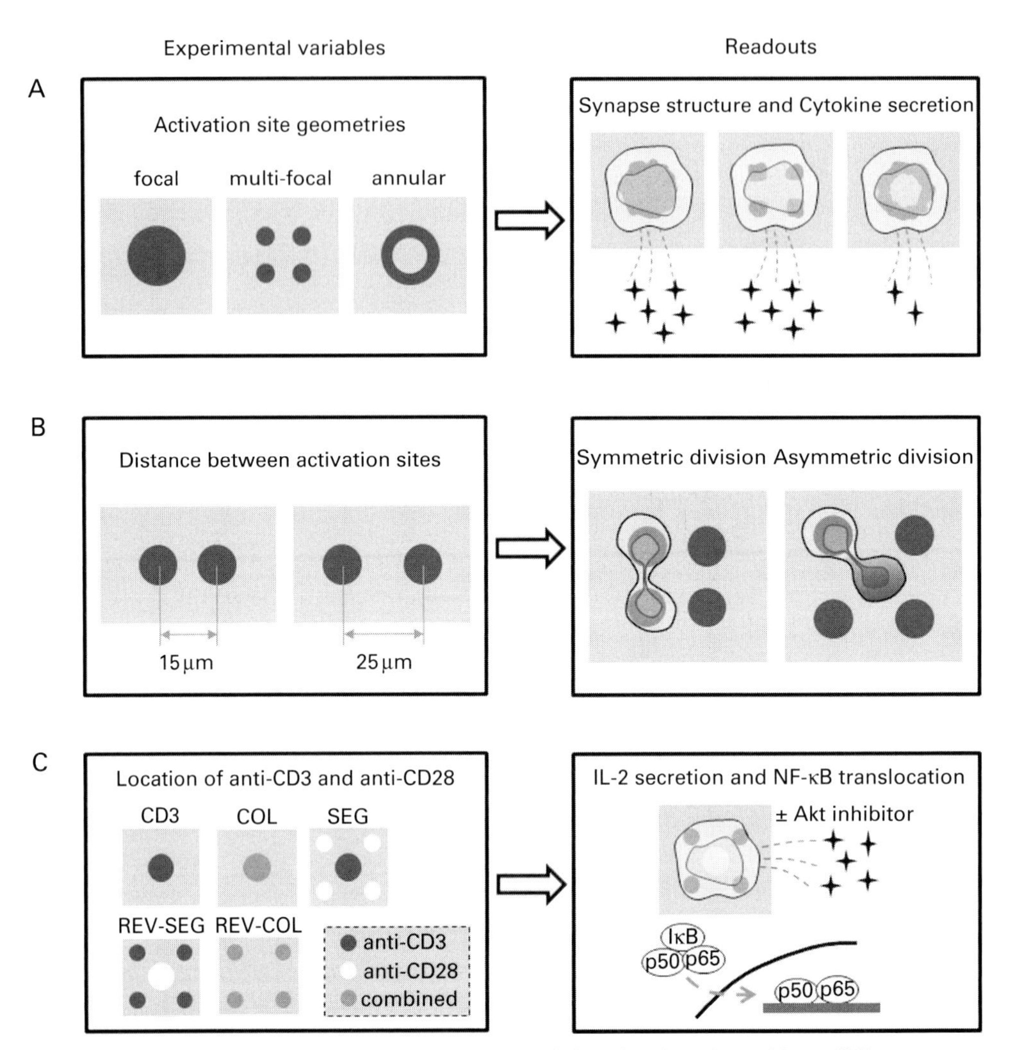

Figure 17.4 **Schematic illustration of various types of ISAs fabricated to address different questions related to T cell activation.**

Based on these characteristics of ISAs, we first asked whether the geometry of the activation sites influenced the structure of ISs formed in T cells. ISAs with three different geometries of activation sites, as schematically shown in Fig. 17.4A, were used. Overall, distributions of TCRs and the θ form of protein kinase C (PKC-θ) in T cells on ISAs were similar to the geometries of the activation sites, thus indicating that the structure of the IS in T cells can be manipulated by micropatterning anti-CD3. Altered synapse formation also affected functional outcomes of T cell activation. T cells on ISAs with annular-shaped activation sites secreted significantly lower amounts of interferon γ (IFN-γ) than T cells on ISAs with focal-shaped activation sites. These results suggest that the spatial distribution of anti-CD3 can control T cell activation by modulating the structures of the IS in T cells.

T cells activated by APCs differentiate into either effect or memory T cells (Fig. 17.1A). Memory formation is one of the important characteristics of adaptive immunity and it allows fast and strong immune responses against repetitive infections, and thus understanding mechanisms of memory T cell formation is critical for vaccine development. Activated T cells can differentiate into diverse subsets of effector or memory T cells (Williams and Bevan 2007; Zhu et al. 2010). While microenvironmental cues determining the differentiation of activated T cells are not yet fully understood, asymmetric division of T cells may be a key initial step for diversification of phenotypes (Chang et al. 2011; Chang et al. 2007). During mitosis of activated T cells, asymmetric partitioning of receptors such as CD4, CD8, LFA-1, and TCR, which are receptors known to polarize to the IS, in two nascent daughter cells occurred. In addition, asymmetric division of T cells depended on the presence of LFA-1 in T cells, which supports stable interaction between T cells and dendritic cells (DCs) (Scholer et al. 2008). These results indicate that the IS plays an essential role in the asymmetric partitioning of molecules. To further identify key factors regulating asymmetric division of T cells, we utilized ISAs (Jung et al. 2014). Mouse primary naïve CD4+ T cells isolated from spleens and lymph nodes of mice were used because naïve T cells are better for differentiation study than the T cell blasts used in the previous study. Compared with T cell blasts, naïve T cells require CD28 co-stimulation for full activation (Sharpe and Freeman 2002), and thus anti-CD28 was co-immobilized in the activation sites along with anti-CD3. First, we checked whether T cells on ISAs underwent asymmetric cell division by examining the distribution of molecules in cytokinetic T cells. Similar to previous ex vivo experiments (Chang et al. 2011; Chang et al. 2007), an asymmetric distribution of TCR, CD3, PKC-ζ, and T-bet was observed in two nascent daughter cells for about half of T cells on the ISAs, suggesting ISAs can induce asymmetric division of T cells. ISAs with various activation site sizes, anti-CD3 surface densities, and inter-activation site distances were then fabricated and frequencies of asymmetric divisions of T cells on different types of ISAs were measured and compared. Interestingly, frequencies of asymmetric divisions significantly increased as the inter-activation site distance decreased, while other variables exhibited minimal effects on the frequency of asymmetric division. Further analysis revealed that asymmetric division mostly occurred when one nascent daughter cell made stable contact with the activation site and the other nascent daughter cell migrated on the adhesion fields during cytokinesis (Fig. 17.4B). Thus, increased inter-activation site distances would increase the probability of only one nascent daughter cell contacting the activation site to form a single IS during cytokinesis, resulting in increased asymmetric cell division. Inter-activation site distances of ISAs can be translated into density of APCs presenting antigens recognized by a given T cell in a lymph node. Typically, self-antigens inducing autoimmune diseases are present in high density, which corresponds to short inter-activation site distance in ISAs and suppresses asymmetric cell division. Considering asymmetric cell division is an important mechanism for memory T cell differentiation, reduction of asymmetric cell division with a high density of APCs would minimize the generation of potentially harmful autoreactive memory T cells.

As mentioned above, naïve T cells require co-stimulatory signals as well as TCR signals for full activation. Similar to the case with TCR ligands, spatial organization of co-stimulatory ligands may have profound effects on T cell activation. Indeed, CD28 has been reported to be spatially segregated from TCR clusters in some experimental settings (Tseng et al. 2005; Yokosuka et al. 2008). To address this problem, the Kam group fabricated ISAs with various configurations of anti-CD3 and anti-CD28 spots, as schematically shown in Fig. 17.4C (Shen et al. 2008). Mouse primary naïve CD4+ T cells isolated from mouse lymph nodes (Shen et al. 2008) and human primary resting CD4+ T cells isolated from peripheral bloods (Bashour et al. 2014b) were used. First, T cells were plated on the ISAs and distributions of receptors were assessed by immunofluorescence microscopy. TCR was concentrated on anti-CD3 spots and CD28 accumulated on anti-CD28 spots for both types of T cells, indicating that the spatial distribution of CD28 as well as that of TCR could be manipulated using micropatterned ligands. Activation of T cells on each type of ISA was then assessed by monitoring nuclear translocation of NF-κB and cytokine secretion. Mouse T cells were better activated when they were on ISAs with spatially segregated anti-CD3 and anti-CD28 (SEG) than when they were on ISAs with co-localized anti-CD3 and anti-CD28 (COL). Interestingly, opposite results were observed for human T cells; human T cells were better activated on COL patterns than SEG patterns. To understand why human T cells and mouse T cells behaved markedly differently from each other, mobility of Lck, a key downstream kinase of TCR signaling (Smith-Garvin et al. 2009; Huppa and Davis 2003), in human T cells and mouse T cells was measured and compared. Interestingly, mobility of Lck in human T cells was significantly lower than in mouse T cells, potentially due to much denser actin network formation in human T cells near IS than in mouse T cells. Therefore, when human T cells are on SEG patterns, phosphorylated Lck on anti-CD3 spots will not be efficiently transported to the anti-CD28 spots, resulting in attenuation of TCR signaling. In contrast, mouse T cells on SEG patterns will not experience significant transport-limited attenuation of TCR signaling because Lck phosphorylated by anti-CD3 spots will be quickly transferred to anti-CD28 spots with minimal de-phosphorylation. These results suggest that crosstalk between signaling mediated by two different receptors can be fine-tuned by the spatial distribution of receptors.

17.3 Protein nanopatterned surfaces

High resolution imaging of T cells interacting with TCR ligands on planar surfaces revealed intriguing nanoscale clustering of receptors and signaling molecules (Dustin and Groves 2012), indicating that ligand clustering on a nanoscale may play a role in T cell activation. To control nanoscale ligand presentation, the Spatz and Dunlop groups utilized protein nanopatterned surfaces (Deeg et al. 2013; Delcassian et al. 2013; Matic et al. 2013). Block copolymer micelle nanolithography (BCML) (Spatz et al. 2000; Glass et al. 2003) was used to generate regular arrays of gold nanoparticles (Au-NPs) with ~10 nm diameter. Then, either anti-CD3 (Delcassian et al. 2013; Matic et al. 2013)

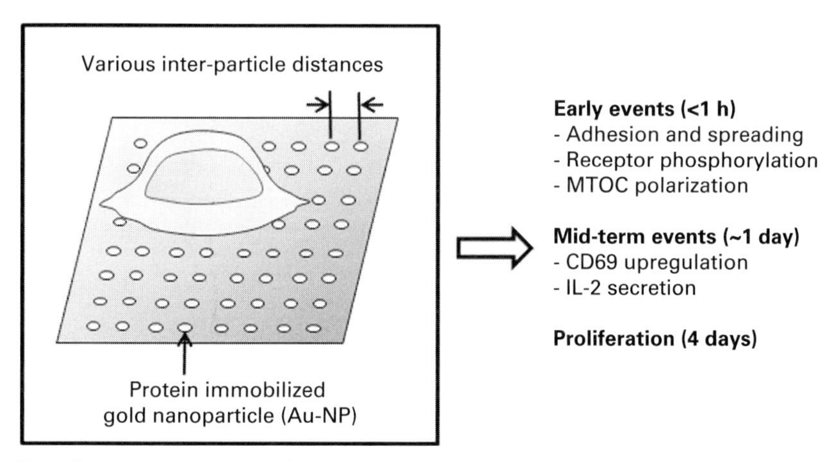

Figure 17.5 **Protein nanopatterned surfaces.**
Schematic illustration of protein nanopatterned surfaces to study the effect of nanoscale ligand presentation on T cell activation.

or antigenic peptide loaded on major histocompatibility complex (p-MHC), which is a native ligand for TCR, (Deeg et al. 2013) was immobilized on Au-NPs. By varying the size of micelles for BCML, distances between Au-NPs were varied, resulting in variation of the ligand density. T cell activation on nanopatterned surfaces was assessed by measuring various events in T cell activation, as summarized in Fig. 17.5. For all cases examined, activation levels of T cells increased as the inter-particle distance between Au-NPs decreased, or as the ligand density increased.

17.4 Nanopatterned supported bilayers

Micro- nanopatterned surfaces described in Section 17.2 and Section 17.3 are based on "immobilized" protein ligands, which are useful in addressing the roles of the spatial distribution of ligands in TCR signaling and T cell activation. However, ligands presented by APCs are "mobile" and dynamically modulated depending on T-APC interactions. To overcome this limitation, supported bilayers – lipid bilayers formed on solid substrates – that allow mobile presentation of ligands have been used to fabricate artificial cell membranes (Sackmann 1996). Indeed, supported bilayers have long been used to study T cell activation (Groves and Dustin 2003), and have been particularly useful in understanding the dynamics of receptors during IS formation at high resolution (Dustin and Groves 2012).

Dynamics of lipids in supported bilayers can be regulated by micro/nanopatterning chromium (Cr) barriers on underlying substrates (Groves and Boxer 2002). The Groves lab incorporated various geometries of nanoscale Cr lines on supported bilayers, presenting p-MHC and ICAM-1 to manipulate the dynamics of TCR microclusters (Fig. 17.6A). High resolution imaging based on supported bilayers revealed that when T cells recognize p-MHC on supported bilayers, TCR microclusters initially generated at

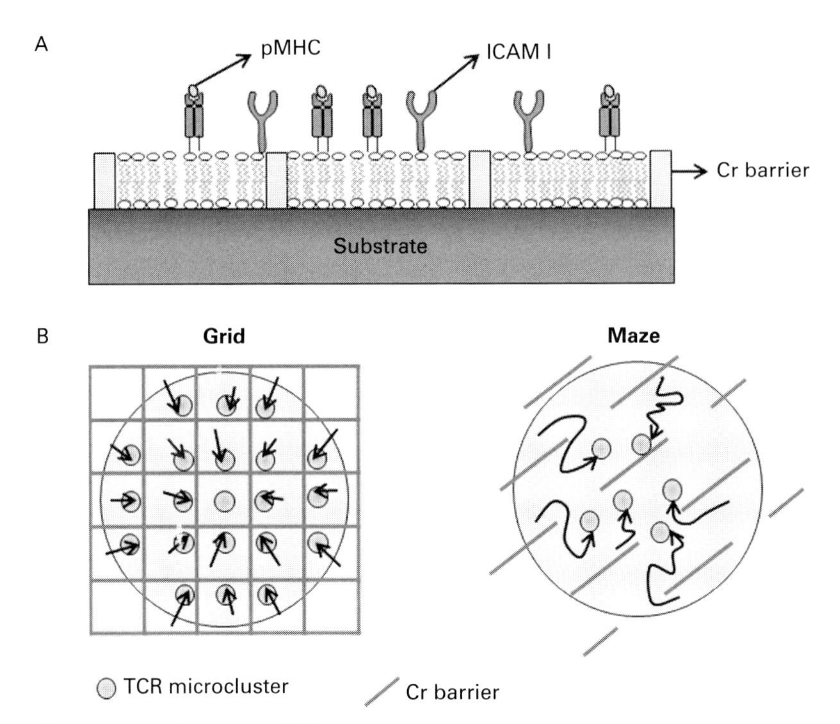

Figure 17.6 **Nanopatterned supported bilayers.**
Schematic illustration of nanopatterned supported bilayer-based antigen presentation (A) and modulated TCR transportation depending on the geometries of Cr barriers (B and C).

the periphery translocated to the center by actin where they coalesce into large TCR clusters and form cSMAC (Campi et al. 2005; Yokosuka et al. 2005; Varma et al. 2006). When grid-structured Cr barriers were inserted into the supported bilayers, locally generated TCR microclusters were trapped within grids, resulting in multifocal synapse formation (Fig. 17.6B) (Mossman et al. 2005). Inhibition of central transportation of TCRs significantly increased initial TCR signaling. In addition, by varying the sizes of grids and fixing the total number of p-MHC, the number of p-MHC molecules within a grid could be controlled (Manz et al. 2011). Since p-MHC molecules within a grid cannot be translocated to other grids, the number of p-MHC molecules interacting with a single TCR microcluster can be precisely controlled. Interestingly, the number of p-MHC molecules within a grid, not the total number of p-MHC interacting with T cells, determined the magnitude of TCR signaling. By varying p-MHC density and quantitatively measuring TCR signaling by the intracellular calcium concentration, the minimum number of p-MHC molecules eliciting TCR signaling was determined.

Nanopatterned supported bilayers were also used to study the dynamics of actin and TCR microclusters in the IS. When supported bilayers with maze-structured Cr nanolines were used, the centripetal flow of TCR microclusters was redirected by obstacles (Fig. 17.6C) (DeMond et al. 2008). In addition to direct observation of TCR

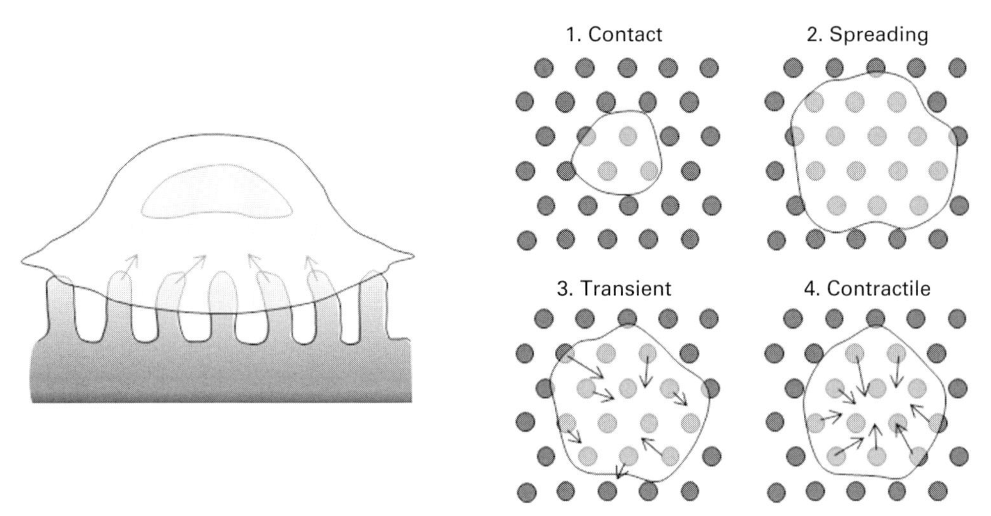

Figure 17.7 **Micropost allays.**
Schematic illustration of a T cell on a micropost array.

microclusters, speckles of GFP-labeled actin were tracked to extract information about actin dynamics (Yu et al. 2010; Babich et al. 2012). By comparing actin dynamics with the positions of TCR microclusters trapped by Cr nanolines, TCR-actin interactions were quantitatively analyzed.

17.5 Micropost arrays

As described in Section 17.4, during the IS formation, centripetal flow generated by cytoskeletons translocate TCR microclusters generated at the periphery to cSMACs. Traction forces exerted by T cells during this process were recently measured by the Kam group by using silicon elastomer-based micropost arrays (Bashour et al. 2014a). Typically, leukocytes generate an order of magnitude weaker force than strongly adhering cells such as fibroblasts and endothelial cells, and thus measurement of forces generated by leukocytes has been technically challenging (Ricart et al. 2011). As shown in Fig. 17.7, T cells contacting micropost arrays coated with anti-CD3 and anti-CD28 rapidly spread over multiple microposts with minimal force generation. Once T cells reached plateau spreading areas, they exerted transient and uncoordinated forces for a while, and eventually generated sustained centripetal contractile forces with ~100 pN/micropost. Spreading and force generation depended on TCR signaling as indicated by the observation that spreading of T cells occurred on micropost arrays coated with anti-CD3 or p-MHC, not on micropost arrays coated with anti-CD28. CD28 signaling appeared to enhance contractility indirectly considering both soluble anti-CD28 and surface immobilized anti-CD28 increased the traction force of T cells on anti-CD3-coated micropost arrays by the same magnitude. Interestingly, the magnitude of forces generated by T cells was not affected by the height of the micropost, which determines the effective stiffness of an individual micropost (Fu et al. 2010).

17.6 Conclusion

In this chapter, various micro- nanofabricated platforms employed to study biochemical and biomechanical mechanisms of T cell activation are described. Micro- nanofabrication has deepened our understanding about the structure and dynamics of the IS formed in T cells over the last decade by enabling us to manipulate spatial organization of ligands and to monitor various biochemical and biomechanical processes (Jung et al. 2013). In addition to addressing fundamental questions in immunology, these technologies can also be potentially useful for T cell-based cancer therapy by allowing augmented ex vivo activation and expansion of cancer-specific T cells isolated from patients (Milone and Kam 2013).

References

Babich, A., Li, S., O'Connor, R. S., Milone, M. C., Freedman, B. D. and Burkhardt, J. K. (2012). "F-actin polymerization and retrograde flow drive sustained PLCγ1 signaling during T cell activation." *The Journal of Cell Biology* **197**: 775–787.

Banchereau, J. and Steinman, R. M. (1998). "Dendritic cells and the control of immunity." *Nature* **392**: 245–252.

Bashour, K. T., Gondarenko, A., Chen, H., Shen, K., Liu, X., Huse, M., Hone, J. C. and Kam, L. C. (2014a). "CD28 and CD3 have complementary roles in T-cell traction forces." *Proceedings of the National Academy of Sciences* **111**: 2241–2246.

Bashour, K. T., Tsai, J., Shen, K., Lee, J.-H., Sun, E., Milone, M. C., Dustin, M. L. and Kam, L. C. (2014b). "Cross talk between CD3 and CD28 is spatially modulated by protein lateral mobility." *Molecular and Cellular Biology* **34**: 955–964.

Brossard, C., Feuillet, V., Schmitt, A., Randriamampita, C., Romao, M., Raposo, G. and Trautmann, A. (2005). "Multifocal structure of the T cell – dendritic cell synapse." *European Journal of Immunology* **35**: 1741–1753.

Campi, G., Varma, R. and Dustin, M. L. (2005). "Actin and agonist MHC-peptide complex-dependent T cell receptor microclusters as scaffolds for signaling." *Journal of Experimental Medicine* **202**: 1031–1036.

Cemerski, S. and Shaw, A. (2006). "Immune synapses in T-cell activation." *Current Opinion in Immunology* **18**: 298–304.

Chang, J. T., Ciocca, M. L., Kinjyo, I., Palanivel, V. R., Mcclurkin, C. E., Dejong, C. S., Mooney, E. C., et al. (2011). "Asymmetric proteasome segregation as a mechanism for unequal partitioning of the transcription factor T-bet during T lymphocyte division." *Immunity* **34**: 492–504.

Chang, J. T., Palanivel, V. R., Kinjyo, I., Schambach, F., Intlekofer, A. M., Banerjee, A., Longworth, S. A., et al. (2007). "Asymmetric T lymphocyte division in the initiation of adaptive immune responses." *Science* **315**: 1687–1691.

Deeg, J., Axmann, M., Matic, J., Liapis, A., Depoil, D., Afrose, J., Curado, S., et al. (2013). "T cell activation is determined by the number of presented antigens." *Nano Letters* **13**: 5619–5626.

Delcassian, D., Depoil, D., Rudnicka, D., Liu, M., Davis, D. M., Dustin, M. L. and Dunlop, I. E. (2013). "Nanoscale ligand spacing influences receptor triggering in T cells and NK cells." *Nano Letters* **13**: 5608–5614.

Demond, A. L., Mossman, K. D., Starr, T., Dustin, M. L. and Groves, J. T. (2008). "T cell receptor microcluster transport through molecular mazes reveals mechanism of translocation." *Biophysical Journal* **94**: 3286–3292.

Doh, J. and Irvine, D. J. (2006). "Immunological synapse arrays: patterned protein surfaces that modulate immunological synapse structure formation in T cells." *Proceedings of the National Academy of Sciences* **103**: 5700–5705.

Dustin, M. L. and Groves, J. T. (2012). "Receptor signaling clusters in the immune synapse." *Annual Review of Biophysics* **41**: 543.

Freiberg, B. A., Kupfer, H., Maslanik, W., Delli, J., Kappler, J., Zaller, D. M. and Kupfer, A. (2002). "Staging and resetting T cell activation in SMACs." *Nature Immunology* **3**: 911–917.

Fu, J., Wang, Y. K., Yang, M. T., Desai, R. A., Yu, X., Liu, Z. and Chen, C. S. (2010). "Mechanical regulation of cell function with geometrically modulated elastomeric substrates." *Nature Methods* **7**: 733–736.

Glass, R., Möller, M. and Spatz, J. P. (2003). "Block copolymer micelle nanolithography." *Nanotechnology* **14**: 1153.

Grakoui, A., Bromley, S. K., Sumen, C., Davis, M. M., Shaw, A. S., Allen, P. M. and Dustin, M. L. (1999). "The immunological synapse: a molecular machine controlling T cell activation." *Science* **285**: 221–227.

Groves, J. T. and Boxer, S. G. (2002). "Micropattern formation in supported lipid membranes." *Accounts of Chemical Research* **35**: 149–157.

Groves, J. T. and Dustin, M. L. (2003). "Supported planar bilayers in studies on immune cell adhesion and communication." *Journal of Immunological Methods* **278**: 19–32.

Guermonprez, P., Valladeau, J., Zitvogel, L., Th Ry, C. and Amigorena, S. (2002). "Antigen presentation and T cell stimulation by dendritic cells." *Annual Review of Immunology* **20**: 621–667.

Hallman, E., Burack, W. R., Shaw, A. S., Dustin, M. L. and Allen, P. M. (2002). "Immature CD4(+)CD8(+) thymocytes form a multifocal immunological synapse with sustained tyrosine phosphorylation." *Immunity* **16**: 839–848.

Huppa, J. B. and Davis, M. M. (2003). "T-cell-antigen recognition and the immunological synapse." *Nature Reviews Immunology* **3**: 973–983.

Irvine, D. J. and Doh, J. (2007). "Synthetic surfaces as artificial antigen presenting cells in the study of T cell receptor triggering and immunological synapse formation." *Seminars in Immunology* **19**: 245–254.

Jung, H.-R., Song, K. H., Chang, J. T. and Doh, J. (2014). "Geometrically controlled asymmetric division of CD4+ T cells studied by immunological synapse arrays." *PloS One* **9**: e91926.

Jung, H.-R., Choi, J. C., Cho, W. and Doh, J. (2013). "Microfabricated platforms to modulate and monitor T cell synapse assembly." *Wiley Interdisciplinary Reviews: Nanomedicine and Nanobiotechnology* **5**: 67–74.

Krummel, M. F., Sjaastad, M. D., Wulfing, C. and Davis, M. M. (2000). "Differential clustering of CD4 and CD3 zeta during T cell recognition." *Science* **289**: 1349–1352.

Luther, S. A. and Cyster, J. G. (2001). "Chemokines as regulators of T cell differentiation." *Nature Immunology* **2**: 102–107.

Manz, B. N., Jackson, B. L., Petit, R. S., Dustin, M. L. and Groves, J. (2011). "T-cell triggering thresholds are modulated by the number of antigen within individual T-cell receptor clusters." *Proceedings of the National Academy of Sciences* **108**: 9089–9094.

Matic, J., Deeg, J., Scheffold, A., Goldstein, I. and Spatz, J. P. (2013). "Fine tuning and efficient T cell activation with stimulatory aCD3 nanoarrays." *Nano Letters* **13**: 5090–5097.

Mempel, T. R., Henrickson, S. E. and von Andrian, U. H. (2004). "T-cell priming by dendritic cells in lymph nodes occurs in three distinct phases." *Nature* **427** 154–159.

Miller, M. J., Safrina, O., Parker, I. and Cahalan, M. D. (2004). "Imaging the single cell dynamics of CD4+ T cell activation by dendritic cells in lymph nodes." *The Journal of Experimental Medicine* **200**: 847–856.

Milone, M. C. and Kam, L. C. (2013). "Investigative and clinical applications of synthetic immune synapses." *Wiley Interdisciplinary Reviews: Nanomedicine and Nanobiotechnology* **5**: 75–85.

Monks, C. R. F., Freiberg, B. A., Kupfer, H., Sciaky, N. and Kupfer, A. (1998). "Three-dimensional segregation of supramolecular activation clusters in T cells." *Nature* **395**: 82–86.

Mossman, K. D., Campi, G., Groves, J. T. and Dustin, M. L. (2005). "Altered TCR signaling from geometrically repatterned immunological synapses." *Science* **310**: 1191–1193.

Murphy, K. M. and Stockinger, B. (2010). "Effector T cell plasticity: flexibility in the face of changing circumstances." *Nature Immunology* **11**: 674–680.

O'Keefe, J. P., Blaine, K., Alegre, M. L. and Gajewski, T. F. (2004). "Formation of a central supramolecular activation cluster is not required for activation of naive CD8(+) T cells." *Proceedings of the National Academy of Sciences of the United States of America* **101**: 9351–9356.

Purtic, B., Pitcher, L. A., van Oers, N. S. C. and Wulfing, C. (2005). "T cell receptor (TCR) clustering in the immunological synapse integrates TCR and costimulatory signaling in selected T cells." *Proceedings of the National Academy of Sciences of the United States of America* **102**: 2904–2909.

Ricart, B. G., Yang, M. T., Hunter, C. A., Chen, C. S. and Hammer, D. A. (2013). "Measuring traction forces of motile dendritic cells on micropost arrays." *Biophysical Journal* **101**: 2620–2628.

Sackmann, E. (1996). "Supported membranes: scientific and practical applications." *Science* **271**: 43–48.

Scholer, A., Hugues, S., Boissonnas, A., Fetler, L. and Amigorena, S. (2008). "Intercellular adhesion molecule-1-dependent stable interactions between T cells and dendritic cells determine CD8+ T cell memory." *Immunity* **28**: 258–270.

Sharpe, A. H. and Freeman, G. J. (2002). "The B7–CD28 superfamily." *Nature Reviews Immunology* **2**: 116–126.

Shen, K., Thomas, V. K., Dustin, M. L. and Kam, L. C. (2008). "Micropatterning of costimulatory ligands enhances CD4+ T cell function." *Proceedings of the National Academy of Sciences* **105**: 7791–7796.

Smith-Garvin, J. E., Koretzky, G. A. and Jordan, M. S. (2009). "T cell activation." *Annual Review of Immunology* **27**: 591.

Spatz, J. P., M Ssmer, S., Hartmann, C., M Ller, M., Herzog, T., Krieger, M., Boyen, H.-G., et al. (2000). "Ordered deposition of inorganic clusters from micellar block copolymer films." *Langmuir* **16**: 407–415.

Thauland, T. J., Koguchi, Y., Wetzel, S. A., Dustin, M. L. and Parker, D. C. (2008). "Th1 and Th2 cells form morphologically distinct immunological synapses." *The Journal of Immunology* **181**: 393–399.

Tseng, S. Y., Liu, M. and Dustin, M. L. (2005). "CD80 cytoplasmic domain controls localization of CD28, CTLA-4, and protein kinase C theta in the immunological synapse." *Journal of Immunology* **175**: 7829–7836.

Varma, R., Campi, G., Yokosuka, T., Saito, T. and Dustin, M. L. (2006). "T cell receptor-proximal signals are sustained in peripheral microclusters and terminated in the central supramolecular activation cluster." *Immunity* **25**: 117–127.

Williams, M. A. and Bevan, M. J. (2007). "Effector and memory CTL differentiation." *Annual Review of Immunology* **25**: 171–192.

Yokosuka, T., Kobayashi, W., Sakata-Sogawa, K., Takamatsu, M., Hashimoto-Tane, A., Dustin, M. L., Tokunaga, M. and Saito, T. (2008). "Spatiotemporal regulation of T cell costimulation by TCR-CD28 microclusters and protein kinase C theta translocation." *Immunity* **29**: 589–601.

Yokosuka, T., Sakata-Sogawa, K., Kobayashi, W., Hiroshima, M., Hashimoto-Tane, A., Tokunaga, M., Dustin, M. L. and Saito, T. (2005). "Newly generated T cell receptor microclusters initiate and sustain T cell activation by recruitment of Zap70 and SLP-76." *Nature Immunology* **6**: 1253–1262.

Yu, C.-H., Wu, H.-J., Kaizuka, Y., Vale, R. D. and Groves, J. T. (2010). "Altered actin centripetal retrograde flow in physically restricted immunological synapses." *PloS One* **5**: e11878.

Zhu, J., Yamane, H. and Paul, W. E. (2010). "Differentiation of effector CD4+ T cell populations." *Annual Review of Immunology* **28**: 445–489.

18 Study of tumor angiogenesis using microfluidic approaches

Yoojin Shin, Sewoon Han, Hyo Eun Jeong, Jeong Ah Kim,
Jessie S. Jeon, and Seok Chung

Tumor angiogenesis is a key regulator of tumor growth and metastasis. Assays allowing the analysis of tumor angiogenesis are an essential tool to elucidate the role played by the tumor microenvironment in regulating tumor angiogenesis. The assays should also be capable of systematically investigating the effects of physiologically relevant, mechanical and chemical stimuli and their synergistic interactions. The high optical resolution of microfluidic assays facilitates three-dimensional studies of cellular morphogenesis. Their versatility can be applied to study the multi-parameter control of angiogenic factors.

18.1 Tumor angiogenesis as a key regulator of tumor growth and metastasis

Tumor angiogenesis is the recruitment of new blood vessels from existing ones and it is an essential step in tumor progression and metastasis. The new blood vessels provide the tumor with nutrients and oxygen and therefore accelerate its growth. They also allow tumor cells to escape from the primary site and enter the blood circulation. However, the blood vessels formed by tumor angiogenesis are immature, with impaired cell-to-cell interactions, abnormal basement membranes, and high permeability. These irregular features accelerate the intravasation of tumor cells, which are then referred to as circulating tumor cells (CTCs). By extravasating from the blood vessels into neighboring tissues, CTCs are the seeds of secondary tumors. Although most of the extravasated tumor cells will become dormant in their new microenvironments, some will differentiate and proliferate to form new sites of micrometastasis regulated by complex microenvironmental interactions with stromal cells, including osteoblasts, osteoclasts, fibroblasts, hematopoietic progenitors, Kupffer cells, and glial cells. The extracellular matrix (ECM) and neighboring blood vessels are additional microenvironmental factors contributing to micrometastasis formation. Highly vascularized tumors therefore generally have a high incidence of metastasis, the interruption of which is being explored as a promising tactic in limiting cancer progression (Weis and Cheresh 2011; Woyach and Shah 2009; Steeg 2006). Judah Folkman was the first to suggest the suppression of tumor angiogenesis as a therapeutic strategy (Folkman 1995). Since then, many anti-angiogenic drugs have been

Acknowledgments: This work was supported by the Human Resources Program in Energy Technology of the KETEP grant from the MTIE (no. 20124010203250) and the National Research Foundation grant funded by MEST (no. 2013R1A2A2A03016122).

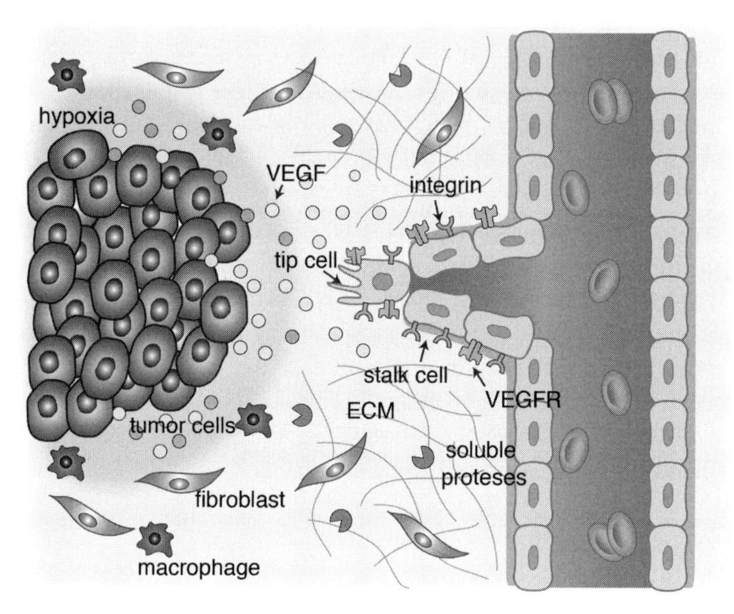

Figure 18.1 **Tumor microenvironment related to tumor angiogenesis.**

developed, although their efficacy has not been fully verified and new regulators of tumor angiogenesis are being intensively pursued (Bettinger et al. 2012).

The tumor microenvironment comprises numerous signaling molecules and pathways that influence the angiogenic response (Weis and Cheresh 2011). Among the soluble factors regulating the angiogenic response are cytokines, growth factors, including members of the fibroblast growth factor (FGF) and vascular endothelial growth factor (VEGF) families, and guidance molecules (Zetter 1998). Tumor cells generate and/or interrupt the local gradient of these and other soluble factors, thereby activating and attracting endothelial cells (ECs) to initiate new vascular growth. Active ECM remodeling by extracellular proteases, including matrix metalloproteinases, accelerate both the angiogenic responses and capillary morphogenesis.

Tumor-associated blood vessels are exposed to various stimuli in their local microenvironments, which induce obvious changes in their structure and function. They respond to stimulatory soluble factors released by the tumor cells and participate in active cell-cell and cell-matrix interactions. The immature, leaky, tortuous vessels result in irregular blood flow and high interstitial fluid pressure inside the tumor (Eilken and Adams 2010). The vascular leakage causes platelet activation and thus increased amount of platelet-derived growth factor (PDGF), which escalates local cellular responses in both tumor and stromal cells. The vessels are therefore located in and also contribute to a microenvironment that promotes tumor progression and stimulates neighboring stromal cells (Rørth 2009).

The complexity of this microenvironment, with its various soluble factors, cell types (stromal, endothelial, immune, and tumor), and ECM, poses a challenge regarding the development of a representative in vitro angiogenesis assay. Angiogenesis results in the formation of a three-dimensional (3-D) tube-like vascular network. It consists of multiple steps: sprouting of selected ECs, outgrowth of the sprouts, lumen formation, and vessel maturation. During sprouting, two types of specialized ECs are present: tip cells form a large number of filopodia whereas stalk cells make up the stalk of the vascular sprouts. The tip cells express high levels of a Notch ligand, delta-like 4 (Dll4), and their phenotypic specialization is very transient, as it is sensitive to microenvironmental factors (Eilken and Adams 2010). Lumen formation and the maturation of stalk cells are driven by collective EC movements. This 3-D process is very difficult to reproduce and therefore difficult to investigate both in vivo and in vitro. In this chapter, we describe the various assays that attempt to reconstitute the complex 3-D process of tumor angiogenesis. Among these, microfluidic approaches have been used to address the precise regulation of microenvironmental factors, including neighboring ECM, chemical gradients, mechanical stimuli, and stromal cells.

18.2 Tumor angiogenesis assays

Typical cell migration assays are limited in their ability to reproduce the complex 3-D structure of newly developing blood vessels and the various microenvironmental factors that facilitate angiogenesis. The capillary-like structures that form on an ECM substrate in the planar tubular network angiogenesis assay (Nakayasu et al. 1992; Shiu et al. 2004) are made up of cells with a polarity that is opposite that of capillary cells in vivo (Fig. 18.2A). The use of EC-coated beads to generate tube-like structures in a 3-D ECM microenvironment allows the formation of stable tube-like structures, but the initial EC seeding surface is rigid and completely different from that formed by the ECM and existing blood vessels. This assay also lacks the fluid-matrix interface that

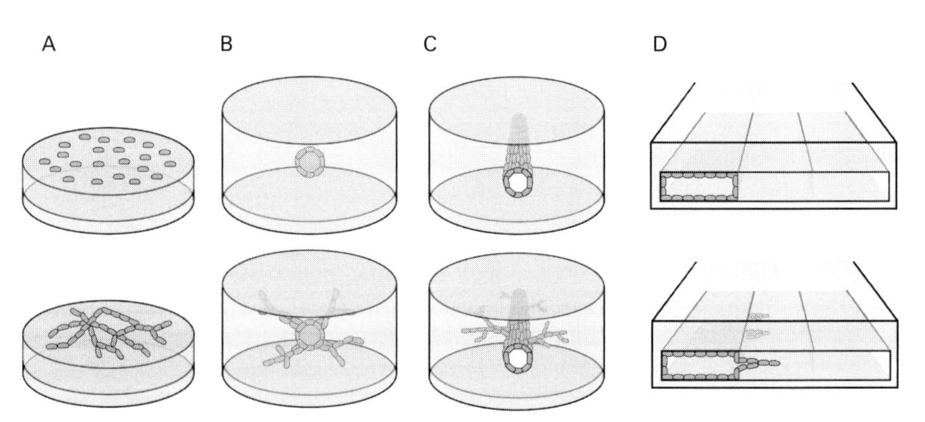

Figure 18.2 **Various cell migration assays.**
(A) Planar tubular network assay on an extracellular matrix (ECM) substrate. (B) Assay consisting of endothelial-cell-coated beads on an ECM. (C) Rapid casting assay of vascular networks in ECM. (D) ECM-incorporating microfluidic assay.

characterizes fluid shear stress and transendothelial flows (Bayless et al. 2000; Ghajar et al. 2006). Moreover, the system is not perfusable and therefore is not applicable to the study of chemokinetic and chemotactic effects (Fig. 18.2B).

A novel angiogenic assay based on the rapid casting technique of patterned vascular networks in ECM offered a novel method to generate perfusable, EC-lined cylindrical tubes (Nguyen et al. 2013; Miller et al. 2012; Zheng et al. 2012) (Fig. 18.2C). The assay was developed to maintain the metabolic functions of engineered tissue constructs (Miller et al. 2012) and was made further applicable for angiogenesis studies (Nguyen et al. 2013; Zheng et al. 2012). In the assay, angiogenic sprouting and new vessel formation are initiated from an original cylindrical vessel surrounded by a 3-D ECM (Nguyen et al. 2013). It can be used to identify inhibitors or promoters of both processes. The assayed cells undergo complex 3-D morphogenesis, with the invasion first of tip cells and then of stalk cells with apical-basal polarity. The lumens and branches that form in the ECM connect neighboring microchannels by perfusable vessels. Mass transfer processes and the long-term stability of the vessel were confirmed, allowing the elucidation of angiogenic interactions with perivascular cells seeded in the collagen bulk (Zheng et al. 2012). However, the handling difficulties and stiffness of the ECM, as the primary structural material supporting the newly formed vasculature, limits the cellular morphologies that develop in vitro.

The continuing need for an assay that reconstitutes the precise microenvironment regulating angiogenesis resulted in the design of novel microfluidic assays that incorporate ECMs in a microfluidic channel (Yeon et al. 2012; Mustonen et al. 1995; Dai et al. 2011; Chung et al. 2009a; Mack et al. 2009; Sudo et al. 2009; Chung et al. 2009b; Jeong et al. 2011b; Shin et al. 2011; Jeong et al. 2011a; Shin et al. 2012; Kim et al. 2012) (Fig. 18.2D). Microfluidic assays capture the complex 3-D tissue physiology of angiogenesis and allow observations of the spatiotemporal regulation that occurs in response to chemical gradients of growth factors, angiogenic promoters and inhibitors, proteins, and small molecules. Various hydrogels have been used in these assays to mimic the ECM invaded by angiogenic sprouts and have consistently allowed cultured ECs to form blood vessels (Yeon et al. 2012; Chung et al. 2009b; Jeong et al. 2011b; Jeong et al. 2011a). Carcinoma cell lines (Chung et al. 2009a), hepatocytes (Sudo et al. 2009), fibroblasts (Mack et al. 2009), and other cell types have also been investigated in microfluidic assays.

18.3 The ECM in tumor angiogenesis assays

18.3.1 ECM partitioning methods

In a paper published by Whiteside's group at Harvard University in 2008, the partitioning of microchannels using one or more microslabs of hydrogel and the construction of adjustable ECM partitions using laminar flow were demonstrated (Wong et al. 2008). The method was flexible in creating a cellular microenvironment that allowed the controlled spatial patterning of the ECM and the application of gradients of soluble factors. The assay was used to culture macrophage-like cells, as a simple model system of intercellular communication, but it was not used to study angiogenesis.

Also in 2008, Kamm's group at MIT reported the design, fabrication, and implementation of a multi-parameter control microfluidic assay with a partitioned ECM (Vickerman et al. 2008). The assay had two parallel microfluidic channels with a central ECM cage that housed the ECM and consisting of a micropillar array. The sterilized ECM was subsequently microinjected by micropipette, and the microfluidic channels were covered with glass coverslips. With the ECM in place, the two-sided media channels were isolated from each other, such that communication could occur only through the ECM that separated them. A VEGF concentration gradient across the ECM that regulated capillary morphogenesis into the matrix was demonstrated and quantified. In 2009, an ECM filling method to partition microchannels and thus enable multicellular co-cultures was proposed by Kamm's group (Chung et al. 2009a) and by Jeon and coworkers at Seoul National University (Huang et al. 2009).

Surface tension, hydrophobic interaction, and the spatial geometry of the microfluidic channel were considered as design parameters to contain the ECM within distinct channels by the filling method. A capillary burst valve model was modified to investigate the effects of different geometric dimensions to optimize ECM partitioning (Huang et al. 2009). A multichannel device with optimized geometry enabled the formation of multilayered ECM gels such that various types of cells could be co-cultured as multi-cellular niches in the assay. A further paper by Kamm's group published in the same year (Chung et al. 2009a) presented a similar concept of ECM partitioning at desired locations. Three independent flow channels were separated by an ECM, which was filled through connecting microchannels. ECs were grown in the center flow channel and a VEGF stimulus was applied to one of the outside channels, forming a stable gradient toward the cultured EC monolayer. This resulted in a sudden and steep gradient near the EC monolayer, which was thus shown to be an effective barrier for molecular diffusion. Capillary morphogenesis into the ECM along the VEGF gradient was monitored and quantified using the microchannel device. The observed changes in the angiogenic response depending on the mechanical properties (e.g., stiffness) of the ECM provided evidence of different modes of capillary morphogenesis. A cancer-cell-associated angiogenic response was also demonstrated in co-cultures with carcinoma cell lines. The type and seeding density of the cells was shown to influence the amount of capillary formation into the ECM. The methods for ECM filling, channel partitioning, and post-EC-seeding procedures have been summarized in a review (Chung et al. 2010) and in a protocol paper (Shin et al. 2012).

18.3.2 Types of ECM

The ECM is a protein-rich entity that supports cells and tissues and controls cell quiescence, survival, growth, and differentiation (Järveläinen et al. 2009). It also provides attachment sites and mechanical support for cells and constitutes both an interstitial matrix and a basement membrane (the thin and fibrous layer underlying the epithelium, mesothelium, and endothelium). Thus, the ECM is not only a regulator but also a structural base for complex, 3-D tube-like structures. The various partitioning methods can be applied to different types of ECMs to create the 3-D microenvironments allowing tumor angiogenesis.

Matrigel, a basement membrane extracted from a murine Engelbreth-Holm-Swarm (EHS) sarcoma, is widely used as an ECM for angiogenesis studies and especially in planar tubular network assays (Fig.18.2A). Matrigel supports the organization of ECs into complex networks, but it has many as yet unidentified proteins that can affect the angiogenic response. Moreover, capillary morphogenesis occurs only on the Matrigel surface, rather than within it (Xu et al. 2009). (Although Dai et al. [2011] reported the directional migration of ECs into Matrigel, their study lacked definitive proof of an

Table 18.1 ECMs used in rapid casting assays of vascular networks in ECM and in ECM incorporating microfluidic assays.

ECM	Assays	Studies	References
Type I collagen	Rapid casting assay	Angiogenesis	(Zheng et al. 2012; Nguyen et al. 2013; Verbridge et al. 2013)
		Blood vessel formation, extravasation	(Chrobak et al. 2006)
	ECM-incorporating microfluidic assay	Cell migration, angiogenesis	(Sudo et al. 2009; Song and Munn 2011; Vickerman and Kamm 2012; Chung et al. 2009a, Mack et al. 2009; Chung et al. 2009b; Jeong et al. 2011b; Jeong et al. 2011a, Shin et al. 2012; Kim et al. 2012)
		Blood vessel formation, intravasation, extravasation	(Zervantonakis et al. 2012; Jeon et al. 2013; Bersini et al. 2014)
		Blood vasculature formation	(Kim et al. 2012; Shin et al. 2011)
Fibrin	ECM-incorporating microfluidic assay	Blood vessel formation	(Yeon et al. 2012; Kim et al. 2013)
		Blood vasculature formation (vasculogenesis)	(Carrion et al. 2010; Whisler et al. 2012; Jeon et al. 2014b; Moya et al. 2013)
		Vasculogenesis, extravasation	(Chen et al. 2013; Jeon et al. 2014a)
Matrigel	ECM-incorporating microfluidic assay	Cell migration, extravasation without blood vessel formation	(Chaw et al. 2007)
		Angiogenesis	(Dai et al. 2011; Mustonen et al. 1995; Bischel et al. 2013)

intact vascular network within the gel.) Collagen and fibrin are other sources of ECM proteins, with the advantages of being a more realistic matrix for capillary formation, their high purity, and extensive remodeling capacity. Collagen is the main structural protein of the various connective tissues in the human body. Type I collagen, which consists of collagen nanofibers, is the most abundant collagen. As noted above, the mechanical properties of type I collagen regulate the dimensionality and structure of the capillaries that form in the collagen matrix. Fibrin is, as its name implies, is a fibrous protein that is polymerized by thrombin during blood clotting and, together with platelets, forms a hemostatic plug over the wound site. Fibrin is also an inducer of capillary tube formation (Chalupowicz et al. 1995). The complex capillary networks that form in fibrin during angiogenesis and vasculogenesis are very stable. However, due to its softness, fibrin can only be used in microfluidic channels as space-filling material; it is not suitable for use as a structural material.

Other ECM proteins, such as hyaluronic acid, fibronectin and alginate, also cannot be used as structural materials because they do not easily form rigid hydrogels even in a microfluidic channel. In previous research, ECM proteins were mixed with type I collagen hydrogel to build a gel structure. In addition to type I collagen, type IV collagen, a basic component of the basement membrane (Maeshima et al. 2000), stabilizes blood vessels (Figg and Folkman 2008). Laminin is another basement membrane component involved in vessel formation and maturation (Thyboll et al. 2002). It has been used to create a vascular microenvironment (Shin et al. 2014) but not to study angiogenesis. For example, a recent report described the use of a type I collagen hydrogel mixed with laminin and fibronectin in a microfluidic device to regulate the differentiation of neural stem cells (Yang et al. 2013). The ECM protein fibronectin plays an important role in promoting EC growth (George et al. 1997). Decorin and thrombospondin have yet to be applied in microfluidic studies; however, because of their ability to inhibit the tumor angiogenic response (Grant et al. 2002; Rodríguez-Manzaneque et al. 2001) they warrant investigation in microfluidic studies.

Table 18.2 Other ECMs used in ECM-incorporating microfluidic assays in complex mixtures with type I collagen.

ECM	Description	Studies	References
Hyaluronic acid	ECM-incorporating microfluidic assay	(Mixed with type I collagen) Cell migration	(Jeong et al. 2011b)
Fibronectin	Endothelial-cell-coated beads in ECM and ECM-incorporating microfluidic assays	(Mixed with type I collagen) Angiogenesis	(Shamloo et al. 2011)
	ECM-incorporating microfluidic assay	(Surface coating) Cell adhesion	(Young et al. 2007)
Alginate	ECM-incorporating microfluidic assay	(Mixed with type I collagen; cells cannot migrate into alginate) Angiogenesis	(Kim et al. 2012)

Angiogenesis is the result of the collective migration of ECs followed by their sprouting and branching. Sprouting is the outgrowth of ECs into the neighboring ECM of existing blood vessels. It is guided by leading tip cells—which have a polarity such that the front end of the cell is free while the back end is attached to stalk cells—that actively interact with the ECM. Stalk cells form the tube-like structures of new blood vessel and, by branching and expanding, a connective network. ECM proteins are therefore essential regulators of both the invasion by tip cells and tube morphogenesis guided by stalk cells. Recently, the role of ECM components in the 3-D migration morphologies of carcinoma cells was examined in detail in vitro using a microfluidic assay (Shin et al. 2013; Shin et al. 2014). Similar approaches can be applied to investigate the role of ECM components in angiogenesis. The mechanical properties of the ECM, including its stiffness, may be another important regulator of capillary morphology during angiogenesis.

18.4 Angiogenic factors in tumor angiogenesis assays

18.4.1 Angiogenic signaling molecules

Angiogenesis is a multi-step process. During their acquisition of angiogenic activities, ECs produce protease to degrade the basement membrane. A local stimulus consisting of a gradient of regulatory signals initiates the migration and proliferation of ECs and their invasion of the neighboring ECM, followed by lumen formation and capillary stabilization. Regulatory signals come from the ECM, stromal cells (fibroblasts, keratinocytes, macrophages, etc.), and tumor cells and they actively direct all aspects of angiogenesis.

Members of the VEGF family are the main promoters of angiogenesis and vasculogenesis. They stimulate EC migration, proliferation, and differentiation and increase the permeability of blood vessels (Yoshida et al. 1996). Placental growth factor and neuropilin-1 also modulate angiogenesis, albeit via mechanisms that are poorly understood (Carmeliet 2000). Members of the FGF and PDGF family also play a role in angiogenesis, by inducing EC migration and proliferation and by recruiting mesenchymal (including smooth muscle cells) and inflammatory (Yoshida et al. 1996; Nissen et al. 2007) cells. Transforming growth factor $\alpha 1$ is an inducer of VEGF synthesis that stimulates EC growth (Ferrari et al. 2006). Similar effectors of angiogenesis include hepatocyte growth factor, insulin-like growth factor-1, and monocyte chemoattractant protein-1 (Carmeliet and Jain 2000).

Angiopoietin 1 (ANG1) tightens the vessel network and has a major role in angiogenesis (Metheny-Barlow and Li 2003), as demonstrated in an ECM-incorporating microfluidic assay. Shin et al. found that a gradient consisting of both ANG1 and VEGF enhanced the connectivity of tip and stalk cells. Capillary sprouts under a VEGF gradient without ANG1 were characterized by a greater number of tip cells but failed to form stable and complex vascular structures. The addition of ANG1 greatly enhanced the number of tip cells that attached to collectively migrating stalk cells (Shin et al. 2011; Nguyen et al. 2013) and increased the connection between these two cell types, thereby regulating the morphogenesis and life cycle of stalk cells. Using an ECM-incorporating microfluidic assay, C. Chen's group at Boston University found that spingoshine-1-phosphate, a major regulator of vascular and immune systems, including angiogenesis and vascular maturation (Vickerman et al. 2008; Farahat et al. 2012; Nguyen et al. 2013), exerted pro-angiogenic effects by stimulating directional filopodial extension (Nguyen et al. 2013). The recent demonstration of hypoxia-driven pathological angiogenesis has important implications for tumorigenesis, wound healing, diabetes, asthma, atherosclerosis, and other diseases. Cells residing deep within tumors are far from blood vessels and thus subject to hypoxia. Abnormal ECM deposition near the tumor site also hinders the delivery of oxygen. Under hypoxic conditions, cells overexpress angiogenic factors such as VEGF and PDGF (Semenza 1998).

Tumor necrosis factor-α is an inhibitor of endothelial growth and may be involved in tumor dormancy (Carmeliet 2000). Angiostatin, endostatin, and interleukins inhibit EC migration. These molecules are therefore being tested in clinical trials in which anti-angiogenic therapies administered to cancer patients. However, residual surviving

Table 18.3 Growth factors analyzed in rapid casting assays of vascular networks in ECM and in ECM incorporating microfluidic assays.

Factor	Description	Functions	References
VEGF	Vascular endothelial growth factor	Physiological regulator of angiogenesis during embryogenesis, skeletal growth, and reproductive functions. Also implicated in the pathological angiogenesis associated with tumor growth.	(Shin et al. 2011; Vickerman et al. 2008; Farahat et al. 2012; Nguyen et al. 2013; Barkefors et al. 2009)
S1P	Spingoshine-1-phosphate	Regulator of angiogenesis, vascular maturation, cardiac development, and immunity.	(Vickerman et al. 2008; Farahat et al. 2012; Nguyen et al. 2013)
ANG1	Angiopoietin 1	Regulates the maintenance and stabilization of mature vessels by promoting interactions between endothelial cells and surrounding supporting cells.	(Shin et al. 2011; Nguyen et al. 2013)
HGF	Hepatocyte growth factor	Stimulates mitogenesis, cell motility, and matrix invasion and thus plays a central role in angiogenesis, tumorigenesis, and tissue regeneration.	(Nguyen et al. 2013)
MCP-1	Monocyte chemotactic protein-1	Responsible for the regeneration of the endothelial layer and angiogenesis.	(Nguyen et al. 2013)
bFGF	Basic fibroblast growth factor	Induces the proliferation of endothelial and smooth muscle cells and inhibits capillary formation.	(Nguyen et al. 2013; Barkefors et al. 2009)

cancer cells that lead to disease recurrence limit the efficacy of this approach (Carmeliet and Jain 2000). The further development of microfluidic angiogenic assays will allow investigations of tumor angiogenesis related to tumor recurrence.

18.4.2 Chemical gradients of angiogenic signaling molecules

Chemical gradients of angiogenic promoters/inhibitors have been simulated and their effects visualized in many types of assays. Visualization of the gradient distribution of VEGF in an ECM-incorporating microfluidic assay was achieved using 40-kDa FITC-dextran (Vickerman et al. 2008; Chung et al. 2009a; Chung et al. 2009b), which has also been applied in the rapid casting assays of vascular networks in ECM (Verbridge et al. 2013) (see Fig. 18.3). A steady-state linear gradient of FITC-dextran is considered to be representative of VEGF gradients. The Chung group at Korea University reported the local accumulation of FITC-dextran near the EC monolayer, due to the reduced permeability of the confluent EC monolayer (Chung et al. 2009a). The Kamm group at MIT used a FITC-dextran intensity profile to characterize endothelial permeability. Experimental and computational simulations of molecular transport have illustrated a sharp drop in the dextran concentration across the EC monolayer because of the latter's barrier effect (Zervantonakis et al. 2012; Jeon et al. 2013). This same method was also used to study the intravasation and extravasation of tumor cells across the EC monolayer.

In previous studies a VEGF gradient was used to study diffusion through the ECM and transport inhibition by the EC monolayer. However, the Chung group at Korea University was the first to include cellular VEGF consumption in a simulation of the distribution of this growth factor in microfluidic channels (Jeong et al. 2011a). The ability of cellular metabolism to perturb growth factor gradients was experimentally verified by monitoring the sprouting response of an EC monolayer into an ECM consisting of type I collagen. The reduced steepness of the gradient could be attributed solely to cellular metabolism.

18.4.3 Interstitial flow

Interstitial flow is determined by the hydrostatic pressure in the capillaries that drives water into the interstitial space and by the osmotic pressure that drives water back into the capillaries, with an additional contribution by the lymphatic system. EC morphogenesis and thus angiogenesis is stimulated by interstitial flow, which directs these cells to form multicellular, branched, lumen-containing vessel networks (Ng et al. 2004). Another potential angiogenic mechanism of interstitial flow is its influence on the establishment of gradients of angiogenic factors, by inducing large changes in their distribution regardless of flow rates, resulting in skewed gradients of these molecules (Helm et al. 2005). Because tumor cells stimulate angiogenesis, which leads to the formation of leaky, structurally abnormal vessels, interstitial flow contributes significantly to tumor microenvironments (Jain et al. 2007).

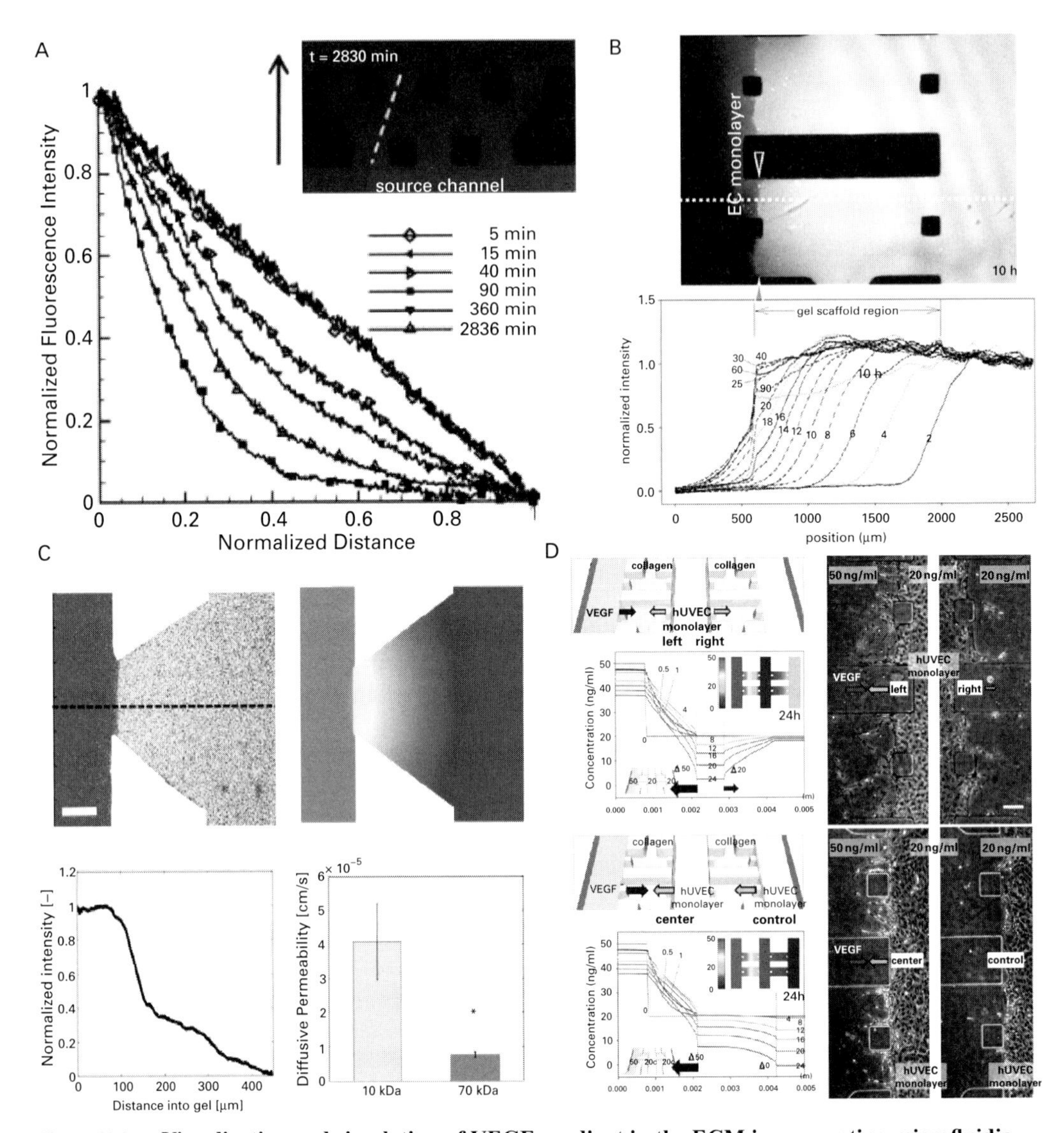

Figure 18.3 **Visualization and simulation of VEGF gradient in the ECM incorporating microfluidic assays.**

(A) Linear gradient of FITC-dextran (Vickerman et al. 2008). (B) Inhibition of the accumulation of FITC-dextran near the endothelial cell (EC) monolayer. (C) Permeability of the EC monolayer as indicated by the accumulation of FITC-dextran. (D) Simulated vascular endothelial growth factor gradient resulting from diffusion, inhibition, and metabolic consumption.

Reproduced from: (A) Chung et al. (2009a); (B) Zervantonakis et al. (2012); (C) and (D) Jeong et al. (2011a).

Based on their ability to mimic tumor microenvironments consisting of stromal cells, ECMs, and blood vessels, microfluidic assays can reveal the complex relationships of convective mass transfer, direct stimulation by interstitial flow, and associated cellular responses in 3-D (Stroock and Fischbach 2010). The direction of interstitial flow was shown to be an important cue in angiogenesis. Thus, basal-to-apical transendothelial flow triggered the transition of ECs from quiescent to invasive, promoted angiogenesis in response to localized vascular endothelial (VE)-cadherin, and increased focal adhesion kinase (FAK) phosphorylation (Vickerman and Kamm 2012). In other research employing microfluidic assays, interstitial flow directed endothelial morphogenesis and sprouting, whereas fluid shear stress attenuated EC sprouting (Song and Munn 2011). ECs integrate signals from fluid forces and local VEGF gradients to become dilated, sprouted, or inactive.

In tumor microenvironment in vivo, the synergistic relationship between interstitial flow effects (convective mass transfer and cellular shear stress) and chemical gradients is complex, involving the ECM, neighboring blood and lymph vessels, tumor and stromal cells, and the response of the immune system (Swartz and Lund 2012). Interstitial flow enhances communication between tumor sites and lymph vessels and can affect the host immune response to tumor growth. It also causes stress-induced changes in stromal cells and the ECM (increased stiffness) and therefore alters immune microenvironments (Swartz and Lund 2012).

18.5 Conclusion

The tumor angiogenesis assays (rapid casting assays of vascular networks in ECM and ECM-incorporating microfluidic assays) described in this review can be exploited to demonstrate the importance of the ECM microenvironment in tumor angiogenesis. They enable systematic investigations into the effects of physiologically relevant mechanical (e.g., interstitial flow) and chemical (e.g., growth factors and their gradients) factors and their synergistic interactions. The microfluidic platform provides high optical resolution in investigations of cellular morphogenesis in 3-D and offers versatility in studies of the multi-parameter control of angiogenesis factors. This platform also enables parallel analyses, and therefore studies with increased throughput (Farahat et al. 2012), in addition to providing stable culture conditions for the quantitative assessment of various chemical and mechanical conditions (Kim et al. 2015).

References

Barkefors, I., Thorslund, S., Nikolajeff, F. and Kreuger, J. (2009). "A fluidic device to study directional angiogenesis in complex tissue and organ culture models." *Lab on a Chip* **9**: 529–535.

Bayless, K. J., Salazar, R. and Davis, G. E. (2000). "RGD-dependent vacuolation and lumen formation observed during endothelial cell morphogenesis in three-dimensional fibrin matrices involves the $\alpha_v \beta_3$ and $\alpha_5 \beta_1$ integrins." *The American Journal of Pathology* **156**: 1673–1683.

Bersini, S., Jeon, J. S., Dubini, G., Arrigoni, C., Chung, S., Charest, J. L., Moretti, M. and Kamm, R. D. (2014). "A microfluidic 3-D in vitro model for specificity of breast cancer metastasis to bone." *Biomaterials* **35**: 2454–2461.

Bettinger, C., Borenstein, J. T. and Tao, S. L. (2012). *Microfluidic Cell Culture Systems*. Oxford: William Andrew.

Bischel, L. L., Young, E. W., Mader, B. R. and Beebe, D. J. (2013). "Tubeless microfluidic angiogenesis assay with three-dimensional endothelial-lined microvessels. *Biomaterials* **34**: 1471–1477.

Carmeliet, P. (2000). "Mechanisms of angiogenesis and arteriogenesis." *Nature Medicine* **6**: 389–396.

Carmeliet, P. and Jain, R. K. (2000). "Angiogenesis in cancer and other diseases." *Nature* **407**: 249–257.

Carrion, B., Huang, C. P., Ghajar, C. M., Kachgal. S., Kniazeva, E., Jeon, N. L. and Putnam, A. J. (2010). "Recreating the perivascular niche ex vivo using a microfluidic approach." *Biotechnology and Bioengineering* **107**: 1020–1028.

Chalupowicz, D. G., Chowdhury, Z. A., Bach, T. L., Barsigian, C. and Martinez, J. (1995). "Fibrin II induces endothelial cell capillary tube formation." *The Journal of cell Biology* **130**: 207–215.

Chaw, K., Manimaran, M., Tay, E. and Swaminathan, S. (2007). "Multi-step microfluidic device for studying cancer metastasis." *Lab on a Chip* **7**: 1041–1047.

Chen, M. B., Whisler, J. A., Jeon, J. S. and Kamm, R. D. (2013). "Mechanisms of tumor cell extravasation in an in vitro microvascular network platform." *Integrative Biology* **5:** 1262–1271.

Chrobak, K. M., Potter, D. R. and Tien, J. (2006). "Formation of perfused, functional microvascular tubes in vitro." *Microvascular Research* **71**: 185–196.

Chung, S., Sudo, R., Mack, P. J., Wan, C.-R., Vickerman, V. and Kamm, R. D. (2009a). "Cell migration into scaffolds under co-culture conditions in a microfluidic platform." *Lab on a Chip* **9**: 269–275.

Chung, S., Sudo, R., Vickerman, V., Zervantonakis, I. K. and Kamm, R. D. (2010). "Microfluidic platforms for studies of angiogenesis, cell migration, and cell-cell interactions." *Annals of Biomedical Engineering* **38**: 1164–1177.

Chung, S., Sudo, R., Zervantonakis, I. K., Rimchala, T. and Kamm, R. D. (2009b). "Surface-treatment-induced three-dimensional capillary morphogenesis in a microfluidic platform." *Advanced Materials* **21**: 4863–4867.

Dai, X., Cai, S., Ye, Q., Jiang, J., Yan, X., Xiong, X., Jiang, Q., et al. (2011). "A novel in vitro angiogenesis model based on a microfluidic device." *Chinese Science Bulletin* **56**: 3301–3309.

Eilken, H. M. and Adams, R. H. (2010). "Dynamics of endothelial cell behavior in sprouting angiogenesis." *Current Opinion in Cell Biology* **22**: 617–625.

Farahat, W. A., Wood, L. B., Zervantonakis, I. K., Schor, A., Ong, S., Neal. D., Kamm, R. D. et al. (2012). "Ensemble analysis of angiogenic growth in three-dimensional microfluidic cell cultures." *PloS One* **7**: e37333.

Ferrari, G., Pintucci, G., Seghezzi, G., Hyman, K., Galloway, A. C. and Mignatti, P. (2006). "VEGF, a prosurvival factor, acts in concert with TGF-β1 to induce endothelial cell apoptosis." *Proceedings of the National Academy of Sciences* **103**: 17260–17265.

Figg, W. and Folkman, J. (2008). *Angiogenesis: An Integrative Approach from Science to Medicine*. New York: Springer.

Folkman, J. (1995). "Angiogenesis in cancer, vascular, rheumatoid and other disease." *Nature Medicine* **1**: 27–30.

George, E. L., Baldwin, H. S. and Hynes, R. O. (1997). "Fibronectins are essential for heart and blood vessel morphogenesis but are dispensable for initial specification of precursor cells." *Blood* **90**: 3073–3081.

Ghajar, C. M., Blevins, K. S., Hughes, C. C., George, S. C. and Putnam, A. J. (2006). "Mesenchymal stem cells enhance angiogenesis in mechanically viable prevascularized tissues via early matrix metalloproteinase upregulation." *Tissue Engineering* **12**: 2875–2888.

Grant, D. S., Yenisey, C., Rose, R. W., Tootell, M., Santra, M. and Iozzo, R. V. (2002). "Decorin suppresses tumor cell-mediated angiogenesis." *Oncogene* **21**: 4765–4777.

Helm, C.-L. E., Fleury, M. E., Zisch, A. H., Boschetti, F. and Swartz, M. A. (2005). "Synergy between interstitial flow and VEGF directs capillary morphogenesis in vitro through a gradient amplification mechanism." *Proceedings of the National Academy of Sciences of the United States of America* **102**: 15779–15784.

Huang, C. P., Lu, J., Seon, H., Lee, A. P., Flanagan, L. A., Kim, H.-Y., Putnam, A. J. et al. (2009). "Engineering microscale cellular niches for three-dimensional multicellular co-cultures." *Lab on a Chip* **9**: 1740–1748.

Jain, R. K., di Tomaso, E., Duda, D. G., Loeffler, J. S., Sorensen, A. G. and Batchelor, T. T. (2007). "Angiogenesis in brain tumours." *Nature Reviews Neuroscience* **8**: 610–622.

Järveläinen, H., Sainio, A., Koulu, M., Wight, T. N. and Penttinen, R. (2009). "Extracellular matrix molecules: potential targets in pharmacotherapy." *Pharmacological Reviews* **61**: 198–223.

Jeon, J. S., Bersini, S., Gilardi, M., Dubini, G., Charest, J. L., Moretti, M. and Kamm, R. D. (2014a). "Human 3-D vascularized organotypic microfluidic assays to study breast cancer cell extravasation." *Proceedings of the National Academy of Sciences*: 201417115.

Jeon, J. S., Bersini, S., Whisler, J. A., Chen, M. B., Dubini, G., Charest, J. L., Moretti, M. et al. (2014b). "Generation of 3-D functional microvascular networks with human mesenchymal stem cells in microfluidic systems." *Integrative Biology* **6**: 555–563.

Jeon, J. S., Zervantonakis, I. K., Chung, S., Kamm, R. D. and Charest, J. L. (2013). "In vitro model of tumor cell extravasation." *PloS One* **8**: e56910.

Jeong, G. S., Han, S., Shin, Y., Kwon, G. H., Kamm, R. D., Lee, S.-H. and Chung, S. (2011a). "Sprouting angiogenesis under a chemical gradient regulated by interactions with an endothelial monolayer in a microfluidic platform." *Analytical Chemistry* **83**: 8454–8459.

Jeong, G. S., Kwon, G. H., Kang, A. R., Jung, B. Y., Park, Y., Chung, S. and Lee, S.-H. (2011b). "Microfluidic assay of endothelial cell migration in 3D interpenetrating polymer semi-network HA-Collagen hydrogel." *Biomedical Microdevices* **13**: 717–723.

Kim, C., Chung, S., Yuchun, L., Kim, M.-C., Chan, J. K., Asada, H. H. and Kamm, R. D. (2012). "In vitro angiogenesis assay for the study of cell-encapsulation therapy." *Lab on a Chip* **12**: 2942–2950.

Kim, C., Kasuya, J., Jeon, J., Chung, S. and Kamm, R. D. (2015). "A quantitative microfluidic angiogenesis screen for studying anti-angiogenic therapeutic drugs." *Lab on a Chip* **15**: 301–310.

Kim, S., Lee, H., Chung, M. and Jeon, N. L. (2013). "Engineering of functional. perfusable 3-D microvascular networks on a chip." *Lab on a Chip* **13**: 1489–1500.

Mack, P. J., Zhang, Y., Chung, S., Vickerman, V., Kamm, R. D. and García-Cardeña, G. (2009). "Biomechanical regulation of endothelium-dependent events critical for adaptive remodeling." *Journal of Biological Chemistry* **284**: 8412–8420.

Maeshima, Y., Colorado, P. C. and Kalluri, R. (2000). "Two RGD-independent αvβ3 integrin binding sites on tumstatin regulate distinct anti-tumor properties." *Journal of Biological Chemistry* **275**: 23745–23750.

Metheny-Barlow, L. J. and Li, L. Y. (2003). "The enigmatic role of angiopoietin-1 in tumor angiogenesis." *Cell Research* **13**: 309–317.

Miller, J. S., Stevens, K. R., Yang, M. T., Baker, B. M., Nguyen, D.-H. T., Cohen, D. M., Toro, E., et al. (2012). "Rapid casting of patterned vascular networks for perfusable engineered three-dimensional tissues." *Nature Materials* **11**: 768–774.

Moya, M. L., Hsu, Y.-H., Lee, A. P., Hughes, C. C. and George, S. C. (2013). "In vitro perfused human capillary networks." *Tissue Engineering Part C: Methods* **19**: 730–737.

mustonen, T. and Alitalo, K. (1995). "Endothelial receptor tyrosine kinases involved in angiogenesis." *The Journal of Cell Biology* **129**: 895–898.

Nakayasu, K., Hayashi, N., Okisaka, S. and Sato, N. (1992). "Formation of capillary-like tubes by vascular endothelial cells cocultivated with keratocytes." *Investigative Ophthalmology & Visual Science* **33**: 3050–3057.

Ng, C. P., Helm, C.-L. E. and Swartz, M. A. (2004). "Interstitial flow differentially stimulates blood and lymphatic endothelial cell morphogenesis in vitro." *Microvascular Research* **68**: 258–264.

Nguyen, D.-H. T., Stapleton, S. C., Yang, M. T., Cha, S. S., Choi, C. K., Galie, P. A. and Chen, C. S. (2013). "Biomimetic model to reconstitute angiogenic sprouting morphogenesis in vitro." *Proceedings of the National Academy of Sciences* **110**: 6712–6717.

Nissen, L. J., Cao, R., Hedlund, E.-M., Wang, Z., Zhao, X., Wetterskog, D., Funa, K., et al. (2007). "Angiogenic factors FGF2 and PDGF-BB synergistically promote murine tumor neovascularization and metastasis." *The Journal of Clinical Investigation* **117**: 2766–2777.

Rodríguez-Manzaneque, J. C., Lane, T. F., Ortega, M. A., Hynes, R. O., Lawler, J. and Iruela-Arispe, M. L. (2001). "Thrombospondin-1 suppresses spontaneous tumor growth and inhibits activation of matrix metalloproteinase-9 and mobilization of vascular endothelial growth factor." *Proceedings of the National Academy of Sciences* **98**: 12485–12490.

Rørth, P. (2009). "Collective cell migration." *Annual Review of Cell and Developmental* **25**: 407–429.

Semenza, G. L. (1998). "Hypoxia-inducible factor 1: master regulator of O_2 homeostasis." *Current Opinion in Genetics & Development* **8**: 588–594.

Shamloo, A., Xu, H. and Heilshorn, S. (2011). "Mechanisms of vascular endothelial growth factor-induced pathfinding by endothelial sprouts in biomaterials." *Tissue Engineering Part A* **18**: 320–330.

Shin, Y., Han, S., Jeon, J. S., Yamamoto, K., Zervantonakis, I. K., Sudo, R., Kamm, R. D. et al. (2012). "Microfluidic assay for simultaneous culture of multiple cell types on surfaces or within hydrogels." *Nature Protocols* **7**: 1247–1259.

Shin, Y., Jeon, J. S., Han, S., Jung, G.-S., Shin, S., Lee, S.-H., Sudo, R., et al. (2011). "In vitro 3D collective sprouting angiogenesis under orchestrated ANG-1 and VEGF gradients." *Lab on a Chip* **11**: 2175–2181.

Shin, Y., Kim, H., Han, S., Won, J., Jeong, H. E., Lee, E. S., Kamm, R. D., et al. (2013). "Extracellular matrix heterogeneity regulates three-dimensional morphologies of breast adenocarcinoma cell invasion." *Advanced Healthcare Materials* **2**: 790–794.

Shin, Y., Yang, K., Han, S., Park, H. J., Seok Heo, Y., Cho, S. W. and Chung, S. (2014). "Reconstituting vascular microenvironment of neural stem cell niche in three-dimensional extracellular matrix." *Advanced Healthcare Materials* **3**: 1457–1464.

Shiu, Y.-T., Weiss, J. A., Hoying, J. B., Iwamoto, M. N., Joung, I. S. and Quam, C. T. (2004). "The role of mechanical stresses in angiogenesis." *Critical Reviews in Biomedical Engineering* **33**: 431–510.

Song, J. W. and Munn, L. L. (2011). "Fluid forces control endothelial sprouting." *Proceedings of the National Academy of Sciences* **108**: 15342–15347.

Steeg, P. S. (2006). "Tumor metastasis: mechanistic insights and clinical challenges." *Nature Medicine* **12**: 895–904.

Stroock, A. D. and Fischbach, C. (2010). "Microfluidic culture models of tumor angiogenesis." *Tissue Engineering Part A* **16**: 2143–2146.

Sudo, R., Chung, S., Zervantonakis, I. K., Vickerman, V., Toshimitsu, Y., Griffith, L. G. and Kamm, R. D. (2009). "Transport-mediated angiogenesis in 3-D epithelial coculture." *The FASEB Journal* **23**: 2155–2164.

Swartz, M. A. and Lund, A. W. (2012). "Lymphatic and interstitial flow in the tumour microenvironment: linking mechanobiology with immunity." *Nature Reviews Cancer* **12**: 210–219.

Thyboll, J., Kortesmaa, J., Cao, R., Soininen, R., Wang, L., Iivanainen, A., Sorokin, L., et al. (2002). "Deletion of the laminin α4 chain leads to impaired microvessel maturation." *Molecular and Cellular Biology* **22**: 1194–1202.

Verbridge, S. S., Chakrabarti, A., Delnero, P., Kwee, B., Varner, J. D., Stroock, A. D. and Fischbach, C. (2013). "Physicochemical regulation of endothelial sprouting in a 3-D microfluidic angiogenesis model." *Journal of Biomedical Materials Research Part A* **101**: 2948–2956.

Vickerman, V., Blundo, J., Chung, S. and Kamm, R. (2008). "Design, fabrication and implementation of a novel multi-parameter control microfluidic platform for three-dimensional cell culture and real-time imaging." *Lab on a Chip* **8**: 1468–1477.

Vickerman, V. and Kamm, R. D. (2012). "Mechanism of a flow-gated angiogenesis switch: early signaling events at cell-matrix and cell-cell junctions." *Integrative Biology* **4**: 863–874.

Weis, S. M. and Cheresh, D. A. (2011). "Tumor angiogenesis: molecular pathways and therapeutic targets." *Nature Medicine* **17**: 1359–1370.

Whisler, J. A., Chen, M. B. and Kamm, R. D. (2012). "Control of perfusable microvascular network morphology using a multiculture microfluidic system." *Tissue Engineering Part C: Methods* **7**: 543–552.

Wong, A. P., Perez-Castillejos, R., Christopher Love, J. and Whitesides, G. M. (2008). "Partitioning microfluidic channels with hydrogel to construct tunable 3-D cellular microenvironments." *Biomaterials* **29**: 1853–1861.

Woyach, J. A. and Shah, M. H. (2009). "New therapeutic advances in the management of progressive thyroid cancer." *Endocrine-Related Cancer* **16**: 715–731.

Xu, X., Yang, G., Zhang, H. and Prestwich, G. D. (2009). "Evaluating dual activity LPA receptor pan-antagonist/autotaxin inhibitors as anti-cancer agents in vivo using engineered human tumors." *Prostaglandins & Other Lipid Mediators* **89**: 140–146.

Yang, K., Han, S., Shin, Y., Ko, E., Kim, J., Park, K. I., Chung, S. et al. (2013). "A microfluidic array for quantitative analysis of human neural stem cell self-renewal and differentiation in three-dimensional hypoxic microenvironment." *Biomaterials* **34**: 6607–6614.

Yeon, J. H., Ryu, H. R., Chung, M., Hu, Q. P. and Jeon, N. L. (2012). "In vitro formation and characterization of a perfusable three-dimensional tubular capillary network in microfluidic devices." *Lab on a Chip* **12**: 2815–2822.

Yoshida, A., Anand-Apte, B. and Zetter, B. R. (1996). "Differential endothelial migration and proliferation to basic fibroblast growth factor and vascular endothelial growth factor." *Growth Factors* **13**: 57–64.

Young, E. W., Wheeler, A. R. and Simmons, C. A. (2007). "Matrix-dependent adhesion of vascular and valvular endothelial cells in microfluidic channels." *Lab on a Chip* **7**: 1759–1766.

Zervantonakis, I. K., Hughes-Alford, S. K., Charest, J. L., Condeelis, J. S., Gertler, F. B. and Kamm, R. D. (2012). "Three-dimensional microfluidic model for tumor cell intravasation and endothelial barrier function." *Proceedings of the National Academy of Sciences* **109**: 13515–13520.

Zetter, P., and Bruce R. (1998). "Angiogenesis and tumor metastasis." *Annual Review of Medicine* **49**: 407–424.

Zheng, Y., Chen, J., Craven, M., Choi, N. W., Totorica, S., Diaz-Santana, A., Kermani, P., et al. (2012). "In vitro microvessels for the study of angiogenesis and thrombosis." *Proceedings of the National Academy of Sciences* **109**: 9342–9347.

19 Neuromechanobiology of the brain

Mechanics of neuronal structure, function, and pathophysiology

Jerel Mueller and William Tyler

This chapter discusses recent progress and future directions regarding mechanobiology as applied to neuronal function. Along with the generation and transduction of mechanical forces by neuronal elements, the influence of mechanical forces on the neuronal membrane, actin, and ion channels is highlighted. Further topics such as cortical folding and traumatic brain injury expand discussion of the role of mechanical forces into a more macroscopic scale. As the mechanical properties of the nervous tissue environment and other mechanical cues influence neural development and contribute to the regulation of endogenous brain function, there is great utility in investigating the mechanical properties of the central nervous system. Through discussion of the role of mechanical forces in neural elements, and early biophysical formulations to understand neural systems that incorporate mechanical analysis, this chapter hopes to encourage expansion of studies and methods investigating mechanobiology applied to the nervous system.

19.1 Introduction

The consideration of mechanical forces on neuronal function has been of great interest over the past several decades and continues to grow while gaining support through its incorporation in system characterization and manipulation. Numerous mechanical events are known to occur in neurons. In axons, for instance, action potentials are accompanied by propagating membrane deformations (volumetric changes). Mechanical impulses have also been recorded at axon terminals during action potential firing and vesicle fusion. Dendritic spines "twitch" and experience rapid actin-mediated contractions in response to synaptic activity. In sensory neuroscience it is broadly recognized that mechanosensitive channels are involved in signal transduction processes in hair cells for hearing and free nerve endings for touch.

The extent to which these cellular-mechanical dynamics influence brain function, however, remains a mystery. This gap in our knowledge probably exists because neuroscientists have not traditionally considered the roles of classical mechanics in brain function. The intent of this chapter is therefore to challenge our current models of neuronal physiology and plasticity, which do not at present account for the cellular mechanics that affect neurons. An overview of the principles by which the neuronal elements endow the brain with mechanical properties will be discussed. Additionally, hypotheses pertaining to how interactions among some of the mechanical features of the brain underlie various aspects of synaptic signaling, neuronal plasticity, and traumatic injury are examined. Furthermore, recent mechanobiological methods and formulations are highlighted to further encourage the investigation of cellular mechanics in neuroscience. Overall, there is a need to expand our consideration of the forces that underlie the mechanical (physical) plasticity of the brain and their consequences for neuronal signaling. By starting to consider the interplay between electrical, chemical, thermal, and mechanical energy, rather than separately compartmentalizing them, fresh insights into nervous system function and dysfunction will likely evolve.

19.2 Mechanical properties of neuronal elements

Elastography uses magnetic resonance or ultrasound approaches to estimate the stiffness of tissues by imaging their responses to sound (shear) waves propagated through the body. Magnetic resonance elastography (MRE) is useful for characterizing and mapping the nonlinear viscoelastic properties of the intact human brain (Kruse et al. 2008; McCracken et al. 2005). MRE has shown that the stiffness of brain regions varies substantially in normal humans (Kruse et al. 2008; McCracken et al. 2005; Zhang et al. 2011) and that these mechanical properties change with age (Sack et al. 2011) and disease state (Murphy et al. 2011; Wuerfel et al. 2010). Understanding the molecular and cellular properties of neurons that underlie these mechanical changes and how they give rise to functional outcomes should serve as focal points for mechanobiological studies of the brain. Owing to the integrated structural and mechanical properties of cells and tissues, forces affecting one of their components can in turn produce tension and strain in others (Eyckmans et al. 2011). A discussion of the plasma membrane, actin, and ion channels with respect to how they generate or are functionally affected by mechanical forces in neurons follows.

19.2.1 The plasma membrane

The phospholipid bilayers of the plasma membranes of neuronal elements give rise to many of the viscoelastic properties of the brain. Plasma membranes are dynamic and experience structural changes across broad time and length scales ranging from nanoscopic (nanosecond and nanometer) to microscopic and mesoscopic (microsecond and millimeter) (Crawford and Earnshaw 1987; Pastor and Feller 1996). Phospholipids have lateral diffusion constants in bilayers on the order of 10^{-12} m^2/s (Almeida and Vaz 1995), they undergo trans-gauche isomerization every 10–20 nanoseconds, and rotate or wobble on nanosecond timescales (Pastor and Feller 1996). Plasma membranes respond to force in a time-varying manner as nonlinear functions of strain, meaning that they are viscoelastic or non-Newtonian fluids (Crawford and Earnshaw 1987). Here, phospholipid bilayers reflect a Maxwell material that shows frequency-dependent changes in tension and viscosity with viscoelastic relaxation times on the order of tens of microseconds (Crawford and Earnshaw 1987).

The deformation of plasma membranes in response to a force can be described by their compression (K_C), area expansion (K_A), and bending (K_B) moduli. In each case, a larger

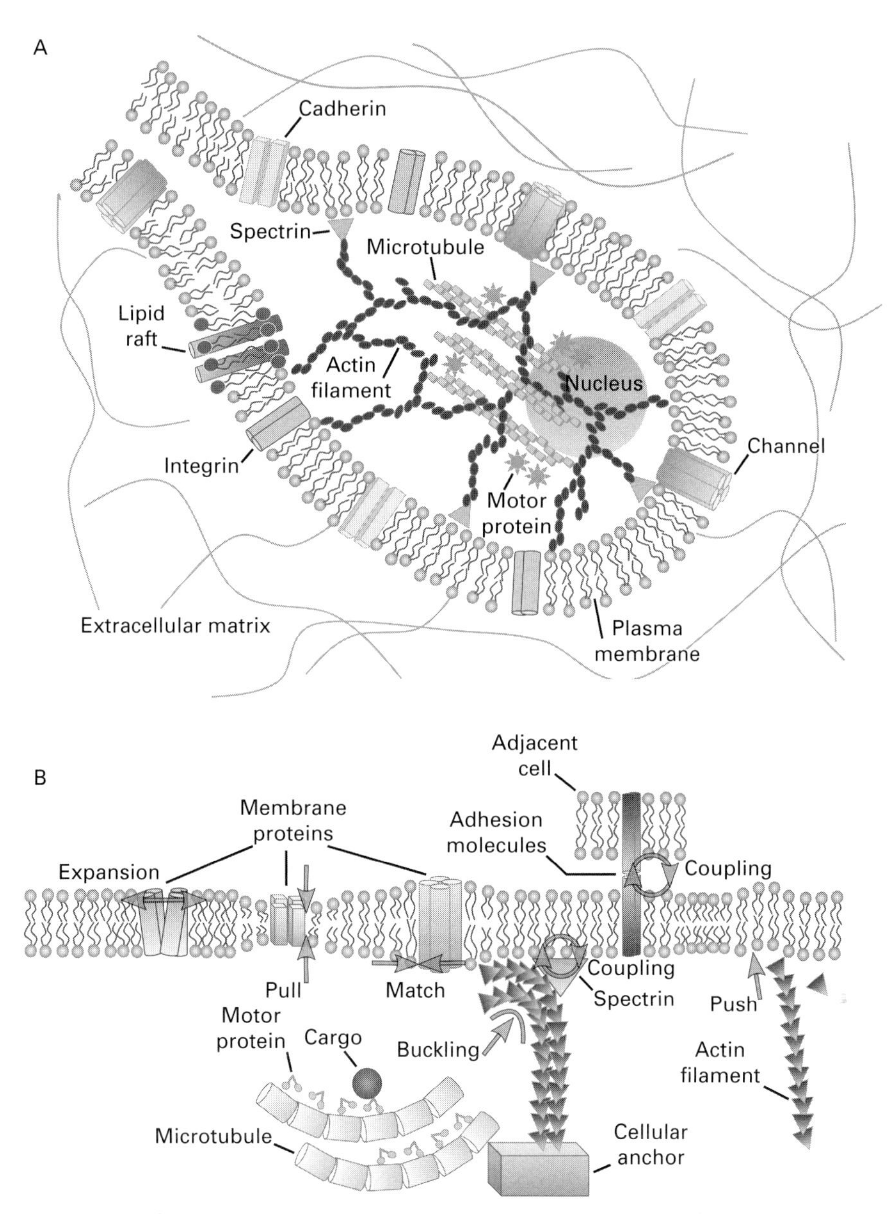

Figure 19.1 **Illustration of components that generate and transduce mechanical forces in neurons.** (A) Components that are affected by mechanical forces include the plasma membrane, ion channels, actin cytoskeleton, microtubules, and extracellular matrix proteins. Each of the illustrated components is known to play significant roles in regulating neuronal function. (B) Expanded illustration of the plasma membrane highlighting various micromechanical forces transduced and sensed by neurons including expansion, pushing, pulling, coupling, and bending forces.

modulus indicates greater resistance to a deformation force, whereas a smaller modulus indicates lower resistance. Membrane deformations can affect the activity of ion channels on millisecond timescales relevant to neuronal activity. The basic viscoelastic properties of plasma membranes can be approximately summarized as $K_B < K_A < K_C$ (Evans and Hochmuth 1978) (meaning that it is most sensitive to bending forces and least sensitive to compression forces). The neuronal plasma membrane's extreme sensitivity to bending deformation mediates its ability to exocytose and recapture vesicles, as well as to respond to protrusive and repulsive forces experienced during growth and motility. The intrinsic viscoelastic properties of neuronal membranes are further governed by cytoskeletal elements, which provide structural tension within a cell.

19.2.2 Actin

Actin fibers form part of the neuronal cytoskeleton, which shows dynamic structural plasticity and functions as a three-dimensional array of force transducers. The polymerization and depolymerization of actin monomers (G-actin) into actin polymers (F-actin) generates mechanical forces that are important for many cellular processes, such as generating cell membrane propulsion and protrusion, counteracting plasma membrane tension and deformation changes during endo- and exocytosis, and acting as molecular tension sensors to regulate numerous aspects of intracellular homeostasis. When polymerizing actin filaments approach a biological load, such as a plasma membrane, they generate pushing forces, and thermal fluctuations enable the continued incorporation of G-actin monomers into F-actin. This actin elongation is thought to resemble a "Brownian ratchet" (Feinman, Leighton, and Sands 1963), as random thermal fluctuations enable a gear-like churning of F-actin polymerization (Peskin, Odell, and Oster 1993). The elongation of F-actin will continue to occur until the counteracting load forces stall polymerization at a thermodynamic limit commonly referred to as the stalling force ($F_{\text{stall}} \approx 1$ pN) (Footer et al. 2007; Hill and Kirschner 1982). Besides stalling, actin can buckle under other forces, allowing it to continue elongating along a boundary (Footer et al. 2007; Hill and Kirschner 1982). F-actin fiber bundles can generate forces of several kPa (nN/μm^2) by contacting surface loads from different angles and continuously undergoing branch formation and elongation (Footer et al. 2007; Parekh et al. 2005). Actin motor proteins (known as myosins), GTP-binding proteins (such as RAS, RAC, RHO, and CDC42), and a host of other proteins can influence forces generated by actin through various mechanisms (Kosztin et al. 2002; Schliewa and Woelkhe 2003).

Actin-generated forces regulate axonal growth-cone dynamics. Growth cones have a low elastic modulus ($E = 106 \pm 21$ Pa; 1 Pa = 1 pN/μm^2) and can generate internal stress on the order of 30 Pa (Bets et al. 2011). As actin polymerizes at the leading edge of growing axons, growth cones begin to form focal adhesions with extracellular matrix (ECM) proteins to navigate their environment. Growth cones are weak (i.e., they are not capable of generating high mechanical stress) and soft (i.e., they are not rigid or stiff) force generators, which render them particularly sensitive to the mechanical properties of their environment (Bets et al. 2011). The dynamics of retrograde flow of F-actin can change abruptly to accelerate the growth of embryonic chick forebrain filopodia when they encounter a substrate stiffness of about 1 kPa (Chan and Odde 2008). These properties might encourage mechanically tuned synapse formation (see Section 19.3.3.2). On the postsynaptic side of synapses, actin is an established regulator of dendritic spine formation and plasticity (Matrus 2000). However, quantitative descriptions of how actin spatially and temporally distributes mechanical forces in spines while working toward such outcomes are lacking.

19.2.3 Ion channels

Pressure, tension, stretch, and stress at a plasma membrane can activate a broad range of mechanosensitive channels (MSCs) in the CNS, as well as in sensory systems (Arnadottir and Chalfie 2010; Hamil and Martinac 2001). The gating mechanisms that underlie MSC activity involve a number of complex membrane deformations and membrane–protein interactions (Hamil and Martinac 2001; Reeves et al. 2008; Suhkarev and Corey 2004). Besides the bulk effects of pressure and tension on plasma membranes, intermolecular mechanical forces are generated when the hydrophobic regions of a protein and a lipid try to constrain themselves to each other's physical lengths (hydrophobic matching). These and other interaction forces arising from protein inclusion in membranes can also modulate ion channel activity (Hamil and Martinac 2001; Reeves et al. 2008; Suhkarev and Corey 2004).

Many polymodal channels from diverse families, including the transient receptor potential (TRP) channels, two-pore domain potassium (K_{2P}) channels, and calcium-activated potassium (BK) channels, are modulated by membrane deformations (Arnadottir and Chalfie 2010; Suhkarev and Corey 2004; Morris 2001). Recent advances in our knowledge of ion channel biophysics indicate that the classic voltage-sensing mechanisms of many transient receptor potential (TRP), for example voltage-gated sodium (Na_V), potassium (K_V), and calcium (Ca_V) channels, are sensitive to mechanical fluctuations in plasma membranes (Reeves et al. 2008; Suhkarev and Corey 2004; Morris 2001).

More evidence for the influence of membrane mechanics on ion channel activity comes from thermodynamic investigations into the mechanisms of action of some anesthetics (see Section 19.4.4). Ketamine and isoflurane are thought to act by increasing the lateral pressure profile of lipids, which alters channel activity (Cantor 1997; Jerebek 2010). Even small membrane deformations with length scales of a few angstroms are sufficient to affect channel behavior (Hamil and Martinac 2001; Reeves et al. 2008). Accepting that micromechanical forces influence channel gating raises important issues for neuroscience – the predominant one being that our conventional understanding of neuronal excitability does not account for cellular-mechanical consequences. How does the realization that many ion channels (including voltage-gated ones) are mechanosensitive affect our comprehension of neuronal activity and plasticity? Confronting this problem seems to represent a particularly difficult challenge, especially as structural changes (tension and stress) at the plasma membrane of neurons

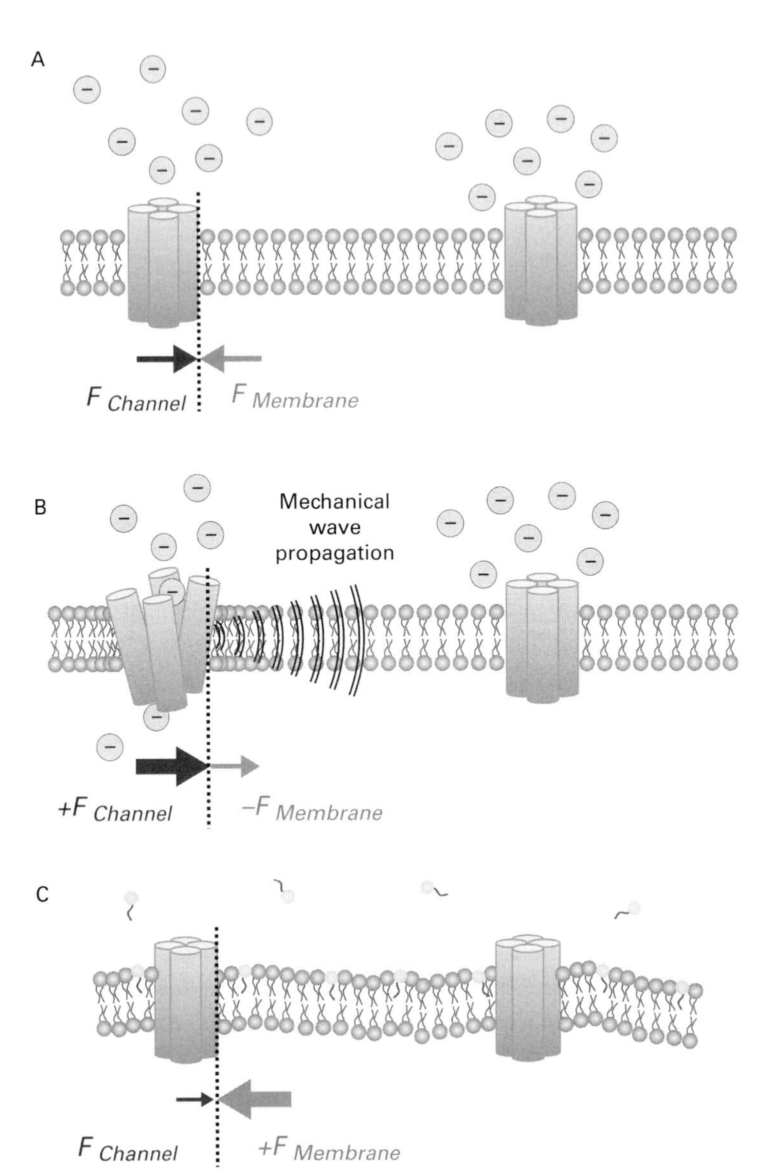

Figure 19.2 **Illustration of the interaction forces between ion channels and the neuronal membrane.** (A) At rest, the ion channels are closed to ion transport and the forces between the phospholipid bilayer and channel protein are equalized. (B) When the channel opens to allow ion transport, phospholipids are displaced by the conformational change of the channel and this produces a mechanical wave that propagates through the viscous phospholipid bilayer membrane towards other channels to possibly influence their behavior. (C) Anesthetic compounds that follow the Meyer-Overton rule will solubilize in the phospholipid bilayer and cause a redistribution of membrane forces that can shift channel conformational equilibrium towards the closed state, as channel opening would require a greater force to overcome that of the anesthetized membrane.

occur continually in the brain. For example, inserting ion channels into a postsynaptic membrane or modulating the rate of vesicle exocytosis would almost certainly lead to dynamic changes in synaptic membrane tension and stress. Addressing the consequences of these mechanical changes for ion channel activity and synaptic signaling will increase our overall understanding of the mechanisms that underlie brain function.

19.3 Role of mechanical forces in the brain

The critical importance of proper intrinsic force distributions in the central nervous system can be underscored by the disorders resultant from abnormalities in cortical folding. The folding of the gyrencephalic cortex is a major mechanical event, and abnormalities in folding have been identified in neurological disorders including autism and schizophrenia (Nordahl et al. 2007; White and Hilgetag 2011). The two major theories currently considered to explain cortical folding are differential expansion (Ronan et al. 2013) and a tension-based theory (VanEssen 1997). Differential expansion proposes that the tangential expansion of cortical regions is the driving force of cortical folding, propelled by the local proliferation of cells and changes in their sizes and connections (Ronan et al. 2013). The tension-based theory hypothesizes that tension along axons in the white matter drives the folding of cortex (VanEssen 1997). However, microdissection arrays have shown that while axons are under significant tension, the patterns of tissue stress are not entirely consistent with the tension-based hypothesis. Tension directed across developing gyri by axons was not found in developing ferret brains (Xu et al. 2010), and observed relaxation of tissue after cutting was suggested to be due to enhanced growth in the gray matter compared to white matter in adult mouse brain (Xu, Bailey, and Taber 2009). Overall, there is currently no theory that can explain all observations related to cortical folding, and proof for either differential expansion or the tension hypothesis is still incomplete. However, this provides a singular example demonstrating a need for an increased emphasis on studying the basic role of mechanical forces in the brain.

19.3.1 Traumatic mechanical injuries

At one end of the spectrum, endogenous micromechanical energy is important for normal brain development and function. At the opposite end of the spectrum, however, substantially greater mechanical forces acting on the brain can result in loss of consciousness, irreversible cognitive dysfunction, progressive neurodegeneration, and even death (Hoge et al. 2008; Meaney and Smith 2011). The deleterious consequences of concussions and traumatic brain injury (TBI) have recently been of central interest to the neuroscience community. However, we know little about the cellular-mechanical consequences of head trauma and how these injuries trigger a host of deleterious molecular signaling pathways, which in turn give rise to the clinical manifestations of TBI (Farkas and Povilshock 2007). Consistent with the cellular-mechanical features of the brain discussed earlier, the evidence suggests that traumatic injuries act at least on protein ion channels, the cytoskeleton, and the plasma membrane.

Diffuse axonal injury resulting in progressive neurodegeneration stems from whiplash-like injuries in which the mechanical activation of tetrodotoxin-sensitive sodium channels ultimately leads to calcium-mediated excitotoxicity (Wolf et al. 2001). Whether a similar phenomenon is associated with head trauma resulting in concussion or TBI is not known, but it warrants investigation. Integrin-mediated activation of RHO could also be an important contributor to diffuse axonal injury after mild TBI (Hemphill et al. 2011). At the level of the plasma membrane, mechanoporation resulting in increased membrane conductance occurs in response to TBI in rats (Farkas, Lifshitz, and Povilshock 2006). Similarly, TBI-induced mechanoporation of the plasma membrane can trigger axon blebbing and focal microtubule disruption (Kilinc, Gallo, and Barbee 2008). Whether mechanoporation per se is a primary consequence of TBI has not been clearly established. It seems likely that disruptions to the functions of the ECM, cytoskeletal elements, and cell membranes after head injury are responsible for the deleterious consequences. For example, intermediate filaments of astrocytes have been shown to protect against mechanical injuries to the brain and spinal cord and to mediate injury responses in the CNS (Pekny and Lane 2007). Further, the breakdown and remodeling of ECM proteins have been implicated in modulating injury responses to brain trauma (Lo, Wang, and Cuzner 2002).

How can an expansion of studies investigating the mechanical properties of the CNS be used to better inform us about the prevention and treatment of TBIs? By gaining a fundamental grasp on the cytomechanical features of brain circuits, neuroscience will be in a better position to address the challenges associated with TBI. At present, we do not fully comprehend the extent to which mechanical energy regulates endogenous brain function. Thus it remains difficult to understand how extreme impact forces can disrupt the natural mechanical properties of the CNS.

19.3.2 Mechanical forces and neural development

During development, the mechanical properties of nervous tissue are prone to alteration, and neurons encounter different mechanical cues depending on their location and developmental stage. Indeed, the elasticity and mechanical properties of the brain has been shown to change across different stages of development (Gefen et al. 2003; Prange and Margulies 2002; Thibault and Margulies 1998). Thus, neurons and growth cones are likely to encounter environments with differing mechanical properties as they migrate in situ. Growth cones are highly motile structures that are the leading tips of developing axons and dendrites, which generate the forces during neuronal growth thru polymerization of actin at their leading edge. In vitro, many neuronal cell types adapt their morphology to the stiffness of their substrate, and as such neuronal growth is likely a mechanical process influenced by the interaction with the mechanical environment in vivo as well. Even between neuronal types, dorsal root ganglion neuron growth cones have been shown to generate significantly greater traction forces compared to hippocampal neuron growth cones (~537 pN vs 71 pN), as determined using traction force microscopy (Koch et al. 2012). Moreover these neuronal types exhibited differential cytoskeletal adaption to substrate stiffness (Koch et al. 2012). Such differences in cytoskeletal mechanics pose the possibility that different forces generated by actin may serve unique mechanical scripts for synapse formation, maturation, and operation in neurons and between their types. Perhaps a mechanical environment, such as the ECM within a given anatomical area can change to optimize the growth dynamics of specific groups of invading axons across distinct stages of development.

Several additional observations seem to provide evidence for such mechanical tuning mechanisms of growth. Neurons plated on top of various geometrically constrained micropatterns revealed neuronal polarization was sensitive to external constraints and that axon polarization was favored along straight lines, such as may be used by newborn neurons extending their axon along preexisting straight structures (Roth et al. 2012). Beyond growth cones and the mechanical environment, mechanical tension along neurites and axons affect network development as well. Applied tension determined axonal specification in undifferentiated minor processes of cultured hippocampal neurons, and could even induce a second axon in an already polarized neuron (Lamoreaux 2002). Furthermore, once a neurite has connected to its target, tension promotes its stabilization and at the same time causes retraction or elimination of collateral neurites (Anava et al. 2009). Supported by these aforementioned primary observations, neuroscience should focus efforts on characterizing changes in traction force, elasticity, or viscosity across different anatomical and cellular regions, levels of activity, and stages of development.

19.3.3 Mechanical forces and neural communication

Mechanical tension along axons contributes not only to neuronal network development, but also to regulation of neural function and communication, and the evidence of the influence of mechanical forces on neural function is numerous. It has recently been shown that mechanical tension within axons plays an essential role in the accumulation of proteins at presynaptic terminals; biochemical signaling and recognition of synaptic partners is not sufficient (Seichen et al. 2009). Presynaptic vesicle clustering at neuromuscular synapses vanished upon severing the axon from the cell body and could be restored by applying tension to the severed end, and further stretching of intact axons could even increase vesicle clustering (Seichen et al. 2009). Furthermore, rest tensions of approximately 1 nN in axons were restored over approximately 15 minutes when perturbed mechanically, implicating mechanical tension as a modulation signal of vesicle accumulation and synaptic plasticity (Seichen et al. 2009). Axonal tension modulates local and global vesicle dynamics (Ahmed et al. 2012), and increased axonal tension from the resting state may induce further actin polymerization and increased clustering via mechanical trapping or interactions between F-actin and vesicles (Seichen et al. 2009).

Regarding the function of neurites in neural networks, actin in dendritic spines has been shown to regulate synapse formation and spine growth (Zito et al. 2004), activity-dependent spine motility (Fischer et al. 1998; Halpin, Hipolito, and Saffer 1998;

Star, Kwiatkowski, and Murthy 2002), and plasticity (Kim and Lisman 1999; Krucker, Siggins, and Halpain 2000; Matus 2000). In gelsolin knockout mice, reduced actin depolymerization has been shown to enhance NMDA-mediated and voltage-gated calcium activity in hippocampal neurons (Furukawa et al. 1997). However, the contribution of mechanical force changes to any of the above observations is not clearly understood. The viscoelasticity of dendritic spines was found to be critical to their function through atomic force microscopy elasticity mapping and dynamic indentation methods (Smith et al. 2007). Through this mechanical characterization, the activity-dependent structural plasticity, metastability, and congestion in the cytoplasm of spines are all gauged by merely a few physically measurable parameters. The degree of which spines are able to remodel and retain stability is determined in large part by viscosity; where soft, malleable spines have properties likely associated with morphological plasticity for learning, and the properties of rigid, stable spines are likely associated with memory retention (Smith et al. 2007). Perhaps the stabilization or destabilization of actomyosin networks produces direct mechanical consequences on synaptic activity by increasing or decreasing plasma membrane tension to coordinate the bending or compression of presynaptic compartments and dendritic spines. Given the dynamic nature of the actin cytoskeleton in the regulation of membrane tension and channel activity, the aforementioned idea seems natural for expanded investigations. Additionally, the contribution of the various other elements composing the cytoskeletal and extracellular matrices besides actin as discussed here can analogously be investigated.

19.3.3.1 Physical coupling at synapses

Synapses comprise discrete compartments coupled by the cytoskeleton, cell adhesion molecules, and extracellular matrix proteins, each of which exerts adhesion forces on synaptic elements. Presynaptic and postsynaptic compartments can change their structures (Schikorski and Stevens 1997) and physiological strengths (Murthy et al. 2001) in a correlated manner. However, we do not know whether mechanical forces transduced by one of these compartments can trigger functional changes in the other. Do dendritic spines transmit mechanical forces to presynaptic compartments to influence neurotransmitter release or recycling? Several observations with respect to synaptic coupling, as well as the influence of presynaptic membrane tension on neurotransmitter vesicle dynamics, support this possibility.

Depending on the cellular-mechanical matching of a synapse (the degree of equality in the stiffness or elasticity between connected presynaptic and postsynaptic compartments), it seems likely that a spine can exert retraction forces, twitching forces or extension forces on its presynaptic partner to trigger functional changes in presynaptic membrane tension or bending (Siechen et al. 2009). If a spine generates a retraction force that is greater than the bending modulus of the presynaptic membrane, but less than the bond-breaking forces of synaptic adhesion molecules, then tension in the presynaptic membrane should be influenced by the spine's retraction. At hippocampal synapses, the motility of a spine is often coupled to the motility of its presynaptic bouton partner, and expansion of one compartment leads to contraction of the other (Umeda, Ebihara, and

Okabe 2005). Presynaptic membrane tension can influence synaptic vesicle organization (Siechen et al. 2009) and modulate the rate of synaptic vesicle exocytosis in an integrin-mediated manner (Chen and Grinnell 1995). In fact, since the seminal descriptions of miniature end plate potentials provided by Fatt and Katz, it has been recognized that mechanical forces can modulate synaptic transmission (Fatt and Katz 1952). These and other observations suggest that mechanical signals can be functionally transmitted in the brain from one synaptic compartment to the other through mechanisms mediated by interactions between the cytoskeleton and cell adhesion molecules. Evidence of similar adhesion-mediated changes between presynaptic and postsynaptic partners has been observed at neuromuscular junctions during muscle growth (Balice-Gordon and Lichtman 1990). Mechanobiological studies in neuroscience should aim to determine to what degree anterograde and retrograde mechanical signaling at synapses participates in development, information transfer, and plasticity.

19.3.3.2 Patterned synapse formation

To ensure optimal growth, cellular processes must sense the mechanical properties of their environment while making necessary adjustments in the traction forces they generate. Although differential modulation of growth cone mechanical properties and ECM stiffness has been shown to regulate synapse formation, the intricacies of the mechanical interplay between these elements are not completely understood. The cellular layers of the rodent hippocampus possess markedly different rigidities (CA1 stratum pyramidale, 0.14 kPa; CA1 stratum radiatum, 0.20 kPa; CA3 stratum pyramidale, 0.23 kPa; and CA3 stratum radiatum, 0.31 kPa) (Elkin et al. 2007). When cultured on substrates with rigidities ranging from 0.5 kPa to 7.5 kPa, hippocampal axons grow faster on softer substrates (Kostic, Sap, and Sheetz 2007). Similarly, neurons from the embryonic spinal cord develop a fivefold higher neurite branch density when grown on soft substrates (0.05 kPa) compared to more rigid ones (0.55 kPa) (Flanagan et al. 2002). Interestingly, the axons of dorsal root ganglion neurons grow significantly more quickly when they are mechanically stretched (Pfister et al. 2004). The initiation of growth, and the growth rate of embryonic chicken forebrain (Fass and Odde 2003) and sensory neurons (Wang et al. 1991), can also be modulated by mechanically applied tension. Expressing differential growth rates as a function of substrate stiffness or growth cone traction force might represent a process for mechanically generating patterned synapse formation.

Molecular mediators of traction force generation cooperate with force-sensing mechanisms to optimize axonal growth dynamics (Chan and Odde 2008). Myosin IIB mediates the generation of growth cone traction forces (Bridgeman et al. 2011); the filopodia of superior cervical ganglion neurons from myosin IIB knockout mice generate significantly less traction force than filopodia from wild-type mice (~660 pN vs 970 pN) (Bridgeman et al. 2011). Both F-actin (Chan and Odde 2008) and integrins (Moore, Roca-Cusacha, and Sheetz 2010) can report substrate rigidity to growing cellular processes, which can in turn optimize their own mechanical properties to govern growth dynamics within their environment. This type of closed-loop feedback system might be functionally relevant in enabling the growth rates of axons to keep up with the hypertrophy of their organism's growing body by responding to mechanical stress cues (Bray 1984). Collectively, the dynamic cellular-mechanical matching principles described above probably provide mechanisms for tuning synapse formation during developmental and adult plasticity. Characterizing changes in growth cone traction force across different anatomical regions, levels of activity, and stages of development should further reveal how cellular-mechanical matching principles influence synapse formation. This seems to be a particularly important issue because the mechanical properties of the brain change during development (Gefen et al. 2003). In addition, cellular-mechanical matching might be important for signaling at gap junctions, neurovascular junctions, and other cell adhesion sites at which cells with different mechanical properties physically interact with one another.

19.4 Formulations in neuromechanobiology

The physiological functions of nervous systems are primarily regarded as being regulated by electrical and chemical driving forces. For instance, we have an intimate portrait of how electrical signaling along axonal fibers is converted to chemical signaling at the synapse between neurons. It is not well understood however how other forces, such as mechanical ones, impart actions upon neural function. Thus, to advance our understanding of how nervous systems operate it is important to develop comprehensive models where electrical, chemical, and mechanical energies are not compartmentalized from one another, but rather cooperate in a synergistic manner to govern neuronal excitability and signaling.

One common approach to mechanobiology involves the application of analysis to the cytoskeletal and extracellular matrices of a cell and determining its associated effects on cellular and molecular processes. One such approach is based on the concept of tensegrity architecture as a simple mechanical model of cell structure to relate cell shape, movement, and cytoskeletal mechanics, as well as the cellular response to mechanical forces (Ingber 2003). Another approach examines interactions between microtubules and actin as basic phenomena behind many fundamental processes, classified as either regulatory or structural interactions (Rodriguez 2003). The structural interactions of neurons lend well to mechanobiological analysis of the forces in the structural matrices as the mechanical forces of structural elements and the mechanical environment have broad implications on neural function.

Regarding the signaling functions of neurons, the Hodgkin-Huxley (HH) model is a bioelectric description of neuronal excitability based on conductance of ion-selective channels and a membrane capacitor, and is the currently accepted model for describing the action potential (Hodgkin and Huxley 1952). However, there are a number of observations related to the action potential that are not electrical or electro-chemical in nature. Several studies have shown the geometric dimensions of nerve fibers change in phase with action potential propagation, exerting forces normal to the membrane surface (Lundstrom 1974; Iwasa and Tasaki 1980; Kim et al. 2007; Tasaki, Iwasa, and

Gibbons 1980; Tasaki and Bytne 1990; Tusaki, Kusano, and Byrne 1989). Additionally, there is a reversible change in heat generation during action potential propagation, where heat released during the first phase of the action potential is compensated by heat uptake in the second phase (Tusaki, Kusano, and Byrne 1989; Abbott, Hill, and Howarth 1958; Zohar 1998; Tasaki and Byrne 1992; Howarth et al. 1975). The HH model however is based on irreversible processes and does not include thermodynamic variables required to sufficiently explain all the physically observed features of a nerve impulse. Despite this shortcoming the equivalent RC circuit formalism of the HH model (Hodgkin and Huxley 1952) has, no doubt, acquired global support through a bewildering number of independent observations over the past 60 years. While separate models accounting for the other non-electrical behaviors observed during the action potential have been proposed, there is not a broadly accepted model unifying electrical, chemical, and mechanical descriptions of the neuronal action potential. Regardless, the coupling of mechanical and electrical energy has seen considerable research and development (e.g., piezoelectricity) and its consideration as applied to the nervous system follows in this section. From these formulations, models and technology for mechanical manipulation and interfacing with the nervous system can be further developed. By starting to consider the interplay between electrical, chemical, and mechanical energy, rather than separately compartmentalizing them, new paradigms for understanding and studying the biophysics of neural systems should advance.

19.4.1 Flexoelectric effect

The flexoelectric effect is a liquid crystal analogue to the piezoelectric effect in solid crystals. Flexoelectricity refers specifically to the curvature dependent polarization of the membrane (Petrov 2006). As opposed to area stretching, thickness compression, and shear deformation in solid crystals, the flexoelectric effect includes the deformation of membrane curvature. This effect is manifested in liquid crystalline membrane structures, as a curvature of membrane surface leads to a splay of lipids and proteins. The molecules would otherwise be oriented parallel to each other in the normal flat state of the local membrane. Similar to piezoelectricity of solids, flexoelectricity is also manifested as a direct and a converse effect, featuring electric field induced curvature. The flexoelectric effect provides a basic mechanoelectric mechanism enabling nanometer-thick biomembranes to exchange responsiveness between electrical and mechanical stimuli. Consideration that cellular membranes possess mechanoelectric properties has raised concerns regarding the possible origin of inductance in early circuit models of the neuronal membrane and giant squid axons (Cole 1941).

Flexoelectricity (current generation from bending) and converse flexoelectricity have been demonstrated in lipid bilayers and cell membranes (Petrov 2006; Petrov 2002) and is likely involved in biological systems. The direct and converse flexoelectric effects have been used to describe the transformation of mechanical into electrical energy by stereocila and the electromotility of outer hair cell membranes for hearing (Petrov 2006; Petrov 2002). Many membrane functions involve the manipulation of membrane curvature (e.g., exocytosis, endocytosis, and cell migration) and the prospects that flexoelectricity is intricately involved in these processes thereby relating membrane mechanics and electrodynamics is likely.

19.4.2 Voltage induced changes in membrane tension

Differences in tension between the intracellular and extracellular interfaces will create changes in membrane curvature, referred to earlier as converse flexoelectricity. Thus, modulation of membrane tension by transmembrane voltage in a neuron will cause movement of the membrane with magnitude and polarity governed by the neuronal membrane stiffness and surface potentials at the membrane interfaces. Such an effect has been observed in real-time using atomic force microscopy (AFM) and voltage clamped HEK293 cells (Zhang, Kelashian, and Sachs 2001). In these studies Zhang et al. (2001) observed that depolarization caused an outward movement of the membrane, with amplitude proportional to voltage. Furthermore, a mathematical model able to predict the membrane tension over a range of surface potentials was developed, explicitly relating the tension in the membrane to the voltage and ionic charges across the membrane (Nygen et al. 2012). Consequently, the change in voltage with a nerve impulse is associated with a change in membrane tension, which will result in an alteration of cell radius to keep pressure constant across the membrane. This offers a mechanism and quantitative description for the observed change in the diameter of nerve fibers during the action potential, as opposed to alternative hypothesized mechanisms such as cell swelling due to water transport (Kim et al. 2007).

19.4.3 Optoelectric and electromechanical coupling

Our early understanding of the phenomena of electrical coupling with the mechanical modification of the neuronal membrane has already begun to yield innovative methods and technologies for interfacing to the nervous system. The modulation of refractive index or thickness of the cell due to transmembrane potential dependent deformations has allowed label free imaging of the membrane potential without the need of organic dyes or optogenetic probes which themselves likely alter membrane dynamics (Oh et al. 2012). By measuring milliradian scale phase shifts in the transmitted light, changes in the membrane potential of individual mammalian cells have been detected using low coherence interferometric microscopy without the use of exogenous labels (Oh et al. 2012). Using this technique, it was also demonstrated that propagation of electrical stimuli in gap junction-coupled cells could be monitored using wide-field imaging. This technique offers the advantages of simple sample preparation, low phototoxicity, and no need for photobleaching. Previous successes in label-free imaging of electrical activity has been possible in invertebrate nerves and neurons, as mammalian cells are smaller, optically transparent, and scatter light significantly less (Oh et al. 2012). While such approaches still require further refinement to enable a resolving power capable of imaging single action potentials, these methods have been able to experimentally confirm that the source of light phase shifts are due to potential-mediated changes in membrane tension, as opposed to swelling due to water transport or electrostriction of the cell membrane (Oh et al. 2012).

Regarding probing the mechanical response of mammalian cells to electrical excitation, AFM is the most commonly used tool for quantifying cellular deformation despite its invasiveness. Recently, piezoelectric nanoribbons have been developed for electro-mechanical biosensing and have demonstrated that cells deflect by 1 nm when 120 mV is applied to the membrane (Nygen et al. 2012). Furthermore, these nanoribbons support previous investigations of cellular electro-mechanics using AFM. Nanoribbons are made using microfabrication techniques, and so can be scaled more readily than AFM probes. Additionally, advances in microfabrication techniques could allow the manufacture of thinner nanoribbons to enhance their sensitivity, and facilitate the electro-mechanical observation of smaller neural structures, such as axons, dendrites, and dendritic spines. It will be interesting to see how further innovative methods and technologies develop to advance our understanding of the electrical coupling with the mechanical modification of the neuronal membrane.

19.4.4 The Meyer-Overton rule

The properties of the phospholipid bilayer membrane determine in part the behavior of various dynamic and relaxation processes of neuronal function. Processes influenced include the propagation and attenuation of mechanical waves, the decay of thermal shape fluctuations, and the translational and rotational diffusion of membrane components (Jeon and Voth 2005). When subjected to lateral stretching or compression, bilayer membranes behave as a viscoelastic material with anisotropy, and this can serve as a mechanism to modulate the state of the membrane, and hence all associated neuronal membrane processes such as channel activity. Thermodynamic investigations of lipid phase transitions have shown that lipid density pulses (sound or mechanical waves) can be adiabatically propagated through lipid monolayers, lipid bilayers, and neuronal membranes to influence fluidity and membrane excitability (Griesbauer, Wixforth, and Schneider 2009; Heimberg 2010; Heimberg and Jackson 2005). Interestingly, recent evidence indicates such sound wave propagation in pure lipid membranes can produce depolarizing potentials ranging from 1 mV to 50 mV with negligible heat generation (Griesbauer, Wixforth, and Schneider 2009), linking mechanical waves in neuronal membranes to changes in transmembrane potentials.

How the properties and changes in density of phospholipid bilayers influence the propagation of mechanical waves and neuronal processes, such as action potential initiation and propagation, is not precisely known. Anesthetics though, make for an interesting case to examine the influence of phospholipid bilayer state on neuronal function. It is known that anesthetics affect various functions of the neurons, including membrane permeability, hemolysis, and the function of ion channels and proteins. The Meyer-Overton rule for anesthetics relates that the critical dose is linearly proportional to the membrane solubility of the anesthetic molecules in the neuronal membrane, independent of the chemical ligand actions of the molecule (Heimberg and Jackson 2007). This rules out specific binding effects based on protein models for the wide variety of anesthetics that follow the Meyer-Overton rule. For example, voltage-gated sodium and potassium channels are slightly inhibited by halogenated alkanes and ethers, but not by xenon and nitrous oxide, despite all these anesthetics following the Meyer-Overton rule (Heimberg and Jackson 2007). It is known however that anesthetics have a pronounced effect on the physical properties of lipid bilayers, such as their lipid melting point phase transitions. This change in physical properties of the membrane can be related to the alteration of neuronal function by anesthetics, providing a mechanism for the alterations in function dependent on their solubility in the membrane and independent of their chemical nature. Such changes in the properties of the neuron's membrane would then influence such mechanisms as flexoelectricity, voltage induced membrane tension, the forces in cytoskeletal and extracellular matrices, ion channels, and all other mechanically sensitive processes coupled to the membrane.

19.5 Conclusions

The brain is a mechanically sensitive organ, the properties of which enable endogenous forces to regulate many aspects of neuronal function. The influence of mechanical energy on the brains of living organisms is omnipresent. For instance, cerebrovascular blood flow accompanying every heartbeat in humans generates forces that can displace brain tissue by tens of micrometers. Nanoscopic changes in plasma membrane stress and tension can influence ion channel activity, synaptic vesicle clustering, neurotransmitter release, and axonal growth cone dynamics. Therefore, we should not disregard the physical mechanics of the nervous systems we study.

Over the past several decades, neuroscience has been dominated by electrophysiological, biochemical, molecular, and genetic studies of brain function. Consequently, the mechanical forces that influence neuronal processes remain largely unexplored. The recent development of methods and devices utilizing mechanical energy to interact with and observe the nervous system represent initial technological advances that capitalizes on the coupling between neuronal function and mechanical forces. The extent to which cellular-mechanical dynamics influence neuronal activity –and effectually the interfacing to the nervous system using mechanical forces – remains largely unexplored. To advance neuroscience and our understanding of the complex nervous system, the compartmentalization of analyses and processes due to electrical, chemical, or mechanical energies in system characterization and manipulation needs to be stepped away from. Knowledge gained from studies of mechanical forces in neurons will probably not disprove our current working models; rather, it should enable us to refine and expand upon them.

In conclusion, the amalgamation of neuroscience and mechanobiology (neuromechanobiology) will provide greater insight into how endogenous mechanical forces physically collide with conventional signal transduction pathways to govern neuronal development and plasticity. Neuromechanobiology will reveal basic brain functions that require the regulated contraction, expansion, pushing, pulling, and relaxing of neurons, glia, and their molecular and cellular components. At a minimum, neuromechanobiology studies will improve our understanding of the primary sequela associated with concussion, traumatic brain injury, and disease states that affect neuronal viscoelasticity.

References

Abbott, B. C., A. V. Hill, and J. V. Howarth. (1958). "The positive and negative heat production associated with a nerve impulse." *Proc R Soc Lond B Biol Sci* **148**(931): 149–187.

Ahmed, W. W., et al. (2012). "Mechanical tension modulates local and global vesicle dynamics in neurons." *Cell Mol Bioeng* **5**(2): 155–164.

Almeida, P. F. F. and W. L. C. Vaz. (1995). "Lateral diffusion in membranes." In *Handbook of Biological Physics: Structure and Dynamics of Membranes – From Cells to Vesicles*, R. Lipowsky and E. Sackmann, eds. Amsterdam: North-Holland, 305–357.

Anava, S., et al. (2009). "The regulative role of neurite mechanical tension in network development." *Biophys J* **96**(4): 1661–1670.

Arnadottir, J. and M. Chalfie. (2010). "Eukaryotic mechanosensitive channels." *Annu Rev Biophys* **39**: 111–137.

Balice-Gordon, R.J. and J.W. Lichtman. (1990). "In vivo visualization of the growth of pre– and postsynaptic elements of neuromuscular junctions in the mouse." *J Neurosci* **10**(3): 894–908.

Betz, T., et al. (2011). "Growth cones as soft and weak force generators." *Proc Natl Acad Sci USA* **108**(33): 13420–13425.

Bray, D. (1984). "Axonal growth in response to experimentally applied mechanical tension." *Dev Biol* **102**(2): 379–89.

Bridgman, P. C., et al. (2001). "Myosin IIB is required for growth cone motility." *Journal of Neuroscience* **21**(16): 6159–6169.

Cantor, R. S. (1997). "The lateral pressure profile in membranes: a physical mechanism of general anesthesia." *Biochemistry* **36**(9): 2339–2344.

Chan, C. E. and D. J. Odde. (2008). "Traction dynamics of filopodia on compliant substrates." *Science* **322**(5908): 1687–1691.

Chen, B. M. and A. D. Grinnell. (1995). "Integrins and modulation of transmitter release from motor nerve terminals by stretch." *Science* **269**(5230): 1578–1580.

Cole, K. S. (1941). "Rectification and inductance in the squid giant axon." *J Gen Physiol* **25**(1): 29–51.

Crawford, G. E. and J. C. Earnshaw. (1987). "Viscoelastic relaxation of bilayer lipid membranes. Frequency-dependent tension and membrane viscosity." *Biophys J* **52**(1): 87–94.

Elkin, B. S., et al. (2007). "Mechanical heterogeneity of the rat hippocampus measured by atomic force microscope indentation." *J Neurotrauma* **24**(5): 812–22.

Evans, E.A. and R.M. Hochmuth, *Current Topics in Membranes and Transport*. 1978. 1–64.

Eyckmans, J., et al. (2011). "A hitchhiker's guide to mechanobiology." *Dev Cell* **21**(1): 35–47.

Farkas, O., J. Lifshitz, and J.T. Povlishock. (2006). "Mechanoporation induced by diffuse traumatic brain injury: an irreversible or reversible response to injury?" *J Neurosci* **26**(12): 3130–3140.

Farkas, O. and J. T. Povlishock. (2007). "Cellular and subcellular change evoked by diffuse traumatic brain injury: a complex web of change extending far beyond focal damage." *Prog Brain Res* **161**: 43–59.
Fass, J. N. and D. J. Odde. (2003). "Tensile force-dependent neurite elicitation via anti-beta 1 integrin antibody-coated magnetic beads." *Biophysical Journal* **85**(1): 623–636.
Fatt, P. and B. Katz. (1952). "Spontaneous subthreshold activity at motor nerve endings." *J Physiol* **117**: 109–128.
Feynman, R. P., R. B. Leighton, and M. Sands, *The Feynman Lectures on Physics*. 1963. 46–49.
Fischer, M., et al. (1998). "Rapid actin-based plasticity in dendritic spines." *Neuron* **20**(5): 847–854.
Flanagan, L. A., et al. (2002). "Neurite branching on deformable substrates." *Neuroreport* **13**(18): 2411–2415.
Footer, M. J., et al. (2007). "Direct measurement of force generation by actin filament polymerization using an optical trap." *Proceedings of the National Academy of Sciences of the United States of America* **104**(7): 2181–2186.
Furukawa, K., et al. (1997). "The actin-severing protein gelsolin modulates calcium channel and NMDA receptor activities and vulnerability to excitotoxicity in hippocampal neurons." *J Neurosci* **17**(21): 8178–8186.
Gefen, A., et al. (2003). "Age-dependent changes in material properties of the brain and braincase of the rat." *J Neurotrauma* **20**(11): 1163–1177.
Griesbauer, J., A. Wixforth, and M. F. Schneider. (2009). "Wave propagation in lipid monolayers." *Biophys J* **97**(10): 2710–2716.
Halpain, S., A. Hipolito, and L. Saffer. (1998). "Regulation of F-actin stability in dendritic spines by glutamate receptors and calcineurin." *J Neurosci* **18**(23): 9835–9844.
Hamill, O. P. and B. Martinac. (2001). "Molecular basis of mechanotransduction in living cells." *Physiol Rev* **81**(2): 685–740.
Heimburg, T. (2010). "Lipid ion channels." *Biophys Chem* **150**(1–3): 2–22.
Heimburg, T. and A. D. Jackson. (2005). "On soliton propagation in biomembranes and nerves." *Proc Natl Acad Sci USA* **102**(28): 9790–9795.
Heimburg, T. and A. D. Jackson. (2007). "On the action potential as a propagating density pulse and the role of anesthetics." *Biophysical Reviews and Letters* **02**(01): 57–78.
Hemphill, M. A., et al. (2011). "A possible role for integrin signaling in diffuse axonal injury." *PLoS One* **6**(7): e22899.
Hill, T. L. and M. W. Kirschner. (1982). "Subunit treadmilling of microtubules or actin in the presence of cellular barriers – possible conversion of chemical free energy into mechanical work." *Proc Natl Acad Sci USA - Biological Sciences* **79**(2): 490–494.
Hodgkin, A. L. and A. F. Huxley. (1952). "A quantitative description of membrane current and its application to conduction and excitation in nerve." *J Physiol* **117**(4): 500–544.
Hoge, C. W., et al. (2008). "Mild traumatic brain injury in U.S. Soldiers returning from Iraq." *N Engl J Med* **358**(5): 453–463.
Howarth, J. V., et al. (1975). "Heat production associated with passage of a single impulse in pike olfactory nerve-fibers." *Journal of Physiology-London* **249**(2): 349–368.

Ingber, D. E. (2003). "Tensegrity I. Cell structure and hierarchical systems biology." *J Cell Sci* **116**(Pt 7): 1157–1173.
Iwasa, K. and I. Tasaki. (1980). "Mechanical changes in squid giant axons associated with production of action potentials." *Biochem Biophys Res Commun* **95**(3): 1328–1331.
Jeon, J. and G. A. Voth. (2005). "The dynamic stress responses to area change in planar lipid bilayer membranes." *Biophys J* **88**(2): 1104–1119.
Jerabek, H., et al. (2010). "Membrane-mediated effect on ion channels induced by the anesthetic drug ketamine." *J Am Chem Soc* **132**(23): 7990–7997.
Kilinc, D., G. Gallo, and K. A. Barbee. (2008). "Mechanically-induced membrane poration causes axonal beading and localized cytoskeletal damage." *Exp Neurol* **212**(2): 422–430.
Kim, C. H. and J. E. Lisman. (1999). "A role of actin filament in synaptic transmission and long-term potentiation." *J Neurosci* **19**(11): 4314–4324.
Kim, G. H., et al. (2007). "A mechanical spike accompanies the action potential in Mammalian nerve terminals." *Biophys J* **92**(9): 3122–3129.
Koch, D., et al. (2012). "Strength in the periphery: growth cone biomechanics and substrate rigidity response in peripheral and central nervous system neurons." *Biophys J* **102**(3): 452–460.
Kostic, A., J. Sap, and M. P. Sheetz. (2007). "RPTPalpha is required for rigidity-dependent inhibition of extension and differentiation of hippocampal neurons." *J Cell Sci* **120**(Pt 21): 3895–3904.
Kosztin, I., et al. (2002). "Mechanical force generation by G proteins." *Proc Natl Acad Sci USA* **99**(6): 3575–3580.
Krucker, T., G. R. Siggins, and S. Halpain. (2000). "Dynamic actin filaments are required for stable long-term potentiation (LTP) in area CA1 of the hippocampus." *Proc Natl Acad Sci USA* **97**(12): 6856–6861.
Kruse, S. A., et al. (2008). "Magnetic resonance elastography of the brain." *Neuroimage* **39**(1): 231–237.
Lamoureux, P., et al. (2002). "Mechanical tension can specify axonal fate in hippocampal neurons." *J Cell Biol* **159**(3): 499–508.
Lo, E. H., X. Wang, and M. L. Cuzner. (2002). "Extracellular proteolysis in brain injury and inflammation: role for plasminogen activators and matrix metalloproteinases." *J Neurosci Res* **69**(1): 1–9.
Lundstrom, I. (1974). "Mechanical wave propagation on nerve axons." *J Theor Biol* **45**(2): 487–499.
Matus, A. (2000). "Actin-based plasticity in dendritic spines." *Science* **290**(5492): 754–758.
Matus, A. (2000). "Actin-based plasticity in dendritic spines." *Science* **290**: 754–758.
McCracken, P. J., et al. (2005). "Mechanical transient-based magnetic resonance elastography." *Magn Reson Med* **53**(3): 628–639.
Meaney, D. F. and D. H. Smith. (2011). "Biomechanics of concussion." *Clin Sports Med* **30**(1): 19–31, vii.
Moore, S. W., P. Roca-Cusachs, and M. P. Sheetz. (2010). "Stretchy proteins on stretchy substrates: the important elements of integrin-mediated rigidity sensing." *Dev Cell* **19**(2): 194–206.

Morris, C. E. (2011). "Voltage-gated channel mechanosensitivity: fact or friction?" *Front Physiol* **2**: 25.
Murphy, M. C., et al. (2011). "Decreased brain stiffness in Alzheimer's disease determined by magnetic resonance elastography." *J Magn Reson Imaging* **34**(3): 494–498.
Murthy, V. N., et al. (2001). "Inactivity produces increases in neurotransmitter release and synapse size." *Neuron* **32**(4): 673–682.
Nguyen, T. D., et al. (2012). "Piezoelectric nanoribbons for monitoring cellular deformations." *Nat Nanotechnol* **7**(9): 587–593.
Nordahl, C. W., et al. (2007). "Cortical folding abnormalities in autism revealed by surface-based morphometry." *J Neurosci* **27**(43): 11725–11735.
Oh, S., et al. (2012). "Label-free imaging of membrane potential using membrane electromotility." *Biophys J* **103**(1): 11–18.
Parekh, S. H., et al. (2005). "Loading history determines the velocity of actin-network growth." *Nat Cell Biol* **7**(12): 1219–1223.
Pastor, R. W. and S. E. Feller. (1996). "Time scales of lipid dynamics and molecular dynamics." *Biol Membr* **1**: 4–29.
Pekny, M. and E. B. Lane. (2007). "Intermediate filaments and stress." *Exp Cell Res* **313**(10): 2244–2254.
Peskin, C. S., G. M. Odell, and G. F. Oster. (1993). "Cellular motions and thermal fluctuations: the Brownian ratchet." *Biophys J* **65**(1): 316–24.
Petrov, A. G. (2002). "Flexoelectricity of model and living membranes." *Biochim Biophys Acta* **1561**(1): 1–25.
Petrov, A. G. (2006). "Electricity and mechanics of biomembrane systems: flexoelectricity in living membranes." *Anal Chim Acta* **568**(1–2): 70–83.
Pfister, B. J., et al. (2004). "Extreme stretch growth of integrated axons." *J Neurosci* **24**(36): 7978–7983.
Prange, M. T. and S. S. Margulies. (2002). "Regional, directional, and age-dependent properties of the brain undergoing large deformation." *J Biomech Eng* **124**(2): 244–252.
Reeves, D., et al. (2008). "Membrane mechanics as a probe of ion-channel gating mechanisms." *Physical Review E* **78**(4): 041901.
Rodriguez, O. C., et al. (2003). "Conserved microtubule-actin interactions in cell movement and morphogenesis." *Nat Cell Biol* **5**(7): 599–609.
Ronan, L., et al. (2013). "Differential tangential expansion as a mechanism for cortical gyrification." *Cereb Cortex* **24**(8): 2219–2228.
Roth, S., et al. (2012). "How morphological constraints affect axonal polarity in mouse neurons." *PLoS One* **7**(3): e33623.
Sack, I., et al. (2011). "The influence of physiological aging and atrophy on brain viscoelastic properties in humans." *PLoS One* **6**(9): e23451.
Schikorski, T. and C. F. Stevens. (1997). "Quantitative ultrastructural analysis of hippocampal excitatory synapses." *J Neurosci* **17**(15): 5858–5867.
Schliwa, M. and G. Woehlke. (2003). "Molecular motors." *Nature* **422**(6933): 759–765.
Siechen, S., et al. (2009). "Mechanical tension contributes to clustering of neurotransmitter vesicles at presynaptic terminals." *Proc Natl Acad Sci USA* **106**(31): 12611–12616.
Smith, B. A., et al. (2007). "Dendritic spine viscoelasticity and soft-glassy nature: balancing dynamic remodeling with structural stability." *Biophys J* **92**(4): 1419–1430.

Star, E. N., D. J. Kwiatkowski, and V. N. Murthy. (2002). "Rapid turnover of actin in dendritic spines and its regulation by activity." *Nat Neurosci* **5**(3): 239–246.
Sukharev, S. and D. P. Corey. (2004). "Mechanosensitive channels: multiplicity of families and gating paradigms." *Sci STKE* **2004**(219): re4.
Tasaki, I. and P. M. Byrne. (1990). "Volume expansion of nonmyelinated nerve fibers during impulse conduction." *Biophys J* **57**(3): 633–635.
Tasaki, I. and P. M. Byrne. (1992). "Heat production associated with a propagated impulse in bullfrog myelinated nerve fibers." *Jpn J Physiol* **42**(5): 805–813.
Tasaki, I., K. Iwasa, and R. C. Gibbons. (1980). "Mechanical changes in crab nerve fibers during action potentials." *Jpn J Physiol* **30**(6): 897–905.
Tasaki, I., K. Kusano, and P. M. Byrne. (1969). "Rapid mechanical and thermal changes in the garfish olfactory nerve associated with a propagated impulse." *Biophys J* **55**(6): 1033–1040.
Thibault, K. L. and S. S. Margulies. (1998). "Age-dependent material properties of the porcine cerebrum: effect on pediatric inertial head injury criteria." *J Biomech* **31**(12): 1119–1126.
Umeda, T., T. Ebihara, and S. Okabe. (2005). "Simultaneous observation of stably associated presynaptic varicosities and postsynaptic spines: morphological alterations of CA3-CA1 synapses in hippocampal slice cultures." *Mol Cell Neurosci* **28**(2): 264–274.
VanEssen, D. C. (1997). "A tension-based theory of morphogenesis and compact wiring in the central nervous system." *Nature* **385**(6614): 313–318.
White, T. and C. C. Hilgetag. (2011). "Gyrification and neural connectivity in schizophrenia." *Dev Psychopathol* **23**(1): 339–352.
Wolf, J. A., et al. (2001). "Traumatic axonal injury induces calcium influx modulated by tetrodotoxin-sensitive sodium channels." *J Neurosci* **21**(6): 1923–1930.
Wuerfel, J., et al. (2010). "MR-elastography reveals degradation of tissue integrity in multiple sclerosis." *Neuroimage* **49**(3): 2520–2525.
Xu, G. et al. (2010). "Axons pull on the brain, but tension does not drive cortical folding." *J Biomech Eng* **132**(7): 071013.
Xu, G., P. V. Bayly, and L. A. Taber. (2009). "Residual stress in the adult mouse brain." *Biomech Model Mechanobiol* **8**(4): 253–262.
Zhang, J., et al. (2011). "Viscoelastic properties of human cerebellum using magnetic resonance elastography." *Journal of Biomechanics* **44**(10): 1909–1913.
Zhang, P. C., A. M. Keleshian, and F. Sachs. (2001). "Voltage-induced membrane movement." *Nature* **413**(6854): 428–432.
Zheng, J., et al. (1991). "Tensile regulation of axonal elongation and initiation." *J Neurosci* **11**(4): 1117–1125.
Zito, K., et al. (2004). "Induction of spine growth and synapse formation by regulation of the spine actin cytoskeleton." *Neuron* **44**(2): 321–334.
Zohar, O., et al. (1998). "Thermal imaging of receptor-activated heat production in single cells." *Biophys J* **74**(1): 82–89.

Index